Advanced Control, Optimization, Stability and Reliability of Microgrids and Power Systems

Advanced Control, Optimization, Stability and Reliability of Microgrids and Power Systems

Guest Editor

Muhammad Akmal

MDPI

Basel • Beijing • Wuhan • Barcelona • Belgrade • Novi Sad • Cluj • Manchester

Guest Editor
Muhammad Akmal
Department of Engineering
and Mathematics
Sheffield Hallam University
Sheffield
UK

Editorial Office
MDPI AG
Grosspeteranlage 5
4052 Basel, Switzerland

This is a reprint of the Special Issue, published open access by the journal *Energies* (ISSN 1996-1073), freely accessible at: https://www.mdpi.com/journal/energies/special_issues/P362725Y4P.

For citation purposes, cite each article independently as indicated on the article page online and as indicated below:

Lastname, A.A.; Lastname, B.B. Article Title. *Journal Name* **Year**, *Volume Number*, Page Range.

ISBN 978-3-7258-7186-5 (Hbk)
ISBN 978-3-7258-7187-2 (PDF)
https://doi.org/10.3390/books978-3-7258-7187-2

Contents

About the Editor

Muhammad Akmal

Muhammad Akmal is currently with Electrical Engineering at Sheffield Hallam University, United Kingdom, where he leads Electrical and Electronic Engineering programmes and major academic initiatives, including program design and accreditations. He joined Sheffield Hallam University in 2018 and previously served as Associate Professor and Assistant Professor at Abu Dhabi University, contributing to programme development and accreditation activities, including ABET and WASC. He holds a PhD in Electric Power Systems from Queen's University Belfast. With over 18 years of international academic and consultancy experience, his research focuses on power systems modelling and simulation, distributed generation integration, smart grids, renewable energy systems, and demand-side management. He has published more than 50 peer-reviewed papers and received the Charles Parsons Energy Research Award from Science Foundation Ireland for his doctoral research. He actively reviews for leading journals and is a Senior Member of IEEE and a Senior Fellow of Advance HE.

Preface

The rapid transformation of modern power systems, driven by high penetration of renewable energy sources, distributed generation, energy storage, and power electronic interfaces, has fundamentally reshaped the operational and dynamic characteristics of electrical networks. While these developments offer significant environmental and economic benefits, they also introduce unprecedented challenges in terms of stability, controllability, protection, optimization, and long-term reliability. Microgrids and actively managed distribution systems, in particular, require advanced analytical tools and intelligent control strategies to operate securely under increasingly uncertain and dynamic conditions.

This Special Issue, "*Advanced Control, Optimization, Stability and Reliability of Microgrids and Power Systems*", was conceived to provide a focused platform for disseminating high-quality research addressing these critical challenges. The collected contributions present innovative theoretical developments, practical control frameworks, data-driven methodologies, and reliability assessment techniques that collectively advance the state-of-the-art.

The editors sincerely thank all authors for their valuable contributions and the reviewers for their rigorous and constructive evaluations, which have ensured the high quality of this Special Issue. It is our hope that this collection will serve as a meaningful reference and inspire further research toward secure, resilient, and sustainable next-generation power systems.

Muhammad Akmal
Guest Editor

energies

MDPI

Article

Stability Analysis of a Multi-Machine Parallel Microgrid Using a Time-Domain Method

Boning Chang and Yifeng Ren *

School of Electrical and Control Engineering, North University of China, Taiyuan 030051, China; 18834070682@163.com
* Correspondence: renyifeng126@126.com

Abstract: Current microgrid research primarily focuses on radial topologies and their control strategies, while exploration of the time-domain dynamic behavior of closed-loop controlled microgrids remains relatively insufficient. This research gap makes it difficult to directly observe and deeply analyze the evolution mechanisms of critical phenomena, such as oscillations and instability, when they occur. Therefore, conducting time-domain analysis on closed-loop structures is crucial for revealing system instability mechanisms and ensuring their safe and stable operation. This paper establishes a state-space model for a closed-loop microgrid structure composed of multiple parallel inverters and conducts time-domain stability analysis under grid-connected operation. First, a mathematical model of the closed-loop microgrid system is constructed using state-space equations. Subsequently, time-domain analysis of small-signal stability is performed on the model. By varying key parameters such as the droop coefficient, the influence patterns on system stability are investigated. The results indicate that the droop control coefficient and LC filter parameters exert the most significant impact on system dynamic characteristics. Simulation experiments validate the correctness and effectiveness of the theoretical model. Finally, the time-domain characteristics of this model were further analyzed and validated through simulations. Results demonstrate that the system maintains robust stability under disturbances even in grid-connected mode.

Keywords: multi-node system; stability analysis; microgrid; time-domain analysis

1. Introduction

In recent years, the problem of environmental pollution caused by traditional fossil fuels has become more and more serious and has become a major issue in today's world economic development. Renewable energy based on wind energy and solar energy provides an effective solution to the current problems [1,2]. The combination of innovation in small distributed generation systems and the technological advancements of power electronic systems has given rise to the concept of future network technologies, such as microgrids. With the advancement of global energy transformation and the 'dual carbon' goal, microgrids have become an important part of the new power system due to their flexibility, low-carbon nature, and reliability. Many forms of distributed generation, such as fuel cells, photovoltaic power generation, and micro turbines, are connected to the grid through power electronic converters [3]. The microgrid can operate in grid-connected mode or stand-alone mode. However, the uncertainty of renewable energy generation has brought a series of problems to the safe and stable operation of microgrids [4,5]. There are a large number of power electronic converters in the microgrid. The increase in various micro-sources will significantly affect the characteristics of the microgrid, and the microgrid exhibits nonlinear characteristics such as low inertia, low damping, and strong time-variability [6]. These problems make the system more prone to oscillation instability [7]. Even if each subunit is stable on its own, the interaction between the subunits will cause stability problems in the system [8]. Therefore, it is necessary to model and analyze the stability of the microgrid.

In the research of microgrids, how to ensure the power quality and system stability in both grid-connected and islanded operation modes has always been one of the key technical challenges. Whether operating in parallel with the main network or independently, there will be some possible problems that can lead to system instability.

In power system networks, voltage and frequency are the two most critical operational parameters [9]. One of the most fundamental and valuable operational modes of a microgrid is islanded operation. In this mode, the microgrid disconnects from the main grid (macrogrid) and operates independently, relying entirely on its internal distributed energy resources—such as photovoltaics, wind turbines, battery energy storage systems, and diesel generators—to provide continuous and reliable power to a local area [10,11]. When the system is unstable, it is often difficult for the system to maintain the stability of voltage and frequency independently [12–14]. Compared with radial microgrids, closed-loop microgrids can significantly mitigate these issues. In the event of a line fault, power can be supplied from the opposite end through the closed-loop path without interrupting critical loads. Furthermore, when a grid-forming control strategy is adopted, the stability and operational reliability of the microgrid can be substantially enhanced.

Grid-Forming Control (GFM) is an advanced control strategy for power electronic converters that enables them to emulate the operational behavior of traditional synchronous generators, thereby autonomously establishing and maintaining grid voltage and frequency. Droop control is one of the most widely used methods in grid-connected inverter control. It emulates the active power-frequency (P–f) and reactive power-voltage (Q–V) droop characteristics of synchronous generators [15]. In power systems with a high penetration of renewable energy, grid-forming inverters—which operate without relying on grid synchronization signals—can autonomously regulate the voltage and frequency. This capability significantly enhances the system stability and resilience to disturbances and helps mitigate stability challenges resulting from weakened grid strength [16,17].

The application in a weak grid environment poses a unique challenge to the stability of grid-connected inverters. A weak grid has the characteristics of low short-circuit capacity and low inertia. Therefore, it is necessary to explore strategies to improve the stability and performance of inverters in this challenging environment [18,19].

In [20], a three-degree-of-freedom rotating pendulum dynamic model integrated with an electromagnetic energy harvesting device was systematically constructed, establishing a comprehensive theoretical framework. However, this work lacked effective validation and practical application of the proposed methods. To address this limitation, the study in [21] focused on verifying the methodology and advancing its engineering applications. While employing the same theoretical approaches—Lagrangian equations and the method of multiple scales—it particularly emphasized validating the high consistency between analytical and numerical solutions, thereby providing a mature methodological reference for analyzing the dynamic behavior and stability of complex systems. Nevertheless, with the continuous increase in renewable energy penetration, modern power systems are progressively evolving into "quasi-dynamic systems" dominated by a large number of power electronic inverters. Ref. [22] estimates the interaction and stability of grid-connected inverters equipped with LC filters. The resonance and oscillation characteristics of multiple inverters are analyzed through simulation. An interactive admittance model is proposed to effectively characterize these interactions in relation to the physical network admittance. Frequency-domain stability analysis of the microgrid model enables early prediction of oscillatory instability and facilitates stability assessment at specific frequency points. However, this paper does not examine time-domain characteristics; as a result, the impact of specific parameters on system stability remains unclear. In practical applications, system stability may be vulnerable to variations in these parameters. In [23], a comprehensive dynamic model of the network is established, with each module represented in state-space form and integrated into the overall system. Sensitivity analysis is conducted on the full network model to evaluate the influence of key parameters on system stability. However, this study is restricted to the dynamic model of a radial network, which is prone to instability in the event of sudden circuit breakages. Although closed-loop networks provide higher power supply reliability and improved stability, they also introduce greater operational and analytical complexity. In [24], a steady-state analysis is carried out for a microgrid system consisting of multiple inverters operating under droop control, and a state-space model is developed to examine the impact of droop coefficients on system stability. The study focuses exclusively on radial microgrid configurations and does not investigate closed-loop microgrids under droop control. Although employing multiple droop-controlled inverters in a closed-loop microgrid can enhance operational stability in practice, it also increases structural and control complexity.

Based on existing research, this paper addresses the challenges of modeling and stability analysis for complex microgrid system structures, with a specific focus on the adaptability of modeling methods and configurations. To overcome the limitations of conventional approaches in handling complex topologies, this paper proposes the application of the state-space method to model loop-structured microgrids composed of multiple parallel inverters. A systematic small-signal stability analysis is conducted, along with a dedicated investigation into the time-domain characteristics of the looped microgrid system. The core strength of the proposed method lies in its ability to reveal underlying mechanisms and its strong generalization capability. By integrating eigenvalue analysis and parameter sensitivity analysis, this study not only identifies key factors influencing system stability but also clearly elucidates the intrinsic mechanisms through which various parameters affect stability and ranks their sensitivity. This provides direct guidance for the optimized design of system parameters. To comprehensively validate stability, the research introduces small-disturbance tests and time-domain analysis during grid-connected operation, establishing a complete verification framework that spans from the frequency domain to the time domain and from theoretical analysis to simulation. Simulation results confirm that the developed model accurately and efficiently characterizes the stability behavior of the system. To demonstrate the general applicability of the proposed method, the modeling and analysis framework was successfully extended to radial microgrids. Comparative experiments show that the method maintains analytical accuracy across different network structures, overcoming the structural limitations of traditional approaches and proving its potential as a universal tool for microgrid stability analysis.

Finally, this work systematically summarizes the quantitative influence of parameter variations on stability and designs simulation experiments to analyze the time-domain characteristics of the system under small disturbances. The results verify the system's capability to suppress daily small load disturbances, thereby validating the rationality and feasibility of the system design from a dynamic performance perspective. The research outcomes provide a theoretically sound and practically viable solution for both academic research and engineering practice in the stability analysis and design of complex microgrids.

2. Materials and Methods

This paper aims to propose a universal stability analysis framework applicable to multiple microgrid structures, followed by a small-signal stability analysis specifically for meshed-structure microgrids. First, a unified state-space modeling approach is established to verify the applicability and generality of the proposed method across different microgrid topologies. On this basis, the influence of multi-parameter variations on the stability of the meshed-structure microgrid system is systematically evaluated, and time-domain stability analysis is further conducted. The overall research procedure is illustrated in Figure 1.

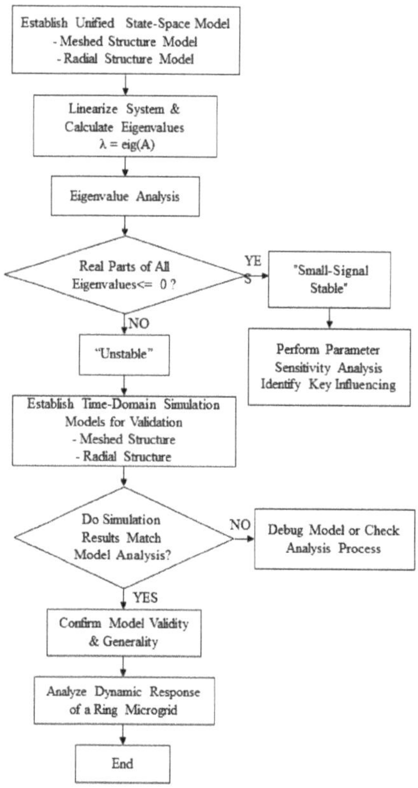

Figure 1. Flowchart or schematic.

This section presents the methodology employed in this study, which comprises two main components: mathematical modeling and stability analysis. For the mathematical modeling part, we develop models for two typical microgrid topologies—radial and meshed structures—and systematically elucidate the differences and interconnections in their modeling approaches, thereby laying a theoretical foundation for subsequent analysis. In terms of stability analysis, the eigenvalue method is adopted to assess the small-signal stability of the system, while parameter sensitivity analysis is conducted to identify key factors influencing system stability.

2.1. Model of Microgrid

Presently, most distributed power sources with power electronic interfaces for renewable energy generation employ three-phase inverters. Therefore, this paper focuses on a microgrid system based on three-phase inverter technology [25,26], as illustrated in Figure 2.

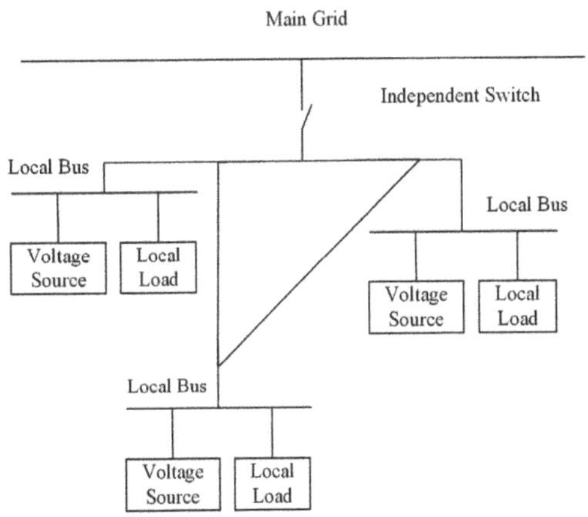

Figure 2. Schematic diagram of micro-grid structure based on the inverter.

The modeling method proposed in this paper divides the whole system into three sub-modules: inverter, network, and load. Each inverter is modeled in its own reference frame, and the rotation frequency of the microgrid system is determined by the rotation frequency of one of the selected inverters. The inverter model includes a droop power controller model, output filter model, LC filter model, and voltage and current controller model. The modules shown in Figure 3 will be explained in the following sections [27].

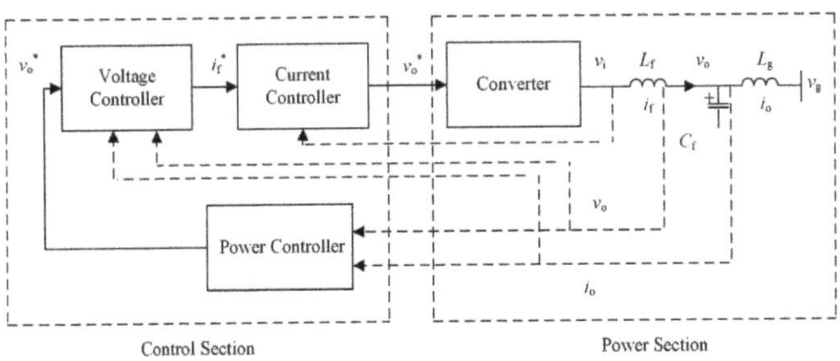

Figure 3. Control block diagram of droop control inverter.

The basic idea of droop control is to simulate the working characteristics of synchronous generators: by reducing the frequency of the output voltage to meet the increase in active power in the load, and by reducing the amplitude of the output voltage to meet the increase in reactive power in the load, the power control loop diagram is shown in Figure 4.

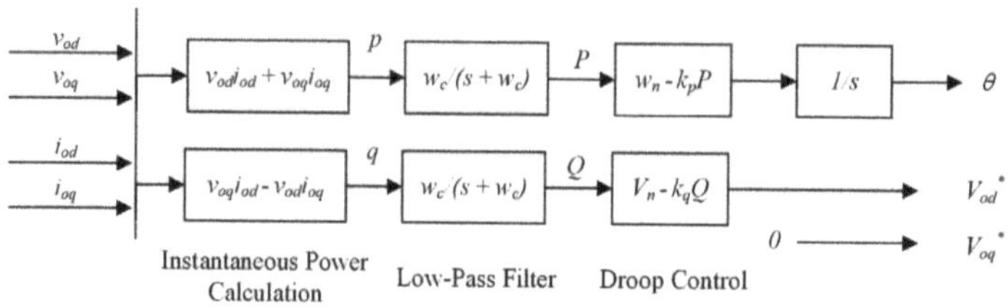

Figure 4. Power control loop diagram.

Based on the established meshed-structure microgrid model, this paper conducts a comparative study with radial-structure microgrids, validating the universality of the proposed modeling method for different network topologies. The detailed research process and corresponding results will be elaborated in the subsequent sections.

2.1.1. Modeling of a Single Inverter

By integrating the state models of each part, the overall small signal model of the droop control inverter can be obtained as follows.

$$[\Delta\dot{x}_{inv}] = A_{inv}[\Delta x_{inv}] + B_{inv}\left[\Delta v_{gdq}\right] + B_{\omega com}[\Delta\omega_{com}] \tag{1}$$

$$\begin{bmatrix}\Delta\omega \\ \Delta i_{odq}\end{bmatrix} = \begin{bmatrix}C_{inv\omega} \\ C_{invc}\end{bmatrix}[\Delta x_{inv}] \tag{2}$$

where

$$x_{inv} = \begin{bmatrix}\delta & P & Q & \varphi_{dq} & \gamma_{dq} & i_{fdq} & v_{odq} & i_{odq}\end{bmatrix}^T \tag{3}$$

$$A_{inv} = \begin{bmatrix} A_p & 0 & 0 & B_p \\ B_{v1}C_{pv} & 0 & 0 & B_{v2} \\ B_{c1}C_{pv}D_{v1} & B_{c1}C_v & 0 & B_{c1}D_{v2} + B_{c2} \\ B_{LCL1}C_{pv}D_{v1}D_{c1} & B_{LCL1}D_{c1}C_v & B_{LCL1}C_v & A_{LCL} + B_{LCL1}(D_{c1}D_{v2} + D_{c2}) \end{bmatrix} \tag{4}$$

$$B_{inv} = \begin{bmatrix} 0 \\ 0 \\ 0 \\ B_{LCL2} \end{bmatrix}, B_{\omega com} = \begin{bmatrix} B_{P\omega com} \\ 0 \\ 0 \\ 0 \end{bmatrix} \tag{5}$$

$$C_{inv\omega} = \begin{bmatrix} C_{P\omega} & 0 & 0 & 0 \end{bmatrix} \tag{6}$$

$$C_{invc} = \begin{bmatrix} 0 & 0 & 0 & \begin{bmatrix} 0 & 0 & I \end{bmatrix} \end{bmatrix} \tag{7}$$

To connect the inverter to the whole system, the output variables need to be converted to the common reference frame. In this case, the output variable of the inverter is expressed as the output current of the vector, and the small signal output current on the common reference frame can be obtained by using the transformation formula, such as Park.

$$[\Delta i_{oDQ}] = [T_S]\left[\Delta i_{odq}\right] + [T_C][\Delta\delta] \tag{8}$$

$$[T_S] = \begin{bmatrix} \cos(\delta_0) & -\sin(\delta_0) \\ \sin(\delta_0) & \cos(\delta_0) \end{bmatrix} \tag{9}$$

$$T_C = \begin{bmatrix} -I_{od}\sin(\delta_0) - I_{oq}\cos(\delta_0) \\ I_{od}\cos(\delta_0) - I_{oq}\sin(\delta_0) \end{bmatrix} \tag{10}$$

Similarly, the input signal of the inverter model is the bus voltage represented on the common reference frame. The bus voltage can be converted into a separate inverter reference frame using inverse transformation, which is given by the following formula:

$$\left[\Delta v_{gdq}\right] = \left[T_S^{-1}\right][\Delta v_{gDQ}] + \left[T_v^{-1}\right][\Delta\delta] \tag{11}$$

$$\left[T_S^{-1}\right] = \begin{bmatrix} -V_{gD}\sin(\delta_0) + V_{gQ}\cos(\delta_0) \\ -V_{gD}\cos(\delta_0) - V_{gQ}\sin(\delta_0) \end{bmatrix} \tag{12}$$

Each independent inverter model has 13 states, 3 inputs, and 2 outputs (except for the inverter with a common reference frame, which has 3 outputs).

$$\begin{gathered} \left[\Delta\dot{x}_{inv}\right] = A_{invi}[\Delta x_{inv}] + B_{invi}\left[\Delta v_{gDQ}\right] + B_{i\omega com}[\Delta\omega_{com}] \\ \begin{bmatrix} \Delta\omega i \\ \Delta i_{oDQi} \end{bmatrix} = \begin{bmatrix} C_{inv\omega i} \\ C_{invci} \end{bmatrix}[\Delta x_{inv}] \end{gathered} \tag{13}$$

$$A_{inv} = \begin{bmatrix} A_{pi} & 0 & 0 & B_{pi} \\ B_{v1i}C_{pvi} & 0 & 0 & B_{v2i} \\ B_{c1i}C_{pvi}D_{v1i} & B_{c1i}C_{vi} & 0 & B_{c1i}D_{v2i} + B_{c2i} \\ B_{LCL1i}C_{pvi}D_{v1i}D_{c1i} + B_{LCL2i}B_x & B_{LCL1i}D_{c1i}C_{vi} & B_{LCL1i}C_{vi} & A_{LCLi} + B_{LCL1i}(D_{c1i}D_{v2i} + D_{c2i}) \end{bmatrix} \tag{14}$$

$$C_{inv\omega i} = \begin{bmatrix} C_{P\omega} & 0 & 0 & 0 \end{bmatrix} \tag{15}$$

$$C_{invc} = \begin{bmatrix} \begin{bmatrix} T_{ci} & 0 & 0 \end{bmatrix} & 0 & 0 & \begin{bmatrix} 0 & 0 & T_{si} \end{bmatrix} \end{bmatrix} \tag{16}$$

$$B_x = \begin{bmatrix} T_v^{-1} & 0 & 0 \end{bmatrix} \tag{17}$$

2.1.2. The Combined Model of All Inverters

The small-signal state-space model of the inverter in the common rotating coordinate system can be obtained via Equations (18) to (19).

$$\left[\Delta\dot{x}_{invi}\right] = A_{invi}[\Delta x_{invi}] + B_{invi}\left[\Delta v_{gDQi}\right] + B_{1\omega com}C_{inv\omega i}[\Delta\omega_{com}] \tag{18}$$

$$\left[\Delta i_{oDQ}\right] = [C_{invci}][\Delta x_{invi}] \tag{19}$$

The rotational coordinate system selected for inverter 1 is regarded as the common rotational coordinate system. This enables the derivation of the state equations for the microgrid inverter network system, comprising all droop-controlled inverter units, as shown in Equation (20).

$$\begin{aligned} \left[\Delta \dot{X}_{INV}\right] &= A_{INV}[\Delta X_{INV}] + B_{INV}\left[\Delta v_{gDQ}\right] \\ \left[\Delta i_{oDQ}\right] &= [C_{INVc}][\Delta X_{INV}] \end{aligned} \tag{20}$$

where

$$[\Delta X_{INV}] = \begin{bmatrix} \Delta x_{inv1} & \Delta x_{inv2} & \Delta x_{inv3} \end{bmatrix} \tag{21}$$

$$\left[\Delta i_{oDQ}\right] = \begin{bmatrix} \Delta i_{oDQ1} & \Delta i_{oDQ2} & \Delta i_{oDQ3} \end{bmatrix} \tag{22}$$

$$A_{INV} = \begin{bmatrix} A_{inv1} + B_{1\omega com}C_{inv\omega 1} & 0 & 0 \\ B_{1\omega com}C_{inv\omega 1} & A_{inv2} & 0 \\ B_{1\omega com}C_{inv\omega 1} & 0 & A_{inv3} \end{bmatrix} \tag{23}$$

$$B_{INV} = \begin{bmatrix} B_{inv1} & 0 & 0 \\ 0 & B_{inv2} & 0 \\ 0 & 0 & B_{inv3} \end{bmatrix} \tag{24}$$

$$C_{INV} = \begin{bmatrix} C_{inv1} & 0 & 0 \\ 0 & C_{inv2} & 0 \\ 0 & 0 & C_{inv3} \end{bmatrix} \tag{25}$$

2.1.3. Network Bus Model

The small-signal model of the busbar line network at nodes i and $i + 1$ in the system may be expressed as given in Equation (26).

$$\Delta \dot{i}_{lineDQi} = A_{NETi}\left[\Delta i_{lineDQi}\right] + B_{1NETi}\left[\Delta v_{gDQ}\right] + B_{2NETi}[\Delta \omega_{com}] \tag{26}$$

$$A_{NETi} = \begin{bmatrix} -\frac{R_{linei}}{L_{linei}} & \omega_0 \\ -\omega_0 & -\frac{R_{linei}}{L_{linei}} \end{bmatrix} \tag{27}$$

$$B_{2NETi} = \begin{bmatrix} I_{lineQi} \\ -I_{lineDi} \end{bmatrix} \tag{28}$$

$$B_{1NET} = \begin{bmatrix} B_{1NET1} \\ B_{1NET2} \\ B_{1NET3} \end{bmatrix}, B_{2NET} = \begin{bmatrix} B_{2NET1} \\ B_{2NET2} \\ B_{2NET3} \end{bmatrix} \tag{29}$$

$$A_{NET} = \begin{bmatrix} A_{NET1} & 0 & 0 \\ 0 & A_{NET2} & 0 \\ 0 & 0 & A_{NET3} \end{bmatrix} \tag{30}$$

The modeling procedures for radial and meshed microgrids are largely similar, with the primary differences stemming from their structural configurations. In a radial structure, inverters are connected pairwise through a single bus in an adjacent manner—for example, the i-th inverter is not directly connected to the (i + 2)-th inverter. In contrast, in a meshed structure, any two inverters can be interconnected; for instance, the i-th inverter may connect not only to the (i + 1)-th but also directly to the (i + 2)-th inverter.

Furthermore, in a meshed configuration, the number of buses equals the number of inverters, and the greater diversity in line connections allows for more flexible mathematical modeling. Even for the same system, the state-space equations may take multiple forms. Additionally, the matrix dimensions of the line models differ between the two structures: the number of line loads in a meshed microgrid equals the number of inverters (i.e., three), whereas in a radial microgrid, it is one less than the number of inverters (i.e., two).

Therefore, when constructing the model, it is essential to clarify the connectivity and sequencing of the inverters and to build the corresponding line mathematical model using the following formulas.

2.1.4. Load Model

The RL load small signal model at the *i*-th node of the system can be expressed in the following (31).

$$\Delta \dot{i}_{loadDQi} = A_{NETi}\left[\Delta i_{loadDQi}\right] + B_{1LOADi}\left[\Delta v_{gDQ}\right] + B_{2LOADi}\left[\Delta \omega_{com}\right] \tag{31}$$

$$A_{LOADi} = \begin{bmatrix} -\frac{R_{loadi}}{L_{loadi}} & \omega_0 \\ -\omega_0 & -\frac{R_{loadi}}{L_{loadi}} \end{bmatrix} \tag{32}$$

$$B_{2LOADi} = \begin{bmatrix} I_{loadQi} \\ -I_{loadDi} \end{bmatrix} \tag{33}$$

$$B_{1LOAD} = \begin{bmatrix} B_{1load1} \\ B_{1load2} \\ B_{1load3} \end{bmatrix}, B_{2LOAD} = \begin{bmatrix} B_{2load1} \\ B_{2load2} \\ B_{2load3} \end{bmatrix} \tag{34}$$

$$A_{LOAD} = \begin{bmatrix} A_{LOAD1} & 0 & 0 \\ 0 & A_{LOAD2} & 0 \\ 0 & 0 & A_{LOAD3} \end{bmatrix} \tag{35}$$

The state-space formulation based on the load model can also be developed using Equation (31). In a meshed topology, each inverter and its local load form an independent unit. Consequently, the modeling methodology and approach are fundamentally the same for both microgrid structures, with the key distinction lying in the flexibility of the model formulation.

2.1.5. A Complete Microgrid Model

The complete microgrid consists of multiple inverters, closed-loop circuit line loads, and local loads of each inverter. The state equation and its small-signal state model can be constructed by combining the models of each subsystem, so the small-signal model of the microgrid system in the common rotating coordinate system is obtained.

$$\begin{gathered} \begin{bmatrix} \Delta \dot{X}_{INV} \\ \Delta \dot{i}_{lineDQ} \\ \Delta \dot{i}_{loadDQ} \end{bmatrix} = A_{MG} \begin{bmatrix} \Delta X_{INV} \\ \Delta i_{lineDQ} \\ \Delta i_{loadDQ} \end{bmatrix} + B_{MG}\left[\Delta v_{gDQ}\right] \\ \left[\Delta i_{oDQ}\right] = C_{MG} \begin{bmatrix} \Delta X_{INV} \\ \Delta i_{lineDQ} \\ \Delta i_{loadDQ} \end{bmatrix} \end{gathered} \tag{36}$$

Even though radial and meshed microgrids share an identical topological shape, the dimensions of their line load matrices differ, leading to a distinct dimensionality in their overall system models.

2.2. Stability Analysis

Eigenvalue analysis is a key means of studying the stability of linearized systems and essentially depends on the distribution characteristics of the system state matrix's eigenvalues in the complex plane. According to the theory of linear systems, the stability of a system can be strictly determined by the sign of the real part of an eigenvalue: if all the eigenvalues have negative real parts and lie the left-half plane, the system is asymptotically stable at the equilibrium point; conversely, if at least one of the eigenvalues has a positive real part, the system is unstable. When the real part of an eigenvalue is zero and its algebraic and geometric multiplicities are equal, the system is in a critical stable state. In practical applications, however, it is generally considered that the system is not robust. Furthermore, the absolute value of the real part of the eigenvalue reflects the attenuation rate of the corresponding mode, and the imaginary part determines the oscillation frequency. Notably, the dominant eigenvalue closest to the imaginary axis plays a pivotal role in the transient response performance of the system. Based on the fifty-one-dimensional microgrid system model that has been established, this paper provides a comprehensive analysis of the system's stability and dynamic characteristics by solving for all the eigenvalues according to the aforementioned distribution rules.

Sensitivity analysis utilizes sensitivity indices to evaluate system stability and is widely used in both conventional power systems and microgrids. In the context of eigenvalue analysis, the sensitivity of eigenvalues to variations in system parameters is examined to quantify their impact on state variables. This information, combined with eigenvalue data, facilitates the rational selection of key controller parameters, such as the power droop coefficient.

3. Results and Discussion

Based on the above analysis, the following simulation verification is conducted.

3.1. Stability Analysis

3.1.1. Eigenvalue Analysis

The eigenvalue analysis method, which extracts key stability indices—such as oscillation frequency, amplitude, and damping ratio—by solving the eigenvalues of the system state matrix, is widely employed in small-signal stability analysis of microgrids due to its conceptual clarity and rigorous criteria [28].

The dimension of the ring-structured microgrid system model is determined by the three parts: the inverter section, the busbar line section, and the load section. In this paper, the microgrid composed of three parallel inverters is taken as the research object. The model dimension of a single inverter is 13, corresponding to thirteen state variables. Therefore, the dimension of the subsystem composed of the three inverters is 39. On this basis, the overall dimension of the system is fifty-one-dimensional, corresponding to 51 eigenvalues. However, the dimensionality of the radial microgrid system model is 49.

Based on the data in Table 1, simulations were performed for the two structural systems, and their operating data at the steady-state working point were obtained (results are shown in Table 2). Figures 5 and 6 show the eigenvalue distributions of the state equations for the microgrid experimental system, calculated using MATLAB 2024a. As shown in the figures, all eigenvalues are located in the left half of the complex plane. According to linear system stability theory [29], this indicates that the microgrid experimental system is stable at the given operating point.

Table 1. Main parameters of the closed-loop microgrid system.

Classification of Parameters	Parameter Symbol	Base Value
System Base Parameters	Rated Capacity, S (kVA) Rated Line-to-Line Voltage, V (V) Rated Frequency, f (Hz)	50 380 50
Inverter Unit	DC-Link Voltage, V_dc (V)	220
LC Filter	Inverter-Side Inductance, L1 (H) Filter Capacitance, Cf (F)	3×10^{-3} 20×10^{-6}
Droop Control	Active Power-Frequency Droop Coefficient, m (rad/W×s) Reactive Power-Voltage Droop Coefficient, n (V/Var)	3.14×10^{-4} 1×10^{-3}
Other parameters	Grid-Side Inductance, L2 (H) Line inductance, Lline (H) Line resistance, Rline (Ω)	7×10^{-3} 3×10^{-5} 4×10^{-5}

After adjusting the key parameters according to the pattern observed in Table 3, these settings were applied to the simulation system. New simulation results were obtained and incorporated into the mathematical model for computation, yielding the new eigenvalue distribution shown in Figures 6 and 7. The figure indicates that the system eigenvalues have shifted overall to the right, with some eigenvalues migrating into the right-half plane. According to eigenvalue analysis theory, this distribution confirms that the system has become unstable.

Table 2. System stability parameter values.

Number of Inverters (i = 1, 2, 3)	Steady-State Operating Point Data for Ring-Structure Microgrids	Steady-State Operating Point Data for Radial Microgrids
Steady-state angular frequency (ω_0/rad·s^{-1})	314.16214	314.16226
The phase difference relative to the common rotating coordinate system (δi/rad)	(0, 0.0066, -5.59×10^{-8})	(0, -4.647×10^{-5}, 3.106×10^{-5})
Output voltage in common rotating coordinate system (V_{gDi}, V_{gQi})/V	(214.95, −13.11) (214.89, −12.95) (−214.94, −13.13)	(308.45, −33.56) (308.22, −33.59) (308.11, −33.60)
Inverter output voltage (V_{odi}, V_{oqi})/V	(214.89, −12.95) (−214.93, −13.09) (214.86, −12.92)	(310.36, 0.159) (310.18, 0.146) (310.11, 0.141)
Inverter output current (I_{odi}, I_{oqi})/A	(214.89, −12.95) (−214.95, −13.11) (214.83, −12.91)	(214.79, −12.15) (214.91, −12.54) (214.96, −12.72)
Inverter-side current (I_{fdi}, I_{fqi})/A	(215.15, 84.31) (214.99, 84.49) (215.07, 84.22)	(214.78, 85.36) (214.85, 84.86) (214.91, −84.81)
Inverter local load current (I_{loadDi}, I_{loadQi})/A	(0.0772, −0.01) (0.0767, −0.01) (0.0767, −0.01)	(0.07118, −0.2517) (0.7112, −0.2516) (0.7110, −0.2516)
Line current (I_{lineDi}, I_{lineQi})/A	(215.34, 0.031) (0, 0) (215.35, 0.032)	(429.167, 0.032) (214.617, 0.033)

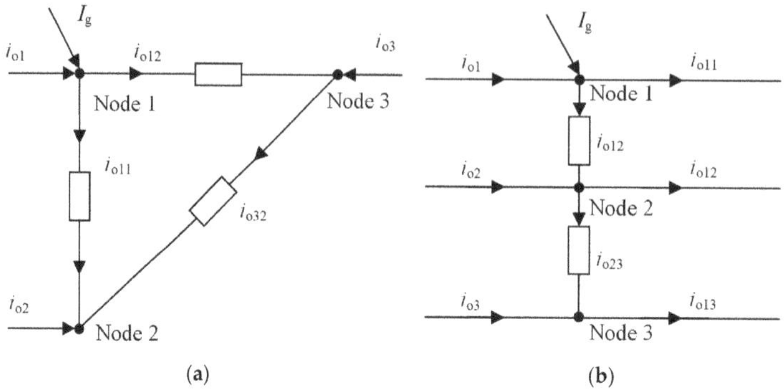

Figure 5. Network Connection Diagram for Two Types of Microgrid Systems. (**a**) Network connection diagram for a ring-structured microgrid system; (**b**) Network connection diagram for a radial-structured microgrid system.

Table 3. Changes in parameters.

Parameters	Kpc	Kpv
Stable Condition	95	3
Unstable Condition	0.95	30

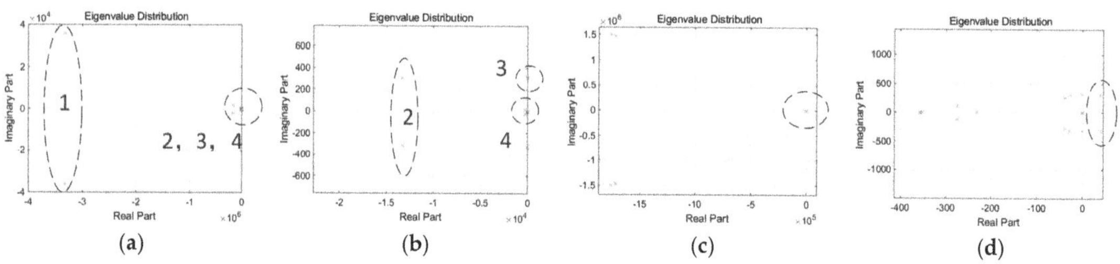

Figure 6. Eigenvalue Distribution of a State-Space Model for Ring-Structured Microgrid. (**a**) Overall View of Eigenvalues under Steady-State Conditions; (**b**) Zoomed-in View of Local Eigenvalues under Steady-State Conditions; (**c**) Overall View of Eigenvalues under Unsteady-State Conditions; (**d**) Zoomed-in View of Local Eigenvalues under Unsteady-State Conditions. The blue × in the diagram represents the eigenvalues.

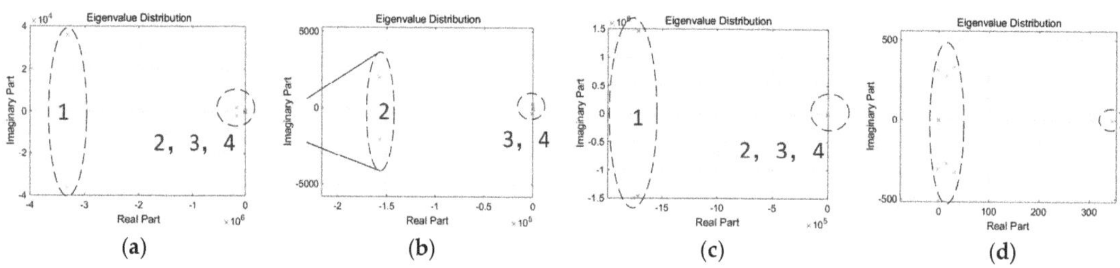

Figure 7. Eigenvalue Distribution of a State-Space Model for Radial-Structured Microgrid. (**a**) Overall View of Eigenvalues under Steady-State Conditions; (**b**) Zoomed-in View of Local Eigenvalues under Steady-State Conditions; (**c**) Overall View of Eigenvalues under Unsteady-State Conditions; (**d**) Zoomed-in View of Local Eigenvalues under Unsteady-State Conditions. The blue × in the diagram represents the eigenvalues.

3.1.2. Sensitivity Analysis

This study primarily focuses on sensitivity analysis for ring microgrids. Figure 6 shows the distribution of all 51 eigenvalues of the system under the initial conditions specified in Table 1. It can be observed that these eigenvalues exhibit a wide range of frequency components, which can be grouped into four distinct clusters.

The further the real part deviates from the imaginary axis, the faster the system's decay rate becomes, enabling quicker stabilization; conversely, the further the imaginary part deviates from the real axis, the higher the system's oscillation frequency becomes. As observed in Figure 6, the high-frequency mode labeled '1' corresponds to the state variables of the inverter's LC filter, exhibiting substantial real and imaginary components. During stable system operation, this mode exerts limited influence on overall stability. However, improper parameter design may induce high-frequency oscillations or resonances, thereby indirectly jeopardizing the system's stable functioning. As can be seen from Figure 6, dual voltage-current control determines the stability at medium and high frequencies. Unlike droop control, it does not directly render the system unstable, but rather determines the magnitude of the system's stability margin. It plays a crucial role in the local stability of the system, such as in the case of droop control numbers '2' and '3'. Droop control governs low-frequency oscillations. An improperly set droop coefficient causes the dominant system eigenvalue to approach the imaginary axis or even enter the right-half plane, thereby triggering low-frequency oscillations or voltage instability. Consequently, droop control represents the most sensitive element in system stability. As indicated by number '4', this mode lies closest to both the imaginary and real axes, rendering it most sensitive to parameter variations.

The distribution of parameter variations shown in Figure 8 indicates that changes in the LC filter parameters (Lf, Cf) and grid-side inductance Lg exert the most significant influence on system stability, potentially causing the system to transition from a stable to an unstable state. These parameters also exert a substantial effect on the oscillation frequency, and droop controller parameters play a pivotal role in system stability. The findings in Figure 8 further validate theoretical analysis: compared to power droop gain, the voltage and current loops primarily influence high-frequency characteristics within the system, while LC filter parameters predominantly affect low-frequency gain. Assessing eigenvalue displacement within identical parameter variation ranges reveals that the droop controller exhibits the highest sensitivity, followed by the LC filter. To better evaluate the impact of the three parameters on system stability, the relevant experimental results are consolidated in Table 4.

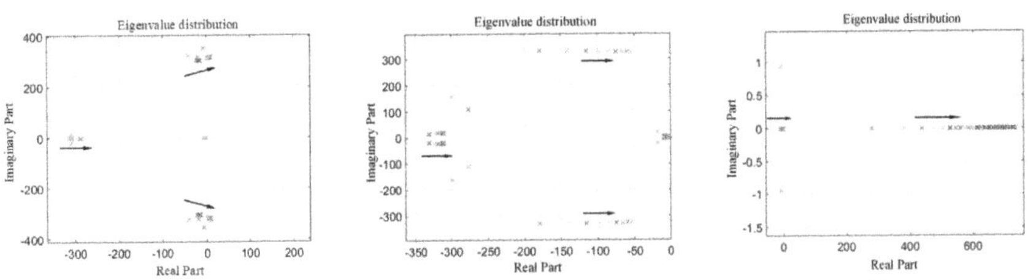

(a) (b) (c)

Figure 8. Changes in eigenvalues with LC parameters, voltage, and current loop coefficient and droop coefficient. (**a**) Changes in eigenvalues with LC parameters; (**b**) changes in eigenvalues with voltage and current loop coefficient; and (**c**) changes in eigenvalues with droop coefficient. The differently coloured "×" marks in the figure indicate the positions of eigenvalues corresponding to parameter variations, while the arrow directions denote the trajectory of eigenvalue shifts as the parameter progressively increases.

Table 4. The Effect of Parameter Variations on System Stability.

Varying Parameter	Variation Amount	Steady-State Result
Controller Gain Kp	+0 +10	Stable Unstable
Controller Gain Km	+0 $+1 \times 10^{-3}$	Stable Unstable
Filter Parameter L	+0 $+10.6 \times 10^{-3}$	Stable Unstable

3.1.3. Verification of the Model

Figures 9 and 10 show the Simulink simulation results for the meshed and radial microgrids, with their parameters corresponding to those in Tables 2 and 3, respectively.

Based on the simulation results in Figures 9 and 10, combined with previous research findings, it can be concluded that when using the parameters from Tables 1 and 2, all eigenvalues of the mathematical model are distributed in the left half-plane, indicating system stability. The stable waveforms in Figure 9a,b, obtained using Table 1 parameters, are consistent with the stability prediction of the theoretical model, thereby achieving mutual validation between the theoretical and simulation models. When parameters causing system instability are applied, eigenvalues appear in the right half-plane while the simulation waveforms exhibit instability, further demonstrating the effectiveness of this stability assessment method.

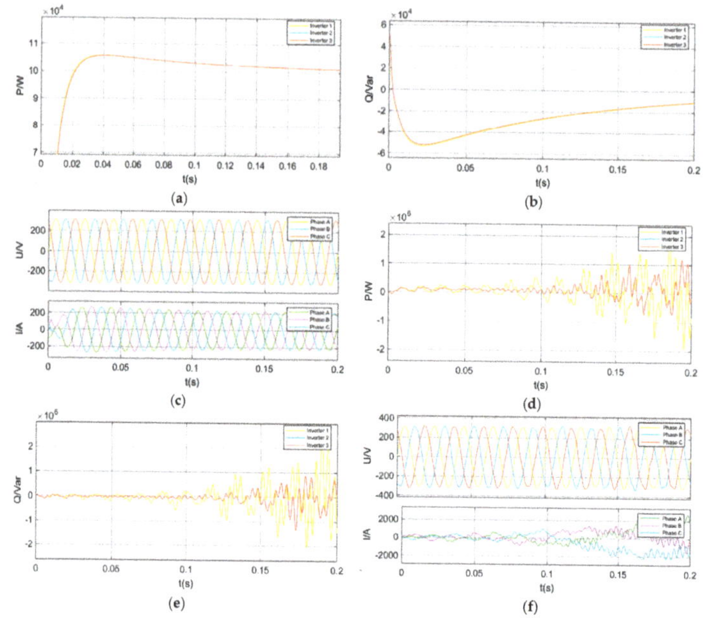

Figure 9. Simulated waveforms of active power, reactive power, voltage, and current output from the ring-structured microgrid structure under stable and unstable operating conditions. (**a**) Waveform of Output Active Power under Stable Conditions; (**b**) Waveform of Output Reactive Power under Stable Conditions; (**c**) Waveforms of Output Voltage and Current under Stable Conditions; (**d**) Waveform of Output Active Power under Unstable Conditions; (**e**) Waveform of Output Reactive Power under Unstable Conditions; (**f**) Waveforms of Output Voltage and Current under Unstable Conditions.

Similarly, Figure 10 presents simulation results for a radial-structured microgrid. By applying the same analytical method used for the mesh structure, mutual confirmation between the theoretical model and simulation results is again achieved, proving the method's applicability to radial-structured microgrids. These results demonstrate that the proposed stability analysis method possesses good versatility across different network configurations.

Thus, the established state-space model enables a rapid and precise assessment of system stability. Furthermore, the results serve to verify that the method is effective for stability analysis in both mesh-structured and radial-structured microgrids, confirming its suitability for the two configurations.

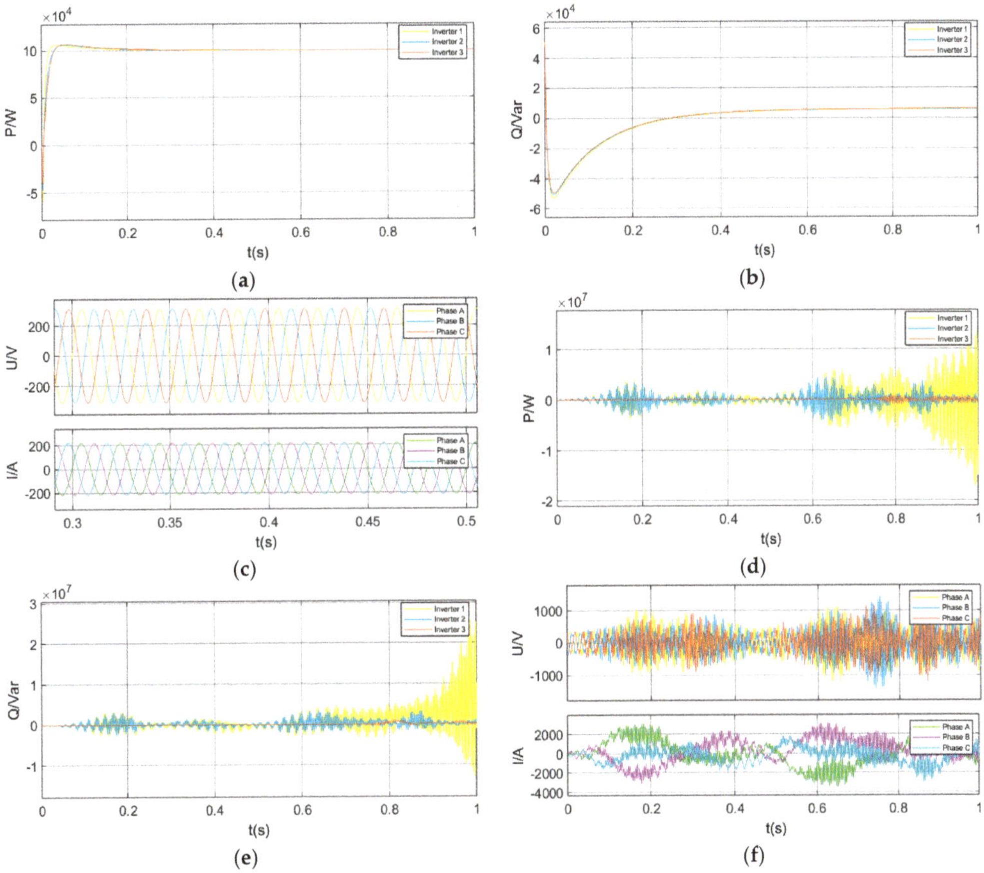

Figure 10. Simulated waveforms of active power, reactive power, voltage, and current output from radial microgrids under stable and unstable operating conditions. (**a**) Waveform of Output Active Power under Stable Conditions; (**b**) Waveform of Output Reactive Power under Stable Conditions; (**c**) Waveforms of Output Voltage and Current under Stable Conditions; (**d**) Waveform of Output Active Power under Unstable Conditions; (**e**) Waveform of Output Reactive Power under Unstable Conditions; (**f**) Waveforms of Output Voltage and Current under Unstable Conditions.

3.2. Time-Domain Stability Analysis

3.2.1. Case 1: A 5% Step Increase in the Active Power Command in Droop Control

In grid-connected mode, a step disturbance signal with an amplitude of 5% of the rated active power is applied at t = 0.5 s. The system dynamic response simulation results are shown in Figure 11. The specific analysis is as follows:

Figure 11a illustrates the variation in the system's active power. As can be observed, the active power exhibits an abrupt increase of 5% at t = 0.5 s. This step change occurs because the command value (P_0) is subjected to a step increase, whereas the inverter's output active power (P) does not respond instantaneously, thereby creating a transient power difference. This difference serves as an additional input to the droop controller, ultimately leading to an increase in the inverter's output power.

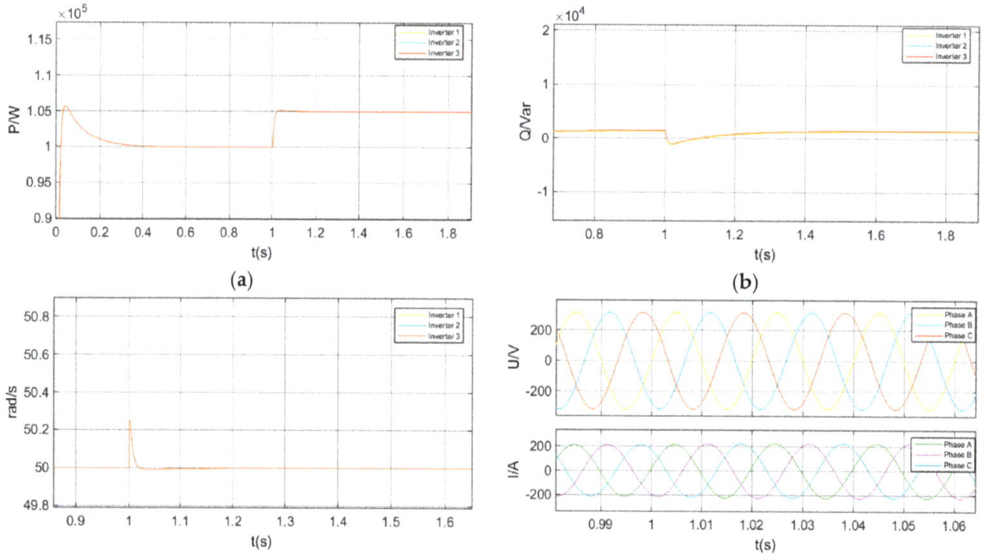

Figure 11. System Response to a 5% Increase in Active Power Command. (**a**) Waveform of Output Active Power; (**b**) Waveform of Output Reactive Power; (**c**) Waveform of Output Frequency; (**d**) Waveforms of Output Voltage and Current.

Figure 11b shows the variation in the system reactive power. It can be observed that the reactive power remains essentially unchanged. The inverter's reactive output is mainly determined by the reactive power command and is independent of active power, so it stays constant. During the transient, a slight dip occurs, but with grid support, it ultimately stabilizes and returns to the initial value. The occurrence of negative reactive power values in Figure 11b is a normal physical phenomenon. This is because during the transient of the step increase in active power command, the inverter's adjustment of output voltage phase and amplitude for maintaining system stability causes significant power coupling. This coupling effect leads to an initial dip in inductive reactive power (Q > 0), briefly bringing it into the capacitive region (Q < 0). The system then establishes a new steady state within 1 to 2 s, after which the reactive power returns to normal.

Figure 11c shows the variation in the system frequency. A brief frequency step can be observed. According to the droop relationship between active power and frequency, power variations cause frequency fluctuations. During dynamic regulation, the phase-locked loop (PLL) and dual-loop controller inside the inverter slightly increase the phase and frequency of the output voltage, resulting in a temporary frequency rise. Due to the strong frequency support provided by the grid in grid-connected mode, the inverter frequency resynchronizes with the grid frequency in steady state and eventually returns to its initial value.

Figure 11d displays the system output voltage and current waveforms. The output voltage remains constant. Based on the control relationship between reactive power and voltage, its steady-state value matches the initial value. The output current increases accordingly. Under the conditions of constant voltage, a 5% increase in active power, and unchanged reactive power, the current amplitude rises in accordance with the apparent power relationship.

3.2.2. Case 2: A 5% Step Decrease in the Reactive Power Command in Droop Control

In grid-connected mode, a step disturbance with an amplitude of −5% of the rated reactive power is applied at t = 0.5 s. The system dynamic response simulation results are shown in Figure 12 with the following specific analysis:

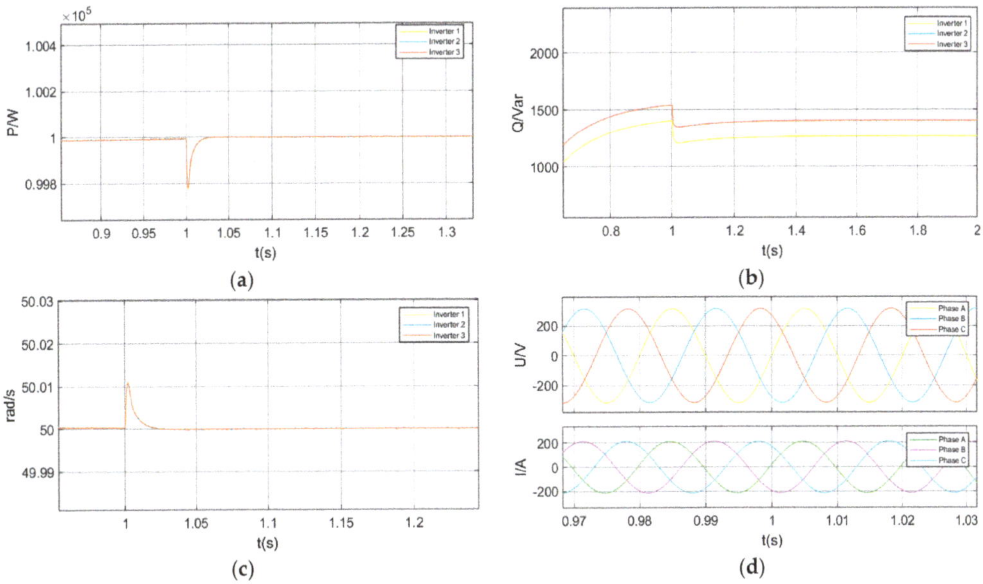

Figure 12. System Response to a 5% Decrease in Reactive Power Command. (**a**) Waveform of Output Active Power; (**b**) Waveform of Output Reactive Power; (**c**) Waveform of Output Frequency; (**d**) Waveforms of Output Voltage and Current.

Figure 12a shows the variation in the system's active power, while Figure 12c shows the change in system frequency. Within the range of small steady-state disturbances, the P–f loop and Q–V loop can be considered decoupled. Therefore, changes in reactive power have little effect on active power. At the same time, the grid frequency is clamped to its rated value. According to the active power–frequency control relationship, the frequency remains essentially constant during disturbances, with any transient fluctuations being negligible.

Figure 12b illustrates the variation in the system's reactive power. At t = 0.5 s, a sudden drop of approximately 5% occurs. This response is triggered by a step decrease in the reactive power setpoint Q_0. Since the actual reactive power output Q of the inverter cannot respond instantaneously, an instantaneous power deviation arises. This deviation is then fed back into the control system, prompting the inverter to reduce its reactive power output.

Figure 12d displays the waveforms of the system's output voltage and current.

Output voltage: Based on the reactive power–voltage control relationship, the inverter control attempts to reduce the output voltage magnitude. However, in grid-connected mode, the grid provides strong voltage support, effectively locking the voltage at the point of common coupling to the grid level. Therefore, the output voltage generally remains stable. Output current: With the voltage essentially unchanged, reactive power reduced by 5%, and active power held constant, the current amplitude should decrease accordingly according to the apparent power relationship. The waveform trend in the figure aligns with this expectation. Although the reduction is not very pronounced, using a cursor measurement tool reveals that the current decreases by approximately 1–2 amperes.

3.2.3. Case 3: A 5% Step Increase in the Active Power Load at the PCC in Droop Control

In grid-connected mode, when a step disturbance with an amplitude of 5% of the power load is applied at the PCC point at t = 0.6 s, the system dynamic response simulation results are shown in Figure 13. The specific analysis is as follows:

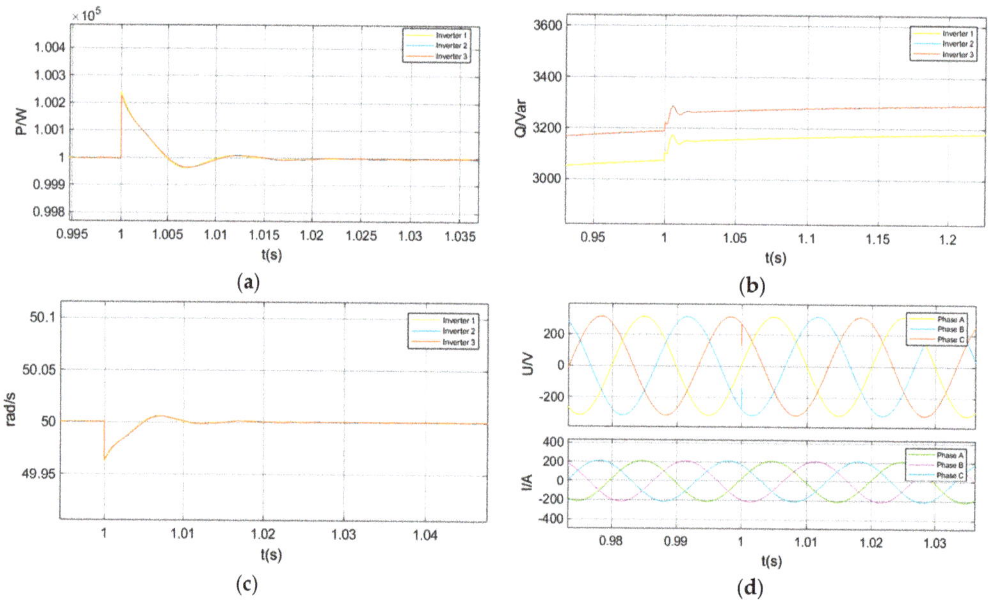

Figure 13. System Response to a 5% Step Increase in Power Load. (**a**) Waveform of Output Active Power; (**b**) Waveform of Output Reactive Power; (**c**) Waveform of Output Frequency; (**d**) Waveforms of Output Voltage and Current.

Figure 13a,c illustrate the variation in active power and frequency in the system, respectively. At t = 0.6 s, the active power demand at the PCC suddenly increases by 5%. To meet this sudden demand, the system draws on its inertia, resulting in a very small and rapid frequency drop. Based on the power-frequency droop characteristic, the local inverters then increase their active power output to counteract this frequency decline. This action reduces both the magnitude and the rate of the frequency drop, thereby helping to prevent false tripping of protection devices. Simultaneously, the main grid, with its substantial inertia and robust speed regulation system, begins injecting additional active power into the PCC. This injection strongly pulls the system frequency back toward and subsequently maintains it at the rated value. At this stage, the overall active power stabilizes at a level higher than its initial value. By using the cursor measurement tool, it can be observed that this net increase is primarily attributable to the additional active power supplied by the main grid.

Figure 13b shows the variation in reactive power in the system. Similarly to the active power profile, the reactive power curve exhibits a step increase at t = 1 s and then stabilizes at a higher level, indicating an increased reactive power demand in the system. The Q-V droop control loop automatically enhances the reactive power output to support the voltage, while the main grid also provides voltage support. The fact that the reactive power does not return to its initial value indicates that the three inverters collectively share the additional reactive load.

Figure 13d displays the system output voltage and current curves. The voltage experiences a very slight dip but quickly recovers and stabilizes. According to the relationship between power, voltage, and current, the increase in active power corresponds to a proportional rise in current, as shown in Figure 13.

The simulation results in Figures 11–13 reveal the system's dynamic response to a small disturbance: after transient oscillations, all observed variables converge to new steady-state values. Notably, these steady-state values are determined by the system operation mode and the specific components affected by the disturbance. In summary, the system demonstrates the ability to maintain stable operation following a minor disturbance.

In summary, the impacts on various system parameters under the three aforementioned scenarios are consolidated in Table 5. The table delineates the dynamic response characteristics of active power, reactive power, frequency, voltage, and current under typical small-disturbance conditions in grid-connected mode. The results demonstrate that the system maintains stability after being subjected to different types of small disturbances, validating the capability of the proposed ring structure to withstand such perturbations.

Table 5. Effects of Small Disturbances under Grid-Connected Mode.

Operating Condition	Active Power (P)	Reactive Power (Q)	Frequency (f)	Voltage and Current (V&C)
+5% Step in Rated Active Power	Increases	Remains unchanged	Rises then recovers	Slightly increases
−5% Step in Rated Reactive Power	Remains unchanged	Decreases	Remains unchanged	Remains unchanged
+5% Step in Power Load	Remains unchanged	Increases	Remains unchanged	Remains unchanged

4. Conclusions

This paper presents a systematic investigation into the stability of closed-loop microgrids, spanning theoretical modeling to simulation verification. By establishing an accurate state-space model supplemented with eigenvalue analysis and parameter sensitivity analysis, it thoroughly reveals the key modes and dominant factors influencing system stability. The research confirms that droop control coefficients and LC filter parameters exert the most significant influence on system dynamics, providing clear theoretical foundations and design guidance for optimizing microgrid controller parameters. Additionally, the validity of the mathematical model was verified through simulation experiments. Further time-domain simulation analyses were conducted, and multiple simulation experiments investigated the dynamic characteristics of the closed-loop structured microgrid under grid-connected operation. The experiments demonstrated that the time-domain approach can clearly reveal the system's dynamic behavior. Furthermore, stability analysis under grid-connected conditions was performed after introducing minor perturbations to the rated power and load power components, confirming that such slight disturbances do not compromise system stability in this mode.

However, faults in power systems (e.g., short circuits, generator tripping, and sudden load changes) constitute large disturbances, rendering small-signal stability analysis based on linearized models inadequate. The introduction of the Lyapunov direct method is essential to accurately assess transient stability under such conditions. Furthermore, the practical applicability of these theoretical studies in engineering remains to be fully investigated. Therefore, future work will focus on the following aspects:

(1) Investigate the transient stability of systems under large disturbances using Lyapunov's direct method. This involves constructing an appropriate Lyapunov function, designing an algorithm for stability region estimation, and evaluating its practical value in engineering applications.
(2) Examine the feasibility of applying small-signal stability analysis methods in engineering. The focus includes developing system models, studying and applying model order reduction techniques, and establishing a closed-loop process for model validation using field measurements.

Author Contributions: Conceptualization, B.C. and Y.R.; methodology, B.C.; validation, B.C. and Y.R.; investigation, B.C. and Y.R.; data curation, B.C. and Y.R.; writing—original draft preparation, B.C.; writing—review and editing, B.C. and Y.R.; visualization, B.C. and Y.R. All authors have read and agreed to the published version of the manuscript.

Funding: This study is not funded by external funding.

Data Availability Statement: The original contributions presented in this study are included in the article. Further inquiries can be directed to the corresponding author.

Conflicts of Interest: The authors declare no conflict of interest.

References

1. Arefifar, S.A.; Ordonez, M.; Mohamed, Y. Energy management in multi-microgrid systems—Development and assessment. In Proceedings of the 2017 IEEE Power & Energy Society General Meeting, Chicago, IL, USA, 17 July 2017; pp. 910–922.
2. Hossein, F.; Fotuhi-Firuzabad, M.; Moeini-Aghtaie, M. Enhancing Power System Resilience Through Hierarchical Outage Management in Multi-Microgrids. *IEEE Trans. Smart Grid.* **2016**, *7*, 2869–2879. [CrossRef]
3. Meng, T.; Lin, Z.; Wan, Y.; Shamash, A. An Output Regulation Approach to Distributed Voltage Regulation of Multiple Coupled Distributed Generation Units in DC Microgrids. In Proceedings of the 2022 American Control Conference (ACC), Atlanta, GA, USA, 8–10 June 2022; pp. 4534–4539.
4. Dong, Z.; Zhao, J.; Wen, F.; Xue, Y. From smart grid to energy Internet: Basic concept and research framework. *Autom. Electr. Power Syst.* **2014**, *38*, 1–11. [CrossRef]
5. Ma, Z.; Zhou, X.; Shang, Y.; Sheng, W. Exploring the Concept, Key Technologies, and Development Model of Energy Internet. *Power Syst. Technol.* **2015**, *39*, 3014–3022.
6. de la Cruz, J.; Wu, Y.; Candelo-Becerra, J.E.; Vásquez, J.C.; Guerrero, J.M. Review of Networked Microgrid Protection: Architectures, Challenges, Solutions, and Future Trends. *CSEE J. Power Energy Syst.* **2024**, *10*, 448–467. [CrossRef]
7. Sui, X.; Tang, Y.; He, H.; Wen, J. Energy-Storage-Based Low-Frequency Oscillation Damping Control Using Particle Swarm Optimization and Heuristic Dynamic Programming. *IEEE Trans. Power Syst.* **2014**, *29*, 2539–2548. [CrossRef]
8. Zeng, Z.; Zhao, R.; Lv, Z.; Ma, L. Impedance Reshaping of Grid-tied Inverters to Damp the Series and Parallel Harmonic Resonances of Photovoltaic Systems. *Proc. CSEE* **2014**, *34*, 4547–4558. [CrossRef]
9. Li, J.; Lü, Z.; Wu, Z.; Liu, H.; Yang, S. Adaptive switching strategy of AC/DC hybrid microgrid operating mode based on the power electronic transformer. *Electr. Power Autom. Equip.* **2020**, *40*, 126–131+138.

10. Sun, Z.; Yang, Z.; Wang, Y.; Xu, H. A Hybrid Islanding Detection Method for Distributed Multi-inverter Systems. *Proc. CSEE* **2016**, *36*, 3590–3597+3378.
11. Gu, X.; Yuan, X.; Zheng, H.; Xiong, W. Review of AC/DC hybrid microgrid stability analysis and its enhancement measures. *Distrib. Util.* **2025**, *42*, 8–20.
12. Guerrero, J.M.; Chandorkar, M.; Lee, T.L.; Loh, P.C. Advanced Control Architectures for Intelligent Microgrids—Part I: Decentralized and Hierarchical Control. *IEEE Trans. Ind. Electron.* **2013**, *60*, 1254–1262. [CrossRef]
13. Xu, Y.; Fang, N. Control strategy research and simulation analysis of an independent optical storage microgrid based on voltage stabilizing control. *Power Syst. Prot. Control.* **2020**, *48*, 67–74.
14. Feng, W.; Sun, K.; Guan, Y.; Josep, M.G.; Xiao, X. An Active Harmonic Grid-Connecting Current Suppression Strategy for Hierarchical Control-Based Microgrid. *Trans. China Electrotech. Soc.* **2018**, *33*, 1400–1409.
15. Li, M.; Zhang, X.; Fu, X.; Geng, H.; Zhao, W. Stability Studies on PV Grid-connected Inverters under Weak Grid: A Review. *Chin. J. Electr. Eng.* **2024**, *10*, 1–19. [CrossRef]
16. Lopes, J.P.; Moreira, C.L.; Madureira, A.G. Defining control strategies for Micro Grids islanded operation. *IEEE Trans. Power Syst.* **2006**, *21*, 916–924. [CrossRef]
17. Liu, X.; Zhang, G.; Pei, W.; Zhu, E.; Zhang, X. Current status and development trends of grid-type converters. *Acta Energiae Solaris Sin.* **2024**, *45*, 101–111.
18. Huang, X.; Shao, Y.; Gong, Y.; Liu, D.; Zhu, X.; Luo, J. Control technology for grid-forming energy storage under unbalanced working conditions of weak current networks. *Therm. Power Gener.* **2024**, *53*, 59–67.
19. Wu, Y.; Xiao, S. Static Voltage Stability Analysis of New Energy Islanded Microgrids Supported by Grid-forming Energy Storage. *Power Syst. Technol.* **2025**. [CrossRef]
20. Amer, T.; Wahba, A.; Abolila, A.; Galal, A.A. Optimizing stability and characteristics of a vibrating rigid body pendulum with energy harvesting device. *J. Low Freq. Noise Vib. Act. Control.* **2025**, *44*, 893–937. [CrossRef]
21. Amer, T.; Arab, A.; Galal, A. On the influence of an energy harvesting device on a dynamical system. *J. Low Freq. Noise Vib. Act. Control.* **2024**, *43*, 669–705. [CrossRef]
22. Lu, M.; Yang, Y.; Johnson, B.; Blaabjerg, F. An Interaction-Admittance Model for Multi-Inverter Grid-Connected Systems. *IEEE Trans. Power Electron.* **2019**, *34*, 7542–7557. [CrossRef]
23. Pogaku, N.; Prodanovic, M.; Green, T.C. Modeling, Analysis, and Testing of Autonomous Operation of an Inverter-Based Microgrid. *IEEE Trans. Power Electron.* **2007**, *22*, 613–625. [CrossRef]
24. Shi, J.; Zheng, Z.; Ai, Q. Modeling of DC micro-grid and stability analysis. *Electr. Power Autom. Equip.* **2010**, *30*, 86–90.
25. He, J.; Yun, W.L. An Enhanced Microgrid Load Demand Sharing Strategy. *IEEE Trans. Power Electron.* **2012**, *27*, 3984–3995. [CrossRef]
26. Wang, S.; Su, J.; Yang, X.; Du, Y. A review of the small signal stability of the microgrid. *Electr. Eng.* **2016**, *11*, 39–45.
27. Liu, H.; Bao, M.; Lu, S.; Duan, Q.; Shentu, L.; Huang, Y. Coordinated operation method of the distribution network and multiple microgrids considering network reconfiguration. *Electr. Power Autom. Equip.* **2025**, *45*, 114–121.
28. Xiang, Y.; Fu, M.; Zhang, A.; Dou, X.; Jiao, Y.; Yang, Y. Small signal stability analysis of AC/DC hybrid microgrid under isolated island operation. *Electr. Power Constr.* **2017**, *38*, 96–105.
29. Zheng, J.; Li, X.; Wang, Y.; Zhu, S.; Wang, X.; Zhu, H. Small-signal stability analysis of a microgrid switching to islanded mode. *Autom. Electr. Power Syst.* **2012**, *36*, 25–32.

energies

MDPI

Article

Active and Reactive Power Coordinated Optimization of Distribution Network–Microgrid Clusters Considering Three-Phase Imbalance Mitigation

Zhenhui Ouyang [1,2,*], Hao Zhong [1,2], Yongjia Wang [1,2], Xun Li [1,2] and Tao Du [1,2]

[1] College of Electrical Engineering and New Energy, China Three Gorges University, Yichang 443000, China; zhonghao022@163.com (H.Z.); 202308580121232@ctgu.edu.cn (Y.W.); 202308580121144@ctgu.edu.cn (X.L.); 17843818774@163.com (T.D.)

[2] Key Laboratory of Hubei Province for Operation and Control of Cascade Hydropower Stations, China Three Gorges University, Yichang 443002, China

* Correspondence: o15327890302@163.com

Abstract: With the continuous increase in the penetration of single-phase microgrids in low-voltage distribution networks (LVDNs), the phase asymmetry of source–load distribution has made the problem of three-phase imbalance increasingly prominent. To address this issue, this paper proposes an active–reactive power coordinated optimization model for distribution network–microgrid clusters considering three-phase imbalance mitigation. The model is formulated within a master–slave game framework: in the upper level, the distribution network acts as the leader, formulating time-of-use prices for active and reactive power based on day-ahead forecast data with the objective of minimizing operating costs. These price signals guide the flexible loads and photovoltaic (PV) inverters of the lower-level microgrids to participate in mitigating three-phase imbalance. In the lower level, each microgrid responds as the follower, minimizing its own operating cost by determining internal scheduling strategies and power exchange schemes with the distribution network. Finally, the resulting leader–follower game problem is transformed into a unified constrained model through strong duality theory and formulated as a mixed-integer second-order cone programming (MISOCP) problem, which is efficiently solved using the commercial solver Gurobi. Simulation results demonstrate that the proposed model fully exploits the reactive power compensation potential of PV inverters, significantly reducing the degree of three-phase imbalance. The maximum three-phase voltage unbalance factor decreases from 3.98% to 1.43%, corresponding to an overall reduction of 25.87%. The proposed coordinated optimization model achieves three-phase imbalance mitigation by leveraging existing resources without the need for additional control equipment, thereby enhancing power quality in the distribution network while ensuring economic efficiency of system operation.

Keywords: low-voltage distribution network; photovoltaic inverter; three-phase imbalance; distributed photovoltaics; multi-microgrid; flexible load

1. Introduction

With the rapid development of renewable energy and the large-scale integration of distributed generation, the traditional operation paradigm of distribution networks is undergoing profound transformation. As an emerging distribution network architecture, the hybrid integration of single- and three-phase microgrids has attracted significant attention in recent years owing to its flexibility and efficiency [1–3]. By embedding single-phase microgrids into three-phase distribution networks, diverse distributed resources can be coordinated and optimally scheduled, thereby improving resource utilization efficiency and alleviating operational pressure on distribution networks. However, this transformation also poses a series of technical challenges related to the operational characteristics of hybrid single–three-phase distribution networks, including three-phase imbalance mitigation, uncertainty management of distributed energy, and optimal scheduling of flexible loads [4–6]. In this context, the coordinated operation between transmission and distribution systems is particularly important. The traditional 'passive' distribution network is gradually evolving to 'active', and the transmission system needs to share flexible resources to jointly deal with systemic problems such as voltage instability and line congestion. For example, the F-channel platform developed by Greece under the framework of 'OneNet' project in Reference [7] realizes the collaborative optimization of transmission and distribution networks in congestion management, frequency and voltage control through artificial intelligence and high-resolution weather forecast, and shows the great potential of cross-voltage level collaborative management in improving system flexibility and power quality.

At present, domestic and foreign scholars have formed a variety of technical routes in the treatment of three-phase imbalance in the distribution network, and have made remarkable progress. The existing achievements can be roughly divided into two categories: one is based on the scheme of external governance equipment. Literature [8] proposes two modes of commutation switch photovoltaic regulation and flexible load demand response to solve the problem of three-phase imbalance caused by distributed photovoltaic access, and optimize new energy consumption and reduce governance costs. Reference [9] proposed a multi-objective optimization model of active distribution network with commutation soft switching based on three-phase four-wire back-to-back voltage source converter topology. Reference [10] proposed an optimization model combining reactive power compensation devices such as on-load voltage regulating transformers and packet switching capacitors. Reference [11] proposed a three-phase four-wire DSTATCOM reference current calculation method based on PR controller and particle swarm optimization to achieve zero-sequence, harmonic and reactive power compensation, and optimize active power fluctuation control. These methods can effectively improve the three-phase balance by directly injecting compensation current or adjusting the network structure, but they usually require additional investment, and the economy needs to be improved. The second is to tap the potential of existing distributed resources, which provides a new way of thinking for the governance model. For example, reference [12] proposed a three-phase unbalance optimization method for distribution network based on iterative linearization, which uses photovoltaic inverter regulation to adjust voltage by optimizing active and reactive power output to reduce the three-phase unbalance degree of distribution network. Reference [13] proposed a voltage-reactive power coordinated control method based on low-voltage distributed photovoltaic inverters to flexibly compensate the reactive power demand of the distribution network. Reference [14] constructed a two-stage optimization model through photovoltaic phase selection switching, inverter reactive power regulation and energy storage active power regulation, which effectively reduced the three-phase unbalance and network loss of the distribution network. These studies achieve unbalanced governance by optimizing control, indicating that existing distributed resources themselves have the potential to participate in system regulation.

Although the above research has made important progress, there are still obvious limitations and room for expansion. First of all, most studies focus on centralized optimal control at the technical level, and lack effective market mechanisms and economic incentives to guide the active participation of the lower microgrid. Secondly, in terms of methodology and research perspective, existing work, such as reference [15], focuses on data-driven black-box optimization, which is good at dealing with uncertainty but difficult to provide verifiable optimal benchmarks in deterministic scenarios. Reference [16] focused on passive safety assessment after failure, rather than pre-prevention through market mechanisms. In addition, in the study of three-phase imbalance control, the existing schemes mostly consider the regulation of active or reactive power in isolation, and fail to form a complete solution to synergistically utilize the potential of the two under the market mechanism.

In view of the above shortcomings, this paper proposes a framework for active and reactive power collaborative optimization of distribution network-microgrid group based on master-slave game and price incentive. The contribution of this paper lies in methodology, the unified MISOCP solution strategy strictly guarantees the existence and feasibility of the equilibrium solution of the game. From the perspective of research, based on 'active prevention', we guide the pre-optimization of resources by constructing a day-ahead market mechanism. On the technical mechanism, based on the master-slave relationship between the distribution network and each microgrid in the electricity market, this paper uses economic incentives to guide the power generation side of each microgrid in the lower layer to use existing resources and load side control for three-phase imbalance management. The distribution network guides the flexible load demand response of each microgrid through the active time-of-use electricity price, and guides the photovoltaic inverters of each microgrid through the reactive power price incentive to actively participate in the three-phase imbalance management. The active and reactive power collaborative optimization model of distribution network-microgrid group considering three-phase imbalance management is established. As the leader of the upper distribution network, based on the day-ahead forecast data, the active and reactive time-of-use electricity prices for the lower microgrids are formulated to minimize the operating costs. Each microgrid in the lower layer acts as a respondent, and formulates a power interaction plan and an internal scheduling scheme designed to minimize its own operating costs based on the electricity price information provided by the upper layer. Through the unified solution of the master-slave game of distribution network-multi-microgrid, the formulation and implementation of the optimal operation strategy are finally realized. Finally, the effectiveness of the proposed model is verified by simulation examples.

The structure of this paper is as follows: In the second section, the collaborative optimization framework of distribution network-microgrid group is constructed, and the master-slave game mechanism based on electricity price incentive is clarified. In the third section, a refined model of distributed resources is established, including reactive power regulation of photovoltaic inverters, flexible load demand response and multi-energy coupling constraints. In the fourth section, a unified solution method of two-level model based on strong duality theory and KKT condition is proposed, which transforms the master-slave game into a mixed integer second-order cone programming problem that can be solved efficiently. In the fifth section, the effectiveness of the model is verified by multi-scenario simulation, and the performance differences in SVG governance, flexible load response and comprehensive coordination are compared, and the influence of photovoltaic uncertainty processing is analyzed. The sixth section discusses the advantages of the proposed method in engineering applications; the seventh section summarizes the core conclusions and proposes future research directions.

2. Construction of Coordinated Optimization Framework

The distribution network employs time-of-use electricity prices for active and reactive power to guide flexible load demand response in downstream microgrids, while using inverter reactive power output to mitigate three-phase imbalance and improve power quality. The basic structure of the system is shown in Figure 1. A, B and C are three single-phase microgrids respectively.

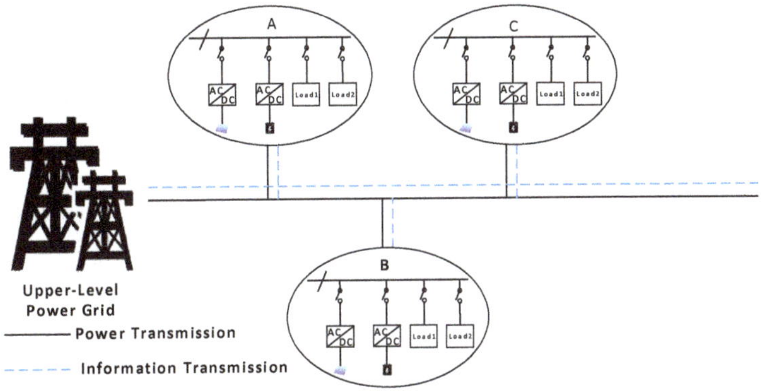

Figure 1. Basic Structure of the Distribution Network–Microgrid Cluster.

This paper proposes a coordinated active and reactive power optimization model for distribution networks with microgrid clusters, considering three-phase imbalance mitigation. First, the distribution network determines the active and reactive power prices and issues them to the downstream microgrids. Each microgrid then develops a power interaction plan with the distribution network and an internal scheduling scheme based on the information provided, and feeds the optimized power curves back to the upstream network. Next, the distribution network updates the issued active power prices and reactive power compensation prices according to the feedback from the microgrids. Finally, the entire model is optimized using a unified solution method, resulting in the formulation and implementation of the optimal operation strategy. The detailed procedure is illustrated in Figure 2.

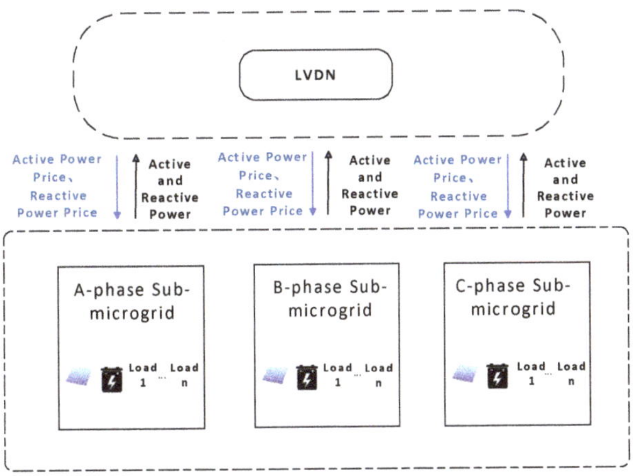

Figure 2. Leader–Follower Game Framework between the Distribution Network and Microgrids.

3. Modeling of Distributed Resources Considering Three-Phase Imbalance Mitigation

3.1. Modeling of Photovoltaic Inverters

The idle capacity of inverters in single-phase microgrid photovoltaic systems is fully utilized to provide reactive power compensation for the distribution network, enabling flexible adjustment of reactive power output and mitigation of three-phase imbalance [17–20]. The maximum reactive power output of a photovoltaic inverter is calculated as follows:

$$\begin{cases} Q_{\mathrm{PV}}^{\max} = \sqrt{S_N^2 - P_{\mathrm{PV}}^2} \\ P_{p,PV}^2(t) + Q_{p,PV}^2(t) \le S_{p,N}^2 \end{cases} \quad (1)$$

where p denotes the phase identifier, taking values from the set {A, B, C}, representing phases A, B, and C, respectively; $P_{p,PV}(t)$ represents the active power output of the PV inverter on phase pat time t; $Q_{p,PV}(t)$ represents the reactive power output of the PV inverter on phase pat time t; $S_{p,N}$ denotes the rated capacity of the inverter on phase p; P_{PV} is the rated active power of the grid-connected PV system; $Q_{\mathrm{PV}}^{\max}$ signifies the maximum reactive power output capability of the inverter.

3.2. Modeling of Flexible Loads

Flexible loads, as key controllable resources, can respond to price signals through load shifting and power curtailment. By guiding their operation with appropriately designed time-of-use electricity prices, flexible loads can not only reduce the system load peak-to-valley difference but also enhance the distribution network's capacity to accommodate distributed photovoltaic generation, thereby mitigating voltage fluctuations and three-phase imbalance to a certain extent.

(1) Price Response Characteristics of Shiftable Loads

$$\begin{cases} P_{\mathrm{tr}}^t = P_{\mathrm{tr0}}^t[1 + \varepsilon_{\mathrm{tr}}^t(\alpha_t - \alpha_t^0)/\alpha_t^0] \\ \sum\limits_{t=1}^{T} P_{\mathrm{tr0}}^t = \sum\limits_{t=1}^{T} P_{\mathrm{tr}}^t \\ \tau_t P_{\mathrm{tr}}^{\min} \le P_{\mathrm{tr}}^t \le \tau_t P_{\mathrm{tr}}^{\max} \end{cases} \quad (2)$$

where P_{tr}^t and P_{tr0}^t represent the shiftable load at node i under the dynamic electricity price and the benchmark electricity price at time t, respectively; $\varepsilon_{\mathrm{tr}}^t$ is the price elasticity coefficient of the shiftable load at node i; α_t and α_t^0 denote the electricity purchase price and the benchmark electricity price at time t, respectively; τ_t is a binary variable (0–1), where $\tau_t = 1$ indicates that the shiftable load transfer occurs at time t; $P_{\mathrm{tr}}^{\max}$ and $P_{\mathrm{tr}}^{\min}$ represent the upper and lower limits of the node power after load transfer, respectively.

(2) Price Response Characteristics of Curtailable Loads

$$\begin{cases} P_{\mathrm{cut}}^t = v_t P_{\mathrm{cut0}}^t[1 + \varepsilon_{\mathrm{cut}}^t(\alpha_t - \alpha_t^0)/\alpha_t^0] \\ \sum\limits_{t}^{t+t_{\mathrm{cut}}^{\max}} (1 - v_t) \ge 1 \\ \sum\limits_{t}^{t+t_{\mathrm{cut}}^{\min}-1} v_t \ge t_{\mathrm{cut}}^{\min}(v_t - v_{t-1}) \end{cases} \quad (3)$$

where P_{cut}^t and P_{cut0}^t represent the curtailable load at node i under the dynamic electricity price and the benchmark electricity price at time i, respectively; v_t is a binary variable (0–1), where $v_t = 1$ indicates that the curtailable load curtailment occurs at time t;

$\varepsilon^t_{\text{cut}}$ is the price elasticity coefficient of the curtailable load at node i; and $t^{\min}_{\text{cut}}$ denote the maximum and minimum allowable duration for load curtailment, respectively.

3.3. Load and Distributed Generation Models

The injection current direction is adopted as the reference direction, and both loads and distributed generation are modeled using constant power models. The models, along with the three-phase and four-wire low-voltage distribution network (LVDN) power flow equations, are described in [14].

4. Bi-Level Optimal Scheduling Model

4.1. Distribution Network Model

4.1.1. Objective Function of the Distribution Network

The distribution network aims to minimize its own operating cost and determines the time-of-use active power prices and reactive power compensation prices for each microgrid. Its objective function is formulated as follows:

$$\min C_1 = \sum_{t=1}^{T}\left[\sum_{p=1}^{3} C^t_{\text{Grid},i,p} + \sum_{p=1}^{3} C^t_{\text{Loss},i,p} + C^t_g\right]\Delta t + \alpha f_c \quad (4)$$

where $p = 1,\ 2,\ 3$ correspond to phases A, B, and C, respectively; $C^t_{\text{Grid},i,p}$ represents the transaction cost between the distribution network and the microgrid at node i on phase p during time period t; $C^t_{\text{Loss},i,p}$ denotes the network loss cost associated with the microgrid at node i on phase p during time period t; C^t_g signifies the cost incurred by the distribution network for purchasing active power from the superior grid during time period t; f_c is the total investment cost of static var generator (SVG); α is a coefficient (value is 0 or 1), 1 is taken when SVG is installed, and 0 is taken when SVG is not installed.

(1) Cost of Purchasing Electricity from Microgrids

$$C^t_{\text{Grid},i,p} = (\delta^t_{buy,i,p}P^t_{buy,i,p} - \delta^t_{\text{sell},i,p}P^t_{\text{sell},i,p} + \delta^t_{\text{Q},i,p}Q^t{}_{i,p}) \quad (5)$$

where $\delta^t_{buy,i,p}$, $\delta^t_{\text{sell},i,p}$, and $\delta^t_{\text{Q},i,p}$ represent the electricity prices at which the distribution network purchases active power from, sells active power to, and purchases reactive power from the microgrid at node i on phase p during time period t, respectively; $P^t_{buy,i,p}$, $P^t_{\text{sell},i,p}$, and $Q^t{}_{i,p}$ denote the active power purchased by the distribution network from, active power sold by the distribution network to, and reactive power purchased by the distribution network from the microgrid at node i on phase p during time period t, respectively.

(2) Network Loss Cost

$$C^t_{\text{Loss},p} = \sum_{l=1}^{n-1}\sum_{p=1}^{3} P_{\text{Loss}l,p}\delta^t_{\text{sell},p} \quad (6)$$

where n represents the total number of nodes in the system network; $l = 1, 2, \cdots, n-1$ indexes the branches; $P_{\text{Loss}l,p}$ represents the power loss on phase p of branch l during time period t.

(3) Cost of Purchasing Electricity from the Superior Grid

$$C^t_g = \delta^t_p P^t_g \quad (7)$$

where δ^t_p denotes the price at which the distribution network purchases active power from the superior grid during time period t; P^t_g represents the active power purchased by the distribution network from the superior grid during time period t.

(4) Cost of SVG

The total investment cost f_c of SVG includes the investment construction cost C_{in} and the operation and maintenance cost C_{om} of SVG. The specific calculation formula is:

$$f_c = C_{in} + C_{om} \tag{8}$$

$$C_{in} = c_{SVG} \sum_{i=1}^{n} S_{SVG,i} \tag{9}$$

$$C_{om} = k_{SVG} C_{in,SVG} \tag{10}$$

where c_{SVG} represents the investment and construction cost per unit capacity of SVG; $S_{SVG,i}$ represents the installation capacity of SVG on the i th node; k_{SVG} represents the annual operation and maintenance cost coefficient of SVG.

4.1.2. Constraints

(1) Electricity Purchase and Sale Constraints

$$0 \leq P^t_{buy,i,p} \leq P^t_{buy,i,p,\max} \tag{11}$$

$$0 \leq Q^t_{\mathrm{buy},i,p} \leq Q^t_{\mathrm{buy},i,p,\max} \tag{12}$$

where $P^t_{buy,i,p,\max}$ is the maximum value of the active power sold to the distribution network by the node i microgrid p in the period t; $Q^t_{\mathrm{buy},i,p,\max}$ is the maximum value of reactive power sold by node i microgrid to distribution network in t period.

(2) Electricity Price Constraints

$$(1-\beta)\delta^t_{\mathrm{f}} \leq \delta^t_{buy,i,p} \leq (1+\beta)\delta^t_{\mathrm{f}}, \forall t, \forall p, \forall i \tag{13}$$

$$(1-\beta)\delta^t_{\mathrm{f}} \leq \delta^t_{sell,i,p} \leq (1+\beta)\delta^t_{\mathrm{f}}, \forall t, \forall p, \forall i \tag{14}$$

where β represents the electricity price adjustment coefficient, and its numerical range is set between 0 and 1. Based on the time-of-use electricity price of the main network, the distribution network can flexibly set the time-of-use purchase and sale electricity price for each micro-network within a certain range, so as to give full play to the incentive role of the time-of-use electricity price in guiding electricity consumption behavior.

(3) Node Voltage Constraints

$$U_{\min,i} \leq |U_{i,p,t}| \leq U_{\max,i} \tag{15}$$

where $U_{\min,i}$ and $U_{\max,i}$ are the minimum and maximum allowable values of node voltage, respectively. According to relevant regulations, the allowable range of voltage offset is −10%~7%.

(4) Branch Current Constraints

$$|I_{ij,t}| \leq I_{\max,ij} \tag{16}$$

where $I_{ij,t}$ represents the actual current carried by the branch ij at time t; $I_{\max,ij}$ represents the maximum allowable current that branch ij can withstand at time t.

(5) Three-Phase Imbalance Constraints

The three-phase imbalance index f_v is defined as the sum of three-phase imbalances at the critical buses of the LVDN. It is calculated as follows:

$$f_v = \sum_{t=1}^{T}\sum_{i=1}^{n} U_{\text{VUF}i,t} \tag{17}$$

where i is the bus (node) number, and $U_{\text{VUF},i,t}$ is the three-phase unbalance degree of the bus (node) i.

The formula of $U_{\text{VUF},i,t}$ is:

$$U_{\text{VUF},i,t} = \left|\frac{V_{\text{neg},i,t}}{V_{\text{pos},i,t}}\right| = \left|\frac{V_{\text{a},i,t} + \alpha^2 V_{\text{b},i,t} + \alpha V_{\text{c},i,t}}{V_{\text{a},i,t} + \alpha V_{\text{b},i,t} + \alpha^2 V_{\text{c},i,t}}\right| \tag{18}$$

where $V_{\text{neg},i,t}$ and $V_{\text{pos},i,t}$ refer to negative sequence voltage and positive sequence voltage, respectively. V_a, V_b, V_c are three-phase voltages of a, b and c, respectively. Parameter $\alpha = 1\angle 120°$.

The three-phase imbalance of the distribution network must remain within the allowable range specified by power system operation standards, that is:

$$U_{\text{VUF},i,t} \le U_{\text{VUF,max}} \tag{19}$$

where $U_{\text{VUF,max}}$ is the maximum allowable value of the three-phase voltage unbalance degree of the common connection point of the distribution network system, and 2% is taken according to the relevant regulations.

4.2. Microgrid Model

4.2.1. Objective Function of the Microgrid

Each microgrid, based on the active power purchase and sale prices and reactive power compensation prices set by the distribution network, aims to minimize its own operating cost and develops both the active power interaction plan with the distribution network and its internal scheduling scheme. Its objective function is formulated as follows:

$$\min C_2 = \sum_{t=1}^{n}\begin{bmatrix} C^t_{\text{w},i,p} + C^t_{\text{fl},i,p} + \\ C^t_{\text{ESS},i,p} + C^t_{\text{PV},i,p} \end{bmatrix}\Delta t \tag{20}$$

where $C^t_{\text{w},i,p}$ is the cost of purchasing electricity from the microgrid to the distribution network during the period; $C^t_{\text{fl},i,p}$ is the flexible load scheduling cost of the microgrid during the period; the $C^t_{\text{ESS},i,p}$ and $C^t_{\text{PV},i,p}$ are the energy storage and photovoltaic operating costs of the microgrid during the period, respectively.

(1) Electricity Purchase Cost

$$C_{\text{w},i,p} = \delta^t_{sell,i,p} P^t_{sell,i,p} - \delta^t_{buy,i,p} P^t_{buy,i,p} - \delta^t_{Q,i,p} {Q^t}_{i,p} \tag{21}$$

(2) Flexible Load Scheduling Cost

$$C_{\text{fl},i,p} = \sum_{t=1}^{T}\sum_{p=1}^{3} \delta^t_{fl,i,p}\left(P^t_{i,p,\text{cut}} + P^t_{i,p,\text{tr}}\right) \tag{22}$$

where $C_{\text{fl},i,p}$ represents the unit cost of the flexible load in the i-node phase microgrid; $P^t_{i,p,\text{cut}}$ and $P^t_{i,p,\text{tr}}$ represent the reducible and transferable load of i-node p-phase microgrid in t period, respectively.

(3) Energy Storage Operating Cost

$$C^t_{\text{ESS},i,p} = w_{\text{es}}\left(P^t_{\text{in,i,p}} + P^t_{\text{out,i,p}}\right) \tag{23}$$

where w_{es} is the operating cost coefficient of energy storage; $P_{in,i,p}^{t}$ and $P_{out,i,p}^{t}$ respectively represent the charging and discharging power of the energy storage of the i-node p-phase microgrid in the t-period.

(4) Photovoltaic Operating Cost

$$C_{PV,i,p}^{t} = w_{pv} P_{pv,i,p}^{t} \tag{24}$$

where w_{pv} is the operating cost coefficient of photovoltaic; $P_{pv,i,p}^{t}$ is the photovoltaic output power of the i-node p-phase microgrid during the period.

4.2.2. Constraints

(1) Power Balance Constraints

$$P_{pv}^{t} + P_{out}^{t} + P_{buy}^{t} = P_{in}^{t} + P_{sell}^{t} + P_{ld}^{t} + P_{fl}^{t}, \forall t \tag{25}$$

where P_{pv}^{t} is the photovoltaic power generation in the period; P_{out}^{t} is the discharge power of energy storage in t period; P_{buy}^{t} is the power value transmitted from the distribution network to the microgrid during the t period; P_{in}^{t} is the charging power of energy storage in t period; P_{sell}^{t} is the power transmitted from microgrid to distribution network in t period; P_{ld}^{t} is the rigid load power in t period; P_{fl}^{t} is the flexible load power in t period.

(2) Energy Storage Constraints

$$SOC^{t+1} = SOC^{t}(1-\sigma) + \frac{\eta P_{in}^{t}\Delta t}{S_{cap}} - \frac{P_{out}^{t}\Delta t}{\eta S_{cap}} \tag{26}$$

$$SOC^{1} = SOC^{n+1} = SOC_{s} \tag{27}$$

$$SOC_{min} \leq SOC^{t} \leq SOC_{max}, \forall t \tag{28}$$

$$0 \leq P_{in}^{t} \leq P_{in,max}, \forall t \tag{29}$$

$$0 \leq P_{out}^{t} \leq P_{out,max}, \forall t \tag{30}$$

where SOC^{t} represents the state of charge of the energy storage system at time t; S_{cap} represents the rated capacity of the energy storage device; SOC_{min} and SOC_{max} are the minimum and maximum charges allowed by the energy storage device, respectively. SOC_{s} is the amount of charge stored in the initial moment; $P_{in,max}$ and $P_{out,max}$ represent the maximum charging power and the maximum discharging power of the energy storage system, respectively. σ denotes the self-discharge rate of energy storage; η denotes the charge–discharge efficiency of energy storage.

(3) Photovoltaic constraints

Considering that the PV output scheduling value is affected by its predicted output, the following constraints are imposed:

$$0 \leq P_{pv}^{t} = [P_{PV1_{dn},t}, P_{PV1_{up},t}] \tag{31}$$

The confidence level ξ is introduced to characterize the satisfaction degree of the interval number constraint. The value range of ξ is [0, 1], then P_{pv}^{t} can be expressed as:

$$P_{pv}^{t} = \xi P_{PV1_{dn},t} + (1-\xi) P_{PV1_{up},t} \tag{32}$$

Specifically, the confidence level ξ characterizes the satisfaction degree of interval number constraints more from the possibility of interval number constraints. If the confidence level ξ is larger, the more it is necessary to ensure that the scheduling value P_{pv}^{t} is within its constraint range. It is considered that $\xi P_{\mathrm{PV1}_{\mathrm{dn}},t} + (1-\xi)P_{\mathrm{PV1}_{\mathrm{up}},t}$ tends to take a small value; if the smaller the confidence level A is, the less likely the interval number constraint is satisfied, and it is considered that $\xi P_{\mathrm{PV1}_{\mathrm{dn}},t} + (1-\xi)P_{\mathrm{PV1}_{\mathrm{up}},t}$ tends to take a large value.

5. Model Solution

The distribution network's master model and the microgrids' follower models are formulated as a leader–follower game. Due to the introduction of time-of-use pricing, the strategies and responses between the leader and followers are highly coupled. Using an iterative method may result in difficulties in ensuring convergence and global optimality. Therefore, a unified solution approach is adopted in this study. By applying strong duality theory and the Karush–Kuhn–Tucker (KKT) conditions, the lower-level responses are equivalently transformed into a set of mixed-integer linear constraints and incorporated into the master model, forming a unified modeling framework. This method effectively preserves the optimality conditions of the lower level, reconstructing the leader–follower game as a mixed-integer second-order cone programming (MISOCP) model. A commercial solver can then be used to achieve integrated solution, ensuring the existence and feasibility of the game-theoretic solution.

Unified Reformulation of the Leader–Follower Model

By applying strong duality theory, the lower-level microgrid models are transformed into mixed-integer linear constraints, which are then incorporated into the master model. Consequently, the distribution network's master model and the microgrids' leader–follower models can be unified and reformulated as a mixed-integer second-order cone programming (MISOCP) problem expressed in compact matrix form, facilitating integrated solution [21], as follows:

$$\mathrm{obj}.C = \min_{y,z} \alpha^{\mathrm{T}} y \tag{33}$$

$$\text{s.t.} \quad Cy = g \tag{34}$$

$$Dy + Ez \leq f \tag{35}$$

$$\|Q_i y + q_i\|_2 \leq c_i^{\mathrm{T}} y + d_i \tag{36}$$

where α is the cost coefficient vector of continuous variables in the objective function; y and z represent continuous optimization variables and discrete optimization variables, respectively. Formula (29), Formula (30) and Formula (31) represent their equality constraints, inequality constraints and second-order cone constraints, respectively.

6. Example Analysis

6.1. Example Setting

The simulation in this study is conducted on a three-phase four-wire low-voltage distribution network (LVDN) whose system topology is shown in Figure 3. The length, impedance, transformer and other parameters of the distribution line are shown in Tables 1 and 2, which are taken from Reference [22], with both loads and distributed photovoltaic generation modeled as constant power. Each single-phase microgrid node is equipped with flexible loads capable of responding to control signals. The rated powers of these flexible loads are 5 kW, 4.2 kW, and 4.6 kW, respectively, with adjustable ranges up to half of their rated power. Single-phase photovoltaic (PV) units and energy storage systems are connected at buses 3, 5, and 7. The rated generation power of the single-phase PV units is 15 kW, and the inverter capacity is 1.1 times the rated active power of the PV units. The inverter cost coefficients are 3.6×10^{-6}, 0.0102, and 29.522, respectively. The energy storage systems have rated capacities of [value], with charge/discharge power limits equal to one-fourth of their rated capacity and an efficiency of 90%. The price elasticity coefficient ε is taken from the reference [23]. c_{SVG} is 200 yuan/kvar, k_{SVG} is 0.05, $S_{SVG,i}$ is 4 kvar. The maximum allowable three-phase voltage imbalance in the distribution network is set to 2%. The aforementioned MISOCP model is implemented in MATLAB R2014b using the YALMIP toolbox and solved with the Gurobi 9.5.2 solver, with the optimization convergence gap set to 1%.

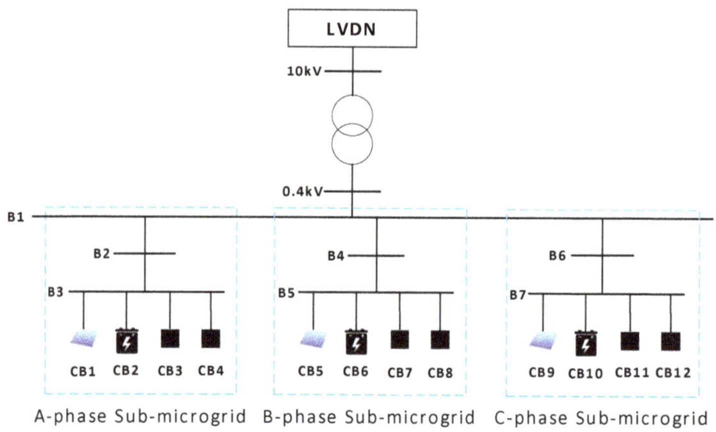

Figure 3. Low-Voltage Distribution Network Topology.

Table 1. Cable and line impedance parameters.

Type	R (Ω/km)	X (Ω/km)	B (μS/km)
N1XE-150	0.206	0.0723	40.8395
AKKJ-150	0.206	0.0628	0
EKKJ-10	1.83	0.091	100.53
AKKJ-240	0.125	0.0565	0

Table 2. Transformer parameters.

Power	500 kVA
Voltage	10/0.4 kV
Connection method	Dyn11
Positive sequence impedance	4.9%
Frequency	50 Hz

The initial electricity price information of the distribution network is presented in Table 3.

Table 3. Initial Electricity Price Information of the Distribution Network.

Scenario	Time	Active Energy Price (Yuan/(kW·h))
Peak Period	8:00–12:00, 18:00–22:00	0.83
Shoulder Period	7:00–8:00, 12:00–18:00	0.5
Off-Peak Period	0:00–7:00, 22:00–24:00	0.17

The baseline load consumption and photovoltaic generation profiles of each microgrid in the three-phase four-wire LVDN are shown in Figure 4. The load demand exhibits a clear "peak–valley–flat" pattern, which, when coordinated with the peak–valley electricity prices in Table 3, can more effectively guide users' demand response behaviors and thereby help balance the loads across the ABC phases of the distribution network.

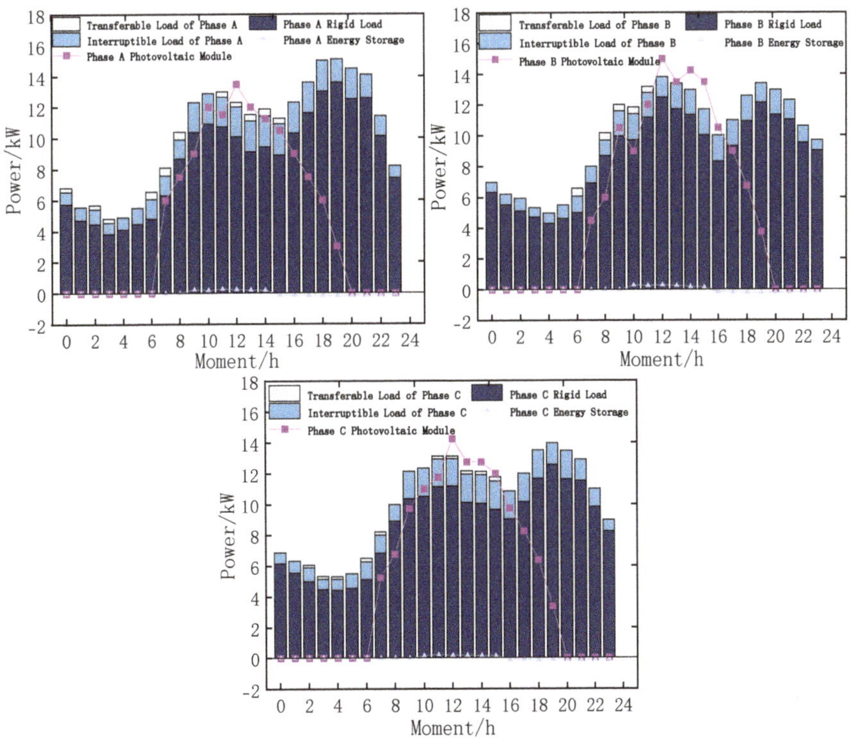

Figure 4. Load and Photovoltaic Generation Profiles of Each Microgrid.

In order to fully verify the comprehensive superiority of the collaborative optimization model proposed in this paper compared with the existing conventional governance methods, this paper sets up three typical scenarios for comparative analysis:

Three comparative scenarios are set in this study. Scenario 1: the distribution network as the main investment in the bus B2, B4, B6 installed capacity of 4 kvar SVG, with the initial load curve and fixed price (all time 0.5 yuan/(kW · h)) to optimize. Scenario 2 considers flexible load demand response, while Scenario 3 considers both flexible load demand response and the reactive power regulation capability of photovoltaic inverters. The three-phase voltage imbalance of the distribution network under these three scenarios is shown in Figure 5, and the simulation results are compared in Table 4.

Table 4. Comparison of Simulation Results.

Scenario	Comprehensive Three-Phase Imbalance/%	Maximum Three-Phase Unbalance/%
Before governance	46.10	3.98
Scenario1	21%	1.6
Scenario2	35.57	3.35
Scenario3	20.23	1.43

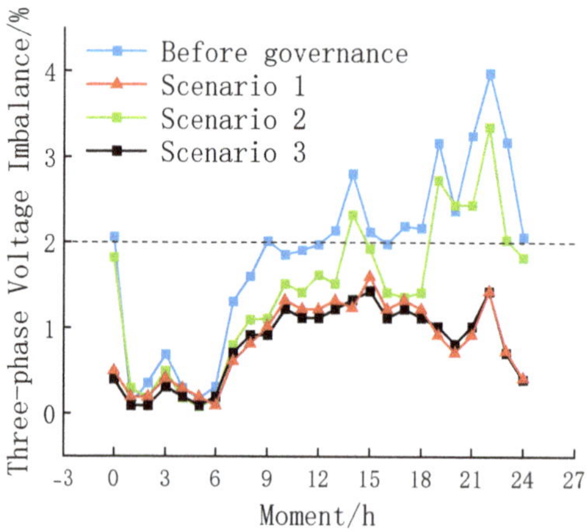

Figure 5. Three-Phase Voltage Imbalance.

6.2. Comparison of Three-Phase Voltage Imbalance

As shown in Figure 5 and Table 4, in Scenario 1, which does not consider flexible load demand response or photovoltaic (PV) inverter reactive power regulation, the three-phase voltage imbalance exhibits significant fluctuations, particularly during the morning (07:00–09:00) and evening (17:00–21:00) peak periods. The imbalance rises rapidly, reaching a maximum of 3.98%, which is well above the limit shown in the figure.

In the first scenario, that is, the distribution network as the main investment in the bus B2, B4, B6 installed capacity of 4 kvar SVG, compared with before treatment, the three-phase voltage imbalance in most of the time has been significantly improved, especially in the morning and evening peak load fluctuations, the imbalance is significantly reduced. The maximum unbalance degree decreases to about 1.6%, and the average value also decreases, indicating that SVG plays an active role in adjusting the three-phase voltage unbalance. In Scenario 2, where flexible loads participate in demand response, the distribution network sets time-of-use electricity prices based on day-ahead load forecasts. These prices guide the optimization of load profiles, prompting users to actively adjust their electricity consumption and thereby balance the three-phase load of the network. Underprice incentives, shiftable loads are transferred to off-peak periods, while curtailable loads are reduced to alleviate peak demand and lower the load peak. Compared with Scenario 1, the three-phase voltage imbalance is significantly improved in most periods, especially during the highly fluctuating morning and evening peaks, with the maximum imbalance reduced to approximately 3.35% and the average value also decreased. This indicates that flexible loads play a positive role in adjusting inter-phase load distribution and mitigating imbalance. However, although Scenario 2 partially balances load distribution across different periods and alleviates some peak-induced imbalance, the adjustment only affects active power, limiting its ability to maintain real-time dynamic voltage balance among the three phases. During periods of high PV output or concentrated asymmetric loads, significant fluctuations in voltage imbalance may still occur.

Scenario 3 further introduces the reactive power regulation capability of PV inverters. PV inverters can flexibly absorb or inject reactive power according to the voltage level of each phase, quickly adjusting node voltages and improving phase-to-phase imbalance. In this scenario, the imbalance curve is not only lower in magnitude than in the previous two scenarios but also smoother. In addition, PV inverters generate active power only during daytime, remaining idle at night. During low-load midnight hours (00:00–06:00), the unused inverter capacity can be utilized to mitigate elevated node voltages caused by low night-time loads, thereby reducing the voltage at each microgrid and decreasing the three-phase voltage imbalance.

These results demonstrate that the proposed method does not require additional power quality equipment; the reactive power compensation capability of PV inverters alone is sufficient to manage three-phase voltage imbalance in the distribution network.

6.3. Power Output of Each Microgrid

Figures 6–8 show the active power output of single-phase sub-microgrids in each scenario. Figures 9 and 10 show the reactive power output of SVG and photovoltaic inverters in each microgrid in Scenario 1 and Scenario 2, respectively. In Scenario 1, the change trend of reactive power output of SVG is similar to that in Scenario 3. Due to the lack of effective price incentive signal, the microgrid only exchanges active power with the distribution network according to its fixed electricity demand. The flexible resources such as energy storage and reactive power capacity of photovoltaic inverters in each microgrid have not been utilized. As a whole, the microgrid has not actively participated in the three-phase unbalanced governance of the distribution network, and its output strategy is relatively single, resulting in poor overall operation economy of the system.

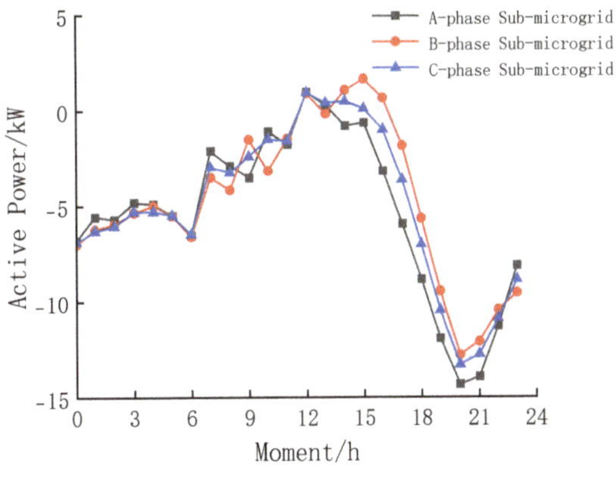

Figure 6. Active Power Purchase and Sale under Scenario 1.

In contrast, the time-of-use electricity price mechanism is introduced in Scenario 2, which makes the output behavior of each microgrid flexible. They will actively respond to the electricity price signal, reduce the load during the peak period of the electricity price or transfer it to the trough period, optimize the electricity cost of the microgrid itself, and smooth the total load curve of the distribution network by 'peak load shifting'. However, the regulation method of this scenario is limited to active power and fails to tap the potential of reactive power regulation of photovoltaic inverters, so the direct effect on improving the voltage and reactive power balance of distribution network is limited.

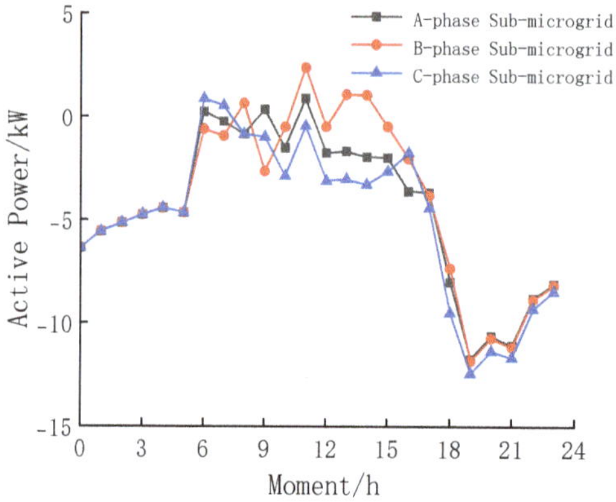

Figure 7. Active Power Purchase and Sale under Scenario 2.

In Scenario 3, from 11:00 to 14:00, the light intensity of each sub-microgrid is sufficient and there is surplus power. Part of the remaining power is stored in the battery, and the other part is sold online to maximize profits. In the rest of the time, almost all the active power is purchased from the distribution network. In the evening of the peak load period, that is, 19:00 to 22:00, the photovoltaic output power is not enough to meet the user's load demand. At this time, energy storage is used as a power supply to reduce the purchase of electricity from the grid during the peak period of electricity price, thereby improving the economy of users' electricity consumption and reducing the peak power supply pressure of the grid. However, the energy storage capacity still cannot fully meet the load demand, resulting in each sub-microgrid purchasing more active power from the distribution network, which significantly reduces the voltage of the distribution network. Since there is basically no light source to generate active power at this time, each sub-microgrid uses the idle capacity of the photovoltaic inverter to issue more reactive power to support the voltage and improve the three-phase voltage imbalance.

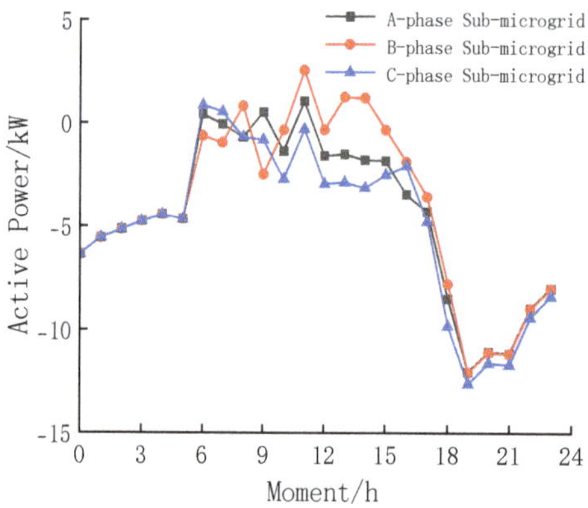

Figure 8. Active Power Purchase and Sale under Scenario 3.

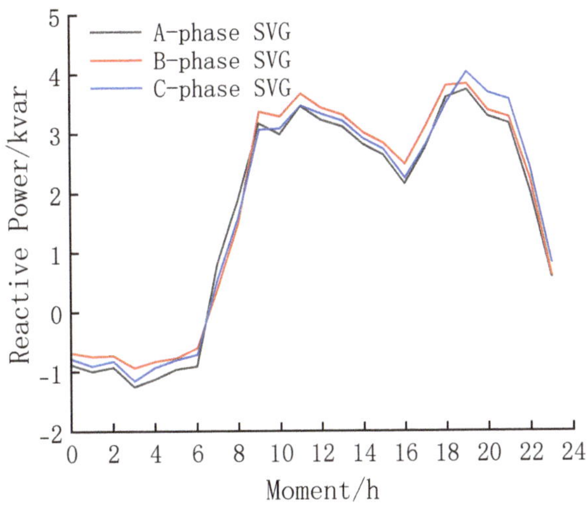

Figure 9. Scenario 1 SVG reactive power output.

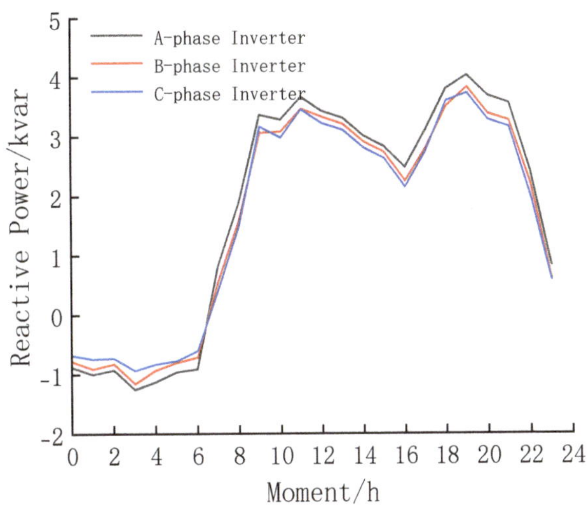

Figure 10. Inverter Power Output under Scenario 3.

It can be observed that, for the distribution network, Scenario 3 not only effectively improves power quality, but also allows each underlying microgrid to generate profit by utilizing the idle reactive power capacity of inverters to supply the distribution network.

6.4. Distribution Network Electricity Price

The electricity prices issued by the distribution network to each microgrid are shown in Figures 11 and 12, while the reactive power prices under Scenario 3 are shown in Figure 13.

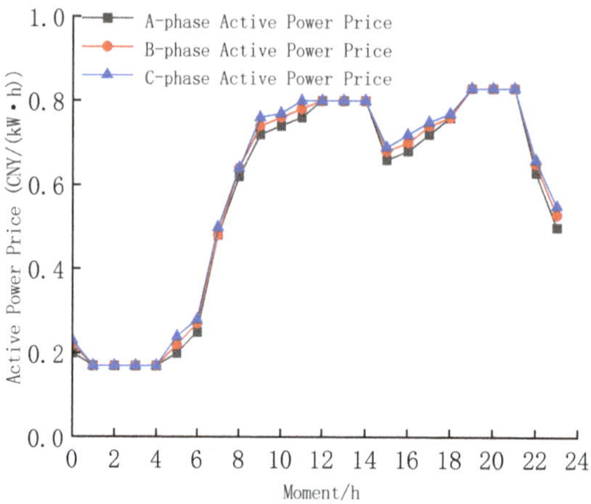

Figure 11. Active Power Electricity Price under Scenario 2.

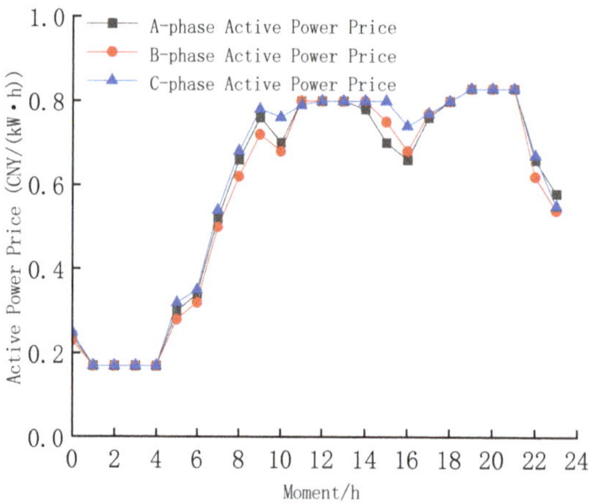

Figure 12. Active Power Electricity Price under Scenario 3.

During 01:00–05:00, the distribution network load is at its off-peak level, and the active power purchased by each microgrid is relatively low, resulting in elevated node voltages. At this time, the generation capacity of the microgrid inverters remains idle, and the reactive power prices issued by the distribution network are relatively high, incentivizing microgrids to provide reactive power to reduce voltages. During 19:00–22:00, the distribution network load reaches its peak, and the microgrids purchase a large amount of active power from the network, causing node voltages to drop and increasing the network's reactive power demand. Consequently, the reactive power prices issued to microgrids rise significantly. Although PV output is low during this period, inverters can utilize their idle capacity to provide local reactive power compensation, improving node voltage levels and reducing line currents. This not only helps mitigate three-phase voltage imbalance in the distribution network but also effectively reduces network loss costs.

From 09:00 to 15:00, PV output in the microgrids is high while loads are near peak, resulting in high inverter active power output and reduced reactive power output. Therefore, appropriate adjustment of reactive power prices is necessary to incentivize microgrids to provide reactive power compensation, balancing network losses, active and reactive power costs, and achieving an overall equilibrium between the distribution network and the underlying microgrids.

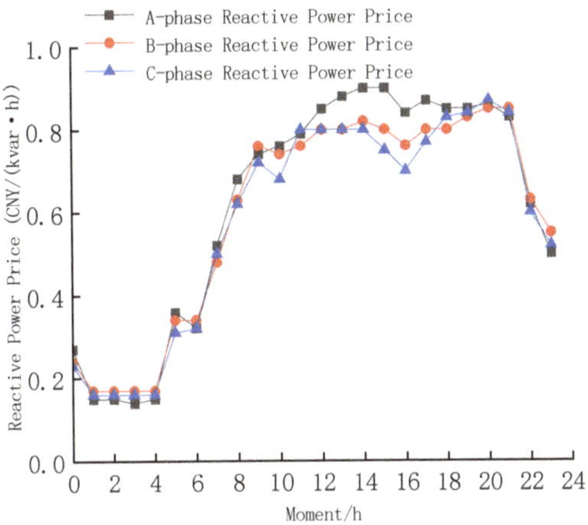

Figure 13. Reactive Power Electricity Price under Scenario 3.

6.5. Economic Analysis

6.5.1. Distribution Network Operating Cost

In Scenarios 1 and 2, the distribution network operating cost consists of the active power purchase cost from the upper-level grid and from each microgrid. In Scenario 3, the cost includes active power purchases from the upper-level grid as well as both active and reactive power purchase costs from the microgrids.

It can be seen from Table 5 that the network loss of the distribution network is reduced in Scenario 1, Scenario 2 and Scenario 3. In Scenario 1, although the distribution network can improve the power quality by installing SVG equipment, its high investment cost leads to a sharp increase in the total operating cost and poor economy. Scenario 2 only uses flexible load demand response. Although it reduces the operating cost of the distribution network, its improvement effect on network loss and imbalance is limited. In Scenario 3, in addition to introducing flexible load demand response, the reactive power regulation capability of microgrid inverters is utilized. By price incentives, microgrids provide reactive power compensation, further reducing three-phase voltage imbalance and network losses in the distribution network. This ensures that the three-phase voltage imbalance of the overall distribution network–microgrid system remains within allowable limits, thereby guaranteeing normal system operation.

Table 5. Distribution Network Operation Cost.

Scenario	Equivalent Network Loss/(kW·h)	Distribution Network Operation Cost/Yuan
Before governance	82.58	45.62
Scenario1	70.34	2452.62
Scenario2	76.96	35.45
Scenario3	60.89	31.82

6.5.2. Operating Cost of Each Microgrid

Table 6 presents the operating costs of each microgrid. It can be observed that, in Scenario 1, due to the absence of a sophisticated pricing mechanism, microgrid power purchases during peak load periods exhibit rigidity, and the energy storage system and inverter reactive power regulation capabilities are not effectively utilized, resulting in significantly reduced system economic efficiency. In Scenario 2, the introduction of flexible load demand response, guided by dynamic electricity pricing, optimizes the load profiles, encourages flexible user-side responses, and facilitates the integration of distributed PV generation, thereby reducing the operating costs of the underlying microgrids. In Scenario 3, the operating costs of the microgrids are noticeably lower than those in Scenarios 1 and 2. This is because Scenario 3 not only incorporates flexible load demand response but also employs reactive power compensation incentives to guide microgrids in fully exploiting the reactive power regulation potential of inverters, allowing microgrids to generate additional revenue and further reduce operating costs.

Table 6. Cost of Each Microgrid.

Microgrid	Scenario 1 Operating Cost/Yuan	Scenario 2 Operating Cost/Yuan	Scenario 3 Operating Cost/Yuan
A-phase sub-microgrid	64.45	65.43	40.12
B-phase sub-microgrid	56.55	56.30	32.16
C-phase sub-microgrid	59.74	68.65	45.66

6.6. The Influence of Photovoltaic Output Uncertainty Processing on the Optimization Results

The model built in this paper represents the uncertainty of photovoltaic output in the form of interval numbers, introduces the confidence level ξ to characterize the credibility of photovoltaic scheduling constraints, and converts the interval number constraints of photovoltaic output into deterministic constraints. In order to reflect the influence of confidence level change on the optimization results, combined with the above analysis, the three-phase unbalance degree of distribution network under different confidence levels is compared and simulated. The results are shown in Tables 7 and 8. It can be seen from the table that with the increase in confidence level ξ, the comprehensive three-phase unbalance degree and equivalent network loss have no obvious linear increase or decrease trend. In the engineering sense, the confidence level ξ reflects the attitude of the regional dispatcher to the photovoltaic prediction interval. The larger the ξ is, the greater the possibility of interval number constraint is, and the more it is necessary to ensure that the photovoltaic output is within the prediction range. From the perspective of mathematical theory, it is considered that the photovoltaic output is too small. In general, the uncertainty of photovoltaic output will cause uncertainty in power quality. However, combined with the results in Tables 7 and 8, it can be found that under different confidence levels ξ, the three-phase unbalance and equivalent network loss are further improved after considering the photovoltaic commutation.

Table 7. Comparison of three-phase unbalance under different confidence levels.

ξ	Three-Phase Unbalance/%
0	20.23
0.2	20.03
0.4	19.2
0.6	18.3
0.8	18.6
1	18.9

Table 8. Equivalent network loss under different confidence levels.

ξ	Equivalent Network Loss/(kW·h)
0	63.89
0.2	64.1
0.4	65.2
0.6	65.4
0.8	63.9
1	63.95

7. Discussion

In this paper, an active-reactive power collaborative optimization model of distribution network-microgrid group based on master-slave game and price incentive is proposed. The simulation results show that the model can not only effectively control the three-phase voltage imbalance within the standard without relying on additional hardware such as SVG, but also significantly improve the overall operation economy of the system, revealing the great superiority of the marketization mechanism in activating the potential of distributed resources.

Compared with the existing governance schemes, this method shows a stronger comprehensive advantage. Compared with the high investment cost of scenario 1 relying on SVG, and the limitation of scenario 2 only relying on flexible load demand response to deal with reactive power problems, the active-reactive power collaborative incentive model proposed in this paper breaks through the bottleneck of single regulation, realizes a more comprehensive mining of distributed resources, and achieves a better balance between governance effect and economy.

From the perspective of engineering application, this model provides a low-cost and high-efficiency governance path for low-voltage distribution networks with high proportion of distributed photovoltaic access. By establishing a reasonable reactive power compensation price mechanism, the lower microgrid photovoltaic inverter can be encouraged to participate in the system voltage support during the idle period, and the idle resources can be converted into adjustment capabilities. This not only improves the power quality, but also promotes the profound transformation of the distribution network operation mode from 'passive response' to 'active guidance' and 'collaborative governance' which has positive practical significance for building a new power system.

8. Conclusions and Future Work

8.1. Conclusions

This study proposes a coordinated active and reactive power optimization model for a distribution network–microgrid system, aiming to mitigate three-phase voltage imbalance in low-voltage distribution networks (LVDNs) by leveraging the reactive power compensation capability of PV inverters and demand response from flexible loads. The main conclusions are as follows:

(1) In this method, the distribution network guides each microgrid to provide reactive power compensation services through reactive power compensation incentives. This strategy can encourage the microgrid to use its own reactive power resources and actively participate in the management of the three-phase imbalance of the distribution network, thereby improving power quality. Compared with the traditional SVG investment scheme, it can achieve the same or even better governance effect, significantly reduce the system operation cost, avoid expensive hardware investment, and show great economic advantages and application potential.

(2) Considering both flexible load demand response and the reactive power regulation capability of PV inverters, the distribution network can more precisely adjust load distribution, reduce voltage imbalance during peak periods, and further balance voltages through reactive power compensation from microgrid PV inverters, effectively reducing three-phase voltage imbalance.

(3) By implementing time-of-use pricing for active and reactive power, the distribution network can guide microgrids to participate in three-phase imbalance mitigation. Without relying on additional power quality devices, the proposed model can reduce the system's maximum three-phase voltage imbalance to 1.43% and decrease the overall imbalance by 25.87%. Through coordinated active and reactive power optimization between the distribution network and microgrids, the total system operating cost is reduced by 20.58%, significantly enhancing both power quality and economic efficiency of the distribution network–microgrid system.

8.2. Future Work

Although the model proposed in this paper has shown good results in dealing with the three-phase unbalance problem of distribution network, there are still some limitations in this study. These points also point out the direction for future research work:

(1) This paper focuses on the uncertainty of photovoltaic output and uses the interval number method to deal with it, but does not model the load fluctuation. The future research work will focus on constructing a robust scheduling strategy considering the joint uncertainty of load and renewable energy. Specifically, distributed robust optimization or data-driven stochastic programming methods can be explored to incorporate the prediction errors of load and photovoltaic into the current master-slave game framework at the same time, so as to formulate a safe and economical scheduling scheme in a more complex fluctuating environment.

(2) At present, the model in this paper is mainly oriented to the day-ahead optimization stage. An important direction in the future is to apply the proposed model to the real-time rolling optimization stage, using ultra-short-term prediction data to stabilize real-time fluctuations and achieve closed-loop optimization of decision-making. In addition, for the scenario of ultra-large-scale microgrid cluster access in the future, the current centralized solution method may face challenges, so the development of more scalable distributed algorithms will be a key research direction.

Author Contributions: Conceptualization, Z.O. and Y.W.; methodology, Z.O.; software, Z.O.; validation, Z.O.; formal analysis, Y.W.; investigation, Y.W.; data curation, H.Z.; writing—original draft preparation, H.Z.; writing—review and editing, X.L.; visualization, T.D.; Supervision, T.D. All authors have read and agreed to the published version of the manuscript.

Funding: This research was funded by Hubei Provincial Natural Science Foundation Joint Fund Project, grant number 2022CFD167.

Institutional Review Board Statement: Not applicable.

Informed Consent Statement: Not applicable.

Data Availability Statement: The data supporting the findings of this study are available from the corresponding author, Zhenhui Ouyang, upon reasonable request.

Conflicts of Interest: The authors declare no conflicts of interest.

References

1. Bian, H.; Ji, K. A Robust Collaborative Optimization of Multi-Microgrids and Shared Energy Storage in a Fraudulent Environment. *Energies* **2025**, *18*, 4635. [CrossRef]
2. Sun, H.; Zou, H.; Jia, J.; Shen, Q.; Duan, Z.; Tang, X. Master–Slave Game Optimization Scheduling of Multi-Microgrid Integrated Energy System Considering Comprehensive Demand Response and Wind and Storage Combination. *Energies* **2024**, *17*, 5762. [CrossRef]
3. Wang, X.; Wang, S.; Ren, J.; Song, Z.; Zhang, S.; Feng, H. Optimizing Economic Dispatch for Microgrid Clusters Using Improved Grey Wolf Optimization. *Electronics* **2024**, *13*, 3139. [CrossRef]
4. Feng, J.; Zhang, X.; Wang, S.; Zhao, W. ADMM-Based Two-Tier Distributed Collaborative Allocation Planning for Shared Energy Storage Capacity in Microgrid Cluster. *Electronics* **2025**, *14*, 3234. [CrossRef]
5. González-Niño, M.E.; Sierra-Herrera, O.H.; Pineda-Muñoz, W.A.; Muñoz-Galeano, N.; López-Lezama, J.M. Exploring Technology Trends and Future Directions for Optimized Energy Management in Microgrids. *Information* **2025**, *16*, 183. [CrossRef]
6. Yang, L.; Hu, Z. Implementation of Dynamic Virtual Inertia Control of Supercapacitors for Multi-Area PV-Based Microgrid Clusters. *Sustainability* **2020**, *12*, 3299. [CrossRef]
7. Zafeiropoulou, M.; Sijakovic, N.; Zarkovic, M.; Ristic, V.; Terzic, A.; Makrygiorgou, D.; Zoulias, E.; Vita, V.; Maris, T.I.; Fotis, G. Development and Implementation of a Flexibility Platform for Active System Management at Both Transmission and Distribution Level in Greece. *Appl. Sci.* **2023**, *13*, 11248. [CrossRef]
8. Shan, G.; Jie, X.; Yinghao, S.; Yi, G. Research on Three-Phase Unbalance of Distribution Networks Considering the Collaboration of Photovoltaic Phase Switching and Flexible Load. *Power Gener. Technol.* **2025**, *46*, 768–777.
9. Min, L.; Lou, C.; Yang, J. Multi-objective Operation Optimization of Active Distribution Network Considering Three-phase Unbalance Regulation of Phase-switching Soft Open Points. *Autom. Electr. Power Syst.* **2023**, *47*, 56–65.
10. Zhao, J.; Liu, X.; Lin, C.; Wei, W. Three phase Unbalanced Reactive Power Optimization for Distribution Systems with a Large Number of Single Phase Solar Generators. *Autom. Electr. Power Syst.* **2016**, *40*, 13–18.
11. Liu, F.; Gao, C.; Zhao, J. Optimization Control Strategy for DSTATCOM in Low-voltage Three-phase Unbalance Distribution Network. *Power Capacit. React. Power Compens.* **2017**, *38*, 58–63.
12. Gao, X.; Gao, Y.; Zhao, J.; Liu, X. Three-phase unbalanced optimization of a distribution network with a high proportion of distributed photovoltaic energy based on a data-driven power flow model. *Power Syst. Prot. Control.* **2024**, *52*, 77–87. [CrossRef]
13. Fu, Y.; Zhou, X.; Su, X. Adaptive and Coordinated Volt/Var Optimization for Unbalanced Active Distribution Networks of Multiple Voltage Levels. *Power Syst. Technol.* **2018**, *42*, 2136–2147. [CrossRef]
14. Chen, J.; Li, Q.; Tang, Y.; Zhang, Y. Three-phase balance optimization of low-voltage distribution network considering photovoltaic phase selection and switching. *Electr. Power Autom. Equip.* **2022**, *42*, 71–78. [CrossRef]
15. Jia, X.; Xia, Y.; Yan, Z.; Gao, H.; Qiu, D.; Guerrero, J.M.; Li, Z. Coordinated operation of multi-energy microgrids considering green hydrogen and congestion management via a safe policy learning approach. *Appl. Energy* **2025**, *401*, 126611. [CrossRef]
16. Zhang, S.; Zhang, X.; Zhang, R.; Gu, W.; Cao, G. N-1 Evaluation of Integrated Electricity and Gas System Considering Cyber-Physical Interdependence. *IEEE Trans. Smart Grid* **2025**, *16*, 3728–3742. [CrossRef]
17. Sefa, I.; Altin, N.; Ozdemir, S.; Balci, S.; Battal, F. A Novel Hybrid-Core Structure for Split-Core Inverter Output Transformers. *J. Electr. Eng. Technol.* **2025**, 1–15. [CrossRef]
18. Liu, Y.-M.; Kuo, C.-C.; Chen, H.-C. Nighttime Reactive Power Optimization for Large-Scale PV Plants: Minimizing Compensation Equipment Investment. *Appl. Sci.* **2025**, *15*, 10748. [CrossRef]
19. Liu, J.; Peng, Y.; Zhang, X.; Ning, Z.; Fan, D. Distributed Online Voltage Control with Feedback Delays Under Coupled Constraints for Distribution Networks. *Technologies* **2025**, *13*, 327. [CrossRef]
20. Tomasz, B.; Marek, N.; Stanisław, P. Power Supply and Reactive Power Compensation of a Single-Phase Higher Frequency On-Board Grid with Photovoltaic Inverter. *Energies* **2022**, *15*, 2563.

21. Fengzhou, S.; Junchao, M.; Miao, Y. A Day-ahead and Intraday Coordinated Energy Management Method for Active Distribution Networks Based on Multi-terminal Flexible Distribution Switch. *Proc. CSEE* **2020**, *40*, 778–789. [CrossRef]
22. Kim, S.P.; Song, S.G.; Park, S.J.; Kang, F.S. Imbalance Compensation of the Grid Current Using Effective and Reactive Power for Split DC-Link Capacitor 3-Leg Inverter. *IEEE Access* **2021**, *9*, 81189–81201. [CrossRef]
23. Zhao, F.; Yu, X.; Du, S.; Guo, C.; Xing, H.; Fang, Y. Assessment on flexibility of a power grid with large-scale wind farm integration considering demand response. *Power Syst. Prot. Control.* **2021**, *49*, 42–51. [CrossRef]

energies

MDPI

Article

An AC and DC Inertia Enhancement Strategy for Bidirectional Grid-Connected Converter in DC Microgrid

Haiguo Tang [1,2], Kehui Zhou [2], Yubo Huang [3,*], Zili Wang [3,4] and Jiajun Fu [3]

1 School of Electronic Information and Electrical Engineering, Shanghai Jiao Tong University, Shanghai 200240, China; homebbs2085@163.com
2 State Grid Hunan Electric Power Company Limited Research Institute, Changsha 410004, China; epzhou@foxmail.com
3 The College of Electrical and Information Engineering, Hunan University, Changsha 410082, China; ziliwang@hnu.edu.cn (Z.W.); jfu11@hnu.edu.cn (J.F.)
4 School of Intelligent Manufacturing, Hunan First Normal University, Changsha 410205, China
* Correspondence: hnus230900514@hnu.edu.cn

Abstract: The virtual synchronous generator (VSG) strategy has good application prospects as an effective measure to improve the grid inertia level. However, VSG strategy cannot stabilize DC voltage and may even affect the stability of DC voltage. To address this issue, this paper proposes an AC and DC inertia enhancement strategy for a bidirectional grid-connected converter (BGC) in a DC microgrid, which is simple and easy to implement, and highly similar to the traditional VSG strategy in terms of control structure. When the load fluctuates, the strategy can better improve DC voltage stability, slow down AC frequency variations, and adjust the active power of the BGC more smoothly. Firstly, the integration of the virtual inertia equations on both sides of the AC and DC is achieved without using the differentiator. Secondly, the proportional relationship between DC voltage and frequency is utilized to simplify the AC virtual inertia equation and improve the control dynamic response speed. Finally, simulation and experimental results validate the effectiveness of the proposed strategy.

Keywords: bidirectional grid-connected converter; DC microgrid; DC voltage control; frequency support; power system stability; virtual capacitance; virtual synchronous generator

1. Introduction

The rapid growth of distributed generation is driving the development of DC microgrids [1]. As a key link between the DC microgrid and the AC grid, the bidirectional grid-connected converter (BGC) plays a crucial role in maintaining the DC bus voltage stability. For DC microgrids, which are rich in new energy sources and transmit energy to the grid, the BGC can emulate synchronous generators (SGs) using the virtual synchronous generator (VSG) strategy to provide inertia support to the grid [2,3].

Most of the work on VSG assumes that the DC side is an ideal voltage source, and for a given AC frequency, only a fixed power can be output that does not vary with the DC voltage. It only simulates the dynamic response characteristics of the AC frequency but does not take into account the dynamic response characteristics of the DC voltage, which cannot provide stability to the DC voltage. Even in order to provide inertial support to the AC system and the energy required for primary frequency regulation, the VSG power will fluctuate widely and exceed the given power reference value [4]. In addition, the renewable energy generation in the microgrid is also unstable, and the output power is highly susceptible to fluctuations. As a low-inertia system dominated by power electronics, once the above two situations occur in a DC microgrid, the wide range of power fluctuations will lead to sharp fluctuations in DC voltage. Although DC microgrids are equipped with active power reserves to stabilize the DC voltage, the cost and other issues limit the active power reserve capacity [5–8]. More seriously, if the VSG power is too high, the energy storage cannot keep the DC voltage stable, and the VSG output power cannot vary with DC voltage, which will break the dynamic balance of the system power, and a serious power shortage will occur, leading to the collapse of the DC microgrid. Therefore, considering the stability of the DC voltage in the AC inertia enhancement strategy is of great significance for the safe and stable operation of DC microgrids.

Currently, the research on AC inertia enhancement strategies considering the stability of DC voltage mainly includes two methods: Direct control of active power by DC voltage and establishing a link between DC voltage and AC frequency.

For the first method, Refs. [9–11] implement the droop control of DC voltage on active power. In [12], the author proposes a virtual synchronous generator error tracking controller strategy, which can maintain the balance of power on both AC and DC sides and suppress the fluctuation of frequency and voltage without normalization processing. In addition, based on the circuit power equation, Ref. [13] proposes to construct the VSG rotor equation based on the nature of the load by directly replacing the active power with the DC voltage or its square to achieve the control of the DC voltage. In [14,15], the authors use the relationship between the square of the DC voltage and the active power to control the DC voltage by using a PI controller. However, the DC microgrid is a low-inertia system, and merely adjusting the power of the converter through DC voltage without providing inertia support for the DC microgrid is not able to suppress DC voltage fluctuations when there is a large power change in the load or power source of the DC microgrid.

For the second method, in [16–19], the authors construct a proportional relationship between AC frequency and DC voltage to achieve the control of DC voltage and AC inertia support. In [20], the author reconstructs the rotor equation from the perspective of energy and uses the internal energy stored in the capacitor as the state variable instead of the DC voltage to establish a proportional relationship with the AC frequency to achieve DC voltage control and AC inertia support. However, Refs. [16–20] only consider the DC voltage and AC frequency, and do not improve the inertia on the DC side.

In [21], the author proposes a VSG-based inertia control strategy by exploiting the property of equal transmitted power on the AC and DC sides of the inverter, which can enhance both the DC inertia and AC inertia. In [22], based on the idea of virtual resistance of the generalized DC electromotive force control method, the author uses the inertia equations of the AC and DC sides to slow down the changes in the DC voltage and AC frequency, and to enhance the inertia level of the AC and DC sides. However, in [21,22], it is detrimental to the system stability to use a differentiator to obtain the rate of change of the DC voltage. Furthermore, Ref. [23] proposes a virtual synchronous control using the dynamic characteristics of the DC capacitor to achieve self-synchronization, establishes a link between the squared value of the DC voltage and the AC frequency to achieve DC voltage control and AC inertia support, and uses a phase-locked loop (PLL) to track the grid frequency. But PLL has been shown to be sensitive to grid impedance variations, which may lead to oscillations if the control parameters are not properly designed, especially under weak grid conditions.

Aiming at the shortcomings of the existing strategies, this paper proposes an AC and DC inertia enhancement strategy using the BGC in a DC microgrid as an example. The main contributions of this paper are summarized as follows:

(1) The proposed strategy addresses the voltage stability problem on the DC side of the VSG. The proposed strategy better suppresses DC voltage fluctuations and slows down DC voltage and frequency variations compared to existing strategies.
(2) Unlike the existing virtual inertia strategies, the proposed strategy does not use a differentiator, which avoids the introduction of high-frequency fluctuations, prevents fluctuations in the control signal, and improves the accuracy of the system control.
(3) The proposed strategy is highly similar to the VSG in terms of control structure, with the addition of only one proportional link. Compared to the existing strategies, the number of integrators is fewer, and the dynamic response is faster.
(4) When the active power shortage occurs in the DC microgrid, the proposed strategy can adjust the active power output of the grid-connected converter more smoothly and suppress the DC voltage oscillation.

This paper is organized as follows: The system architecture is presented in Section 2. The proposed AC and DC inertia enhancement strategy is described in detail in Section 3. The stability analysis is presented in Section 4. In addition, the simulation and experimental results are discussed in Section 5. The conclusions are given in Section 6.

2. System Configuration

A DC microgrid configuration with BGC is shown in Figure 1, which mainly consists of photovoltaic (PV), energy storage, DC load, and BGC, and is connected to the AC grid through BGC. By controlling the BGC, power can flow in both directions between the DC microgrid and the AC grid. The PV adopts maximum power point tracking (MPPT) technology, the energy storage adopts voltage-current double closed-loop control of the DC bus voltage, and the DC load adopts constant power control. P_{pv}, P_L, P_b, and P_o are the active power of PV, DC load, energy storage, and BGC, respectively.

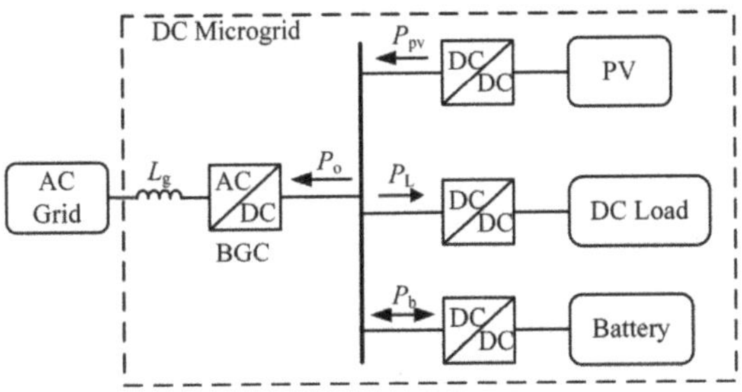

Figure 1. Diagram of DC microgrid configuration with BGC.

Figure 2 shows the BGC topology, where U_{dc} is the DC bus voltage, and C_{dc} is the parallel capacitance on the DC microgrid bus. L_f and C_f are the BGC output filter inductance and filter capacitance, R_L and R_c are the parasitic resistances of the filter inductance and the filter capacitance, and L_g is the equivalent impedance of the grid. e_a, e_b, and e_c are the three-phase voltages of the midpoint of the bridge arm of the converter, $i_{L,abc}$ and $i_{o,abc}$ are the three-phase current of the AC side inductor and the BGC output three-phase current, $V_{o,abc}$ and $V_{g,abc}$ are the three-phase voltage at the two ends of the BGC output capacitor and the three-phase voltage of the grid, respectively.

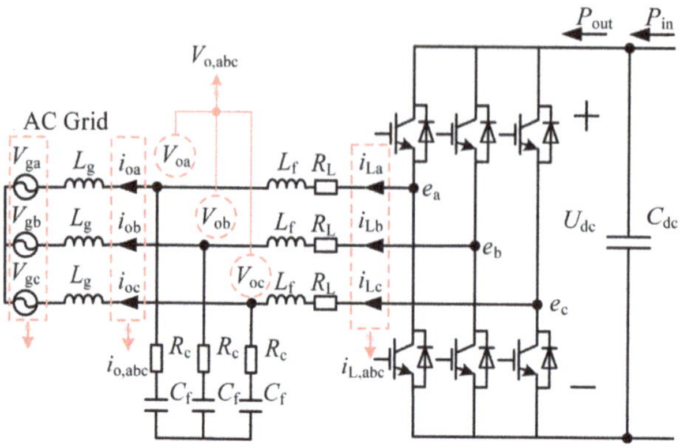

Figure 2. Diagram of BGC topology.

3. Proposed Inertia Enhancement Strategy

For the stability problem of the DC side of the VSG, the existing strategy utilizes the DC voltage as well as its rate of change to control the active power reference value of the VSG in order to enhance the stability of the DC side. The proposed strategy improves the VSG strategy by utilizing the dynamic characteristics of the DC capacitor, and the DC voltage acts directly on the virtual frequency to ensure the stability of the DC side voltage.

The overall block diagram of the proposed strategy is shown in Figure 3, which mainly includes an active power control loop and a reactive power control loop. The reactive power control loop is the same as the conventional VSG strategy. The active power control loop consists of a proportional and a virtual capacitor loop.

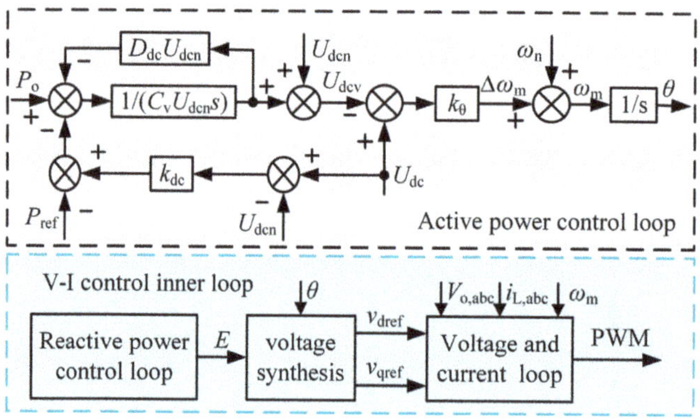

Figure 3. Overall control block diagram of the proposed AC and DC inertia enhancement strategy.

3.1. Proportional Loop

The active power control equation for the VSG strategy can be expressed as

$$P_{ref} - P_o - D_{ac}\omega_n(\omega_m - \omega_n) = J\omega_n \frac{d\omega_m}{dt} \tag{1}$$

where J is the virtual rotational inertia, P_{ref} is the power reference value, D_{ac} is the VSG damping coefficient, and ω_m and ω_n are the virtual and rated angular frequency of the VSG, respectively.

Neglect the damping in the VSG and compare it with the dynamic characteristic equation of the DC capacitor, as follows:

$$P_{\text{ref}} - P_{\text{o}} = J\omega_{\text{n}}\frac{d\omega_{\text{m}}}{dt},\ P_{\text{in}} - P_{\text{out}} = C_{\text{dc}}U_{\text{dc}}\frac{dU_{\text{dc}}}{dt} \tag{2}$$

where P_{in} and P_{out} are the DC capacitor input and output active power. From (2), we can see that the dynamic characteristics of the DC capacitor are similar to the active power equation of the VSG in form. Citing the idea in [17], we construct a proportional relationship for the DC voltage and virtual angular frequency, which can be expressed as

$$\Delta\omega_{\text{m}} = \omega_{\text{m}} - \omega_{\text{n}} = k_{\theta}(U_{\text{dc}} - U_{\text{dcref}}) \tag{3}$$

where $\Delta\omega_{\text{m}}$ is the deviation of the virtual angular frequency with the rated angular frequency, k_{θ} is the proportional factor, and U_{dcref} is the DC voltage reference value.

3.2. Virtual Capacitor Loop

The transfer of energy from the DC side to the AC side is taken as the positive direction, and the active power control equation for the virtual capacitor (VC) strategy can be expressed as

$$P_{\text{o}} - P_{\text{ref}} - DU_{\text{dcn}}(U_{\text{dcv}} - U_{\text{dcn}}) = C_{\text{v}}U_{\text{dcn}}\frac{dU_{\text{dcv}}}{dt} \tag{4}$$

where C_{v} is the virtual capacitance, D is the damping coefficient, and U_{dcv} and U_{dcn} are the virtual capacitance voltage and the DC nominal voltage.

To enhance the inertia on the DC side, the virtual capacitor voltage U_{dcv} is then used as the DC voltage reference value U_{dcref}, which realizes the improvement of the VSG strategy. Finally, DC voltage-power droop control is added to the active power control loop. Substituting (3) into (4) yields

$$\begin{aligned} &P_{\text{ref}} + k_{\text{dc}}(U_{\text{dc}} - U_{\text{dcn}}) + DU_{\text{dcn}}(U_{\text{dc}} - U_{\text{dcn}}) - P_{\text{o}} \\ &= J\omega_{\text{n}}\tfrac{d\omega_{\text{m}}}{dt} + D_{\text{ac}}\omega_{\text{n}}(\omega_{\text{m}} - \omega_{\text{n}}) - C_{\text{v}}U_{\text{dcn}}\tfrac{dU_{\text{dc}}}{dt} \end{aligned} \tag{5}$$

$$J = C_{\text{v}}U_{\text{dcn}}/k_{\theta}\omega_{\text{n}},\ D_{\text{ac}} = DU_{\text{dcn}}/k_{\theta}\omega_{\text{n}} \tag{6}$$

where k_{dc} is the DC voltage-power droop coefficient. According to (5), we can see that the proposed strategy not only enhances the inertia of the AC and DC sides but also provides AC and DC damping effects. Among them, the inertia level and damping effect on the AC side are improved through the AC terms $J\omega_{\text{n}}\frac{d\omega_{\text{m}}}{dt}$ and $D_{\text{ac}}\omega_{\text{n}}(\omega_{\text{m}} - \omega_{\text{n}})$, while those on the DC side are improved through the DC terms $C_{\text{v}}U_{\text{dcn}}\frac{dU_{\text{dc}}}{dt}$ and $DU_{\text{dcn}}(U_{\text{dc}} - U_{\text{dcn}})$.

Observing the active power control structure of the proposed strategy, we can know that the proposed strategy is highly similar to the VSG strategy in terms of structure, with the addition of only one proportional loop. Compared with the existing inertial strategy, the proposed strategy does not use differentiators, and the fluctuation of DC voltage can directly affect the value of the virtual frequency; the response speed to the change of DC voltage is much faster than the existing strategy. Therefore, the proposed strategy can provide a stronger DC voltage response speed with the same AC/DC inertia and damping effects as the existing inertia strategy, and it is not affected by the high-frequency clutter introduced by the differentiator.

4. Stability Analysis

Considering that the response speed of the power outer loop in the proposed strategy is slower than that of the voltage-current control inner loop, this paper ignores the effect of the inner loop on the power loop when analyzing the power characteristics. The active power injected into the AC network through the BGC is shown as

$$P_o = \frac{3EV_g}{2X_g}\sin\delta \tag{7}$$

where E is the virtual excitation electromotive force magnitude, V_g is the AC grid voltage magnitude, X_g is the equivalent impedance between the BGC and the grid, and δ is the phase angle difference between the BGC output voltage and the grid voltage. According to [10], the small signal model of (7) can be expressed as

$$\Delta P_o = \frac{3EV_g}{2X_g}\Delta\delta \tag{8}$$

Therefore, the small signal model of BGC is shown in Figure 4. The small signal closed-loop transfer function of the strategy proposed in this paper is

$$\Delta P_o = \frac{U_{dcn}(C_v s + D) + k_{dc}}{U_{dcn}s(C_v s + D)/k + 1}\Delta U_{dc} + \frac{1}{U_{dcn}s(C_v s + D)/k + 1}\Delta P_{ref} \tag{9}$$

where

$$k = \frac{3EV_g k_\theta}{2X_g} \tag{10}$$

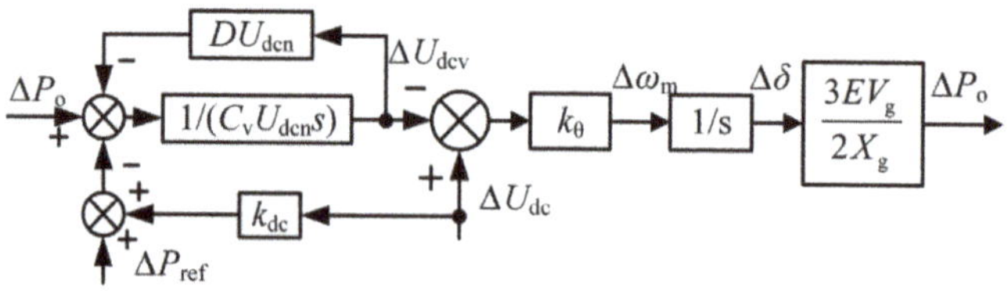

Figure 4. Small signal model of BGC.

Then, according to (9), the closed-loop pole distribution is plotted under different AC virtual inertia J, DC damping D, and DC virtual capacitance C_v, as shown in Figure 5. According to Figure 5a, when J and D are constant, with the increase in virtual capacitance, the stronger the DC inertia, the pair of conjugate poles moves toward the imaginary axis, and the system dynamic response is slower.

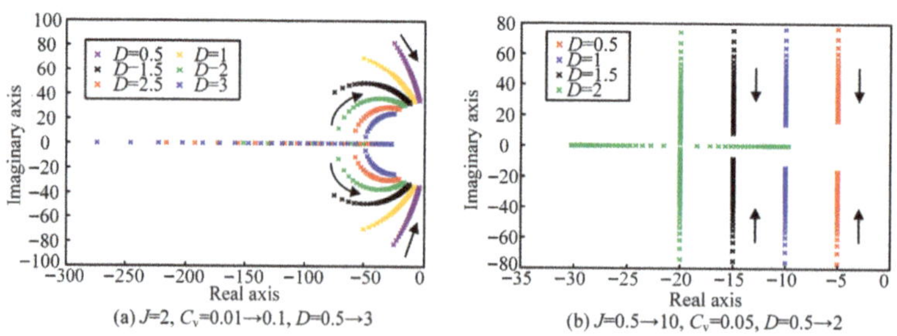

Figure 5. Pole distribution of closed-loop transfer function.

From Figure 5b, when C_v and D are unchanged, as J increases, the closer the pair of conjugate poles is to the negative real axis, the stronger the AC inertia, the system damping is gradually enhanced, and the system active power output overshoot is smaller. When J and C_v are unchanged, as D is increased, the poles gradually move away from the imaginary axis and approach the negative real axis, system damping is gradually enhanced, and system active power output overshoot is smaller.

Figure 6 shows the unit-step response of $\Delta P_o/\Delta P_{ref}$. As we can observe from Figure 6a, as C_v increases, the output active power overshoot is larger, and the system damping ratio gradually becomes smaller. From Figure 6b, as J increases, the output active power overshoot is smaller, and the system damping ratio gradually becomes larger.

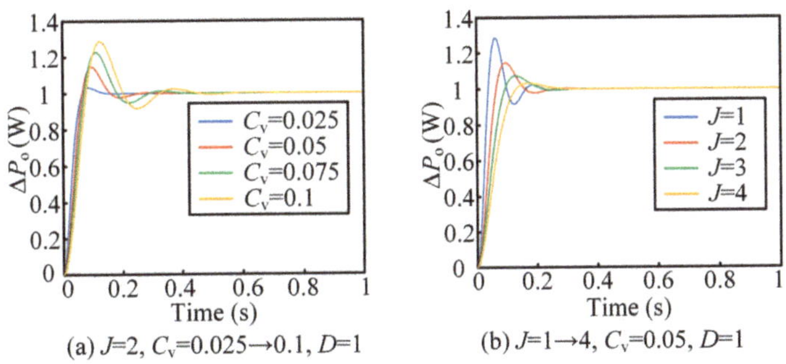

Figure 6. Unit-step responses of $\Delta P_o/\Delta P_{ref}$.

Furthermore, considering that the strategy proposed in this paper is applied to a DC microgrid, analyzing the stability of this strategy alone has limitations. Therefore, an overall analysis of the DC microgrid is necessary. In this paper, impedance modeling of each module in the DC microgrid is conducted to analyze the stability of the DC microgrid.

The output power of BGC can be expressed as

$$P_o = U_{dc}\left(i_{dc} - C_{dc}\frac{dU_{dc}}{dt}\right) \tag{11}$$

Write the state variable in (11) as the sum of the steady state quantity and the small perturbation:

$$\begin{aligned} U_{dc} &= U_{dcn} + \Delta U_{dc} \\ i_{dc} &= I_{dcn} + \Delta i_{dc} \\ P_o &= P_{ref} + \Delta P_o \end{aligned} \tag{12}$$

Substituting (12) into (11), neglecting the quadratic perturbation term, and simplifying and performing Laplace transformation, the small-signal equation can be obtained as

$$\Delta P_o = (I_{dcn} - sC_{dc}U_{dcn})\Delta U_{dc} + U_{dcn}\Delta i_{dc}, \tag{13}$$

According to [24], transforming (13), the input impedance Z_{BGC} on the DC side of the BGC is found as

$$Z_{BGC} = \frac{\Delta U_{dc}}{\Delta i_{dc}} = \frac{U_{dcn}}{\Delta P_o/\Delta U_{dc} - I_{dcn} + sC_{dc}U_{dcn}} \tag{14}$$

Combining (9) and (14), the input impedance of the DC side of BGC can be obtained. Based on [25,26], impedance modeling of each module of the system is developed. Based on impedance analysis [27,28], the individual modules of the DC microgrid are categorized into two types: Bus Voltage Controlled Converter (BVCC) and Bus Current Controlled

Converter (BCCC). Based on the analysis of the control strategies of each module, it can be known that PV, load, and BGC belong to BCCC, while the energy storage belongs to BVCC. Therefore, the DC microgrid impedance ratio, i.e., the system open-loop transfer function T_{dc}, is

$$T_{dc} = \frac{Z_{BVCC}}{Z_{BCCC}} = \frac{Z_B}{Z_{BGC} \parallel Z_{PV} \parallel Z_L} \quad (15)$$

The distribution of the system's closed-loop poles under different load powers is plotted, as shown in Figure 7. According to Figure 7, when the load power changes, all the closed-loop poles of the system are distributed in the left half-plane, and there are no poles in the right half-plane. Therefore, the DC microgrid system can operate stably under the strategy proposed in this paper.

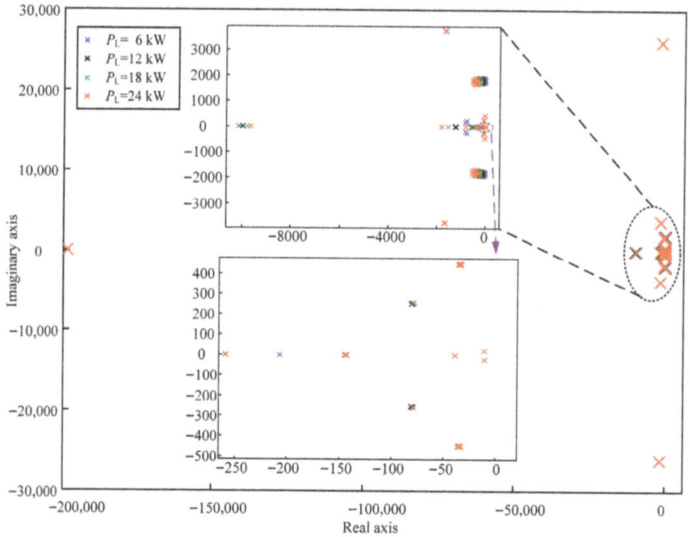

Figure 7. System closed-loop pole distribution when the system load changes.

5. Simulation and Experimental Verification

In order to verify the effectiveness of the proposed strategy, this paper builds a simulation model on MATLAB 2023a/Simulink and a semi-physical simulation on the real-time laboratory (RT-LAB), which uses the DSP28335 to control the OP5600. The experimental platform, based on the RT-LAB, is shown in Figure 8. The system structure is shown in Figure 1, and the rated power of PV, energy storage, BGC, and DC load are all 24 kW, all switch frequencies are 10 kHz, the rated frequency of the grid is 50 Hz, the line voltage is 380 V, and the DC voltage-power droop coefficient is 1.6 kW/V. The parameters of the proposed control strategy are shown in Table 1.

Table 1. Parameters of the proposed strategy.

Parameter	Value	Parameter	Value
J (kg·m^2)	2	D	1
U_{dcn} (V)	800	C_{dc} (mF)	0.5
C_v (F)	0.05	L_g (mH)	0.1
L_f (mH)	0.4	C_f (μF)	10
R_L (Ω)	0.04	R_c (Ω)	0.01

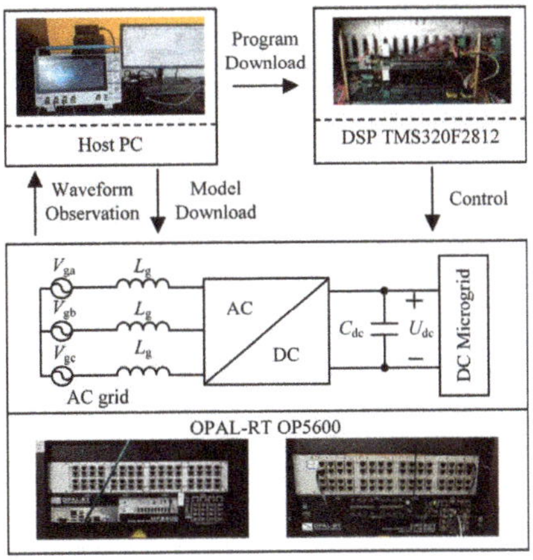

Figure 8. The experimental platform based on the RT-LAB.

In order to verify the inertia-enhancing ability of the proposed strategy on both AC and DC sides and the DC voltage stabilizing ability, simulated and experimental tests are carried out under AC-side and DC-side disturbances, respectively. The proposed strategy is compared with VSG control, PI control [12], and virtual inertia control [29] in simulation and experiment.

5.1. Simulation Verification

(1) Case 1: Simulation of AC side inertia enhancement test. Figure 9 shows the simulation results of DC voltage, frequency, and BGC active power when the active power reference value and the grid frequency are changed under the control of four strategies. The load is 10 kW at the beginning, and the active power reference value is set to 24 kW, which is set to 0 kW at 1 s and 24 kW at 2 s; the grid frequency decreases to 49.9 Hz at 3 s, and the grid frequency returns to 50 Hz at 4.5 s.

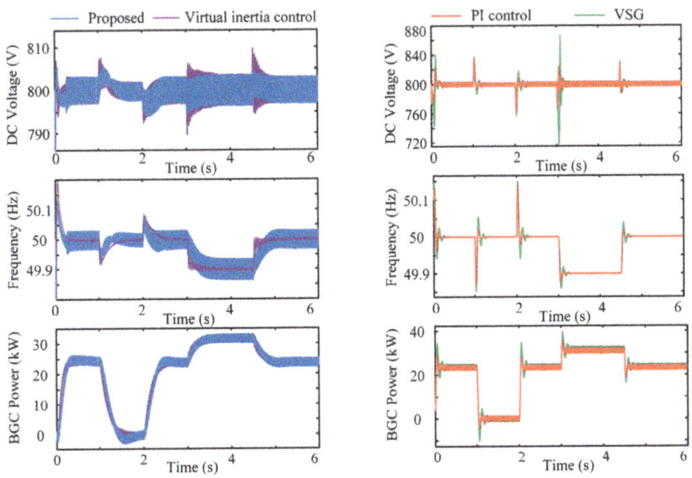

Figure 9. Simulation diagrams of the proposed strategy and existing strategies under AC side disturbances: (1) Proposed control (2) PI control [12] (3) Virtual inertia control [29] (4) VSG.

The simulated waveforms show that all four control methods can realize power control and increase the active power output to realize frequency support when the grid frequency is reduced. Table 2 shows the fluctuations of DC voltage when the active power reference value and the grid frequency change in simulation. The DC voltage fluctuation under the VSG control and [12] reach more than 30 V when the active power reference varies. The DC voltage fluctuation under the VSG control even reaches 82 V when the grid frequency varies. [12] reduces the voltage fluctuation in comparison to the VSG control, but it also has about 30 V fluctuation. The DC voltage of [29] fluctuates below 10 V, and there are small oscillations during grid frequency changes. However, the DC voltage fluctuation under the proposed strategy is kept below 5 V, the output active power has no obvious oscillation, and the change rate of frequency is even smaller.

Table 2. Fluctuation of DC voltage in Simulation Case 1.

Variable	Scenario	1 s	2 s	3 s	4 s
DC voltage (V)	Proposed	5.2	5.9	5.1	4.5
	[29]	7.6	7.5	10.4	9.8
	[12]	35.6	38.4	33.4	29.8
	VSG	37.6	42.9	82	30.9

(2) Case 2: Simulation of DC side inertia enhancement test. Figure 10 shows the simulation results of DC voltage and BGC active power under the control of four strategies for DC microgrid PV light intensity and load variation. The strategies are tested to see whether they have a stabilizing effect on the DC voltage through the DC side disturbances.

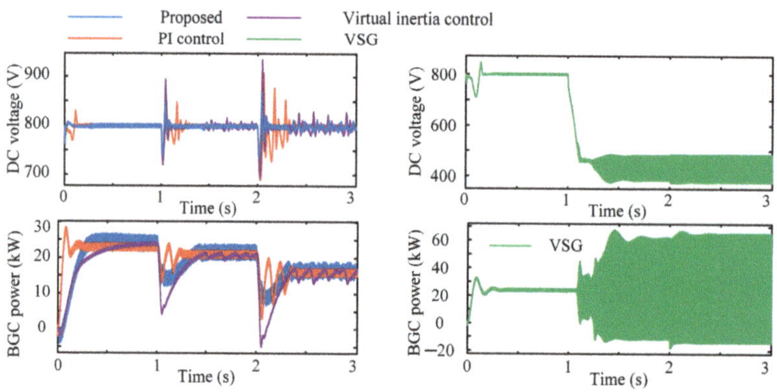

Figure 10. Simulation diagrams of the proposed strategy and existing strategies under DC side disturbances: (1) Proposed control (2) PI control [12] (3) Virtual inertia control [29] (4) VSG.

The active power reference value is set to 24 kW, the load power is 15 kW, and the PV output is 24 kW. The PV light intensity is reduced at 1 s to make the PV output less than 15 kW. At this time, since the energy storage is rated at 24 kW, the DC voltage will not be able to maintain stability if the BGC active power is kept constant at 24 kW. From the DC voltage and active power waveforms under VSG control, we can observe that when there is an active power deficit in the system, the DC voltage under the VSG strategy continues to drop to 400 V, and there is a large oscillation in the BGC output active power. So, we can see how much the VSG strategy affects the degree of voltage stability on the DC side.

In the proposed strategy, Refs. [12,29] can stabilize the DC voltage by adjusting the active power of the BGC according to the power deficit of the DC microgrid. Table 3 demonstrates the DC voltage fluctuations as the DC microgrid load and PV output power vary in simulation. The proposed strategy and [29] quickly stabilize the DC voltage, although there are DC voltage fluctuations of nearly 50 V and 100 V, respectively. The DC voltage of [12] appears to fluctuate for 0.4 s, with a peak value of 50 V. At 2 s, the DC load power rises to 20 kW, and the DC microgrid power deficit widens further. The proposed strategy and [29] quickly stabilize the DC voltage, whereas [12] shows a large fluctuation that lasts for 0.4 s and peaks at 108 V. In summary, compared to the VSG strategy [12,29], the proposed strategy can adjust the BGC active power more smoothly when there is an active power deficit in the system.

Table 3. Fluctuation of DC voltage in Simulation Case 2.

Variable	Scenario	1 s	2 s
DC voltage (V)	Proposed	48.1	67.1
	[29]	95	135
	[12]	56	108.3

5.2. Experimental Verification

(1) Case 1: AC-side inertia enhancement test experiment. Figure 11 shows the experimental results under Case 1, where the experimental parameters and AC-side perturbation settings are the same as the simulation. Figure 11a–d shows the experimental waveforms under the proposed strategy, PI control strategy [12], Virtual inertia strategy [29], and the VSG strategy, respectively.

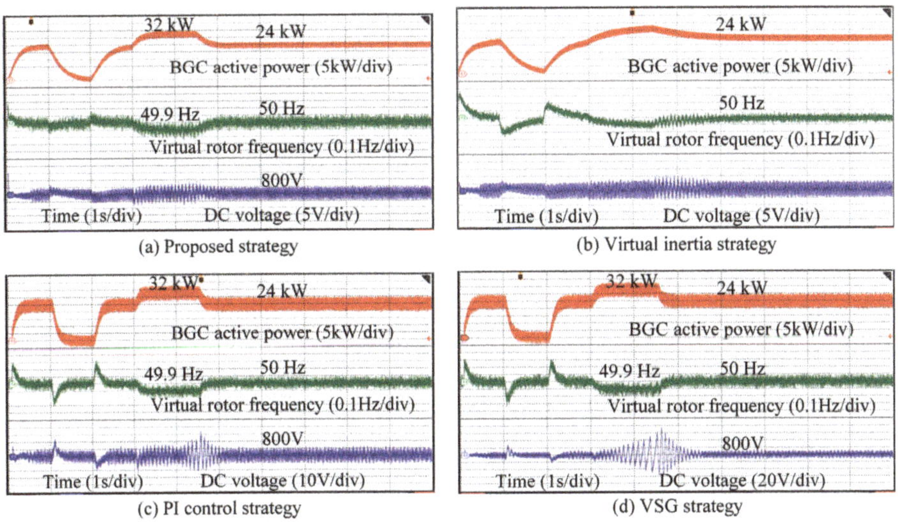

(a) Proposed strategy (b) Virtual inertia strategy (c) PI control strategy (d) VSG strategy

Figure 11. Experimental results of the proposed strategy and existing strategies under AC side disturbances.

The experimental waveforms show that all four control methods can achieve power control and inertia improvement on the AC side. Table 4 shows the fluctuations of DC voltage when the active power reference value and the grid frequency change in the experiment. The DC voltage fluctuation is kept below 10 V under the proposed strategy and [29]; Ref. [12] and the VSG strategy show some amplitude of voltage oscillation, with oscillation peaks of 30 V and 70 V, respectively.

Table 4. Fluctuation of DC voltage in Experiment Case 1.

Variable	Scenario	1 s	2 s	3 s	4 s
DC voltage (V)	Proposed	7	6	8	9
	[29]	8	7	8	11
	[12]	26	21	23	32
	VSG	32	20	21	76

Therefore, compared with the VSG control [12,29], the proposed strategy effectively slows down the DC voltage and frequency variations and ensures the stability of the system control.

(2) Case 2: DC-side inertia enhancement test experiment. Figure 12 shows the experimental results under case 2, where the experimental parameters and DC side disturbance settings are consistent with the simulation. Figure 12a–d shows the experimental waveforms under the proposed strategy, PI control strategy [12], Virtual inertia strategy [29], and the VSG strategy, respectively. Table 5 demonstrates the DC voltage fluctuations as the DC microgrid load and PV output power vary in the experiment.

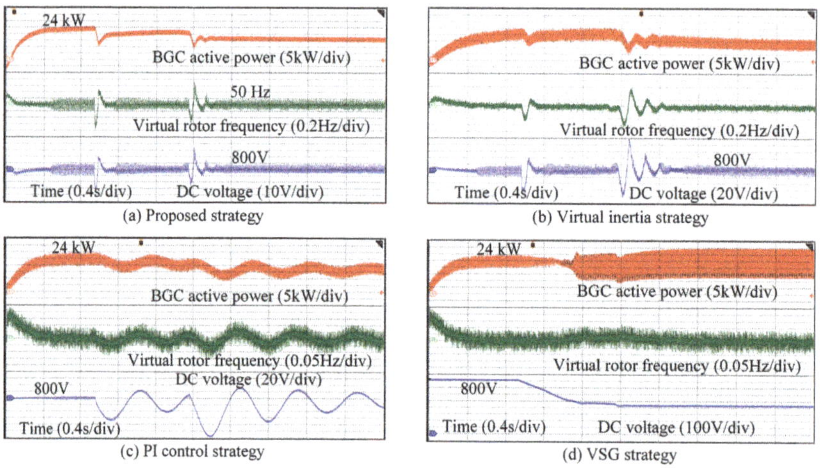

Figure 12. Experimental results of the proposed strategy and existing strategies under DC side disturbances.

Table 5. Fluctuation of DC voltage in Experiment Case 2.

Variable	Scenario	1 s	2 s
DC voltage (V)	Proposed	35	36
	[29]	68	100
	[12]	70	116

At 1 s, according to Figure 12d, we can see that the VSG strategy does not have the ability to keep the DC voltage stable and even affects the DC voltage stabilization. Once the energy storage is unable to stabilize the DC voltage, it will lead to system collapse, the DC voltage will continue to drop to 400 V, and the BGC output power will oscillate.

According to Figure 12c, we can observe that the DC voltage of [12] fluctuates with a peak value of 70 V at 1 s and gradually stabilizes. At 2 s, the DC load power rises to 20 kW, the active power deficit of the DC microgrid further expands, and the DC voltage of [12] fluctuates more, with a peak value of 116 V, but still gradually stabilizes. It can be seen that the ability of PI control in [12] to steady the DC voltage is not strong, and when the active power of the system has a large fluctuation, the response speed of [12] is slow, it cannot stabilize the DC voltage quickly, and the DC voltage fluctuates greatly.

According to Figure 12b, at 1 s and 2 s perturbations, the DC voltage of [29] shows some oscillations with fluctuation amplitudes of 68 V and 100 V, respectively. According to Figure 12a, we can observe that the DC voltage of the proposed strategy fluctuates 30 V under 1 s and 2 s perturbations and remains stable quickly. In summary, the proposed strategy can better stabilize the DC voltage, effectively improve the inertia level of the DC side, and control with higher stability.

6. Conclusions

In this paper, an AC and DC inertia enhancement strategy for a bidirectional grid-connected converter is proposed without the use of the differentiator to solve the voltage stability problem on the DC side of the VSG and to provide inertia support for the AC and DC sides. To begin with, the AC and DC virtual inertia equations are combined using virtual capacitance and the proportionality between DC voltage and frequency. Secondly, the proposed strategy is analyzed, and the advantages of this strategy are discussed. A small signal model is developed to analyze the reliability of the proposed strategy control. The system impedance model is established, which proves the stability of the DC microgrid system under the proposed strategy. Finally, the effectiveness of the proposed strategy is verified through simulation and experiment.

Author Contributions: Investigation, H.T. and J.F.; Methodology, H.T. and Y.H.; Software, K.Z.; Supervision, K.Z. and Z.W.; Validation, Z.W.; Writing—original draft, H.T., Y.H. and K.Z.; Writing—review and editing, H.T. and Y.H. All authors have read and agreed to the published version of the manuscript.

Funding: This work was supported in part by the State Grid Hunan Electric Power Science and Technology Project under Grant 5216A524001S, in part by the Hunan Provincial Natural Science Foundation of China under Grant 2024JJ6172, Changsha Natural Science Foundation of China under Grant kq2402091, and in part by the Open Fund of State Key Laboratory of High-Efficiency and High-Quality Conversion for Electric Power (Hunan University) (2024KF005).

Data Availability Statement: The original contributions presented in this study are included in the article. Further inquiries can be directed to the corresponding author.

Conflicts of Interest: Authors Haiguo Tang and Kehui Zhou were employed by the company State Grid Hunan Electric Power Company Limited Research Institute. The remaining authors declare that the research was conducted in the absence of any commercial or financial relationships that could be construed as a potential conflict of interest.

References

1. Guerrero, J.M.; Chandorkar, M.; Lee, T.-L.; Loh, P.C. Advanced control architectures for intelligent microgrids—Part I: Decentralized and hierarchical control. *IEEE Trans. Ind. Electron.* **2013**, *60*, 1254–1262. [CrossRef]
2. Zhong, Q.-C.; Weiss, G. Synchronverters: Inverters that mimic synchronous generators. *IEEE Trans. Ind. Electron.* **2011**, *58*, 1259–1267. [CrossRef]
3. Ashabani, S.M.; Mohamed, Y.A.-R.I. New family of microgrid control and management strategies in smart distribution grids—Analysis, comparison and testing. *IEEE Trans. Power Syst.* **2014**, *29*, 2257–2269. [CrossRef]
4. Chen, J.R.; Liu, M.Y.; Milano, F.; O'Donnell, T. Adaptive virtual synchronous generator considering converter and storage capacity limits. *CSEE J. Power Energy Syst.* **2022**, *8*, 580–590.
5. Liu, J.; Hou, Y.; Guo, J.; Liu, X.; Liu, J. A cost-efficient virtual synchronous generator system based on coordinated photovoltaic and supercapacitor. *IEEE Trans. Power Electron.* **2023**, *38*, 16219–16229. [CrossRef]

6. Zhong, C.; Li, H.; Zhou, Y.; Lv, Y.; Chen, J.; Li, Y. Virtual synchronous generator of PV generation without energy storage for frequency support in autonomous microgrid. *Int. J. Electr. Power Energy Syst.* **2022**, *134*, 107–343. [CrossRef]
7. Zhang, B.; Zhao, P.; Zhao, J. Research on control strategy of two-stage photovoltaic virtual synchronous generator with variable power point tracking. *Energy Rep.* **2022**, *8*, 283–290. [CrossRef]
8. Ma, Y.; Cao, W.; Yang, L.; Wang, F.; Tolbert, L.M. Virtual synchronous generator control of full converter wind turbines with short-term energy storage. *IEEE Trans. Ind. Electron.* **2017**, *64*, 8821–8831. [CrossRef]
9. Citro, C.; Al-Numay, M.; Siano, P.; Alhelou, H.H. Experimental evaluation of virtual synchronous generators with high regulation performance to enhance AC–DC grid interconnections. *Int. J. Electr. Power Energy Syst.* **2024**, *158*, 109964. [CrossRef]
10. Wang, R.; Chen, L.; Zheng, T.; Mei, S. VSG-based adaptive droop control for frequency and active power regulation in the MTDC system. *CSEE J. Power Energy Syst.* **2017**, *3*, 260–268. [CrossRef]
11. Li, C.; Li, Y.; Cao, Y.; Zhu, H.; Rehtanz, C.; Hager, U. Virtual synchronous generator control for damping DC-side resonance of VSC-MTDC system. *IEEE J. Emerg. Sel. Top. Power Electron.* **2018**, *6*, 1054–1064. [CrossRef]
12. Lyu, Z.; Yi, J.; Song, J.; Liu, B. Coordinated optimal control and dispatching operation of grid-connected AC/DC hybrid microgrid group. *Electr. Power Syst. Res.* **2023**, *221*, 109–430. [CrossRef]
13. Ashabani, M.; Mohamed, Y.A.-R.I. Novel comprehensive control framework for incorporating VSCs to smart power grids using bidirectional synchronous-VSC. *IEEE Trans. Power Syst.* **2014**, *29*, 943–957. [CrossRef]
14. Leon, A.E.; Mauricio, J.M. Virtual synchronous generator design to improve frequency support of converter-interfaced systems. *IEEE Trans. Energy Convers.* **2024**, *39*, 862–871. [CrossRef]
15. Leon, A.E.; Mauricio, J.M. Virtual synchronous generator for VSC-HVDC stations with DC voltage control. *IEEE Trans. Power Syst.* **2023**, *38*, 728–738. [CrossRef]
16. Chen, D.; Xu, Y.; Huang, A.Q. Integration of DC microgrids as virtual synchronous machines into the AC grid. *IEEE Trans. Ind. Electron.* **2017**, *64*, 7455–7466. [CrossRef]
17. Khajehoddin, S.A.; Karimi-Ghartemani, M.; Ebrahimi, M. Grid-supporting inverters with improved dynamics. *IEEE Trans. Ind. Electron.* **2019**, *66*, 3655–3667. [CrossRef]
18. Yang, R.; Shi, G.; Cai, X.; Zhang, C.; Li, G.; Liang, J. Autonomous synchronizing and frequency response control of multi-terminal DC systems with wind farm integration. *IEEE Trans. Sustain. Energy* **2020**, *11*, 2504–2514. [CrossRef]
19. Shi, K.; Chen, Y.; Xu, P.; Sun, Y.; Liu, M.; Wang, X. A generalized DC electromotive force control strategy for bi-directional interface converters in a standalone hybrid microgrid. *IEEE Trans. Energy Convers.* **2023**, *38*, 1155–1165. [CrossRef]
20. Yang, R.; Shi, G.; Zhang, C.; Li, G.; Cai, X. Internal energy based grid-forming control for MMC-HVDC systems with wind farm integration. *IEEE Trans. Ind. Appl.* **2023**, *59*, 503–512. [CrossRef]
21. Zhang, L.; Wang, X.; Zhang, Z.; Cui, Y.; Ling, L.; Cai, G. An adaptative control strategy for interfacing converter of hybrid microgrid based on improved virtual synchronous generator. *IET Renew. Power Gener.* **2022**, *16*, 261–273. [CrossRef]
22. Ren, M.; Sun, X.; Sun, Y.; Shi, K.; Xu, P. A virtual inertial control strategy for bidirectional interface converters in hybrid microgrid. *Int. J. Electr. Power Energy Syst.* **2023**, *153*, 109–388. [CrossRef]
23. Huang, L.; Xin, H.; Wang, Z.; Wu, K.; Wang, H.; Hu, J.; Lu, C. A virtual synchronous control for voltage-source converters utilizing dynamics of DC-link capacitor to realize self-synchronization. *IEEE J. Emerg. Sel. Top. Power Electron.* **2017**, *5*, 1565–1577. [CrossRef]
24. Amin, M.; Molinas, M.; Lyu, J.; Cai, X. Impact of power flow direction on the stability of vsc-hvdc seen from the impedance nyquist plot. *IEEE Trans. Power Electron.* **2017**, *32*, 8204–8217. [CrossRef]
25. Lu, X.; Sun, K.; Guerrero, J.M.; Vasquez, J.C.; Huang, L.; Wang, J. Stability enhancement based on virtual impedance for dc microgrids with constant power loads. *IEEE Trans. Smart Grid* **2015**, *6*, 2770–2783. [CrossRef]
26. Lee, J.-S.; Lee, G.-Y.; Park, S.-S.; Kim, R.-Y. Impedance-based modeling and common bus stability enhancement control algorithm in DC microgrid. *IEEE Access* **2020**, *8*, 211224–211234. [CrossRef]
27. Zhang, X.; Ruan, X.; Tse, C.K. Impedance-based local stability criterion for dc distributed power systems. *IEEE Trans. Circuits Syst. I Regul. Pap.* **2015**, *62*, 916–925. [CrossRef]
28. Zhang, L.; Ruan, X. Control schemes for reducing second harmonic current in two-stage single-phase converter: An overview from DC-bus port-impedance characteristics. *IEEE Trans. Power Electron.* **2019**, *34*, 10341–10358. [CrossRef]
29. Wang, J.; Huang, W.T.; Tai, N.L.; Yu, M.D.; Li, R.; Zhang, Y. A bidirectional virtual inertia control strategy for the interconnected converter of standalone AC/DC hybrid microgrids. *IEEE Trans. Power Syst.* **2024**, *39*, 745–754. [CrossRef]

energies

MDPI

Article

Reliability Analysis of CVT Online Monitoring Device Based on Bayesian Network

Xu Chen [1], Haomiao Zhang [1], Chao Zhang [1], Yinzhe Xu [1], Yu Yan [1], Yuntao Zhao [2,*], Xuhui Chen [2] and Rui Ren [2]

1 State Grid Ningxia Marketing Service Center (State Grid Ningxia Metrology Center), Yinchuan 750001, China; chenxu89@126.com (X.C.); zhang197802@126.com (H.Z.); zc198802041@126.com (C.Z.); xuyz98@163.com (Y.X.); yany20000507@126.com (Y.Y.)

2 School of Artificial Intelligence and Automation, Wuhan University of Science and Technology, Wuhan 430041, China; 202304704079@wust.edu.cn (X.C.); 1724830028@wust.edu.cn (R.R.)

* Correspondence: zhyt@wust.edu.cn

Abstract: To address the challenges of equipment reliability assessment in the context of intelligent power systems, especially the shortcomings of traditional methods in dealing with multi-factor coupling and uncertain fault inference of CVT (capacitive voltage transformer) online monitoring devices, this study proposes a reliability analysis method based on Bayesian networks (BNs). The research aims to evaluate the reliability of CVT online monitoring devices, identify key risk factors, and optimize maintenance strategies. Firstly, a Bayesian network reliability model is constructed for the CVT online monitoring device, defining key influencing factors such as environmental factors and component quality as network nodes, and establishing conditional probability dependency relationships between nodes. Subsequently, the MATLAB R2021b simulation platform was used to simulate the system's operating status under different combinations and scenarios. The experimental results indicate that the combination of high-temperature and high-humidity environments has the most significant impact on reliability; among the component factors, the failure of the data acquisition and processing unit has the greatest impact on system reliability; wiring process issues pose a greater threat to reliability than mechanical fixing issues; and regular maintenance can significantly improve system reliability. This method validates the effectiveness of Bayesian networks in dynamic reliability analysis of CVT online monitoring devices, which can accurately locate high-risk factors and support maintenance decision optimization.

Keywords: Bayesian network; online monitoring device; reliability analysis; voltage transformer

1. Introduction

As power systems become increasingly intelligent and complex, voltage transformer monitoring is critical to ensuring their safe operation [1,2]. Therefore, reliability analysis of monitoring devices has become a key component in maintaining system safety [3–7].

Traditional reliability analysis methods, mostly based on physical models [8–10], struggle to effectively handle fault reasoning under multi-factor coupling, dynamic environments, and uncertain conditions.

In recent years, Bayesian networks (BNs) [11–13] have been widely applied in fields involving uncertainty due to their strong probabilistic reasoning capabilities and intuitive graphical modeling advantages, which use probabilistic algorithms as the basis for reasoning and handling uncertainty [14,15]. Internationally, significant progress has been made in applying Bayesian networks to reliability analysis of power equipment. Huo et al. [16] proposed an application method of Bayesian networks (BN) in power system reliability assessment; Qiu et al. [17] developed an evaluation framework integrating Bayesian networks for failure assessment of electric energy metering equipment; and Ramadan H. et al. [18] optimized the health index calculation model of power transformers by combining neural networks with Bayesian reasoning, providing new ideas for dynamic reliability assessment. Domestic scholars, such as Zhang Zhaochuang [19], Liao Caibo [20], and Fan Huifang [21], applied Bayesian networks to transformer fault analysis; Chen Yunhao et al. [22] realized dynamic prediction of fault probabilities for current transformers based on Bayesian networks, verifying their applicability in complex equipment; and Tong Weiyan et al. [23] proposed a hybrid method for reliability analysis of complex systems by combining fuzzy theory with Bayesian networks, further enhancing the model's ability to handle uncertain information. Additionally, Zhao Miao [24] revealed the dynamic characteristics of insulation monitoring parameters for voltage transformers through simulation studies, laying a foundation for the application of Bayesian networks in multi-physics field coupling analysis. Based on the logical relationship between component and system reliability, combined with test fault data, Wang Qing et al. [25] proposed a Bayesian evaluation method for electric energy meter reliability integrating reliability prediction. Zhao Qicheng et al. [26] developed a Bayesian state assessment method for dry-type air-core reactors.

Therefore, this paper proposes a reliability analysis method based on Bayesian networks to address the limitations of traditional physical models in handling multi-factor coupling and uncertain fault reasoning. By integrating the MATLAB simulation platform and through a dynamic probability reasoning mechanism, the model's adaptability to complex environmental changes is enhanced. A multi-level causal dependency network is constructed to identify key risk factors (such as specific component defects and weak links in the process), overcoming the limitations of traditional methods in risk traceability. Compared with assessment systems relying on physical equations, this method demonstrates higher efficiency in dealing with uncertainties and decision optimization, providing quantifiable support for engineering practice and reducing empirical costs.

2. Bayesian Networks

A Bayesian network is a graphical model composed of nodes and directed edges, where each node represents a random variable, and edges represent conditional dependencies between variables. The goal of a Bayesian network is to infer the probability of unknown variables using known evidence, thereby evaluating system reliability.

Its core idea is to express dependencies between variables through conditional probability distributions and perform reasoning and updating using Bayes' theorem. The mathematical basis of Bayesian networks is Bayes' theorem [27], using the following Equation (1):

$$P(H|E) = \frac{P(E|H)P(H)}{P(E)} \tag{1}$$

where $P(H|E)$ is the posterior probability of hypothesis H given evidence E, $P(E|H)$ is the likelihood of observing evidence E under hypothesis H, $P(H)$ is the prior probability of hypothesis H, and $P(E)$ is the marginal probability of evidence E.

As shown in Figure 1,The probability distribution of each node in a Bayesian network depends on its set of parent nodes, i.e., $P(X_1 X_2, \ldots X_n) = \prod_{i=1}^{n} P(X_i|Pa(X_i))$, where X_i is a node in the network, and $Pa(X_i)$ is the set of parent nodes of X_i. This formula indicates that the joint probability distribution in the network can be expressed as the product of the conditional probabilities of each node.

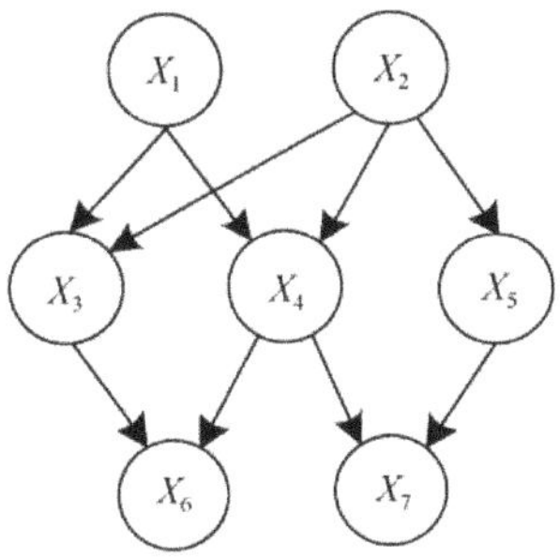

Figure 1. Bayesian network.

By modeling the conditional probability distributions of each node, Bayesian networks can establish dependencies between multiple variables. During reasoning, the probability distribution of other unknown nodes can be derived from observed data of known nodes.

3. Reliability Analysis Based on Bayesian Networks

Reliability analysis of online monitoring devices using Bayesian networks can evaluate the health status of equipment operation. The health status of equipment is typically influenced by multiple factors. In a Bayesian network, each factor is represented as a node, and directed edges between nodes denote conditional dependencies among these factors. System reliability analysis involves inferring relationships between these variables to assess equipment reliability. The evaluation process diagram is shown in Figure 2.

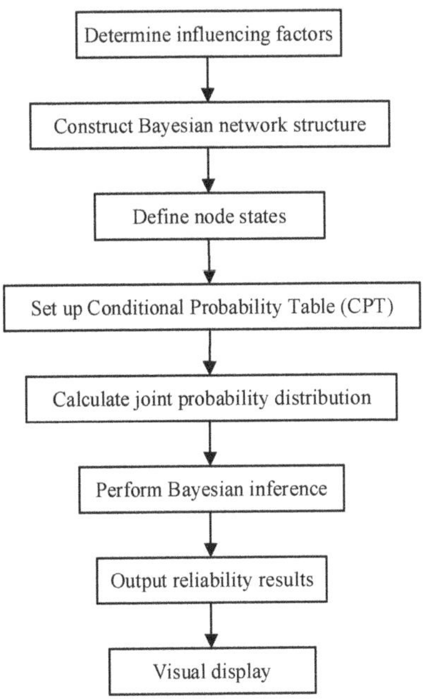

Figure 2. Bayesian network experiment flowchart.

(1) Construction of equipment health status model

When constructing a health status model for online monitoring devices, it is first necessary to define various operating parameters of the device as nodes in the network and determine dependencies between these nodes using historical data. The Bayesian network model for equipment health status can be expressed as Equation (2):

$$P(T,H,U,S) = P(T)P(H|T)P(U|H)(S|U) \tag{2}$$

where T represents temperature, H represents humidity, U represents voltage, and S represents the equipment health status. Temperature, humidity, and voltage, as crucial environmental variables, directly affect the dielectric properties and insulation performance of equipment, and also have an impact on the health status of the equipment. The conditional probability distributions, such as $P(H|T)$, describe the relationship between temperature and pressure; $P(U|H)$ represents the dependency between humidity and voltage; and finally, $P(S|U)$ indicates the influence of voltage on the health status of the equipment.

(2) Fault modes and fault probabilities

Conditional probabilities under different fault modes are established using historical data and equipment design information. The occurrence probabilities of these fault modes are inferred through Bayesian networks.

Let F be the fault state of the equipment, T be temperature, and U be voltage. The probability of equipment failure can be expressed by Equation (3):

$$P(F|T,U) = \frac{P(T,U|F)P(F)}{P(T,U)} \tag{3}$$

where $P(T,U|F)$ is the conditional probability of temperature and voltage under the fault state, $P(F)$ is the prior probability of failure occurrence, and $P(T,U)$ is the marginal probability of temperature and voltage.

(3) Reasoning and decision support

Reasoning through Bayesian networks infers equipment health status based on real-time sensor data. For example, if the temperature and voltage of the equipment reach certain preset thresholds, Bayesian reasoning can calculate the posterior probability of equipment failure. Assuming the temperature and current of the equipment are observed, the posterior probability of equipment failure can be expressed as Equation (4):

$$P(F|T=t,U=u) = \frac{P(T=t,U=u|F)P(F)}{P(T=t,\ U=u)} \tag{4}$$

In this way, Bayesian networks can provide health status assessments of online monitoring devices based on current observed data.

4. Experimental Verification

Bayesian networks are based on probabilistic graphical models, consisting of nodes and directed edges, representing conditional dependencies between variables. Therefore, it is first necessary to identify key factors affecting the reliability of CVT online monitoring devices. Each factor is then treated as a node in the network, and causal relationships between them are established. This paper will use the CVT online monitoring device from Ningxia Company of State Grid at Yinchuan China to conduct the following experiments.

(1) Environmental factors

First, environmental factors such as temperature and humidity are used as nodes to establish relationships. The parent nodes are environmental factors, and the child node is reliability. Each parent node has different states, and the reliability node depends on the conditional probabilities of these two factors. Next, it is necessary to determine the conditional probability table (CPT) for each node. The experimental steps are as follows:

First, the network structure and probability parameters are defined, and a Bayesian network structure with three nodes (temperature, humidity, and reliability) is constructed. Then, the joint probability is calculated to derive the marginal probability of reliability, and results are visualized via histograms and tables.

The reliability analysis results of capacitive voltage transformers are shown in Figures 3 and 4.

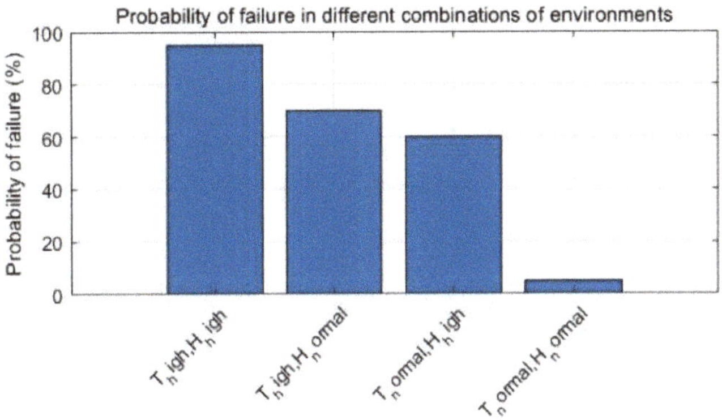

Figure 3. Probability of failure due to environmental factors.

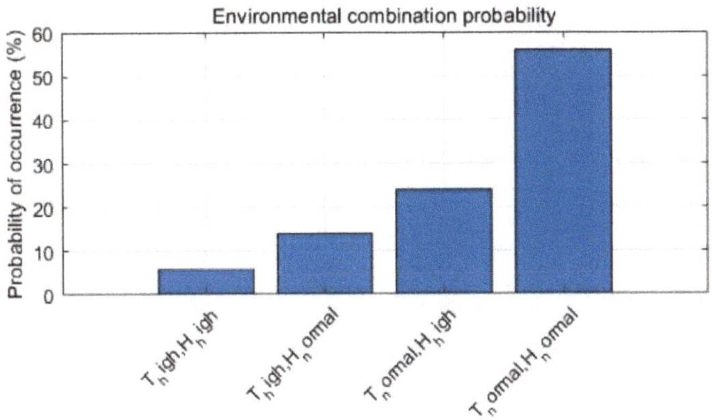

Figure 4. Environmental combination probability.

Analysis of Table 1 results: Overall fault probability: 32.70%; overall normal probability: 67.30%. The most dangerous combination is high temperature + high humidity, with a 95% fault probability. In a high-temperature and high-humidity environment, the failure probability of the CVT online monitoring device reaches 95%, mainly due to the combined effect of high temperature and high humidity. High temperature accelerates the aging of electronic components and reduces insulation performance, while high humidity causes water vapor to penetrate into the sensors and circuit boards, leading to corrosion, short circuits, or signal drift.

Table 1. Reliability analysis results.

Environmental Combination	Combination Probability	Fault Probability
T_high, H_high	6.00%	95.00%
T_high, H_normal	14.00%	70.00%
T_normal, H_high	24.00%	60.00%
T_normal, H_normal	56.00%	5.00%

(2) Component quality

Three core components of the CVT online monitoring device, namely the sensor unit, communication and transmission module, and data acquisition and processing unit, are selected as nodes to establish relationships. These three factors serve as parent nodes, and reliability is the child node. The probabilities of each state are calculated through joint probabilities, and then the conditional probability table for each node is determined. The experimental steps are as follows:

First, define the prior probabilities of each node (each node is independent, and the joint probability is the product of their respective probabilities). Define the conditional probability table (CPT) as the conditional probability of the reliability node R. Enumerate all possible combinations of parent node states, calculate the probability of each combination, and visualize the results, as shown in Figure 5.

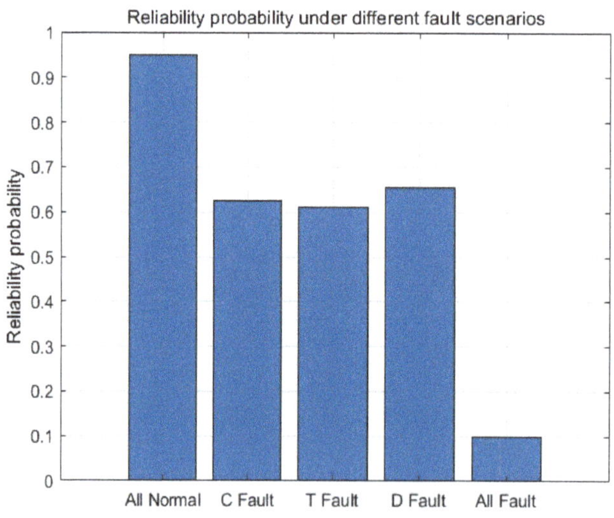

Figure 5. Reliable probability of failure of different components.

Analysis of Table 2 results: The overall normal reliability probability is 95%. The most dangerous combination is all faults, with a 90% fault probability. When the data acquisition and processing Fault (D Fault) fails, the system reliability significantly decreases. Its impact is greater than that of the sensor Fault (C Fault) and the communication module (T Fault). The failure of this Fault will directly cause the entire monitoring system to lose its "decision-making center". Moreover, when all three components fail simultaneously, the system reliability drops sharply to 10%.

(3) Installation process

First, two important processes in the installation of CVT online monitoring devices, wiring process and mechanical fixing, are used as parent nodes, and reliability is the child node. The wiring process has two states: good and poor; mechanical fixing has two states: normal and faulty. The reliability state of the entire system is determined by these factors. The experimental steps are as follows:

Construct a Bayesian network containing three nodes: wiring process, mechanical fixing, and reliability, where the reliability node is affected by the first two. Then calculate the conditional probabilities when each factor fails. Next, enumerate all possible combinations, calculate the joint probabilities, and sum them to obtain the marginal probability of the reliability node. Finally, visualize the results as shown in Figures 6 and 7.

Table 2. List of component quality reliability.

Quality Combination	Reliability Probability
All Normal	95%
C Fault	62.5%
T Fault	61%
D Fault	62.5%
All Fault	10%

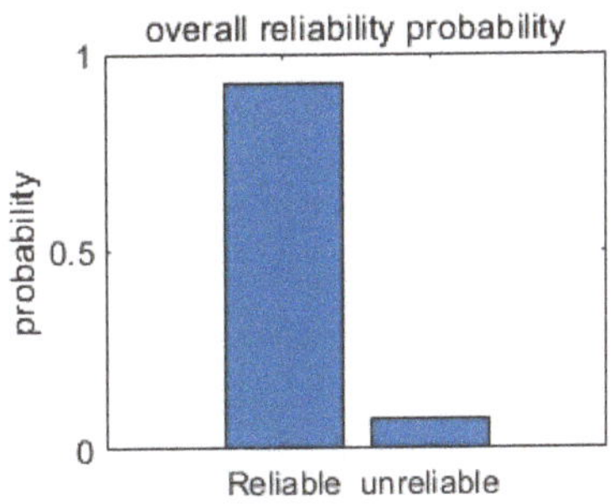

Figure 6. Overall reliability probability.

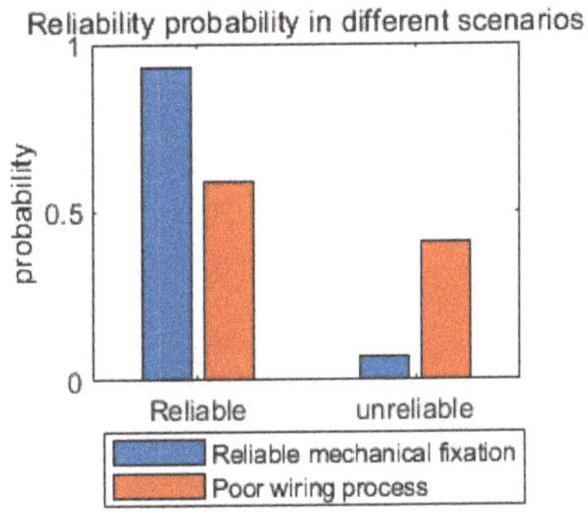

Figure 7. Reliability probabilities in different scenarios.

Analysis of Table 3 results: The overall normal probability is 92.72%. The wiring process has a greater impact on reliability than mechanical fixing.

Table 3. List of installation process reliability.

Scenario Description	Reliability Probability	Unreliability Probability
Overall	92.72%	7.27%
Reliable mechanical fixing	93.25%	6.75%
Poor wiring process	59%	41%

(4) Maintenance management

Regular maintenance and no maintenance are set as parent nodes, and reliability is the child node. The reliability state of the entire system is determined by maintenance management. It is necessary to determine the influence weight of each parent node on system reliability or the specific values of joint probabilities. The experimental steps are as follows:

First, maintenance management, as a parent node, directly affects reliability. Then, determine the states of each node. Maintenance management has two states: "regular maintenance" and "no maintenance"; reliability has two states: "good" and "poor". Next, set the conditional probability table. Reliability under different maintenance states is shown in Figure 8.

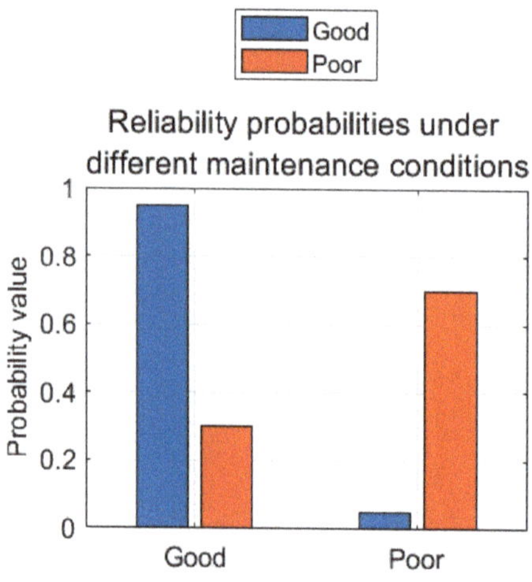

Figure 8. Reliability under different maintenance conditions.

Table 4 shows that when regular maintenance is implemented, the probability of the device being in a "good" state is as high as 95%, and the probability of being in a "poor" state is only 5%. Without maintenance, the probability of the device being in a "good" state drops sharply to 30%, while the probability of being in a "poor" state significantly increases to 70%. These results strongly demonstrate the extreme importance of regular maintenance in ensuring the reliability of CVT online monitoring devices. Compared with the on-site maintenance records of the State Grid in Ningxia region, the actual proportion of "good" is 94.9%, which is only 0.1% lower than the predicted value of 95%. In the un-maintained areas, the proportion of "poor" reaches 71.1%, which is 1.1% higher than the predicted value of 70%. This confirms that the Bayesian model has high accuracy and practical effectiveness.

Table 4. Maintain the reliability list of status.

Maintenance Status	Reliability Probability		Unreliability Probability	
	Predicated Value	Actual Value	Predicated Value	Actual Value
Good	95%	94.9%	5%	4.2%
Poor	30%	29.3%	70%	71.1%

(5) Transformer's own error

The error magnitude of the transformer itself is set as the parent node, and reliability is the child node. The reliability state of the entire system is determined by maintenance management. It is necessary to determine the influence weight of each parent node on system reliability or the specific values of joint probabilities. The experimental steps are as follows:

First, the key variables here are the error characteristics and reliability state of the transformer. The error characteristic is a parent node, and reliability is a child node, as error characteristics affect reliability. Next, determine the states of the variables. Error characteristics may have different states: "low error", "medium error", and "high error". Reliability is divided into "good" and "bad". Then, define the conditional probability table (CPT) as shown in Figure 9.

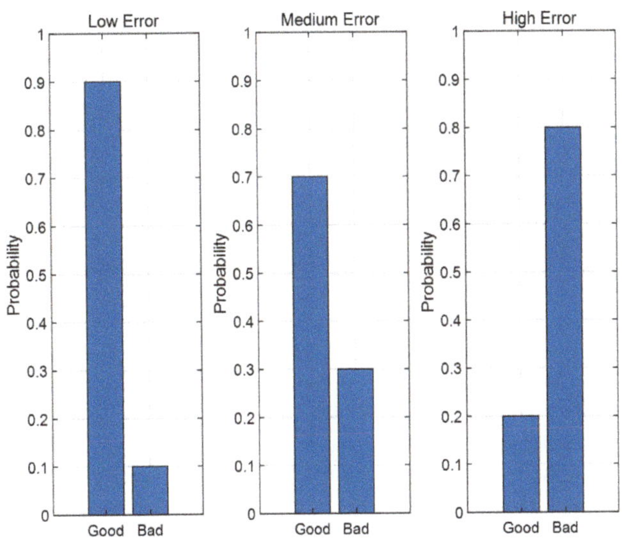

Figure 9. Reliability in case of error.

Table 5 shows that when in the "low error" state, the probability that the monitoring device outputs a "reliable" result is the highest. This indicates that the monitoring device has high credibility in its output results when measuring healthy or well-functioning CVTs. As the error increases, the probability that the monitoring device outputs a "reliable" result decreases to 75%. This means that when the CVT itself begins to have minor issues, the accuracy of the monitoring device's diagnostic results may decrease due to the influence of abnormal signals from the device itself, with a certain risk of misjudgment or missed judgment. When in the "high error" state, the probability that the monitoring device outputs a "reliable" result drops sharply to 40%. This indicates that severe faults or performance degradation of the CVT itself will greatly interfere with the judgment ability of its online monitoring device.

Table 5. List of transformer error reliability.

Error State	Reliability Probability	Unreliability Probability
Low	90%	10%
Medium	70%	30%
High	20%	80%

5. Conclusions

With the intelligent development of power systems, traditional reliability analysis methods struggle to effectively solve the problem of fault reasoning for CVT online monitoring devices under multi-factor coupling, dynamic environments, and uncertain conditions.

This paper proposes a reliability analysis method based on Bayesian networks to dynamically evaluate the health status of online monitoring devices, identify key risk factors, and optimize maintenance strategies. A Bayesian network model for CVT online monitoring devices is constructed, where key variables such as environmental factors, component quality, installation process, and maintenance management are defined as network nodes, and conditional probability dependencies are established. Multi-scenario experiments are conducted using the MATLAB platform to simulate system behavior under different environmental combinations (temperature and humidity), component states (sensors/communication/data processing units), process levels (wiring/mechanical fixing), and maintenance strategies. Dynamic updates and predictions of fault probabilities are realized through Bayesian reasoning. The results show that the combination of high-temperature and high-humidity environments has the most significant impact on reliability; among component factors, failures of the data acquisition and processing unit have the greatest impact on system reliability. The proposed Bayesian network model can not only identify the key reliability threats but also quantify their impact; regular maintenance can increase the system reliability by 65 percentage points (from 30 to 95%), while high-temperature and high-humidity environments will increase the failure probability to 95%. By supporting data-driven decision-making in the complex power grid environment, this method can improve operational efficiency and reduce unplanned downtime. Wiring process issues pose greater risks to reliability than mechanical fixing problems; and regular maintenance can significantly improve system reliability. This study verifies the effectiveness of Bayesian networks in reliability analysis, proposes a Bayesian network modeling method for multi-factor coupling, solves the limitations of traditional models in uncertain reasoning, and provides a new paradigm for complex system reliability theory.

Author Contributions: Conceptualization, R.R. and Y.Z.; methodology, X.C. (Xu Chen) and X.C. (Xuhui Chen); software, Y.Y. and X.C. (Xuhui Chen); validation, Y.X., C.Z. and R.R.; formal analysis, H.Z.; investigation, X.C. (Xu Chen); resources, Y.Y.; data curation, H.Z.; writing—original draft preparation, X.C. (Xu Chen); writing—review and editing, C.Z.; visualization, H.Z.; supervision, X.C. (Xuhui Chen); project administration, Y.Z.; funding acquisition, R.R. All authors have read and agreed to the published version of the manuscript.

Funding: This study was supported by the Science and Technology Project of State Grid Ningxia Electric Power Co., Ltd. (No. 5229YX240005).

Data Availability Statement: Restrictions apply to the availability of these data. Data were obtained from State Grid Ningxia Marketing Service Center (State Grid Ningxia Metrology Center) and are available with the permission of State Grid Ningxia Marketing Service Center (State Grid Ningxia Metrology Center).

Conflicts of Interest: The authors declare that this study received funding from the Science and Technology Project of State Grid Ningxia Electric Power Co., Ltd. (No. 5229YX240005). The funder had the following involvement with the study: The funder participated in the study design.

Abbreviations

The following abbreviations are used in this manuscript:

CVT	capacitive voltage transformer
BN	Bayesian networks
CPT	conditional probability table

References

1. Xu, Y.; Li, H.; Zhang, C. Quantitative evaluation method for voltage transformer error based on optimal modeling and deep belief network. *Electr. Meas. Instrum.* **2024**, *61*, 55–62+77.
2. Han, H.A.; Zhang, Z.; Wang, H.N.; Li, H.B.; Xue, J.L.; Shao, L. Online metering performance evaluation of capacitor voltage transformer based on principal component analysis. *Power Autom. Equip.* **2019**, *39*, 201–206.
3. Li, X.; Tian, F.; Guo, Y. Exploration of key technologies for health management and intelligent operation and maintenance of power equipment in new power systems. *Power Syst. Technol.* **2023**, *47*, 3710–3727.
4. Li, N.; Zhang, W.; Wang, H. Reliability verification method for intelligent energy meters in high dry heat field environments. *Electr. Meas. Instrum.* **2024**, *61*, 192–198.
5. Xue, G.; Wang, Y.; Yu, H.; Wei, L.; Qian, Z. Reliability study of SF6 gas density meter based on constant stress accelerated life test. *Electr. Meas. Instrum.* **2024**, *61*, 194–202.
6. Zhu, Q.Y.; Zhu, L.P.; Wang, Z.Z.; Zhang, X.; Li, Q.; Han, Q.; Yang, Z.; Qin, Z. Hybrid triboelectric-piezoelectric nanogenerator assisted intelligent condition monitoring for aero-engine pipeline system. *Chem. Eng. J.* **2025**, *519*, 165121. [CrossRef]
7. Cao, J.; Lin, Y.; Fu, X.P.; Wang, Z.; Liu, G.; Zhang, Z.; Qin, Y.; Zhou, H.; Dong, S.; Cheng, G.; et al. Self-powered overspeed wake-up alarm system based on triboelectric nanogenerators for intelligent transportation. *Nano Energy* **2023**, *107*, 108150. [CrossRef]
8. Kou, H.; An, Z.; Ma, Q. Fatigue reliability assessment of wind turbine blades based on performance degradation data. *Acta Energiae Solaris Sin.* **2017**, *38*, 3174–3179.
9. Li, K.; Wang, Y.; Zhao, C.C.; Hao, Y.J.; Liu, Z.J.; Hu, B.K. Reliability simulation analysis of leakage protection module based on performance degradation. *Electr. Meas. Instrum.* **2023**, *60*, 64–70+77.
10. Chaturvedi, A.; Sarma, M.; Chaturvedi, S.K.; Bernstein, J. Performance assessment and RUL prediction of power converters under the multiple components degradation. *Microelectron. Reliab.* **2023**, *144*, 114958–114971. [CrossRef]
11. Qian, C.; Hua, F. Research on the reliability of hydraulic systems based on dynamic bayesian networks. *Mach. Tool Hydraul.* **2023**, *51*, 219–224.
12. Zhou, Z.; Dong, D.; Zhou, J. Application of bayesian networks in reliability analysis. *Syst. Eng. Theory Pract.* **2006**, *6*, 95–100.
13. Wang, Q.; Chen, J.; Ni, Y.; Xiao, Y.; Liu, N.; Liu, S.; Feng, W. Application of Bayesian networks in reliability assessment: A systematic literature review. *Structures* **2025**, *71*, 108098–108105. [CrossRef]
14. Das, S.; Das, S.; Chakraborty, A. Bayesian neural network based probability density evolution approach for efficient structural reliability analysis. *Comput. Struct.* **2025**, *315*, 107807. [CrossRef]
15. Chang, Y.; Wu, X.; Zhang, C.; Chen, G.; Liu, X.; Li, J.; Cai, B.; Xu, L. Dynamic Bayesian networks based approach for risk analysis of subsea wellhead fatigue failure during service life. *Reliab. Eng. Syst. Saf.* **2019**, *188*, 454–472. [CrossRef]
16. Huo, L.; Zhu, Y.; Fan, G. Reliability assessment of power systems by Bayesian networks. In Proceedings of the International Conference on Power System Technology, Beijing, China, 13–17 October 2002; Volume 2, pp. 876–879.
17. Qiu, W.; Tang, Q.; Yao, W.; Qin, Y.; Ma, J. Probability Analysis for Failure Assessment of Electric Energy Metering Equipment Under Multiple Extreme Stresses. *IEEE Trans. Ind. Inform.* **2020**, *99*, 3762–3771. [CrossRef]
18. Ramadan, H.; Eldin, H.A.; Elserougi, A. Health index calculation of power transformer using different neural network structures with efficiency evaluation of the calculated indices. *J. Electr. Eng. Technol.* **2025**, *20*, 3041–3052. [CrossRef]
19. Zhang, Z.C.; Wang, J.G.; Xia, J.H.; Wen, Y.C.; Weng, L.T.; Ma, Z.F.; Yang, H.K.; Dou, J.Y. Research on Transformer Fault Classification Method Based on Bayes CNN. *New Technol. Electr. Energy* **2024**, *43*, 18–26.

20. Liao, C.B.; Jiang, Z.H.; Yang, J.X.; Shao, J.; Wang, T.L.; Li, X. Real time warning method and interpretability analysis of transformer health status. *High Volt. Technol.* **2024**, 1–11. [CrossRef]
21. Fan, H.F.; Xian, R.C.; Wang, T.; Gao, H.P.; Chen, L.; Zhang, B.Q. Application of improved naive bayes model in fault location of power transformers. *Gaoya Dianqi* **2023**, *59*, 190–197.
22. Chen, Y.H.; Peng, Z.Y.; Yang, K.Y.; Luo, L.Y.; Shi, Y.Q.; Li, Z.; Wang, H.Z. Probability prediction and analysis of current transformer faults based on Bayesian networks. *Yunnan Electr. Power* **2022**, *50*, 74–78+83.
23. Tong, W.Y.; Lin, H.; Liu, Y.Z.; Shi, Y.H. Research on reliability analysis of complex systems based on fuzzy bayesian networks. *Manuf. Autom.* **2023**, *45*, 148–153+174.
24. Zhao, M. Simulation study on inter turn insulation monitoring parameters of dry-type voltage transformers. *Power Capacit. React. Power Compens.* **2023**, *44*, 59–66.
25. Wang, Q.; Wang, P.X.; Wang, T.T.; Zhang, Q.; Zhu, H.X. Research on bayesian evaluation method of electric energy meter reliability based on fusion reliability prediction. *Electr. Meas. Instrum.* **2024**, *61*, 209–216.
26. Zhao, Q.C.; Zhang, B.; Hong, J.Q.; Han, G.W.; Han, W.F.; Chen, F. State evaluation method for dry-type air core reactors based on weighted naive Bayes. *Therm. Power Gener.* **2023**, *52*, 144–152.
27. Shankar, P.M. Linear algebra and bayes' theorem in probability. *Coll. Math. J.* **2025**, *56*, 49–59. [CrossRef]

energies

MDPI

Article

Discrete Adaptive Nonswitching Reaching Law Algorithm for Sliding Mode Control of a Grid-Following Inverter

Albert Sawiński, Piotr Leśniewski * and Piotr Chudzik

Institute of Automatic Control, Faculty of Electrical, Electronic, Computer and Control Engineering, Lodz University of Technology, 90-924 Łódź, Poland; albert.sawinski@dokt.p.lodz.pl (A.S.); piotr.chudzik@p.lodz.pl (P.C.)

* Correspondence: piotr.lesniewski@p.lodz.pl

Abstract: This paper extends one of the nonswitching-type reaching laws for discrete-time sliding mode control. The control task under consideration is the regulation of the grid current of the grid-following power inverter. A mathematical model of a plant is presented as an example of a microgrid. This system contains a T-type inverter, LCL filter, DC source, power grid connection and control system. The system tests were performed in a simulation environment. First, the methods known in the literature for implementing continuous sliding mode control are presented for the described problem, including stability analysis and implementation. Secondly, the well-known discrete sliding mode control algorithm based on the nonswitching reaching law type is discussed. The main part of this article consists of a proposed modification to the above algorithm. We consider the use of two separate regulation mechanisms: an adaptation of a specific control law parameter responsible for limiting the control signal and a mechanism for reducing the steady-state error. The aim of these procedures is to increase the quality of control, which, in turn, leads to an increase in the quality of energy transmitted in grids. The stability analysis is presented, as well as the simulation results. Finally, the results of all methods are compared and discussed, with conclusions drawn.

Keywords: sliding mode control; discrete-time systems; control laws; robust control; adaptive control; inverters; power systems; modeling; current regulators; computer simulation

1. Introduction

Algorithms that are currently implemented in control systems are subject to very dynamic developments. New, often advanced and sophisticated, methods are constantly emerging which, due to their specificity, can be used in specific and dedicated solutions. One of the most interesting and rapidly developing areas of control systems is the broadly understood sliding mode control, which is known for its ability to provide the control plant with a high degree of robustness to disturbances, and its ease of implementation. In addition, sliding mode control, with its nonlinear operation because of the use of equivalent control and an appropriate reaching law, makes it possible to control the dynamics of the system both in the static state of operation and during the process of reaching the set point, which is not necessarily ensured in the case of the most commonly used classic PID controller. The basis of sliding mode control is the proposal of a proper hyperplane, on which—or in the vicinity of which—a representative system point shall move [1–4]. In the first case, a continuous time domain controller is described, and the representative point movement is called sliding motion. The second case regards a controller in a discrete time domain with quasi-sliding motion of a representative system point [1,4]. Hence, one can define two distinct phases of the sliding control process: the reaching phase, when the operating point is moving towards the hyperplane, and the sliding phase, during which a sliding or quasi-sliding motion is present [1,4]. In the latter phase, the system order is reduced and its dynamics depend only on the proposed form of the hyperplane [1,4,5]. Sliding mode control methods guarantee the system a high level of robustness to external disturbances and model inaccuracies, which will definitely be present in real systems [1,2]. There are many different types of sliding mode control designs, such as Terminal Sliding Mode Control (TSMC), Hybrid Sliding Mode Control (HSMC), Nonsingular Terminal Sliding Mode Control (NTSMC), etc. [6–9]. Sliding mode control is being used more and more often in fields such as electric drive control [10,11], unmanned aerial vehicle (UAV) control [12], power electronics control [9,13], inventory resource control and management [14–17], etc. It is particularly noteworthy that various control strategies are implemented specifically in the field of power electronics [18–24]. One can see that the discussed control method might be used in cases of complex process control and in strongly nonlinear systems that are subjected to an impact resulting from a wide range of disturbance types [1,5]. In this paper, a problem related to power electronics is discussed, namely the issue of the current grid control with the use of a T-type three-phase bidirectional grid-following inverter that is synchronized with the power grid. This work involves the use of a proposed sliding mode control method, which is an extended and modified version of one of the more interesting approaches of developing a so-called nonswitching reaching law [5,15]. This concept is an interesting avenue for the development of advanced control algorithms due to the increasing number of power electronic devices, so-called microgrids, and renewable energy sources, as well as the increasing requirements for the quality of energy transmitted by these devices and systems [25–31]. To implement and test the methods proposed in this paper, simulation studies were carried out using the MATLAB R2025a Simulink environment.

2. Microgrid System Model

The considerations presented in the article were carried out on a grid-following converter model connected to a local power network model. The proposed system consists of the following elements: a DC voltage source, a three-phase T-type inverter, an LCL filter, and the simplified model of the power grid, which is the main load of the system. All of the energy system elements and control systems were also implemented as mathematical models in Matlab Simulink. A simplified diagram of the considered system is shown in Figure 1.

2.1. Power System Structure

The main part of the system is the grid-following inverter. A three-phase three-level inverter in the T topology was selected for the tests. This type of topology allows a simple implementation of three-level output voltages, thanks to which the voltage signal being the output from the device gains a greater degree of flexibility [32,33]. The topology diagram of the T-type inverter is shown in Figure 2.

The proposed inverter works as a current source [34,35]. This type of converter should be connected to the power grid or to a microgrid containing a grid-forming inverter that can define the voltage and frequency conditions of the local grid. The inverter is synchronized to the grid using a Phase-Locked Loop (PLL) system [25,27,29,30,36–40]. The task of this device is to provide the current transferred in both directions, which is hereinafter referred to as the grid current. The DC source on the DC voltage side of the inverter can be an energy storage or; for example, a photovoltaic panel system [25,30,41]. The grid-following inverter is connected to the power grid using an LCL filter because of the discrete nature of the inverter output voltage. This type of filter is widely known and used in similar applications due to its simplicity, as well as the fact that with the appropriate selection of parameters, it is possible to ensure satisfactory filtration results at a low implementation cost [30,42–46].

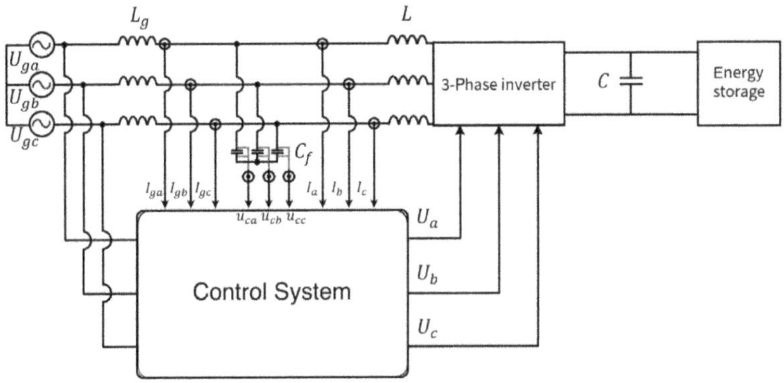

Figure 1. Microgrid system simplified scheme.

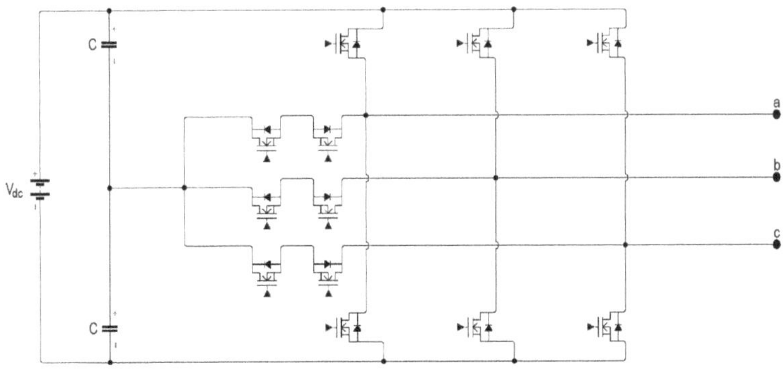

Figure 2. T-type inverter topology scheme. The individual phases are marked with the letters a, b, and c.

2.2. Mathematical Model of the System

Capacitive and inductive elements are essential for describing the dynamics of the control system. The number of these elements dictates the order of mathematical equations describing the dynamics of the system. Therefore, the LCL filter used plays a very important role in the mathematical description. In the case under discussion, the parameters of this filter constitute the main basis for developing a mathematical model of the system [9,30]. Therefore, it is important to propose an appropriate control algorithm to provide protection against imperfections occurring in this mode, because deviations from the correctly selected filter parameter values can cause the occurrence of signal disturbances in the form of oscillations of higher harmonics [42,44,47]. This is strongly related to the fact that the power grid has its own inductance, which is not constant, from which follows the need to use a dedicated solution in the field of control.

Using the LCL-type filter discussed above, it is possible to propose the following equations describing the plant's dynamics:

$$\begin{cases} L\frac{di_k}{dt} = dU_{dc} - u_{ck} \\ C_f\frac{du_{ck}}{dt} = i_k - i_{gk} \\ L_g\frac{di_{gk}}{dt} = u_{ck} - U_{gk}, \end{cases} \tag{1}$$

where

- L—filter inductance;
- L_g—grid inductance;
- C_f—filter capacity;
- i_k—inverter output current of a given phase;
- u_{ck}—capacitive filter voltage of a given phase;
- i_{gk}—grid current of a given phase;
- d—voltage modulator duty cycle;
- U_{dc}—energy storage voltage (inverter DC voltage source);
- U_{gk}—grid voltage of a given phase;
- k—phase designation.

A simplified diagram of this model is visible in Figure 3.

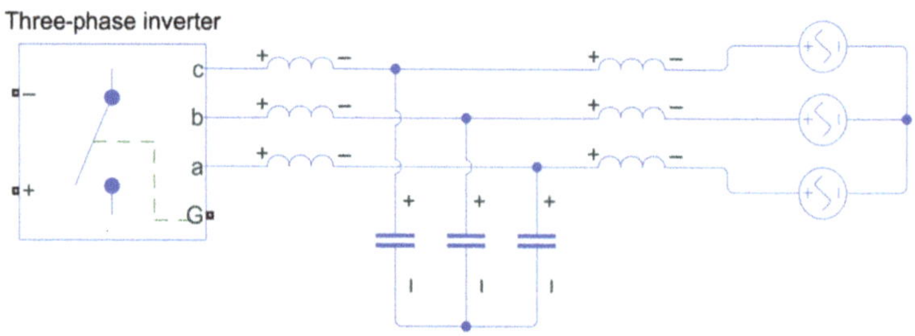

Figure 3. A three-phase simplified model diagram.

2.3. *Transformation of the Model Using dq0 Method*

In order to simplify the control process by introducing two constant controlled signals instead of three phase signals, the dq0 transformation was used [25,41,48–51]. The use of this operation also allows for the implementation of a simple PLL synchronization system. After applying the discussed transformation, the following form of state equations can be obtained:

$$\begin{cases} L\frac{d}{dt}\begin{bmatrix} i_d \\ i_q \end{bmatrix} = U_{dc}\begin{bmatrix} d_d \\ d_q \end{bmatrix} - \begin{bmatrix} u_{cd} \\ u_{cq} \end{bmatrix} - L\omega\begin{bmatrix} -i_q \\ i_d \end{bmatrix} \\ C_f\frac{d}{dt}\begin{bmatrix} u_{cd} \\ u_{cq} \end{bmatrix} = \begin{bmatrix} i_d \\ i_q \end{bmatrix} - \begin{bmatrix} i_{gd} \\ i_{gq} \end{bmatrix} - C_f\omega\begin{bmatrix} -u_{cq} \\ u_{cd} \end{bmatrix} \\ L_g\frac{d}{dt}\begin{bmatrix} i_{gd} \\ i_{gq} \end{bmatrix} = \begin{bmatrix} u_{cd} \\ u_{cq} \end{bmatrix} - \begin{bmatrix} U_{gd} \\ U_{gq} \end{bmatrix} - L_g\omega\begin{bmatrix} -i_{gq} \\ i_{gd} \end{bmatrix}, \end{cases} \tag{2}$$

where

- i_d, i_q—inverter output current vector components;
- i_{gd}, i_{gq}—grid current vector components;
- u_{cd}, u_{cq}—capacitive filter voltage vector components;
- U_{gd}, U_{gq}—grid voltage vector components;
- d_d, d_q—voltage modulator duty cycle vector components;
- ω—pulsation.

It is noticeable that additional terms appeared in the equations, resulting from cross-couplings created in the system after the transformation.

The block diagram of the obtained dynamic system is visible in Figure 4. The use of the transformation allows one to consider the problem of regulation of the discussed system from the power perspective—component d is responsible for active power, and q for reactive power. This work focuses mainly on the problems of controlling the grid current in a wide range for component d, while the aim of regulation of component q is always to reduce the current of this component to the value 0, which is also a problem often discussed [52].

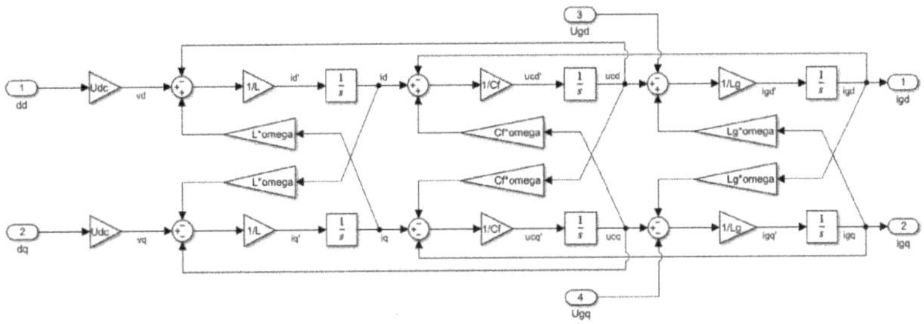

Figure 4. Scheme of the model transformed using dq0 method.

It is therefore a sixth-order system whose state equations can be represented in matrix form:

$$\begin{bmatrix} \dot{i_d} \\ \dot{u_{cd}} \\ \dot{i_{gd}} \\ \dot{i_q} \\ \dot{u_{cq}} \\ \dot{i_{gq}} \end{bmatrix} = \begin{bmatrix} 0 & -\frac{1}{L} & 0 & \omega & 0 & 0 \\ \frac{1}{C_f} & 0 & -\frac{1}{C_f} & 0 & \omega & 0 \\ 0 & \frac{1}{L_g} & 0 & 0 & 0 & \omega \\ -\omega & 0 & 0 & 0 & -\frac{1}{L} & 0 \\ 0 & -\omega & 0 & \frac{1}{C_f} & 0 & -\frac{1}{C_f} \\ 0 & 0 & -\omega & 0 & \frac{1}{L_g} & 0 \end{bmatrix} \begin{bmatrix} i_d \\ u_{cd} \\ i_{gd} \\ i_q \\ u_{cq} \\ i_{gd} \end{bmatrix} + \begin{bmatrix} \frac{U_{dc}}{L} & 0 \\ 0 & 0 \\ 0 & 0 \\ 0 & \frac{U_{dc}}{L} \\ 0 & 0 \\ 0 & 0 \end{bmatrix} \begin{bmatrix} d_d \\ d_q \end{bmatrix}. \quad (3)$$

In order to design control systems, the system described by the matrix Equation (3) was divided into two smaller systems, which leads to the need of an appropriate controller for a given vector component. This action significantly simplifies the further design of controllers, and the overall dynamics of the system can still be taken into account using equivalent control laws.

For the d component of the vector, the matrix state equations take the following form:

$$\begin{bmatrix} \dot{i_d} \\ \dot{u_{cd}} \\ \dot{i_{gd}} \end{bmatrix} = \begin{bmatrix} 0 & -\frac{1}{L} & 0 \\ \frac{1}{C_f} & 0 & -\frac{1}{C_f} \\ 0 & \frac{1}{L_g} & 0 \end{bmatrix} \begin{bmatrix} i_d \\ u_{cd} \\ i_{gd} \end{bmatrix} + \begin{bmatrix} \frac{U_{dc}}{L} \\ 0 \\ 0 \end{bmatrix} d_d. \quad (4)$$

Similarly, the matrices for the q component can be determined.

2.4. Simulation Model

The simulation model was implemented in the MATLAB Simulink environment using elements of the specialized Simscape library. All devices were built from scratch, without using ready-made models, e.g., for an inverter. This model is therefore a realization close to reality, which can be configured in detail. A discrete solver with a time step of 1×10^{-7} was selected.

3. Implementation of Known SMC Algorithms

In order to compare the proposed control method with known literature methods, two algorithms will be first presented and described, which have already been developed in the literature and adapted to the discussed control problems. The main interest concerns the so-called reaching laws, which can have a significant impact on the control process and its quality [4,53–57]. The first selected method is the classic sliding mode control with a discontinuous (switching) part, proportional reaching law and equivalent control. The second algorithm is HSMC—hybrid control, in which the proportionally reaching law is switched with a power reaching law, and the discontinuous control has a variable amplitude.

3.1. Sliding Variable Selection

The first, and most important step in sliding mode control algorithm design is the appropriate proposal of the form of the hyperplane and its coefficients. As mentioned above, the dynamics of the system in the sliding phase will be governed solely by the adopted hyperplane. There are several methods for the appropriate selection of these coefficients, such as pole placement method, quadratic optimization, etc., but also using the dead-beat method [5,15]. In the implementation examples of sliding mode control in the grid current control system using a three-phase inverter known from the literature, the method of selecting these parameters is based on manual selection of values that are experimentally proven to provide stability of the system [30]. An additional difficulty in using some analytical methods is the fact that the state variables are not phase variables. In the two discussed cases, the manual method was used, but its stability was confirmed by using the state matrix described in Section 2, which is presented in the next subsection. Therefore, the following two sliding variables were selected:

$$s_d = c_1(i_d - i_{d_{set}}) + c_2(u_{cd} - u_{cd_{set}}) + c_3(i_{gd} - i_{gd_{set}}), \tag{5}$$

$$s_q = c_1(i_q - i_{q_{set}}) + c_2(u_{cq} - u_{cq_{set}}) + c_3(i_{gq} - i_{gq_{set}}), \tag{6}$$

where

- s_d, s_q—sliding variables;
- c_1, c_2, c_3—sliding variable coefficients, where $c_1 > 0, c_2 > 0, c_3 > 0$;
- $i_{d_{set}}, u_{cd_{set}}, i_{gd_{set}}, i_{q_{set}}, u_{cq_{set}}, i_{gq_{set}}$—desired values of the state variables.

3.2. Stability Analysis

In order to investigate the stability of the system, the state feedback with the equivalent control substitution was applied, which led to obtaining a closed matrix of the system with the equivalent control applied. The stability of the system was investigated based on determining the poles of this system depending on the applied coefficients of the sliding variable, the combination of which was checked to a certain extent [58]. On this basis, a graphical visualization of the three-dimensional space was obtained, showing stable combinations of hyperplane parameters. The values of the slowest modes were presented. The aforementioned visualization is visible in Figure 5. The following sliding variable cofefficient values were selected: $c_1 = 0.5$, $c_2 = 1$, and $c_3 = 1$.

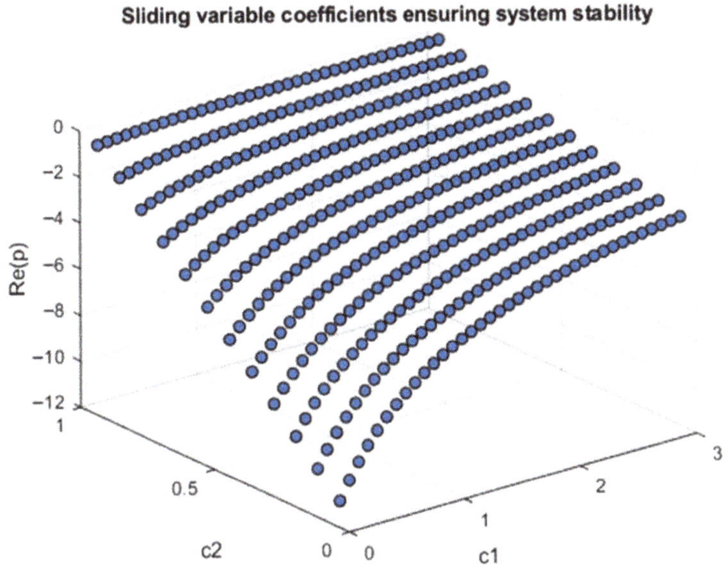

Figure 5. Graphical representation of sliding coefficient stability for $c_3 = 1$.

3.3. Implementation of a Classic SMC

The structure of the classical controller is based on a discontinuous part, implemented using the sign function and equivalent control, which results from the dynamics of the plant. Additionally, the so-called reaching law has been taken into account in the form of a very simple, proportional part of the control law. The controllers for both components of the vectors are designed in an analogous way; therefore, the derivations of equations for the d component only have been presented.

In order to determine the equivalent control, Lyapunov stability analysis can be used, with the following Lyapunov function [9,30]:

$$V_d = \frac{1}{2} s_d^2, \tag{7}$$

will be reduced to the form of the well-known stability condition of a sliding-mode control system:

$$\dot{V_d} = s_d \dot{s_d} < 0. \tag{8}$$

The controlled variable in the case of the discussed issue is the output voltage of the inverter, which can be presented as the dependence of the control signal on the source voltage on the DC side of the inverter and the duty cycle [9,30]:

$$\begin{bmatrix} v_d \\ v_q \end{bmatrix} = U_{dc} \begin{bmatrix} d_d \\ d_q \end{bmatrix}, \tag{9}$$

where

- v_d, v_q—components of the inverter output voltage vector.

Finally, the form of equivalent control for the d component can be presented in the following form [9]:

$$\begin{aligned} u_{d_{eq}} &= -\frac{L}{U_{dc}}\left(\frac{-u_{cd}+L\omega i_q}{L}-i_{d_{set}}\right) \\ &-\frac{L}{c_1c_2U_{dc}}\left(\frac{i_d-i_{gd}+C_f\omega u_{cq}}{C_f}-u_{cd_{set}}\right) \\ &-\frac{L}{c_1c_3U_{dc}}\left(\frac{-u_{cd}+L\omega i_{gq}}{L}-i_{d_{set}}\right). \end{aligned} \tag{10}$$

Taking into account the obtained equivalent part in the control law, the following full form of the control [9] can be obtained:

$$u_d = -k_1\,\mathrm{sgn}(s_d) - k_2 s_d + u_{d_{eq}}, \tag{11}$$

where

- u_d—control law for the d component;
- k_1, k_2—positive gains for corresponding control law parts.

Such a control algorithm ensures Lyapunov asymptotic stability, which is shown in the following equation:

$$\dot{V}_d = s_d(-k_1\,\mathrm{sgn}(s_d) - k_2 s_d) < 0. \tag{12}$$

Similar derivations apply to the q component.

3.4. Implementation of a Hybrid SMC

One of the interesting approaches is the use of so-called hybrid controllers, in which a certain part of the control law changes depending on the fulfillment of certain conditions [8,9]. In the case presented in this subsection, this is a change in the type of the reaching rule, the switching of which takes place at the representative point, which allows to avoid step changes in the control signals [9]. In the discussed case, two types of reaching rules are switched: proportional and power. The power reaching rule is used in the regions of state space far from the hyperplane, which allows to reduce the duration of the reaching phase. In the area closer to the hyperplane, the proportional part is activated, since it tends to provide better results with lower values of the sliding variable [8,9]. Additionally, the amplitude of the discontinuous part is variable and depends on the value of the sliding variable and the appropriate parameter. Such a procedure is intended to reduce the influence of chattering, which has been confirmed in the literature [9,12].

Again, it is possible to apply the methods used for the implementation and analysis of classical SMC, only changing the reaching law and the discontinuous control [9]:

$$u_d = -k_1|s_d|^{1-\alpha}\,\mathrm{sgn}(s_d) - k_2 H(s_d,\alpha) + u_{d_{eq}}, \tag{13}$$

where

- α—an additional parameter for hybrid control law;
- $H(s_d,\alpha)$—an additional function to implement hybrid control type, expressed by equation:

$$\begin{cases} H(s_d, \alpha) = |s_d|^{1+\alpha} \operatorname{sgn}(s_d) & \text{for } |s_d| > 1, \\ H(s_d, \alpha) = s_d & \text{for } |s_d| \le 1. \end{cases} \quad (14)$$

The presented control law again ensures asymptotic stability in the Lyapunov sense:

$$\dot{V}_d = s_d \Big(-k_1 |s_d|^{1-\alpha} \operatorname{sgn}(s_d) - k_2 H(s_d, \alpha) \Big) < 0. \quad (15)$$

3.5. Results Presentation

In order to initially compare both methods, a very simple test scenario case consisting of setting a constant signal value of 5 A is proposed for the desired value of the d component of the grid current vector.

The results for both methods are presented below. Two types of waveforms are presented for the d component of the grid current vector: for a short simulation time to show the beginning of the control process and for a longer simulation time to demonstrate the steady state behavior. Additionally, sliding variable waveforms for each controller case are visualized. Figures 6 and 7 present graphs of the d components of the grid current vector and their given signals for the case of using the classic SMC from Equation (11) and the hybrid HSMC described by Equation (13), respectively. As mentioned earlier, this is the first type of waveform that concerns a short simulation time in order to visualize the dynamic stage itself, i.e., the reaching phase.

The waveforms visible in Figures 8 and 9 allow to observe sliding variable signals. One can see that for the first case, i.e., for the classic sliding mode control approach, the impact of the chattering phenomenon resulting from discontinuous control presence. On the other hand, the second figure mentioned shows that this impact may be indeed reduced using hybrid reaching law and variable amplitude of the switching control part.

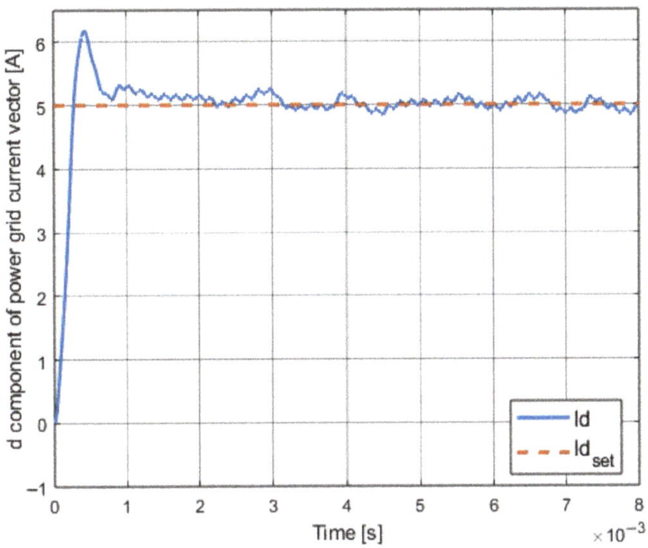

Figure 6. Grid current and its desired value waveforms for the d component. Classic SMC approach implementation used.

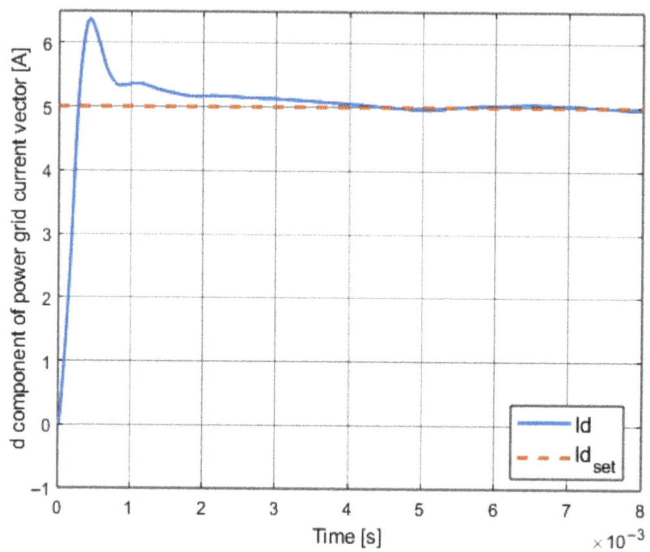

Figure 7. Grid current and its desired value waveforms for the d component. Hybrid SMC approach implementation used.

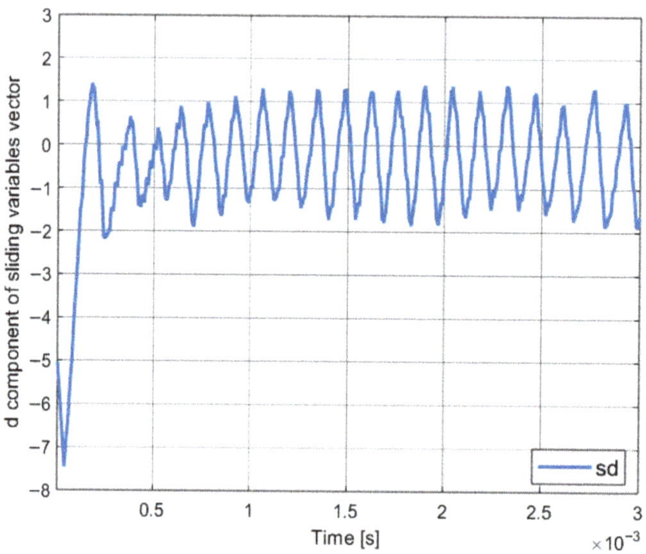

Figure 8. Sliding variable waveform for the d component. Classic SMC approach implementation used.

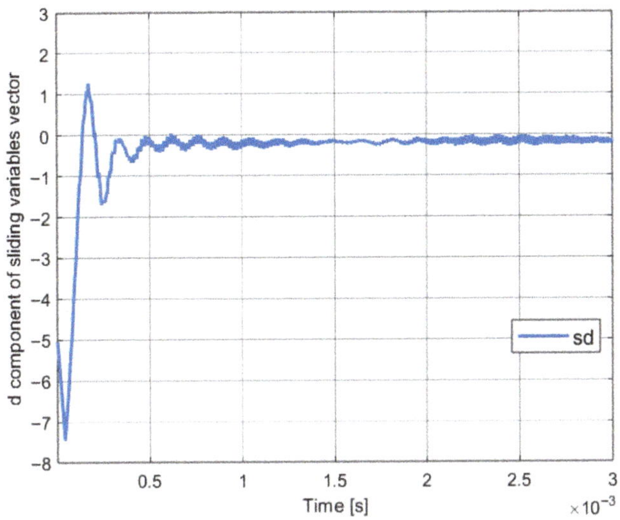

Figure 9. Sliding variable waveform for the d component. Hybrid SMC approach implementation used.

The observable oscillations resulting from the presence of discontinuous control, which are shown in Figure 10, were significantly reduced by means of hybrid control, and specifically by using the implementation of variable amplitude discontinuous control, which is visible in Figure 11. As mentioned earlier, only the oscillations resulting from the operation of the inverter keys remained (6th harmonic). This is a state close to the expected one. Reducing the amplitude value of these oscillations even to a greater extent is one of the goals of using more sophisticated methods presented in the subsequent chapters of the work.

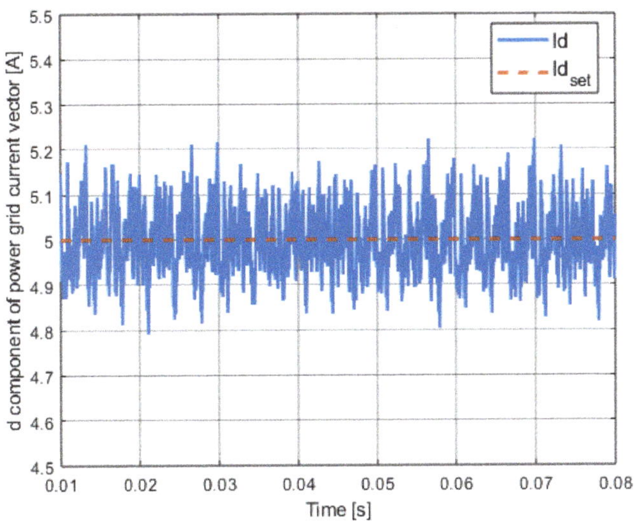

Figure 10. Grid current and its desired value waveforms for the d component. Classic SMC approach implementation used.

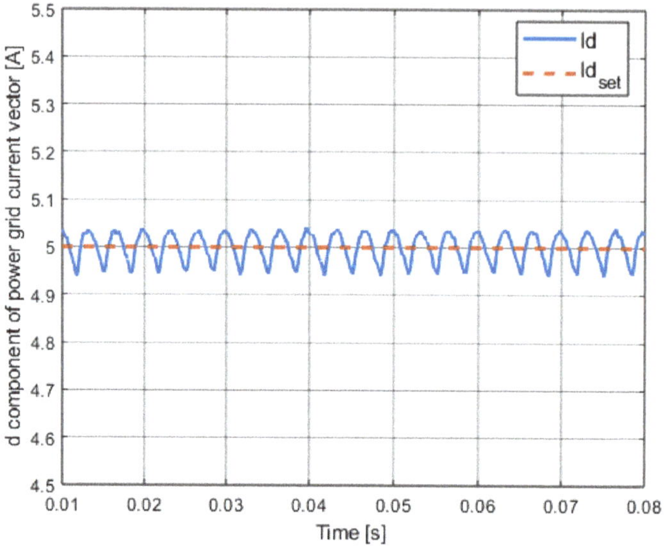

Figure 11. Grid current and its desired value waveforms for the d component. Hybrid SMC approach implementation used.

3.6. Conclusions About Known Solutions

In both cases of the implementation of control algorithms based on sliding methods, the system turned out to be stable and the results were satisfactory. The use of the classical algorithm meets the basic requirements regarding the quality of control; however, a visible drawback is the presence of the chattering phenomenon, which is reduced to the presence of low-amplitude and high-frequency oscillations in the controlled signal of the system [1,4]. The hybrid control algorithm allows for obtaining better results by using the appropriate switching of reaching law and the variable amplitude of discontinuous control, thanks to which the impact of the chattering phenomenon on the system is significantly reduced. However, the proposed solutions also have a number of disadvantages. Firstly, the lack of possibility of using known analytical methods for selecting the coefficients of sliding variables—some methods are not applicable due to the previously mentioned fact of non-phase state variables, and, for example, the method based on dead-beat is not possible to implement due to the generation of too high signal values in the system, which shall lead to instability. It would be possible to use the pole placement method, where the starting point would be dead-beat control, but this is still a manual operation to some extent. The effect of manual selection of hyperplane parameters is most likely the overshoot visible in Figure 6. Secondly, these are continuous methods, and a better solution would be to use discrete control assuming the use of digital systems. Another problem is the use of certain quantities that are not fully analytically confirmed in the literature, but are present in the control laws in the form of gains. It is true that in the case of continuous controllers, the gain of discontinuous control should be greater than the total values of disturbances acting on the system [2]. In order to reduce the impact of the mentioned control system weaknesses on the system, a more advanced sliding control method is proposed in the next chapter.

4. Nonswitching-Type Reaching Law Implementation

In this section, a method for designing a nonswitching reaching rule based on [5] is presented. Then, this method is implemented in the system and its operation is checked. This algorithm belongs to the category of discrete controllers; hence, the system has been discretized. Interestingly, in this method there is no discontinuous switching part included; hence, the chattering phenomena problem should be completely eliminated. This method is designed to use a sliding hyperplane obtained with the dead-beat approach [5,16,55,56]. The selection of the sliding variable coefficients based on the dead-beat approach allows obtaining a hyperplane that shall provide the fastest error reduction, and the reaching law itself is designed in such a way as to appropriately limit the control signal, which could reach very high values without this limitation. The algorithm also allows for taking into account compensation for external disturbances and model imperfections using estimated maximum values of the deviations of these phenomena from the average values.

4.1. Dead-Beat Approach

In order to use the nonswitching reaching law method with the dead-beat hyperplane design approach, the system model was discretized. Two reduced matrices were used, which greatly simplified the discretization process. The discrete state matrix for the d component took the following form:

$$\boldsymbol{A_d} = \begin{bmatrix} A_{d_{11}} & A_{d_{12}} & A_{d_{13}} \\ A_{d_{21}} & A_{d_{22}} & A_{d_{23}} \\ A_{d_{31}} & A_{d_{32}} & A_{d_{33}} \end{bmatrix}. \tag{16}$$

Due to the size of the above matrix, its individual elements are presented in Appendix A. The presented matrix will also be quoted later in the context of introducing the adaptation mechanism to the discussed control method.

To obtain the coefficients of the sliding variable using the dead-beat approach, one must determine the vector $\boldsymbol{c}^T$ from the following equation:

$$\det\left(\begin{bmatrix} z & 0 & 0 \\ 0 & z & 0 \\ 0 & 0 & z \end{bmatrix} - \left(\begin{bmatrix} 1 & 0 & 0 \\ 0 & 1 & 0 \\ 0 & 0 & 1 \end{bmatrix} - \boldsymbol{B_d}\left(\boldsymbol{c}^T\boldsymbol{B_d}\right)^{-1}\boldsymbol{c}^T\right)\boldsymbol{A_d}\right) = z^3 \tag{17}$$

where

- $\boldsymbol{B_d}$—discrete input matrix;
- $\boldsymbol{c}^T$—sliding variable coefficients vector, $\boldsymbol{c}^T = [c_1\ c_2\ c_3]$.

Determining the discussed coefficients is greatly facilitated by dividing the entire matrix system of equations into two simpler ones—this resulted in obtaining two third-order systems, hence the matrix for each component is compared to z^3.

4.2. Control Law Implemenation

To determine the control law based on the nonswitching reaching law type, it is necessary to define sliding variables in the discrete domain:

$$s_d(k) = c_1[i_d(k) - i_{d_{set}}(k)] + c_2[u_{cd}(k) - u_{cd_{set}}(k)] + c_3[i_{gd}(k) - i_{gd_{set}}(k)], \tag{18}$$

where

- $s_d(k)$—sliding variable for the d component,
- $i_d(k), u_{cd}(k), i_{gd}(k)$—discrete state variables,

- $i_{d_{set}}(k), u_{cd_{set}}(k), i_{gd_{set}}(k)$—desired discrete state variables signals.

For this chosen definition of the sliding variable, the control law based on the non-switching reaching law type has the following form:

$$u_{d_{NRL}}(k) = -\left(c^T B_d\right)^{-1}\left\{-[1 - q[s_d(k)]]s_d(k) + c^T A_d x_d(k) - c^T x_{d_{set}}(k)\right\} + \left(c^T B_d\right)^{-1}\{-F_{1d} - S_{1d}\}, \tag{19}$$

where

- $u_{d_{NRL}}(k)$—control law based on nonswitching reaching law type;
- $q[s_d(k)]$—reaching law function; $q[s_d(k)] = \frac{s_{0d}}{|s_d(k)|+s_{0d}}$; where s_{0d}—a positive constant;
- F_{1d}—average value of the disturbances impact on the system;
- S_{1d}—average value of the model inaccuracy impact on the system.

The control law presented above, together with an analogous control signal for the q component, has been implemented in the simulation environment. It should be noted that the equivalent control is largely implemented by state feedback; however, the existence of cross-couplings feedback is ignored, but it can be taken into account by using the appropriate constants that refer to the model inaccuracies and that are included in the control law.

Due to the use of a hyperplane designed with the dead-beat approach, three separate simulation models with different levels of advancement and complexity were designed:

- Mathematical model—The simplest model, which was built using the matrix equations presented earlier. It concerns only one component of the grid current vector. Only basic Simulink elements were used in the design of this system. This model is helpful in verifying control based on the dead-beat method.
- Model with ideal voltage sources—This is a model that takes into account both components of the state variable vectors. For this reason, it contains dq0 transformation systems and a synchronization system. The LCL filter and the power grid are implemented using elements of the Simscape library, which allows for the implementation of electrical elements in a very precise way. Reference voltages are implemented using ideal voltage sources, which allows for testing a fairly advanced version of the system in the absence of control signal constraints and for any output voltage levels.
- Full model, i.e., model with modulator and inverter—This is the most advanced version of the model, which extends the previous model with modulator and inverter, which are also implemented using electrical elements of the Simscape library. Due to the mentioned elements, there are restrictions on the control signal in the system, and the T-type inverter—as a three-level inverter—can realize three voltage levels at the output: positive, zero and negative. This is the model used to conduct series of tests to generate results presented in this paper, also including tests of classic and hybrid control, which were described in the previous chapters.

4.3. Stability Analysis

Based on [5] it is possible to ensure the stability of the system using such a control law by fulfilling the following condition:

$$s_{0d} > S_{2d} + F_{2d}, \tag{20}$$

where

- S_{2d}—maximum permissible deviation of the value of the impact of model inaccuracy on the system from the nominal value;
- F_{2d}—maximum permissible deviation of the value of the impact of external disturbances on the system from the nominal value.

Both sets of parameters: above S_{2d} and F_{2d} and also mentioned in (19) S_{1d} and F_{1d} refer, respectively, to the total effect of the model uncertainty and external disturbances on the system:

$$\tilde{S_d}(k) = \mathbf{c}^T \Delta[\mathbf{x_d}(k)], \tag{21}$$

where

- $\Delta[\mathbf{x_d}(k)]$—a model uncertainty representation.

$$\tilde{F_d}(k) = \mathbf{c}^T \mathbf{f}(k), \tag{22}$$

where

- $\mathbf{f}(k)$—an external disturbance.

First, the control algorithm based on the dead-beat method (without nonswitching reaching law) was tested on a discrete mathematical model. After achieving success, tests were carried out on the second level of model advancement, also obtaining a positive result. The full technical model could not work correctly (in the context of simulation) due to the limitations of the inverter and the DC source, which led to exceeding the limits of the control signals and, as a result, to instability. Taking into account the full control law from (19) allowed to limit the control values in an expected way, so that the control task could be realized using the full technical model that was closest to reality.

In the case of control process discussed in this paper it is possible to introduce assumptions regarding above-described effects:

(1) The total effect of the model uncertainty results from the lack of inclusion of the cross-couplings in the control algorithm and also results from all phenomena resulting from the use of advanced electronic components in simulation environement, whose full mathematical description would be too complex, i.e., parameters of transistors, passive elements, parasitic values, etc.
(2) The total impact of disturbances acting on the system can be assumed arbitrarily to some extent, taking into account additionally the experimentally observed oscillations resulting from the operation of transistors, focusing on the value of their amplitude based on Figure 11. Since there is no external disturbance applied, it is possible to use the discussed signal in this manner.

In the first case, the signal may be specifed by considering the impact of the components added in (2) in relation to (1). To obtain values from (19) and (20), which refer to the average values and maximum deviations, assumptions regarding the test scenario must be made and then these signals can be measured.

Taking into account that

$$S_{1d} = \frac{S_{Ud} + S_{Ld}}{2}, \quad F_{1d} = \frac{F_{Ud} + F_{Ld}}{2}, \tag{23}$$

$$S_{2d} = \frac{S_{Ud} - S_{Ld}}{2}, \quad F_{2d} = \frac{F_{Ud} - F_{Ld}}{2}, \tag{24}$$

where

- S_{Ud}—upper bound of $\tilde{S}_d(k)$;
- S_{Ld}—lower bound of $\tilde{S}_d(k)$;
- F_{Ud}—upper bound of $\tilde{F}_d(k)$;
- F_{Ld}—lower bound of $\tilde{F}_d(k)$.

It is possible to redirect the analysis of the discussed impacts to determine, through appropriate assumptions, their maximum and minimum values.

One can see that missing control law elements, resulting from the lack of inclusion of the cross-couplings, can be found in the full equivalent control from (10). Considering observations mentioned above, the following form of (21) may be proposed for $(c^T B_d) \neq 0$:

$$\tilde{S}_d(k) = -\frac{\frac{L\omega}{U_{dc}}\left(i_q(k) + \frac{u_{cq}(k)}{c_1 c_2} + \frac{i_{gq}(k)}{c_1 c_3}\right)}{\left(c^T B_d\right)^{-1}}. \tag{25}$$

The obtained equation (25) describes the total effect of one vector component (q) to another one (d). Based on that it is possible to propose appropriate assumptions or deliver experimental data regarding quantities presented in (24). As for (22), without any known external disturbance impacting the system, the best solution is to take into consideration observed oscillations amplitude.

4.4. Obtained Results

Similarly to Section 3.5, the results of the simulation tests are presented below, maintaining the same form. The control law parameter (s_{0d}) was selected arbitrary based on (20) condition. The waveform shown in Figure 12 differs significantly from those presented in Figures 6 and 7—there is no overshoot, and it is also possible to observe a completely different waveform characteristic. The same conclusions apply to results presented in Figure 13 when compared with those from Figures 8 and 9. In turn, in Figure 14 one can see that, although the amplitude of oscillations resulting from the operation of the inverter keys decreased, a significant steady-state error appeared. The obtained results are described in more detail in the next subsection.

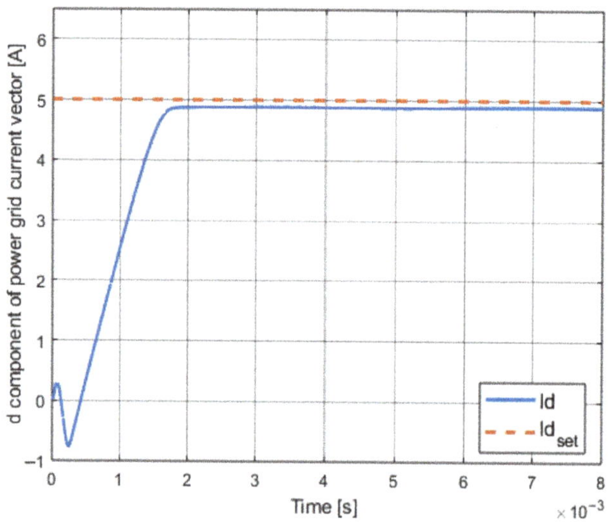

Figure 12. Grid current and its desired value waveforms for the d component. Nonswitching reaching law approach implementation used.

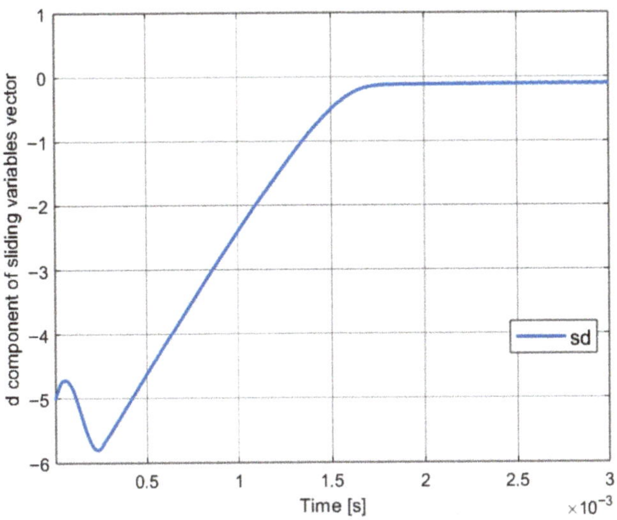

Figure 13. Sliding variable waveform for the d component. Nonswitching reaching law approach implementation used.

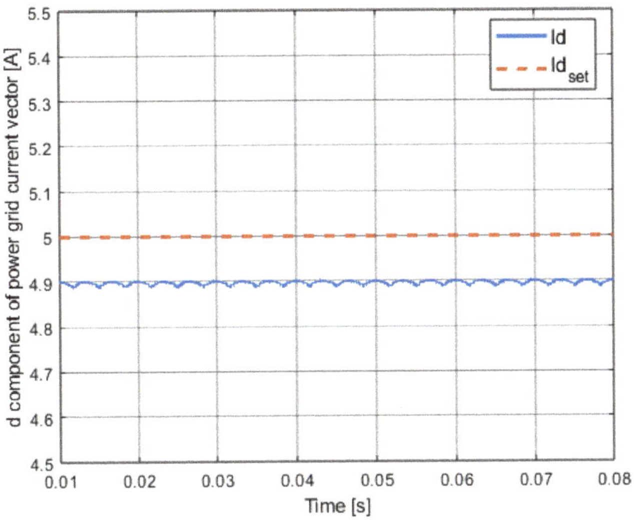

Figure 14. Grid current and its desired value waveforms for the d component. Nonswitching reaching law approach implementation used.

4.5. Conclusions About the Method

The applied solution allowed for the obtainment of a very good quality signal of the controlled value. It is free of oscillations and noise visible for the classical method, and there is no overshoot. The reaching phase is characterized by an almost linear characteristic of the grid current of the d component. The similar linearity itself can be a serious advantage in many different implementations and fields, especially since the set signal is constant.

It is worth mentioning that the characteristic of the sliding variable waveform, visible in Figure 13, is significantly different from those previously obtained (Figures 8 and 9). The signal in this case is smooth and mostly close to a linear shape during reaching phase. No chattering is visible, but non-zero error in steady-state can be observed.

It is possible to point out several aspects of the control law in question that could be changed or modified, namely

a. In the case of the above results, the parameter s_{0d} is based entirely on the condition presented in (20), i.e., its value was assumed arbitrarily in the range specified by the condition. This parameter directly affects the gain of the control signal and thanks to it, it is possible to inhibit the dead-beat control to a greater or lesser extent. Hence, its appropriate selection for the assumed operating conditions could improve the quality of the control process. This problem is the genesis of the introduced adaptation mechanism, which will be presented in the next chapter.
b. There is a steady-state error in the system, which can be seen in Figure 14. This error in long-term processes may turn out to be a factor significantly reducing the potential quality of control. This problem, in turn, is the basis for using the method of reducing the steady-state error proposed in Section 6.

5. Introducing Adaptation Mechanism

The presented control law described by Equation (19) assumes the use of a sliding variable resulting from the dead-beat method. It follows that the key aspect in such a control algorithm is to propose a model with high accuracy of plant representation. The methods presented above for limiting the value of the control signal to the maximum control value can be extended to take into account the occurrence of changes in specific model parameters within a certain assumed range. One of the proposed solutions for selecting the parameter s_{0d} may be to link this parameter to a certain maximum value of the control signal. In the case of using an inverter with a modulator, the range of the control signal is $[-1, 1]$, so the maximum absolute value of the control, which is equal to 1, can be used as follows:

$$u_{d_{NRL}}(0) \leq |u_{d_{max}}|, \tag{26}$$

where

- $u_{d_{max}}$—maximum allowable control.

Using the condition given in (26) and based on Equation (19), it is possible to present the following condition regarding the parameter s_{0d}:

$$s_{0d} \leq \frac{\left(\frac{c^T x_{d_{set}}(0) - |u_{d_{max}}|(c^T B_d) - c^T A_d x_d(0)}{s_d(0)} + 1\right)|s_d(0)|}{\frac{c^T x_{d_{set}}(0) - |u_{d_{max}}|(c^T B_d) - c^T A_d x_d(0)}{s_d(0)}}. \tag{27}$$

Based on this condition, it is possible to determine the appropriate control parameter, which ensures that the signal value at the first moment of time will not exceed the maximum value. This means that it is possible to use the dead-beat sliding hyperplane justifiably. Moreover, the efficiency of the control system can be maintained by implementing a fast and stable regulation process.

To confirm the reasonableness of the above assumptions, a simulation test was performed again, this time for the newly determined parameter s_{0d}, in the Section 5.2.

5.1. Description of Adaptation Aspects

Using the method for determining the range of the non-switching control parameter, it is possible to propose a certain modification of the control based on the adaptation mechanism. Taking into account the dependencies between the parameter s_{0d} and the maximum value of the control signal, state variables and setpoint signals contained in the inequality (27), it is possible to introduce regular updates of the value of this parameter so as not to exceed the control limits, while maintaining the parameter value calculated for the maximum control value.

An additional aspect worth considering is the limitation resulting from the nature of the object itself. In this case, it is possible to narrow the range of the control law parameter based on certain assumed ranges of changes in the grid parameters, i.e., the main load of the system. In the discussed case, the most important thing will be to take into account the variation in the grid capacitance and inductance. However, this method can also be used for any other plant in which a variation in certain parameters, loads or disturbances may be expected.

For the following conditions:

$$L_g \neq L_{g\delta}, \tag{28}$$

$$C_f \neq C_{f\delta}, \tag{29}$$

where

- $L_{g\delta}$—new inductance value related to the grid load changes;
- $C_{f\delta}$—new capacity value related to the grid load changes.

It is possible to propose a new state matrix in which the changed grid parameters are taken into account. First, however, an indication what exact impact a change in the value of a given parameter may have on other parameters is needed:

- an increase in the grid inductance value leads to the need to increase the value of the control signal for a constant s_{0d};
- an increase in the capacitance value leads to the need to increase the value of the control signal for a constant s_{0d};
- an increase in the value of the parameter s_{0d} leads to an increase in the value of the control signal (the larger s_{0d}, the less dead-beat is counteracted).

$$\boldsymbol{A_{d\delta}} = \begin{bmatrix} A_{d_{11\delta}} & A_{d_{12\delta}} & A_{d_{13\delta}} \\ A_{d_{21\delta}} & A_{d_{22\delta}} & A_{d_{23\delta}} \\ A_{d_{31\delta}} & A_{d_{32\delta}} & A_{d_{33\delta}} \end{bmatrix}. \tag{30}$$

where

- $\boldsymbol{A_{d\delta}}$—discrete state matrix with changed inductance and capacity (individual elements are presented in Appendix B).

Hence, one can conclude that in order to maintain the control signal value within a certain limited range, it is possible to use an additional limitation of the parameter s_{0d} due to the assumed maximum values of inductance and capacitance. This is because it is the increase in these values that leads to exceeding the control signal limits. Hence, the values introduced in Equations (28) and (29) actually refer to the maximum values. Reducing either of these parameters will lead to a decrease in the control value, but within a safe range. Therefore, the parameter s_{0d} has been additionally limited due to the assumed maximum values of the grid parameters in such a way that the maximum control value is achieved for the maximum possible values of the changed grid parameters.

If, in accordance with the previous statements, an increase in the value of the grid parameters leads to an increase in the value of the necessary control, then the action of the adaptation mechanism should consist of decreasing the value of the parameter s_{0d} in order to reduce this control and maintain it within the assumed range or to adopt its assumed value.

By using a constant value of the parameter s_{0d}, it is possible to determine the difference in the control value for two different state matrices:

$$\delta u_{d_{max}} = -\left(\boldsymbol{c}^T\boldsymbol{B}_d\right)^{-1}\boldsymbol{c}^T[\boldsymbol{A}_d - \boldsymbol{A}_{d\delta}]\boldsymbol{x}_d(0), \tag{31}$$

However, since the intended goal is to introduce adaptation of the parameter s_{0d}, the range of the control value should always end at the maximum value resulting from the plant's specificity. Hence, it is possible to update the inequality (27) to take into account the assumed range of grid parameter variability:

$$s_{0d\delta} \leq \frac{\left(\frac{\boldsymbol{c}^T\boldsymbol{x}_{d_{set}}(0) - |u_{d_{max}}|\left(\boldsymbol{c}^T\boldsymbol{B}_d\right) - \boldsymbol{c}^T\boldsymbol{A}_{d\delta}\boldsymbol{x}_d(0)}{s_d(0)} + 1\right)|s_d(0)|}{\frac{\boldsymbol{c}^T\boldsymbol{x}_{d_{set}}(0) - |u_{d_{max}}|\left(\boldsymbol{c}^T\boldsymbol{B}_d\right) - \boldsymbol{c}^T\boldsymbol{A}_{d\delta}\boldsymbol{x}_d(0)}{s_d(0)}}. \tag{32}$$

where

- $s_{0d\delta}$—changed control parameter taking into account the assumed range of grid parameter variability.

The inequality (32) is an improved form of the inequality (27) that takes into account the grid variability that affects the model. This is very important when using the state matrix feedback based on the dead-beat method.

Based on the assumed largest and smallest necessary changes in state variables resulting from the current and desired states, it is possible to determine the limits of the signal being a variable representation of the parameter $s_{0d\delta}$. These limits should take into account the previously described assumptions regarding the range of changes in the grid and filter parameters.

For $\max\left\{-\boldsymbol{c}^T\boldsymbol{A}_d\boldsymbol{x}_d(0) + \boldsymbol{c}^T\boldsymbol{x}_{d_{set}}(0)\right\} = \phi_{max}$:

$$s_{0d\delta_{min}} = \frac{\left(\frac{\phi_{max} - |u_{d_{max}}|\left(\boldsymbol{c}^T\boldsymbol{B}_d\right)}{s_d(0)} + 1\right)|s_d(0)|}{\frac{\phi_{max} - |u_{d_{max}}|\left(\boldsymbol{c}^T\boldsymbol{B}_d\right)}{s_d(0)}}, \tag{33}$$

where

- ϕ_{max}—an auxiliary constant expressing maximum value of the respective control law parts;
- $s_{0d\delta_{min}}$—a constant defining the lower limit of the control law parameter value with respect to the value of ϕ_{max}.

For $\min\left\{-\boldsymbol{c}^T\boldsymbol{A}_d\boldsymbol{x}_d(0) + \boldsymbol{c}^T\boldsymbol{x}_{d_{set}}(0)\right\} = \phi_{min}$:

$$s_{0d\delta_{max}} = \frac{\left(\frac{\phi_{min} - |u_{d_{max}}|\left(\boldsymbol{c}^T\boldsymbol{B}_d\right)}{s_d(0)} + 1\right)|s_d(0)|}{\frac{\phi_{min} - |u_{d_{max}}|\left(\boldsymbol{c}^T\boldsymbol{B}_d\right)}{s_d(0)}}, \tag{34}$$

where

- ϕ_{min}—an auxiliary constant expressing minimum value of the respective control law parts;
- $s_{0d\delta_{max}}$—constant defining the upper limit of the control law parameter value with respect to the value of ϕ_{max}.

Using the inequality (20) presenting the proof of stability, it is possible to present the following condition:

$$s_{0d\delta_{max}} \geq s_{0d\delta_{min}} > S_{2d} + F_{2d}. \tag{35}$$

5.2. Adaptation Mechanism Implementation

By adopting the above-mentioned assumptions described by equations and inequalities (26)–(35), it is possible to propose the following mechanism for adapting the parameter $s_{0d\delta}(k)$:

$$s_{0d\delta}(k) = \begin{cases} s_{0d\delta_{max}} & \text{for} \quad s_{0d\delta}(k) \geq s_{0d\delta_{max}} \\ \dfrac{\left(\frac{\phi(k)-|u_{d_{max}}|\left(c^T B_d\right)}{s_d(k)}+1\right)|s_d(k)|}{\frac{\phi(k)-|u_{d_{max}}|\left(c^T B_d\right)}{s_d(k)}} & \\ \text{for} \quad s_{0d\delta}(k) \in \left(s_{0d\delta_{min}}; s_{0d\delta_{max}}\right) & \\ s_{0d\delta_{min}} & \text{for} \quad s_{0d\delta}(k) \leq s_{0d\delta_{min}}, \end{cases} \tag{36}$$

where

- $s_{0d\delta}(k)$—a signal representing variable version of control law parameter;
- $\phi(k)$—an auxiliary signal, $\phi(k) = \boldsymbol{c}^T \boldsymbol{x}_{d_{set}}(k) - \boldsymbol{c}^T \boldsymbol{A}_{d\delta}$.

By implementing the proposed signals$s_{0d\delta}(k)$, the new control law will take the following form:

$$\begin{gathered} u_{d_{ANRL}}(k) = -\left(\boldsymbol{c}^T \boldsymbol{B}_d\right)^{-1} \\ \left\{-[1 - q_\delta[s_d(k)]]s_d(k) + \boldsymbol{c}^T \boldsymbol{A}_d \boldsymbol{x}_d(k) - \boldsymbol{c}^T \boldsymbol{x}_{d_{set}}(k)\right\} \\ +\left(\boldsymbol{c}^T \boldsymbol{B}_d\right)^{-1} \{-F_{1d} - S_{1d}\}, \end{gathered} \tag{37}$$

where

- $u_{d_{ANRL}}(k)$—control law based on adaptive nonswitching reaching law type;
- $q_\delta[s_d(k)] = \frac{s_{0d\delta}(k)}{|s_d(k)|+s_{0d\delta}(k)}$—modified reaching law part.

An important original assumption of the control law based on the nonswitching reaching law type was that the constant parameter used should be positive. The signal introduced instead of the constant should also be positive; therefore, a protection against reaching a negative value was added, which could be used for small values of the difference between the grid current and the set current, i.e., in the operating region close to the sliding hyperplane. Therefore, an updated form of Equation (36) was proposed:

$$s_{0d\delta}(k) = \begin{cases} s_{0d\delta_{max}} \quad \text{for} \quad s_{0d\delta}(k) \geq s_{0d\delta_{max}} \\ \frac{\left(\frac{\phi(k)-|u_{d_{max}}|\left(c^T B_d\right)}{s_d(k)}+1\right)|s_d(k)|}{\frac{\phi(k)-|u_{d_{max}}|\left(c^T B_d\right)}{s_d(k)}} \\ \text{for } s_{0d\delta}(k) \in \left(s_{0d\delta_{min}}; s_{0d\delta_{max}}\right) \wedge |s_d(k)| \geq 1 \\ s_{0d\delta}(k-1) \quad \text{for} \quad |s_d(k)| < 1 \\ s_{0d\delta_{min}} \quad \text{for} \quad s_{0d\delta}(k) \leq s_{0d\delta_{min}}. \end{cases} \tag{38}$$

The equation marked with number (38) is still applicable to the control law given by Equation (37).

5.3. Obtained Results

This subsection presents the simulation test results. Again, the same test scenario was used and the signal waveforms were recorded. It should also be mentioned that the lower limit of the auxiliary signal $\phi(k)$ proposed in (36) is 1 A and the upper limit is 100 A. The below presented waveforms demostrate the operation of the implemented adaptation mechanism—it may not be visible at first glance and will be more exposed through the presentation of quality measures in Section 7, but it is noticeable that the regulation time is shorter for the adaptive case in Figure 15 than for the case without the adaptation mechanism in Figure 12, which occurred without exceeding the maximum value of the control signal.

Comparing the results shown in Figure 16 to the case of using nonswitching reaching law algorithm with lack of adaptation mechanism in Figure 13, it should be mention that, in addition to the shorter reaching phase time mentioned earlier, the initial value changes (up to 0.25 milliseconds) were also reduced in terms of amplitude.

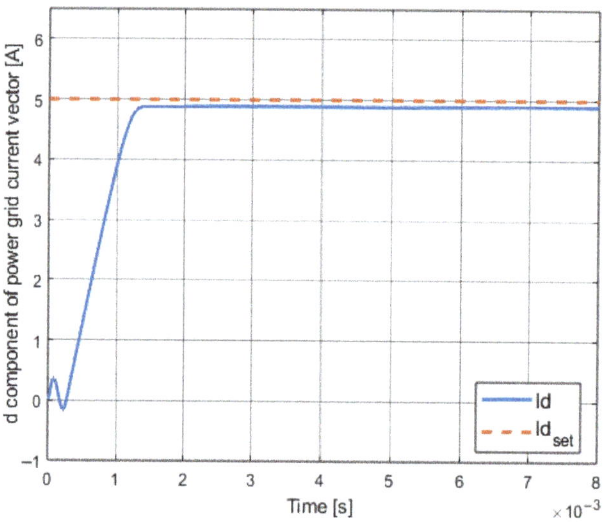

Figure 15. Grid current and its desired value waveforms for the d component. Adaptive nonswitching reaching law approach implementation used.

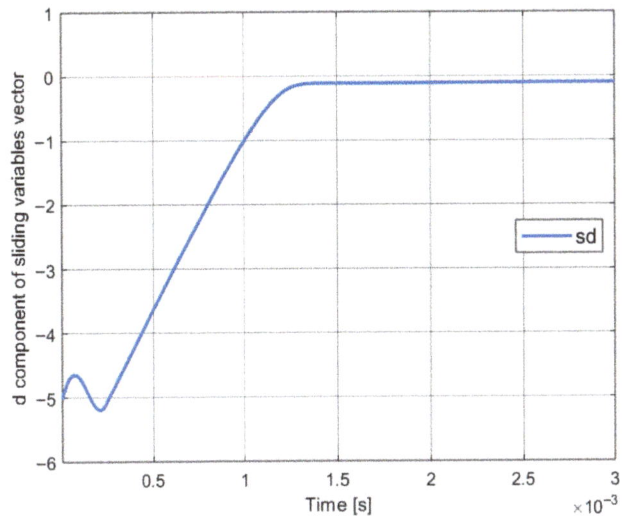

Figure 16. Sliding variable waveform for the d component. Adaptive nonswitching reaching law approach implementation used.

6. Introducing Steady Error Reduction Mechanism

The use of the adaptation mechanism allowed the system to adapt to various setpoint states and dynamic conditions, which significantly decreased the control time; however, one drawback still remains, which is better shown by rescaling the results from Figure 14, namely the non-zero steady-state error. This section presents the proposed concept that shall allow minimizing the steady-state error while using the discussed nonswitching reaching law type in the form with adaptation.

6.1. Description of the Proposed Method

Methods of reducing the steady-state error in the sliding-mode control theory often come down to the use of methods that take into account the calculation of the integral of the controlled quantity error or selected state variable. This element is usually included in the sliding variable [59–61].

In the case discussed in this paper, the topic concerns an advanced and specific control law based on the use of an appropriately limited dead-beat, therefore in the reaching phase (in the dynamic state) the reaching law should not be modified in any way by implementing an integral in the control law. Using an additional element in the sliding variable, which refers to the dead-beat coefficients through the hyperplane coefficients, is not correct due to the fact of modifying this hyperplane without referring to the dead-beat. In practice, in dynamic states the system may turn out to be unstable. Hence, the method proposed in the next subsection should only be activated interchangeably with the adaptation mechanism, i.e., adaptation operates in the dynamic state, and the steady-state error reduction is activated in the steady-state with a non-zero error presence. Such an implementation is possible because in the steady state adaptation is not crucial, and in the dynamic state the steady-state error reduction mechanism does not lead to an improvement in the quality of control, but quite the opposite.

6.2. Implementation of the Mechanism

The steady-state error reduction algorithm cooperating with the adaptive control law from Equation (37) can be described as follows, by modifying Equation (18):

$$s_d(k) = \begin{cases} c_1\left[i_d(k) - i_{d_{set}}(k)\right] + c_2\left[u_{cd}(k) - u_{cd_{set}}(k)\right] \\ +c_3\left[i_{gd}(k) - i_{gd_{set}}(k)\right] \\ \text{for } \psi(k) = 0 \\ \\ c_1\left[i_d(k) - i_{d_{set}}(k)\right] + c_2\left[u_{cd}(k) - u_{cd_{set}}(k)\right] \\ +c_3\left[i_{gd}(k) - i_{gd_{set}}(k)\right] \\ +\gamma T \sum_{j=k_s}^{k}\left[i_{gd}(j) - i_{gd_{set}}(j)\right] \\ \text{for } \psi(k) = 1, \end{cases} \quad (39)$$

for which, together with the condition $k \geq 1$:

$$\psi(k) = \begin{cases} 0, \text{ for} \\ \left(\frac{\left|i_{gd_{set}}(k) - i_{gd_{set}}(k-1))\right|}{T} > i_{gd_{set_{\delta max}}} \vee |s_d(k)| \geq 1\right) \\ \wedge \psi(k-1) = 1 \\ \\ 1, \text{ for} \\ \frac{\left|i_{gd}(k) - i_{gd}(k-1))\right|}{T} < i_{gd_{\delta max}} \wedge |s_d(k)| < 1 \\ \wedge \frac{\left|i_{gd_{set}}(k) - i_{gd_{set}}(k-1))\right|}{T} \leq i_{gd_{set_{\delta max}}} \wedge \psi(k-1) = 0 \\ \\ \psi(k-1) \text{ for other cases,} \end{cases} \quad (40)$$

where

- $i_{gd_{set_{\delta max}}}$—maximum permissible value of the change in the set value of the d component of the grid current vector;
- $i_{gd_{\delta max}}$—maximum value of the change in the d component of the grid current vector, below which it is assumed that the steady state is present;
- γ—gain of control based on the steady-state error reduction mechanism, $\gamma > 0$;
- T—sampling period ($T = 20 \mu s$);
- k_s—a discrete moment of time for which the activation condition of the mechanism is met;
- $\psi(k)$—a logical variable enabling the mathematical notation of the implementation of the activation and deactivation mechanism of the steady-state error reduction algorithm; $\psi(0) = 0$ (mechanism disabled by default).

Considering modifications proposed in (39) and (40) it is important to mention that providing the system with a sliding variable reduction to zero shall not necessarily lead to reduction of the individual errors included in this variable - only the linear combination of these errors shall tend to zero. However, including an error reduction mechanism based on discrete integral calculation precisely for one controlled signal, which is in this case grid current vector d component, allows to achieve the expected results, i.e., to obtain a zero value of the specific error.

The following element from Equation (39):

$$\gamma T \sum_{j=k_s}^{k} [i_{gd}(j) - i_{gd_{set}}(j)], \tag{41}$$

is always equal to zero at the time of mechanism activation and while it is active, its value also tends to zero—right after starting to reduce the steady-state error–due to its construction based on the integral.

6.3. Obtained Results

An analogous set of results was again generated for the same test scenario, this time also incorporating the steady-state error mechanism in the control law.

The control law based on the nonswitching reaching law type from Equation (19) was extended by two mechanisms: adaptation of the key control parameter and reduction in the steady-state error.

The proposed control law described by Equations (37)–(40) allowed for the obtainment of the following effects: the short control time obtained in the case of using the adaptation mechanism alone was maintained, which was visible in Figure 15, but additionally the problem of the presence of the non-zero steady-state error, visible in Figures 14 and 17, was eliminated, which, thanks to the applied method from Equation (39), ceased to be present in the grid current signal, which is visible in Figures 18 and 19 (in terms of the sliding variable) and in particular in Figure 20. All the positives obtained by using the previous techniques were additionally preserved. The detailed summary of results and the summary are placed in the next section.

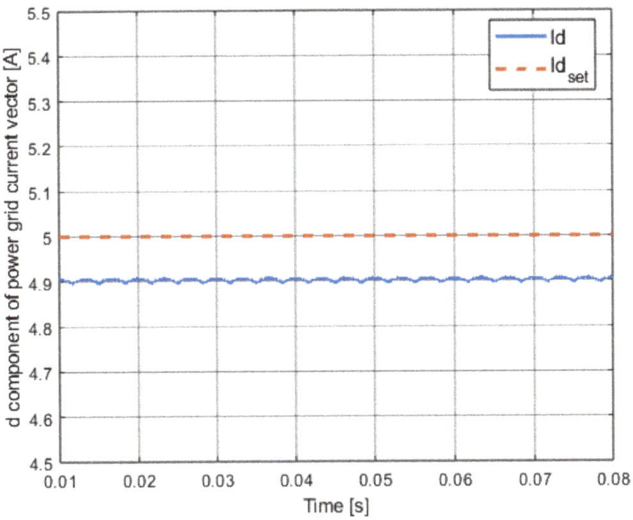

Figure 17. Grid current and its desired value waveforms for the d component. Adaptive nonswitching reaching law approach implementation used.

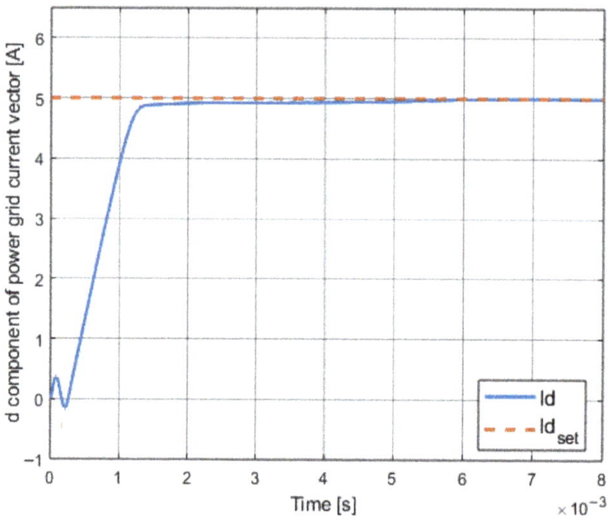

Figure 18. Grid current and its desired value waveforms for the d component. Adaptive integral nonswitching reaching law approach implementation used.

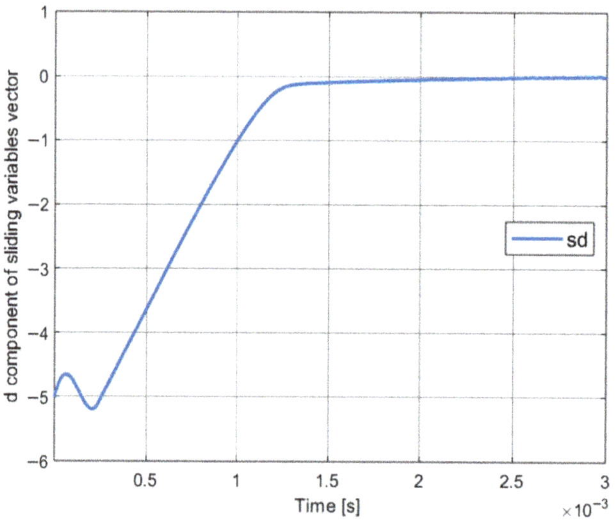

Figure 19. Sliding variable waveform for the d component. Adaptive integral nonswitching reaching law approach implementation used.

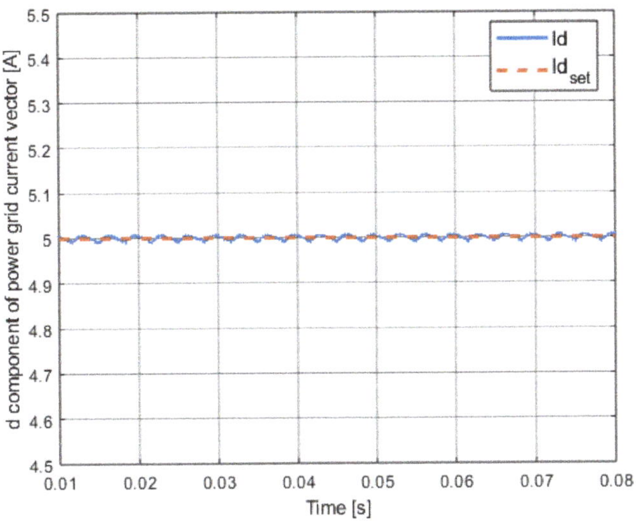

Figure 20. Grid current and its desired value waveforms for the d component. Adaptive integral nonswitching reaching law approach implementation used.

7. Comparative Analysis of the Presented Methods

Based on the test scenario described in Section 3.5, a series of simulation tests were performed, the results of which are visualized in the graphics from Figures 6–20. Using the measured data, the following quality metrics were determined: ISE and IAE for the entire time range of the test scenario, as well as ISE, IAE, minimum and maximum error for the steady state.

7.1. Control Measures Summary

The collected data were prepared for comparison in the form of appropriate tables, which contain the values of the aforementioned quality measures assigned to the given type of control algorithm used. The steady state was defined as the state from 0.01 s to the end of the simulation duration (0.08 s). The steady state study therefore concerns 3.5 periods of the sinusoidal grid current signal.

Table 1 presents the IAE and ISE control measures for all five control methods presented in this publication. The main goal of this test scenario was to initially examine the quality of control considering the dynamic state and the steady state of a specified duration.

Table 1. ISE and IAE control measures summary.

Control Method	ISE	IAE
Classic SMC	3.6939×10^{-3}	6.2885×10^{-3}
Hybrid SMC	3.5924×10^{-3}	3.7845×10^{-3}
Nonswitching Reaching Law (NRL)	2.2296×10^{-2}	1.3434×10^{-2}
Adaptive Nonswitching Reaching Law (ANRL)	1.4890×10^{-2}	1.1379×10^{-2}
Adaptive Integral Nonswitching Reaching Law (AINRL)	1.4157×10^{-2}	4.2585×10^{-3}

The results of the steady-state tests, which is the most important aspect in this work, are included in Tables 2 and 3. The obtained results are described in detail in the next subsection.

Table 2. Minimum and maximum error values summary.

Control Method	MIN (Steady-State)	MAX (Steady-State)
Classic SMC	-2.2278×10^{-1}	2.0677×10^{-1}
Hybrid SMC	-4.0321×10^{-2}	6.1065×10^{-2}
NRL	9.8367×10^{-2}	1.1198×10^{-1}
ANRL	9.1041×10^{-2}	1.0353×10^{-1}
AINRL	-6.6823×10^{-3}	8.4445×10^{-3}

Table 3. ISE and IAE control measures summary for steady-state case.

Control Method	ISE (Steady-State)	IAE (Steady-State)
Classic SMC	4.0190×10^{-4}	4.3989×10^{-3}
Hybrid SMC	6.1030×10^{-5}	1.8120×10^{-3}
NRL	7.6287×10^{-4}	7.3035×10^{-3}
ANRL	6.5887×10^{-4}	6.7884×10^{-3}
AINRL	1.3919×10^{-6}	2.7554×10^{-4}

7.2. Test Results Analysis

The use of the control algorithm based on the control law (37)–(40) resulted from an attempt to introduce a method that should lead to a significant improvement in the quality of control in the context of the dynamic state and the steady state by using a sliding hyperplane with an analytically justified form. Theoretical aspects and assumptions of this method indicated the possibility of completely reducing the impact of the chattering phenomenon on the system and obtaining a characteristic of the waveform, the specificity of which may be desired in certain control tasks in power electronics, and most importantly completely eliminates the problem of overshoot.

Based on the data from Table 1 it is possible to state that the quality of algorithms based on (19) is lower than for the classic solution and the better hybrid solution in the context of proposed control measures. This is due to the specific trajectory of the representative point in the case of using control laws based on (19), which from the perspective of calculating the values of quality measures may indeed seem less qualitative. However, its advantages should be taken into account, i.e., the previously mentioned almost linear signal growth without overshoot, for which the maximum control values resulting from the limitations of the three-phase inverter are not exceeded. Moreover, it is important to observe that the dynamic state in the control processes related to the discussed issue constitutes a significant minority of the entire process. The change of the grid current signal controlled in this way is also to a much lesser extent burdened with complications and consequences resulting from more chaotic and oscillatory, and often noisy waveforms. To discuss the most important and key advantages of the proposed control laws, it is necessary to refer first to the steady-state case.

The first issue concerning the waveforms of the grid current in the steady state is the range of oscillation values caused by the operation of electronic switches in the inverter. The use of the classic SMC from (11) additionally introduced oscillations visible in Figure 10 to the system, which was reduced using the control law from (13), obtaining the result visible in Figure 11, which was largely successful. One of the main goals of introducing the method based on the nonswitching reaching law type was to ensure an even greater reduction in the described phenomena—the nonswitching rule by definition eliminates the chattering problem, while the remaining oscillations should be reduced to an even greater extent than for (13) due to the analytically justified selection of the sliding hyperplane parameters and the control law parameters in (19). Apart from the presence of the steady-state error, visible in Figure 14, it is possible to conclude that the amplitude of the occurring unfavorable oscillations was reduced in comparison to the hybrid algorithm. The problem of the occurrence of the steady-state error was eliminated by using the laws (39)–(40), and the results presented in Table 2 confirm that the algorithm described by the quoted equations allows to narrow down the range of the amplitude of the mentioned oscillations many times more in comparison to other methods used.

The data included in Table 3 shows that the control quality in the case of using the last proposed algorithm from (37)–(40) is significantly higher than in the case of the other methods. This is due to the use of the adaptation mechanism leading to a reduction in the control time within the control constraints, but also the lack of chattering and, above all, the total disappearance in the steady-state error, which was the main disadvantage of the basic version of the algorithm from (19). Therefore, the best results were obtained for the case of the final form of the algorithm based on the nonswitching reaching law type, while maintaining the advantages concerning the shape of the characteristic in dynamic states and the narrowest range of quasi-sliding motion. These advantages are also visible when comparing results from Figures 8, 9, 13, 16 and 19—by using nonswitching reaching law control methods it is possible to ensure the system with a smooth and stiff sliding variable signal waveform that is close to linear shape in its characteristics. Two introduced algorithms provide even better results by shortening the settling time and by reducing the steady-state error to zero.

8. Conclusions

This paper presents an attempt to adapt one of the sliding-mode algorithms known in the theory of control laws to be used in a precisely modeled system representing a real case of the control problem. First, methods known from the literature that were successfully applied to the case of grid-following inverter control were presented and implemented. The results obtained by using these methods were presented, and then the advantages and disadvantages of these solutions were indicated. The next step was the introduction of a more sophisticated sliding mode control method known from the literature, which was supposed to potentially eliminate the disadvantages of the methods mentioned earlier, i.e., the lack of a full analytical approach, overshoots, oscillations, etc. It was possible to implement the nonswitching reaching law algorithm from [5] using the dead-beat sliding hyperplane parameters. For example, this algorithm allowed for the elimination of chattering and obtain a waveform shape that can be a significant advantage in many technological processes. In the next step, the disadvantages of this solution were indicated and two methods were proposed that could potentially eliminate these disadvantages: control parameter adaptation and the steady-state error reduction mechanism. These algorithms were then described and implemented, and the test results confirmed that they provide the highest quality of regulation, including the quality of energy transmitted through the grid. The two methods were also synthesized so that they could cooperate and not interfere with each other. The limitations of the system were also taken into account, as well as the potential variability of the power grid. The sliding mode control methods proposed in this paper turned out to be a very interesting extension of known methods. The control quality of the introduced algorithms was experimentally confirmed, and their advantages were discussed in detail. The possibility of implementing specific methods known in theory to practical problems, in this case related to power electronics, was confirmed. The developed methods can be also utilized in other control problems, i.e., wherever various changes in setpoints and the occurrence of a non-zero steady-state error are expected, the presence of which in real objects is often certain. The continuous development of devices, including power electronics, as well as the increasing requirements for devices, make the development of the field of control algorithms very dynamic, as mentioned in the introduction.

Author Contributions: Conceptualization, A.S., P.L. and P.C.; methodology, A.S., P.L. and P.C.; software, A.S.; validation, A.S., P.L. and P.C.; formal analysis, A.S., P.L. and P.C.; investigation, A.S.; resources, A.S.; data curation, P.L. and P.C.; writing—original draft preparation, A.S.; writing—review and editing, A.S., P.L. and P.C.; visualization, A.S.; supervision, P.L. and P.C.; project administration, P.C. All authors have read and agreed to the published version of the manuscript.

Funding: This research received no external funding.

Data Availability Statement: The data presented in this study are available on request from the corresponding author.

Conflicts of Interest: The authors declare no conflicts of interest.

Appendix A. Discrete State Matrix Elements

$$A_{d_{11}} = \frac{2L + L_g e^{T\sqrt{\frac{-L-L_g}{C_f L L_g}}} + L_g e^{-T\sqrt{\frac{-L-L_g}{C_f L L_g}}}}{2(L+L_g)}, \tag{A1}$$

$$A_{d_{12}} = \frac{C_f L_g \left(e^{T\sqrt{\frac{-L-L_g}{C_f L L_g}}} - e^{-T\sqrt{\frac{-L-L_g}{C_f L L_g}}} \right)}{2\sqrt{-C_f L L_g (L+L_g)}}, \tag{A2}$$

$$A_{d_{13}} = \frac{2L_g + L_g e^{T\sqrt{\frac{-L-L_g}{C_f L L_g}}} + L_g e^{-T\sqrt{\frac{-L-L_g}{C_f L L_g}}}}{2(L+L_g)}, \tag{A3}$$

$$A_{d_{21}} = \frac{\left(e^{T\sqrt{\frac{-L-L_g}{C_f L L_g}}} - e^{-T\sqrt{\frac{-L-L_g}{C_f L L_g}}} \right)(L+L_g)}{2C_f(L+L_g)}, \tag{A4}$$

$$A_{d_{22}} = \frac{e^{T\sqrt{\frac{-L-L_g}{C_f L L_g}}} - e^{-T\sqrt{\frac{-L-L_g}{C_f L L_g}}}}{2}, \tag{A5}$$

$$A_{d_{23}} = \frac{\left(e^{-T\sqrt{\frac{-L-L_g}{C_f L L_g}}} - e^{T\sqrt{\frac{-L-L_g}{C_f L L_g}}} \right)(L+L_g)}{2C_f(L+L_g)}, \tag{A6}$$

$$A_{d_{31}} = \frac{2L + L e^{T\sqrt{\frac{-L-L_g}{C_f L L_g}}} + L e^{-T\sqrt{\frac{-L-L_g}{C_f L L_g}}}}{2(L+L_g)}, \tag{A7}$$

$$A_{d_{32}} = \frac{C_f L \left(e^{T\sqrt{\frac{-L-L_g}{C_f L L_g}}} - e^{-T\sqrt{\frac{-L-L_g}{C_f L L_g}}} \right)}{2\sqrt{-C_f L L_g (L+L_g)}}, \quad (A8)$$

$$A_{d_{33}} = \frac{2L_g + Le^{T\sqrt{\frac{-L-L_g}{C_f L L_g}}} + Le^{-T\sqrt{\frac{-L-L_g}{C_f L L_g}}}}{2(L+L_g)}. \quad (A9)$$

Appendix B. Modified Discrete State Matrix Elements

$$A_{d_{11\delta}} = \frac{2L + L_{g\delta}e^{T\sqrt{\frac{-L-L_{g\delta}}{C_{f\delta} L L_{g\delta}}}} + L_{g\delta}e^{-T\sqrt{\frac{-L-L_{g\delta}}{C_{f\delta} L L_{g\delta}}}}}{2(L+L_{g\delta})}, \quad (A10)$$

$$A_{d_{12\delta}} = \frac{C_{f\delta} L_{g\delta} \left(e^{T\sqrt{\frac{-L-L_{g\delta}}{C_{f\delta} L L_{g\delta}}}} - e^{-T\sqrt{\frac{-L-L_{g\delta}}{C_{f\delta} L L_{g\delta}}}} \right)}{2\sqrt{-C_{f\delta} L L_{g\delta} (L+L_{g\delta})}}, \quad (A11)$$

$$A_{d_{13\delta}} = \frac{2L_{g\delta} + L_{g\delta}e^{T\sqrt{\frac{-L-L_{g\delta}}{C_{f\delta} L L_{g\delta}}}} + L_{g\delta}e^{-T\sqrt{\frac{-L-L_{g\delta}}{C_{f\delta} L L_{g\delta}}}}}{2(L+L_{g\delta})}, \quad (A12)$$

$$A_{d_{21\delta}} = \frac{\left(e^{T\sqrt{\frac{-L-L_{g\delta}}{C_{f\delta} L L_{g\delta}}}} - e^{-T\sqrt{\frac{-L-L_{g\delta}}{C_{f\delta} L L_{g\delta}}}} \right)(L+L_{g\delta})}{2C_{f\delta}(L+L_{g\delta})}, \quad (A13)$$

$$A_{d_{22\delta}} = \frac{e^{T\sqrt{\frac{-L-L_{g\delta}}{C_{f\delta} L L_{g\delta}}}} - e^{-T\sqrt{\frac{-L-L_{g\delta}}{C_{f\delta} L L_{g\delta}}}}}{2}, \quad (A14)$$

$$A_{d_{23\delta}} = \frac{\left(e^{-T\sqrt{\frac{-L-L_{g\delta}}{C_{f\delta} L L_{g\delta}}}} - e^{T\sqrt{\frac{-L-L_{g\delta}}{C_{f\delta} L L_{g\delta}}}} \right)(L+L_{g\delta})}{2C_{f\delta}(L+L_{g\delta})}, \quad (A15)$$

$$A_{d_{31\delta}} = \frac{2L + Le^{T\sqrt{\frac{-L-L_{g\delta}}{C_{f\delta} L L_{g\delta}}}} + Le^{-T\sqrt{\frac{-L-L_{g\delta}}{C_{f\delta} L L_{g\delta}}}}}{2(L+L_{g\delta})}, \quad (A16)$$

$$A_{d_{32\delta}} = \frac{C_{f\delta} L \left(e^{T\sqrt{\frac{-L-L_{g\delta}}{C_{f\delta} L L_{g\delta}}}} - e^{-T\sqrt{\frac{-L-L_{g\delta}}{C_{f\delta} L L_{g\delta}}}} \right)}{2\sqrt{-C_{f\delta} L L_{g\delta} (L+L_{g\delta})}}, \quad (A17)$$

$$A_{d_{33\delta}} = \frac{2L_{g\delta} + Le^{T\sqrt{\frac{-L-L_{g\delta}}{C_{f\delta} L L_{g\delta}}}} + Le^{-T\sqrt{\frac{-L-L_{g\delta}}{C_{f\delta} L L_{g\delta}}}}}{2(L+L_{g\delta})}. \quad (A18)$$

References

1. Bartoszewicz, A. Conventional sliding modes in continuous and discrete time domains. In Proceedings of the 2017 18th International Carpathian Control Conference (ICCC), Sinaia, Romania, 28–31 May 2017; pp. 588–593.
2. Utkin, V.; Shi, J. Integral sliding mode in systems operating under uncertainty conditions. In Proceedings of the 35th IEEE Conference on Decision and Control, Kobe, Japan, 13 December 1996; Volume 4, pp. 4591–4596.
3. Zenteno-Torres, J.; Cieslak, J.; Dávila, J.; Henry, D. Sliding Mode Control with Application to Fault-Tolerant Control: Assessment and Open Problems. *Automation* **2021**, *2*, 1–30. [CrossRef]
4. Latosiński, P.; Adamiak, K. Discrete-Time Sliding Mode Control Strategies—State of the Art. *Energies* **2024**, *17*, 4564. [CrossRef]
5. Bartoszewicz, A.; Lesniewski, P. New Switching and Nonswitching Type Reaching Laws for SMC of Discrete Time Systems. *IEEE Trans. Control Syst. Technol.* **2016**, *24*, 670–677. [CrossRef]
6. Jiang, Y.; Wang, Q.; Dong, C. A reaching law based neural network terminal sliding-mode guidance law design. In Proceedings of the 2013 IEEE International Conference of IEEE Region 10 (TENCON 2013), Xi'an, China, 22–25 October 2013; pp. 1–5.
7. Xi, L.; Yang, S.; Dong, H.; Qi, X. Terminal Sliding Mode Control for Robotic Manipulator Based on Combined Reaching Law. In Proceedings of the 2019 International Conference on Computer Network, Electronic and Automation (ICCNEA), Xi'an, China, 27–29 September 2019; pp. 436–441.
8. Leng, J.; Ma, C. Sliding Mode Control for PMSM Based on a Novel Hybrid Reaching Law. In Proceedings of the 2018 37th Chinese Control Conference (CCC), Wuhan, China, 25–27 July 2018; pp. 3006–3011.
9. Sawiński, A.; Chudzik, P.; Tatar, K. Smc algorithms in t-type bidirectional power grid converter. *Energies* **2024**, *17*, 2970. [CrossRef]
10. Szewczyk, F. Sliding Mode Control Restart Strategy for Rotating Induction Machine Using Time-Varying Switching Line. *Przegląd Elektrotechniczny* **2024**, *1*, 21–26. [CrossRef]
11. Mozayan, S.M.; Saad, M.; Vahedi, H.; Fortin-Blanchette, H.; Soltani, M. Sliding Mode Control of PMSG Wind Turbine Based on Enhanced Exponential Reaching Law. *IEEE Trans. Ind. Electron.* **2016**, *63*, 6148–6159. [CrossRef]
12. Sawiński, A.; Chudzik, P.; Tatar, K. Cascade Sliding Mode Control for Linear Displacement Positioning of a Quadrotor. *Sensors* **2025**, *25*, 883. [CrossRef] [PubMed]
13. Rajanna, G.S.; Nagaraj, H.N. Comparison between sigmoid variable reaching law and exponential reaching law for sliding mode controlled DC-DC Buck converter. In Proceedings of the 2013 International Conference on Power, Energy and Control (ICPEC), Dindigul, India, 6–8 February 2013; pp. 316–319.
14. Latosinski, P.; Bartoszewicz, A.; Lesniewski, P. A new reaching law based DSMC for inventory management systems. In Proceedings of the 2017 21st International Conference on System Theory, Control and Computing (ICSTCC), Sinaia, Romania, 19–21 October 2017; pp. 372–377.
15. Bartoszewicz, A.; Lesniewski, P. Non-switching type reaching law with application to congestion control in connection-oriented communication networks. In Proceedings of the 2014 European Control Conference (ECC), Strasbourg, France, 24–27 June 2014; pp. 1211–1216.
16. Bartoszewicz, A.; Maciejewski, M. Dead-beat sliding mode control of perishable inventories with transportation losses and multiple suppliers. In Proceedings of the 2013 European Control Conference (ECC), Zurich, Switzerland, 17–19 July 2013; pp. 3209–3215.
17. Ignaciuk, P.; Bartoszewicz, A. Sliding Mode Dead-Beat Control of Perishable Inventory Systems With Multiple Suppliers. *IEEE Trans. Autom. Sci. Eng.* **2012**, *9*, 418–423. [CrossRef]
18. Nicola, M.; Nicola, C.-L. Improved Performance in the Control of DC-DC Three-Phase Power Electronic Converter Using Fractional-Order SMC and Synergetic Controllers and RL-TD3 Agent. *Fractal Fract.* **2022**, *6*, 729. [CrossRef]
19. Bouguerra, Z. Comparative study between PI, FLC, SMC and Fuzzy sliding mode controllers of DFIG wind turbine. *J. Renew. Energies* **2023**, *26*, 209–223. [CrossRef]
20. Kalvinathan, V.; Chitra, S. Power Optimization in Hybrid Renewable Energy Standalone System using SMC-ANFIS. *Adv. Electr. Comput. Eng.* **2022**, *22*, 69–78. [CrossRef]
21. Liu, S.; Liu, X.; Jiang, S.; Zhao, Z.; Wang, N.; Liang, X.; Zhang, M.; Wang, L. Application of an Improved STSMC Method to the Bidirectional DC–DC Converter in Photovoltaic DC Microgrid. *Energies* **2022**, *15*, 1636. [CrossRef]
22. Kumar, N.; Saha, T.K.; Dey, J. Design and analysis of a double band hysteresis SMC for cascaded inverter-based PV system. In Proceedings of the 2016 IEEE Region 10 Conference (TENCON), Singapore, 22–25 November 2016; pp. 898–902.
23. Lyu, S.; Zheng, L.; Song, J.; Liu, Z.; Tian, M. A Repetitive Sliding Mode Control for Household Mircogrid Inverter Operating in Islanding Mode. In Proceedings of the 2018 IEEE International Power Electronics and Application Conference and Exposition (PEAC), Shenzhen, China, 4–7 November 2018; pp. 1–6.
24. Zheng, W.; Li, S.; Wang, J.; Wang, Z. Sliding-mode control for three-phase PWM inverter via harmonic disturbance observer. In Proceedings of the 2015 34th Chinese Control Conference (CCC), Hangzhou, China, 28–30 July 2015; pp. 7988–7993.

25. Bouzid, A.M.; Guerrero, J.M.; Cheriti, A.; Bouhamida, M.; Sicard, P.; Benghanem, M. A survey on control of electric power distributed generation systems for microgrid applications. *Renew. Sustain. Energy Rev.* **2015**, *44*, 751–766. [CrossRef]
26. Rocabert, J.; Luna, A.; Blaabjerg, F.; Rodríguez, P. Control of Power Converters in AC Microgrids. *IEEE Trans. Power Electron.* **2012**, *27*, 4734–4749. [CrossRef]
27. Gong, X.; Wang, X. A novel Koopman-inspired method for the secondary control of microgrids with grid-forming and grid-following sources. *Appl. Energy* **2023**, *333*, 120631. [CrossRef]
28. Guo, W.; Mu, L. Control principles of micro-source inverters used in microgrid. *Prot. Control Mod. Power Syst.* **2016**, *1*, 5. [CrossRef]
29. Mohiuddin, S.M.; Qi, J. A Unified Droop-Free Distributed Secondary Control for Grid-Following and Grid-Forming Inverters in AC Microgrids. In Proceedings of the 2020 IEEE Power & Energy Society General Meeting (PESGM), Montreal, QC, Canada, 2–6 August 2020; pp. 1–5.
30. Hao, X.; Yang, X.; Xie, R.; Huang, L.; Liu, T.; Li, Y. A fixed switching frequency integral resonant sliding mode controller for three-phase grid-connected photovoltaic inverter with lcl-filter. In Proceedings of the 2013 IEEE ECCE Asia Downunder, Melbourne, VIC, Australia, 3–6 June 2013; pp. 793–798.
31. Monica, P.; Kowsalya, M.; Subramanian, K. Droop reference based decoupling coupling control of NPC inverter for islanded microgrid. In Proceedings of the 2017 Innovations in Power and Advanced Computing Technologies (i-PACT), Vellore, India, 21–22 April 2017; pp. 1–5.
32. Rąbkowski, J.; Kopacz, R. Extended T-type Inverter. *Power Electron. Drives* **2018**, *3*, 55–64. [CrossRef]
33. Meenakshi, V.; Kavitha, M.; Antony, A.S.M.; Kaveri, V.V.; Radhika, S.; Immanuel, D.G. T-Type Three Phase Inverter with Grid Connected System used in Renewable Energy Resources. In Proceedings of the 2023 International Conference on Sustainable Computing and Data Communication Systems (ICSCDS), Erode, India, 23–25 March 2023; pp. 929–936.
34. Gao, X.; Zhou, D.; Anvari-Moghaddam, A.; Blaabjerg, F. Stability Analysis of Grid-Following and Grid-Forming Converters Based on State-Space Modelling. *IEEE Trans. Ind. Appl.* **2024**, *60*, 4910–4920. [CrossRef]
35. Babu, V.; Roselyn, P.; Chinnaiyan, N.; Sundaravadivel, P. Development of Grid-Forming and Grid-Following Inverter Control in Microgrid Network Ensuring Grid Stability and Frequency Response. *Electronics* **2024**, *13*, 1958. [CrossRef]
36. Fu, X.; Sun, J.; Huang, M.; Tian, Z.; Yan, H.; Iu, H.H.; Hu, P.; Zha, X. Large-Signal Stability of Grid-Forming and Grid-Following Controls in Voltage Source Converter: A Comparative Study. *IEEE Trans. Power Electron.* **2021**, *36*, 7832–7840. [CrossRef]
37. Du, W.; Tuffner, F.K.; Schneider, K.P.; Lasseter, R.H.; Xie, J.; Chen, Z.; Bhattarai, B. Modeling of Grid-Forming and Grid-Following Inverters for Dynamic Simulation of Large-Scale Distribution Systems. *IEEE Trans. Power Deliv.* **2021**, *36*, 2035–2045. [CrossRef]
38. Jasim, A.; Jasim, B. Grid-forming and grid-following based microgrid inverters control. *Iraqi J. Electr. Electron. Eng.* **2022**, *18*, 111–131. [CrossRef]
39. Sadeque, F.; Sharma, D.; Mirafzal, B. Power-Sharing between Grid-Forming and Grid-Following Inverters. In Proceedings of the 2021 IEEE 22nd Workshop on Control and Modelling of Power Electronics (COMPEL), Cartagena, Colombia, 2–5 November 2021; pp. 1–5.
40. Gluščević, J.; Janda, Ž.; Dragosavac, J.; Ristić, L. Enhancing Stability of Grid-Following Inverter for Renewables. In Proceedings of the 2023 22nd International Symposium on Power Electronics (Ee), Novi Sad, Serbia, 25–28 October 2023; pp. 1–6.
41. Hasan, M.M.; Chowdhury, A.H. Performance Analysis of a DQ0 Controlled Grid Forming Inverter for Grid Connected Photovoltaic System: A Case Study. In Proceedings of the 2022 4th International Conference on Sustainable Technologies for Industry 4.0 (STI), Dhaka, Bangladesh, 17–18 December 2022; pp. 1–6.
42. Li, S.; Zhou, S.; Li, H. Harmonic Suppression Strategy of LCL Grid-Connected PV Inverter Based on Adaptive QPR_pc Control. *Electronics* **2023**, *1*, 2282. [CrossRef]
43. Huang, J.; Zhao, Y.; Wang, J.; Zhang, P. A Hybrid Active Damping Strategy for Improving the Adaptability of LCL Converter in Weak Grid. *Electronics* **2023**, *13*, 144. [CrossRef]
44. Ahsan, E.E.; Shobug, M.A.; Tanim, M.M.H.; Reza, M.H. Harmonic Distortion Reduction of Transformerless Inverter's Output Voltage Using 5-Level Single-Phase Inverter and LCL Filter. In Proceedings of the 2020 2nd International Conference on Advanced Information and Communication Technology (ICAICT), Dhaka, Bangladesh, 28–29 November 2020; pp. 251–256.
45. Choi, W.; Lee, W.; Sarlioglu, B. Effect of grid inductance on grid current quality of parallel grid-connected inverter system with output LCL filter and closed-loop control. In Proceedings of the 2016 IEEE Applied Power Electronics Conference and Exposition (APEC), Long Beach, CA, USA, 20–24 March 2016; pp. 2679–2686.
46. Wang, B.; Zhang, X.; Song, C.; Cao, R. Research on the Filters for Dual-Inverter Fed Open-End Winding Transformer Topology in Photovoltaic Grid-Tied Applications. *Energies* **2019**, *12*, 2338. [CrossRef]

47. Dehghani, M.; Mardaneh, M.; Shafiei, M.H. Sliding mode control for load harmonics compensation and PV voltage regulation in a grid-tied inverter through a single-stage MPPT. In Proceedings of the 2020 28th Iranian Conference on Electrical Engineering (ICEE), Tabriz, Iran, 4–6 August 2020; pp. 1–6.
48. de Brito, M.A.G.; Junior, L.G.; Canesin, C.A. Three-phase tri-state integrated boost inverter with special space vector and dq0 control. In Proceedings of the International Symposium on Power Electronics Power Electronics, Electrical Drives, Automation and Motion, Sorrento, Italy, 20–22 June 2012; pp. 644–648.
49. Schonardie, M.F.; Coelho, R.F.; Schweitzer, R.; Martins, D.C. Control of the active and reactive power using dq0 transformation in a three-phase grid-connected pv system. In Proceedings of the 2012 IEEE International Symposium on Industrial Electronics, Hangzhou, China, 28–31 May 2012.
50. Schonardie, M.F.; Ruseler, A.; Coelho, R.F.; Martins, D.C. Three-phase grid-connected pv system with active and reactive power control using dq0 transformation. In Proceedings of the 2010 9th IEEE/IAS International Conference on Industry Applications—INDUSCON 2010, Sao Paulo, Brazil, 8–10 November 2010; pp. 1–6.
51. Vinoliney, A.; Lydia, M.; Levron, Y. Optimal grid integration of renewable energy sources with energy storage using dq0 based inverter controller. In Proceedings of the 2020 International Conference on Inventive Computation Technologies (ICICT), Coimbatore, India, 26–28 February 2020; pp. 966–972.
52. Kumar, N.; Buwa, O. A Review on Reactive Power Compensation of Distributed Energy System. In Proceedings of the 2020 7th International Conference on Smart Structures and Systems (ICSSS), Chennai, India, 23–24 July 2020; pp. 1–6.
53. Lesniewski, P. Discrete time reaching law based sliding mode control: A survey. In Proceedings of the 2018 22nd International Conference on System Theory, Control and Computing (ICSTCC), Sinaia, Romania, 10–12 October 2018; pp. 734–739.
54. Li, Y.; Liu, L. The research of the sliding mode control method based on improved double reaching law. In Proceedings of the 2018 Chinese Control And Decision Conference (CCDC), Shenyang, China, 9–11 June 2018; pp. 672–675.
55. Bartoszewicz, A.; Lesniewski, P. New reaching law for quasi-sliding mode control of discrete time systems. In Proceedings of the 52nd IEEE Conference on Decision and Control, Firenze, Italy, 10–13 December 2013; pp. 2881–2887.
56. Bartoszewicz, A. A new reaching law for sliding mode control of continuous time systems with constraints. *Trans. Inst. Meas. Control* **2015**, *37*, 515–521. [CrossRef]
57. Bartoszewicz, A.; Lesniewski, P. Reaching Law Approach to the Sliding Mode Control of Periodic Review Inventory Systems. *IEEE Trans. Autom. Sci. Eng.* **2014**, *11*, 810–817. [CrossRef]
58. Tatar, K.; Chudzik, P.; Leśniewski, P. Sliding mode control of buck dc–dc converter with lc input filter. *Energies* **2023**, *16*, 6983. [CrossRef]
59. Drazenovic, B.; Milosavljevic, C.; Veselic, B. Discrete-time chattering-free integral sliding mode control of continuous-time descriptor systems. In Proceedings of the 2014 13th International Workshop on Variable Structure Systems (VSS), Nantes, France, 29 June–2 July 2014; pp. 1–6.
60. Veselic, B.; Perunicic-Drazenovic, B.; Milosavljevic, C. Integral Sliding Manifold Design for Linear Systems With Additive Unmatched Disturbances. *IEEE Trans. Autom. Control* **2016**, *61*, 2544–2549. [CrossRef]
61. Veselic, B.; Drazenovic, B.; Milosavljevic, C. Optimal discrete-time integral sliding manifold design for linear systems subjected to a class of unmatched disturbances. In Proceedings of the 2015 International Workshop on Recent Advances in Sliding Modes (RASM), Istanbul, Turkey, 9–11 April 2015; pp. 1–6.

energies

MDPI

Review

Current Status, Challenges and Future Perspectives of Operation Optimization, Power Prediction and Virtual Synchronous Generator of Microgrids: A Comprehensive Review

Ling Miao [1,*], Ning Zhou [1], Jianwei Ma [1], Hao Liu [1], Jian Zhao [1], Xiaozhao Wei [1] and Jingyuan Yin [2,3]

1 State Grid Henan Electric Power Research Institute, Zhengzhou 450052, China; zhounepri@sina.com (N.Z.); hermit_m71@163.com (J.M.); liuhaoepri@139.com (H.L.)
2 Institute of Electrical Engineering, Chinese Academy of Sciences, Beijing 100190, China; yinjingyuan@mail.iee.ac.cn
3 School of Electronic, Electrical and Communication Engineering, University of Chinese Academy of Sciences, Beijing 101408, China
* Correspondence: lingmiao11@outlook.com; Tel.: +86-0371-8779-2986

Abstract: With the increasing prominence of the energy crisis and environmental problems, microgrid technology has received widespread attention as an important technical means to improve the stability and reliability of new energy access. Focusing on the latest development of microgrid operation control technology, this paper combs and summarizes the related research at home and abroad, including the key technologies of microgrid optimization operation, power prediction and virtual synchronous active support control technology, and points out their advantages and limitations. First, this review describes the concept and structure of microgrids, including components such as distributed power sources, energy storage devices, energy conversion devices and loads. Then, the microgrid optimization operation technologies are analyzed in detail, including energy management optimization algorithms for efficient use of energy and cost reduction. Focusing on microgrid power forecasting techniques, including wind energy and PV power forecasting and load forecasting, the contributions and impacts of different power forecasting methods are summarized. Furthermore, the inverter control strategies and the stability mechanism of the virtual synchronous generator (VSG) active support control technology are investigated. Finally, synthesizing domestic and international microgrid development experience, this review summarizes the current state-of-the-art technologies, analyzes the advantages and limitations of these key technologies (including optimization scheduling, power prediction and VSG-based active support control) and highlights the necessity of their continuous improvement to provide a solid foundation for promoting the widespread application and sustainable development of microgrid technology.

Keywords: microgrid; operation optimization; power prediction; virtual synchronous generator

1. Introduction

In recent years, the rapid economic development is accompanied by the increasing demand for energy, and the world is gradually facing problems such as the shortage of conventional fossil fuels and environmental pollution. With advances in small generators, energy storage devices and power electronics and the gradual rise in fuel costs, building large, centralized power plants in many cities is not an optimal solution. Therefore, the microgrid is replacing centralized power generation as a new power system [1]. The microgrid has a high penetration rate of renewable energy, in particular wind and solar, which accounts for 91% of the global renewable energy generation capacity [2,3].

With the continuous improvement of the comprehensive utilization of renewable energy technology, wind power and photovoltaic (PV) power generation, as emerging energies in many countries, play an important role in the strategic energy structure, promoting the rapid development of the industry. Since 2000, the installed capacity of wind power and photovoltaic in the world and China have shown a doubling growth [4,5]. The global cumulative and new installed wind power capacity from 2015–2024 is shown in Figure 1 [6], and the global cumulative and new installed solar PV capacity from 2015–2024 is shown in Figure 2 [7]. Data from the Global Wind Energy Council's (GWEC) Global Wind Power Development Report 2024 show that between 2015 and 2024, the cumulative installed capacity of global wind power increased from 433 GW to 1167 GW, with a compound annual growth rate of 11.12%. In 2024, the global new installed wind power capacity reached 155 GW, while China's cumulative installed wind power capacity exceeded 500 GW, and its cumulative installed solar PV capacity reached 886 GW. China's wind power capacity accounted for over 50% of the global cumulative total, and its solar PV capacity represented 44.54% of the global cumulative installed capacity. Clean energy is projected to exceed 85% of the global energy mix by 2050.

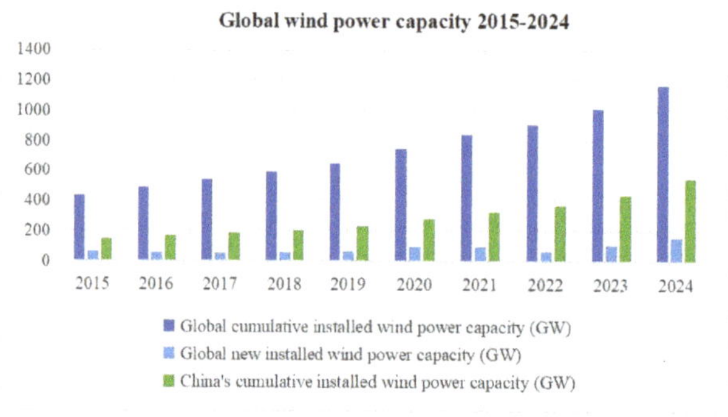

Figure 1. Global cumulative and new installed wind power capacity from 2015–2024.

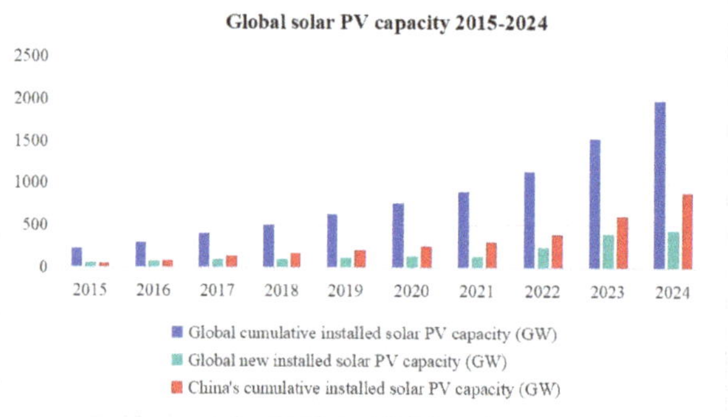

Figure 2. Global cumulative and new installed solar PV capacity from 2015–2024.

The inherent flexibility of the power system can accommodate a certain amount of intermittent renewable energy, but as it enters the medium to high or extremely high ratios, the system will face multiple challenges in different time scales, such as stability control, operation and planning. With the large-scale application of renewable energy, the penetration rate is increasing, and the continuous development of intelligent technology and the construction of a new type of power system are accelerating. In order to ensure the stable and reliable operation of renewable energy grid-connected power generation, microgrids have become an important solution to the problem of the power systems in some countries. Among them, the current status and development trends of microgrid operation optimization technology, the research progress and trends of microgrid power prediction technology, virtual synchronous control technology and other key technologies have attracted significant attention from academia and the industry and have played a key role in promoting the innovation of microgrid technology [8–10]. In recent years, many countries, such as the United States, China, Japan and Europe, have carried out the research of microgrid technology and made important progress, successfully addressing theoretical problems related to microgrid operation, protection, economy and so on [11–13].

The renewable energy microgrid is an important construction object to promote the green development of new power systems. With the increasing application of intermittent renewable energy sources in the new-generation microgrids, the share of renewable energy in energy supply is gradually increasing. This decarbonization transition from fossil fuels to renewable energy not only reduces the controllability of microgrid power output but also introduces uncertainty. Both wind and solar power renewable energies have inherent characteristics such as randomness, intermittency and volatility. In addition, wind power has anti-peaking characteristics. Load demand can also exhibit stochastic behavior, and load demand and renewable generation usually have different peak and valley characteristics. These uncertainties lead to certain difficulties in the stable operation and optimal dispatch of the microgrid after the renewable energy is connected [14]. Therefore, the most central problems of microgrids are the optimized scheduling and power quality control of the distributed power sources, the distributed energy storage, the loads and the grid [15]. The optimal scheduling of microgrids is beneficial to improve the consumption capacity of distributed energy, such as wind power and solar PV power, in the distribution grid, increase the utilization rate of renewable energy and reduce the losses and operating costs of the distribution grid. The source-load-storage multi-type, flexible resources are fully coordinated and interactive in the microgrid-source, load, and storage are no longer an independent operation, but are combined through the power grid framework to realize the interaction of the coupled system. The participation of different types of flexible resources greatly enhances the interaction capabilities among the source-side renewable energy, energy storage system and users in the process of microgrid optimization. Shifting load demand from peak to trough of electricity consumption improves the consumption of the distributed power supply, while the flexible resources on the energy storage side are fully utilized. By absorbing excess renewable generation and releasing electricity at peak times, shifting renewable energy output is realized on a time scale. In addition, the accuracy of wind power, solar PV power, and load power prediction in microgrids constrains the optimal dispatch of the microgrid and affects the full utilization of renewable energy. Therefore, accurate power prediction is an important research direction for renewable energy microgrid energy management systems (EMSs) [16]. Therefore, this paper introduces and summarizes the latest theoretical results of the operation optimization strategy and the power prediction of microgrids in detail.

The microgrid can operate either in grid-connected mode or independently (islanded mode). Regardless of the operating mode, the system requires appropriate control strategies to ensure the reliability and stability of the power supply, and hierarchical control is widely used in power systems [17,18]. The hierarchical control structure of the microgrid is shown in Figure 3. Optimal scheduling calculations of the microgrid occur at the upper decision-making layer (third layer), which regulates the power flow direction of the microgrid and the power grid, solves the optimal scheduling problem to calculate the power of each distributed source and sends it to the second layer control to ensure the stability and economy of microgrid operation. The second layer control regulates the output voltage amplitude and frequency of the inverter through the control signals sent from the given values of the first layer control of the distributed power supply and realizes the balance of the power and stabilization of the main grid system. The signal collected by the first layer control is the local signal of the controller, which generally refers to the power, current and voltage control of the converter [19].

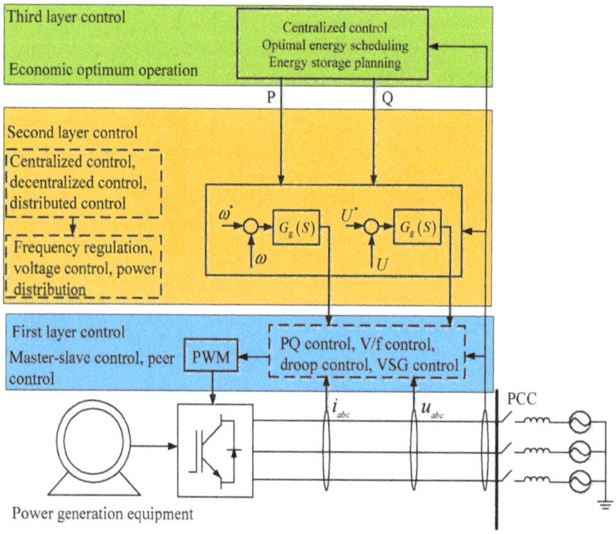

Figure 3. Hierarchical control structure of the microgrid.

Therefore, in renewable energy microgrids, the control technology of inverters is particularly important, as it serves as the medium for connecting distributed generation to the grid and determines the static and dynamic response performance and stability of the microgrid system. The main functions of the inverter are to realize microgrid connection and disconnection, active and reactive power decoupling, terminal voltage stabilization and regulation of the power balance between power supply and load. Currently, the mainstream control strategies for microgrid inverters include the following: PQ control, V/f control, droop control and virtual synchronous generator (VSG) control [20].

PQ control is a common strategy for grid-connected inverter operation, also known as constant power control, which ensures that the output active and reactive power of the inverter are constant and consistent with the given value. Due to the particularity of its control, it can realize decoupling between the output active and reactive power [21]. However, PQ control cannot be used as the main control method in islanded mode because it cannot provide voltage and frequency support to the grid. Therefore, in a microgrid with master–slave control, the distributed power supply, which is the master control unit when grid-connected, is controlled by PQ, and the load changes are followed by the master control unit. When the microgrid changes from grid-connected to islanded, the master control unit changes from PQ mode to V/f control. V/f control is a common strategy for off-grid operation of inverters, also known as constant voltage and constant frequency control. This control ensures that the output voltage and frequency of the inverter are constant, its output power is determined by the load and the external characteristics can be equated to a voltage source [22]. Because it can provide voltage and frequency support, it is very suited for islanded operation, but it requires a sufficient amount of energy to meet load demand, so the capacity of the distributed power supply is relatively large, and it usually requires a backup power supply. Droop control can adjust the output voltage, frequency and power of the inverter, similar to the principle of a traditional synchronous generator; it mainly includes Q-V, P-f regulation of forward control and P-V, Q-f regulation of reverse control. Due to the established linear relationships between active power and frequency (P-f) and between reactive power and voltage (Q-V), the droop control can realize the power equalization without communication. However, droop control does not provide inertia and damping to the system. Virtual synchronous generator (VSG) technology simulates the power and voltage control algorithms of a conventional synchronous generator and introduces inertia and damping characteristics, which can slow down the oscillation of frequency and power and improve the anti-interference performance of microgrids. Currently, there are two main VSG-controlled inverters, voltage type virtual synchronous generator and current type virtual synchronous generator [23]. VSG technology is a new generation of new energy power generation technology that turns new energy from passive regulation to active support. It is an effective way to enable new energy power generation to have the ability to support the power grid with inertia support, primary frequency regulation and active voltage regulation. It is the exploration and practice of solving the problems faced by the double-high power system [24,25]. Therefore, this paper provides a detailed introduction and summary of the latest theoretical results of VSG active support control technology.

Microgrids have been widely used in industrial parks, islands and remote areas due to their flexible and efficient characteristics. Many countries in the world have established their own microgrid demonstration projects.

The United States has made important achievements in microgrid development and is in a leading position in microgrid demonstration projects, technical concepts and key technologies; the policy environment and market mechanisms and research institutions and industrial chains. The United States has the largest number of microgrid demonstration projects in the world, which cover a variety of application scenarios and capacities from a few kilowatts to several megawatts. Among them, some large-scale microgrid projects include the Smart Microgrid Demonstration Project and the Solar City Demonstration Project. At the same time, the United States government and various state governments have given attention to and supported the development and application of microgrids by introducing a series of policies and regulations, such as solar investment tax credits, energy liberalization and power market reform. These policies provide a favorable policy environment and market mechanism for the development and application of microgrids. In addition, the United States has numerous research institutions and industrial chains, such as the National Renewable Energy Laboratory, the Solar Energy Research Institute and the Microgrid Industry Association, which provide important support and guarantees for the research and practice of microgrids [26].

Europe is one of the regions in the world where microgrid development is relatively active. In Europe, intelligent control and integrated management are used to improve the operating efficiency and reliability of microgrids. At present, the European Union has produced some research results in the areas of distributed power modeling and island interconnection. The Demotec microgrid structure of the Solar Energy Technology Institute at Kassel University in Germany and the Cyclades Islands of the Aegean Sea in Greece are typical representatives of microgrid demonstration projects [27].

In recent years, the number of microgrid demonstration projects in Japan has increased year by year to more than 100, covering a variety of application scenarios, such as urban, rural and island. The capacity of these demonstration projects ranges from tens of kilowatts to hundreds of megawatts. The combination of decentralized control and centralized management is used to control and manage the microgrid in Japan, which improves its operational efficiency and reliability. In addition, Japan has promoted some new technologies and applications, such as virtual power plants and the energy internet, which have very promising development prospects [28–30].

In order to improve the structure of the power grid, increase the efficiency of new energy generation and promote the implementation of the dual-carbon strategy, China has also invested a lot of money and energy in the field of microgrids, with the strong support of governmental departments, and has built a variety of microgrid demonstration projects for different environments. For example, small offline microgrids for towns and residential school scenarios have been constructed in Qinghai Province to improve the economy and reliability of the grid. A grid-connected microgrid project for industrial parks has been constructed in Nanjing, Jiangsu Province, to provide a reliable power supply for the parks. Through the implementation of microgrid demonstration projects, some experiences and results have been obtained. With the growing maturity of microgrid technology, it has become an important part of the smart grid and energy internet.

As an important interface between renewable energy and large-grid, the microgrid is an effective way to solve the problem of new energy consumption and improve the overall energy efficiency of the system, and it is also an important direction of future power grid research and development. A microgrid has more efficient, flexible and reliable operation control, which can ensure the stable, efficient and low-carbon operation of the smart grid. A microgrid has broad application prospects in the future energy field. Therefore, it is necessary to carry out more in-depth research on the new power system for microgrids to meet the needs of future smart grids. This paper introduces the latest theoretical results of microgrid key technologies, such as operation optimization strategy, power prediction and VSG active support control technology, and aims to show the current application status of operation optimization strategy, power prediction and VSG active support control technology in microgrid systems, so as to strengthen the connection between theoretical work and industry applications, expand the horizon of future research and provide some references for microgrid-related research.

However, a comprehensive review simultaneously addressing operation optimization, power prediction and virtual synchronous generator (VSG) technologies, particularly their latest advancements and interdisciplinary synergies, is lacking. To bridge this gap, this review aims to (1) Systematically synthesize recent breakthroughs in microgrid operation optimization strategies; (2) Critically evaluate state-of-the-art power forecasting methodologies for renewable sources and loads and (3) Analyze emerging VSG control paradigms for active grid support. The graphic paper is represented in Figure 4. The subsequent structure of this paper is as follows: Section 2 examines optimization strategies for microgrid operations and classifies the energy management algorithms of microgrid and their applications. Section 3 assesses power prediction techniques, comparing statistical, machine learning and hybrid models for wind/PV generation and load forecasting. Section 4 explores VSG control technologies, focusing on stability mechanisms and adaptive control innovations. Finally, conclusions are presented in Section 5.

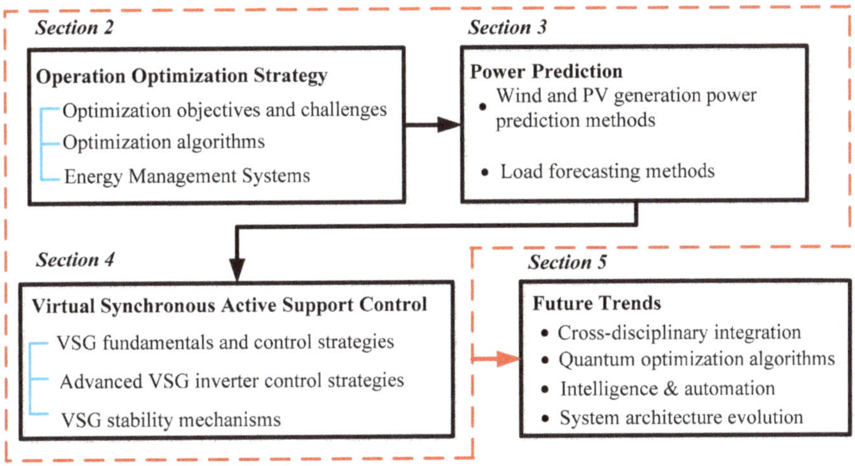

Figure 4. Graphical representation in the review in this paper.

2. Research Status of Operation Optimization Strategy of Microgrids

2.1. Optimization Objectives and Challenges

Different from the traditional big-grid dispatching, a large number of renewable energy sources are connected to the microgrid system. The inherent strong fluctuations and uncertainty of renewable energy output make the constraints, objective functions and dispatching strategies of energy management and the optimal dispatching of microgrids diverse and complex, and the energy management and optimal dispatching of microgrids become more complex. The process of energy management and optimal dispatching is to forecast the renewable energy output power and load power according to a large number of historical data, combine various constraints, environmental information, market rules and other factors to formulate the energy management and optimal dispatching production plan for the microgrid and make real-time adjustments and corrections to realize the optimal dispatching of safety, economy and environmental protection of the microgrid in the actual operation process. The microgrid has various types of power sources, different control methods and variable and complex operation modes, so the stable control of the microgrid itself and the energy management research of distributed power sources and energy storage systems are important. With the continuous development of control technology, the use of advanced control strategies and energy management techniques to improve the performance of the microgrid has become a major trend [31]. The microgrid energy management system (EMS) is an advanced energy management method that integrates monitoring, operation and dispatching. It can optimize and coordinate the operation of energy storage and distributed power sources, so as to achieve efficient and stable operation of the power system and give full play to the advantages of new energy [32]. The energy optimization management of the microgrid can coordinate the multi-side demand of source–load–storage, improve the utilization efficiency of intermittent energy and reduce carbon emissions and operating costs, so as to improve the comprehensive economic benefits of the microgrid but also reduce the negative impact on system stability. It is key to achieving efficient and stable operation of the microgrid [33–35].

The microgrid structure includes renewable energy sources such as wind power and PV power generation, energy storage systems, power dispatching systems, EMS, SCADA (Supervisory Control and Data Acquisition) systems and gas turbines, where the energy storage systems and gas turbines are mainly used to balance the intermittency and volatility of renewable power generation, as shown in Figure 5. The maximum power point tracking (MPPT) control is used for wind power and PV power generation. Gas turbines, as backup power sources, often use power control to regulate their output power. The energy storage system and the bidirectional converter system adopt voltage closed-loop control to maintain the constant DC bus voltage by regulating the power direction and size, absorbing the residual energy in the microgrid. Additionally, a battery management system (BMS) is equipped to monitor the battery pack state of charge (SOC). In the EMS, the net load power is the difference between the user load and the renewable energy output power, which can be obtained in real time by measuring the outlet voltage and current of the battery energy storage system and the generator system. On this basis, with the goal of improving power generation economy and limiting the fluctuation of the battery SOC, the gas turbines output power is determined according to the net load power and the real-time measured battery SOC to realize the dynamic distribution of the net load between the battery storage system and the generator system [36].

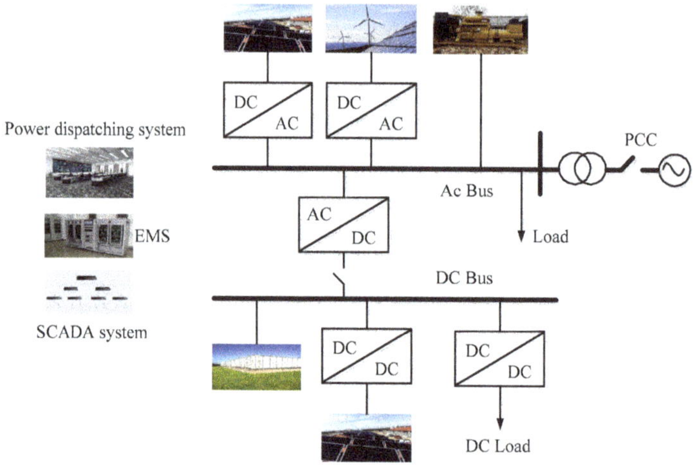

Figure 5. AC/DC hybrid microgrid component structure.

2.2. Classification and Review of Optimization Algorithms

Through the analysis of the optimal operation of the microgrid, it can be learned that the optimal operation of microgrids usually involves multiple objectives, such as operating costs, pollutant gas emissions, profits and other factors [37–39]. If the optimal operation of the microgrid ignores the impact on the environment and only considers the operating profit or operating cost, it will certainly lead to the massive emission of pollutants. However, compared with the consideration of cost and profit, if the objective function only considers the pollutant emission of the microgrid, the model is simpler. Therefore, when establishing the operation model of the microgrid, it is necessary to consider the operation cost and the treatment cost [40–42]. Multi-objective optimization of the microgrid is the key to solving the dispatching problem, and the establishment of a multi-objective optimization model is an important research direction for the long-term healthy operation of microgrids.

At present, the research on optimal operation algorithms for microgrids mainly includes the following aspects: rule-based control strategies, model-based control strategies, cooperative control strategies, optimal control algorithms and artificial intelligence-based and heuristic algorithm control strategies [43]. Rule-based control strategy is developed according to experience and rules, which meet the basic microgrid requirements but lack flexibility and adaptability. Zhang et al. proposed a multi-timescale energy management strategy based on rule-based, which considers different timescales and proposes a real-time scheduling energy management strategy that can achieve the economic optimum for economic scheduling and ensure the stability of the system, but the strategy is less adaptable to complex nonlinear systems [44]. The control strategy based on model prediction models and predicts the output units and loads of the microgrid by establishing a microgrid model, and uses the model to optimize control, which can improve the operation efficiency and performance of the microgrid. Li et al. proposed a control strategy for microgrid EMSs based on model predictive control, to achieve dynamic optimization and dispatching control by modeling and predicting the load of the microgrid [45]. The control strategy based on cooperative control realizes the cooperation between the components of the microgrid through cooperative control to improve the efficiency and performance of the microgrid. Yue et al. proposed establishing a distributed energy unit control scheme based on a multi-agent system. Each distributed energy unit is controlled by a primary unit control agent, while a secondary distributed cooperative control agent manages the second level, so as to realize the cooperative control and optimization of distributed energy. However, this method is highly dependent on information and communication technology [46].

Under the dual pressures of energy and environment, the requirements for the economy and energy-saving and emission reduction performance of microgrid generation are constantly increasing. Using optimal control theory to optimize the energy distribution has gradually become a research hotspot in the field of new energy technology in recent years. The optimal control algorithm is simple to program and can quickly obtain the optimal solution in the offline. Control strategies based on the optimal control algorithm use certain parameters of the microgrid as independent variables and find the optimal solution of the generalized function under the constraints of the allowable control range. There are two main operation optimization strategies based on optimal control theory: Dynamic Programming (DP), proposed by American scholar Bellman, and Pontryagin's Minimum Principle (PMP), proposed by Soviet scholar Pontryagin. The DP and the PMP methods are analytical in nature and are effective in solving variational problems with closed set constraints on control.

In DP, Khalid et al. built a battery simulation model to optimize the charging and discharging cycling process of the battery through a DP algorithm to achieve the lowest operating cost with known load and generation capacity [47]. However, the DP strategy easily leads to dimensional catastrophe due to its huge and complex operations. In order to solve this problem, Venayagamoorthy et al. proposed a neural network-based adaptive DP algorithm model based on the output power of the distributed power of the microgrid system and the forecasted data of load demand; the evaluation function in the DP is processed by using the J-function estimator of the feed-forward neural network, which reduces the computational complexity in the optimization process [48]. Regarding the problem that the DP strategy has difficulty handling large-dimensional data, Das et al. established an islanded microgrid model based on Markov decision-making and proposed an approximate DP to achieve energy optimization with minimum operating cost [49]. Cheng combined DP with fuzzy control to solve the supply–demand imbalance problem in microgrids due to the instability of the clean energy [50]. The energy management algorithm based on DP is complex, and the power generation of clean energy and the power consumption of loads need to be known in advance, so it is less applicable in real-time controls. The optimal control algorithm based on the PMP is mainly used to solve continuous or discontinuous nonlinear problems. It can take a parameter of the island microgrid model as a variable and obtain an optimal solution by constantly adjusting the costate variable. However, it is an offline algorithm overly relying on artificial experience, leading to difficulties in covariate selection [51,52].

For the shortcomings of traditional optimization algorithms in solving the multi-objective optimization process, researchers at home and abroad have continuously explored, improved and proposed various intelligent optimization algorithms [53,54]. The control strategy of the heuristic algorithms is to imitate the algorithm summarized by the law of nature and give the feasible solution according to the example problem. Torkan et al. applied a multi-objective genetic algorithm to deal with the uncertainty of load and renewable energy to achieve the minimum cost and minimum greenhouse gas emissions [55]. In order to select the optimal size of the microgrid components, Li et al. proposed a genetic algorithm to find the minimum microgrid cost [56]. In addition, many researchers have used Simulated Annealing (SA) to solve multi-objective optimization problems. Hafez et al. used SA optimization tools to solve the battery scheduling optimization problem for residential microgrids [57]. The particle swarm optimization (PSO) algorithm is also a widely used algorithm. Wang et al. established an economic power dispatch model for multiple microgrids by comprehensively considering many factors, such as generation cost, discharge cost, power purchase cost, power sales revenue and environmental cost and solved the economic power dispatch model by using the PSO algorithm [58]. Sharmistha et al. used the Grey Wolf Optimizer (GWO) algorithm in order to minimize the energy cost of the microgrid and make better use of renewable energy sources [59]. Wang et al. developed an operational optimization model and optimized it using the Moth Flame Optimization algorithm in order to obtain the minimum operational cost [60]. Pesaran et al. proposed a hybrid PSO algorithm to enhance particle diversity by combining the crossover and mutation operators of genetic algorithms with the original particle swarm algorithm [61]. Ref. [62] adopted a newly developed crow search algorithm (CSA) to optimize the energy management of the microgrid to reduce the generation cost and pollutant emissions. The CSA mimics the memory of crows and the strategies of hiding and chasing food. For the energy management of the microgrid, several practical complexities are considered, such as valve-point load effects, joint economic emission scheduling using the price penalty factor method and modeling of renewable energy and energy storage systems. The obtained results are then compared with many different soft computing techniques, such as genetic algorithms and PSO, to demonstrate the effectiveness of the proposed algorithm. Ref. [63] employed the Self-Adaptive Comprehensive Differential Evolution (SACDE) algorithm for solving the Economic Load Dispatch (ELD) and Combined Economic Emission Dispatch (CEED) problems to achieve optimal power usage in isolated microgrids. This method has strategically superior effectiveness compared to the price penalty factor technique. Ref. [64] presented an improved Lévy optimization algorithm for energy management of a renewable solar/wind microgrid. The microgrid has multiple diesel generators and is suitable for off-grid remote communities. The main objective is to solve an economic emission scheduling problem with a price penalty factor to minimize the energy cost and emission level. The enhanced heuristic Lévy optimization algorithm is used to improve the searchability of the optimal solution compared to traditional arithmetic algorithms. The Lévy optimization algorithm is used for the management of the microgrid and compared with other heuristic optimization algorithms. Results show that the Lévy algorithm achieves significant cost savings compared to other algorithms, such as arithmetic algorithms and the crow search algorithm (CSA), hybrid improved grey wolf optimizer (HIGWO), internal search algorithm (ISA), cuckoo search algorithm (CS), particle swarm optimization (PSO) and ant colony optimization (ACO).

Some scholars have combined multiple heuristics to greatly improve the adaptability and feasibility of the algorithms. The Hybrid Improved Grey Wolf Optimizer (HMGWO) proposed in ref. [65] was used for economic dispatch and emission dispatch of the microgrid. According to the wolf hunting strategy, the population-based sine cosine algorithm strategy is combined with the crow's position update method to form a robust hybrid algorithm. The proposed economic emission scheduling method was compared with the existing price penalty factor method (PPF) and fractional programming (FP) method to solve the joint economic emission scheduling problem on three dynamic test systems, and a better optimization scheme between power generation cost and pollutant emission was found. A new heuristic algorithm called the Arithmetic Optimization Algorithm (AOA) was proposed in ref. [66], which exploited the distributional behavior of the main arithmetic operators (multiplication, division, subtraction and addition) in mathematics in order to perform the optimization process over a wide search space. The performance of the AOA was examined through 29 benchmark functions and several real-world engineering design problems, and the performance, convergence behavior and computational complexity of the proposed AOA were analyzed in different scenarios. The AOA method is used to find the best solution, which sometimes gets stuck in a local optimum during the search process [67,68]. This leads to longer search times to find the optimal solution. In addition, it requires a more comprehensive evaluation when the optimal solution is distributed over a wider range of possibilities. In ref. [69], an augmented algorithm combining the Lévy optimization algorithm and the AOA was proposed to improve the efficiency of the AOA and successfully applied to different engineering problems. The proposed method enhances the AOA's exploration and exploitation capabilities and is used to optimize the operation of a renewable energy microgrid with multiple diesel generators. Ref. [70] proposed an intelligent energy management system (SEMS) based on artificial intelligence-embedded FPGA, which used two multi-objective optimization algorithms, Gorilla Army Optimization Algorithm (GTO) and Reptile Search Algorithm (RSA), to solve the optimization problem. The proposed Smart Energy Management System (SEMS) included two levels of control to achieve optimal management and operation of the isolated microgrid. The first layer used FPGA as the central controller, and the second layer was based on the optimized operation and management of the isolated microgrid to formulate a coordinated operation strategy to optimize the coordinated use of backup power. Ref. [71] developed a hybrid algorithm using the crow search algorithm and JAYA to conduct the combined economic emission dispatch (CEED) for four power systems with and without renewable energy source participation, respectively. Both the price penalty factor (PPF) and fractional programming (FP) methods were used to solve the CEED of all four test systems, and they were analyzed with the goal of minimizing the emission of harmful and toxic gases into the atmosphere.

2.3. Energy Management Systems and Case Studies

New energy, as the main distributed power source to access the grid, has many defects, such as high volatility and instability. The main role of energy management is to ensure the economic operation of the system under the premise of safety and reliability of the microgrid. Solving the problems of intermittency, voltage and frequency fluctuations of distributed power sources is the key to realizing the coordinated management of electric energy and maintaining a dynamic balance [72]. With the rapid development of the microgrid, the EMS faces greater challenges [73]. Although the capacity of the microgrid is not as high as that of the big grid, its EMS is equally complex and requires the integration of several factors, such as variations in natural conditions, charge/discharge control logic, load fluctuations, operating costs and pollution emission costs. [74]. Meanwhile, the SOC of the battery is an important operating condition parameter, and exceeding the SOC limit will affect the long-term operation of the microgrid. The EMS is needed to help the microgrid integrate battery SOC and achieve desirable fuel economy and pollution emission levels. Due to the stochastic nature of the environment, the efficient operation of the microgrid EMS depends to a large extent on the real-time performance of the control algorithm [75–77].

The EMS objectives of the microgrid mainly include the following aspects: minimizing the operating cost, generation cost, voltage deviation and pollutant emission and maximizing the comprehensive benefit. Jung et al. and Vaka et al. minimized the operating cost of the microgrid by optimizing the energy management strategy from the perspective of considering the operating life and the capacity of the energy storage system [78,79]. Fouladfar et al. proposed a multi-objective EMS based on demand response and dynamic pricing. In addition to reducing the market clearing price and increasing the producer profit, it also focuses on reducing the emission level of generating units, and the improvement effect is good [80]. Li et al. developed a mathematical model of an islanded microgrid based on an island with the NSGA-II algorithm, which fully considered the power balance constraints of the microgrid and the operating power of storage batteries and diesel generators to optimize the operation of the micropower [81]. Yong et al. established a model of the lowest power generation cost and the least pollutant emission in the microgrid and balanced and compromised the two goals by using different control strategies to optimize the objective function in the grid-connected microgrid project [82].

Alanazi et al. used variable step size and fixed step size methods to search the optimal value of the microgrid multi-objective function to achieve the minimum voltage deviation and minimum carbon dioxide emissions [83]. Shuai et al. designed a reinforcement learning strategy optimization model framework in order to accurately predict renewable energy generation and realize online dispatch. The Monte Carlo method is used to make the learning model obtain the best decision, so as to achieve the purpose of economic and reliable operation of the microgrid [84]. Ali et al. proposed a novel reinforcement learning approach that utilized a distributed approach to teach the medium to interact with a single microgrid environment and used a global agent to search for a system cost-optimal solution for the microgrid group. However, this type of strategy requires continuous reinforcement learning upfront and a large amount of training data, which may become computationally expensive or less efficient in finding the least-cost solution [85].

The operation optimization strategy of the microgrid is undergoing a remarkable transformation, evolving from its previous single-objective, static-model framework into a more complex and advanced multi-objective, collaborative, dynamic, self-adaptive operation mode. With the continuous advancement of technology and the increasing demand on power systems, the future direction of development will require breaking through the bottlenecks in efficiency and robustness of existing algorithms. In order to achieve this goal, it is necessary to combine emerging technologies, such as artificial intelligence, big data analysis and the Internet of Things (IoT), to build a comprehensive "source–network–charge–storage" intelligent collaborative system. Such a system will be able to achieve efficient allocation and optimal management of resources, so as to provide solid technical support and guarantee the construction and development of new power systems.

3. Research Status of Microgrid Power Prediction

The power output of wind power and PV in microgrids is affected by natural conditions and has large fluctuations and randomness, while the load in microgrids also has the disadvantage of large fluctuations. These lead to large errors in the power prediction of source–load–storage. Therefore, improving the accuracy of wind–PV power prediction and load prediction in the microgrid, and combining with the energy storage system for complementary power generation to optimize the economic and environmental benefits of the microgrid is of great significance to its reliable and economic operation [86,87].

3.1. Research Status of Wind and PV Generation Power Prediction Methods

The wind and PV power prediction methods are divided into physical methods and statistical methods according to the different principles of prediction models. Statistical methods are the most widely used and can be further divided into the time-series method, (e.g., Auto Regressive Integrated Moving Average—ARIMA), machine learning, the deep learning method and the artificial neural network method. Researchers have studied wind power for a long time, and specialized wind power generation prediction systems have been established in Denmark, Germany, the United Kingdom and China, as shown in Table 1. Table 2 shows the classification of solar PV generation methods.

Table 1. Power prediction method of wind power generation.

Prediction Model	Core Technology/Algorithm	Prediction Horizon	Key Strengths/ Focus Area	Typical Input Data	Reference
Prediktor	Physical modeling (NWP + wake modeling)	Short-Medium	Physical accuracy, terrain effects	NWP, terrain data	Louka et al. [88]
WPPT	AR models, transfer functions (Statistical)	Short-Term	Computational efficiency, simplicity	Historical power, simple meteo	Cutler et al. [89]
Zephry	Mesoscale modeling (Phys) + Kalman filter (Stat)	Short-Term	Hybrid approach, data assimilation	NWP, SCADA data	Smith et al. [90]
Previento	CFD modeling (Phys) + Statistical post-proc	Short-Term	High spatial resolution, local effects	Detailed NWP, site specifics	Giebel et al. [91]
Sipreolico	Multi-scale meteo + ML + Data assimilation	Multi-scale	Multi-scale integration, optimization	Multi-source meteo, power data	Lobo et al. [92]
AWPPS	Fuzzy neural network (AI)	Short-Term	Handling uncertainty, non-linearity	Historical power, meteo data	Wang et al. [93]
LocalPred-RegioPred	Phys model + ML correction + Cross-scale	Multi-scale	Regional adaptation, error correction	NWP, regional power, ML features	Banakar et al. [94]
WEPROG-MSEPS	Deep integration phys models + Stat learning	Short-Medium	Robustness, hybrid performance	High-res NWP, historical stats	Pope et al. [95]
AWPT	Fuzzy neural network (AI)	Short-Term	Pattern recognition, adaptability	Historical power, meteo data	De Giorgi et al. [96]
GH-FORECASTER	Adaptive regression (Stat/AI)	Short-Term	Adaptability to changing conditions	Real-time data streams	Pirjan et al. [97]
ANEMOS	Hybrid model	All, esp. Extreme	Extreme weather forecasting	NWP, ensemble forecasts	Kariniotakis et al. [98]
WPP	Physical + statistical method	Short-Medium	General purpose, flexibility	NWP, historical data	Zhang et al. [99]
WPFS	B/S structure, cross-platform	Up to 144 h (Medium)	Practical deployment, long horizon	NWP, operational constraints	Aggarwal et al. [100]
NSF3100	BP-ANN (AI) + Refined techniques	Short-Term	Statistical learning, data refinement	Historical power, processed meteo	Wang et al. [101]

Table 2. Power prediction classification of solar PV generation.

Classification Angle	Predictive Classification	Core Methodology/ Algorithm Examples	Key Characteristics/Applications	Reference
Methodological approach	Physical method	Sky imagery, irradiance modeling, PV cell physics	Based on first principles, good for clear sky	Roberts et al. [102]
	Statistical method	ARIMA, Regression (linear, nonlinear)	Simpler, relies on historical patterns	Shan et al. [103]
	AI/Deep learning method	ANN (BP, RNN, LSTM, GRU), SVM, Ensemble Methods (RF, GBDT)	Handles non-linearity, complex patterns, big data	Wai et al. [104]
	Hybrid method	e.g., Phys + Stat, Phys + AI, AI ensemble	Combines strengths, aims for robustness and accuracy	Voyant et al. [105]
Forecasting horizon	Ultra short-term prediction	Minutes to 6 h	Grid balancing, real-time dispatch	Wang et al. [106]
	Short-term prediction	6 h to 72 h	Unit commitment, market bidding	Yao et al. [107]
	Medium-term prediction	3 days to 2 weeks	Maintenance planning, resource assessment	Wang et al. [108]
	Long-term prediction	Weeks to years	Investment planning, policy making	Liu et al. [109]
Forecast output type	Point forecast	Single expected value	Most common, simple	Van der Meer et al. [110]
	Probabilistic forecast	CDF, quantiles	Quantifies uncertainty, risk-aware decisions	Sanjari et al. [111]
	Interval forecast	Prediction intervals (e.g., PIs)	Provides range of likely values	Kodaira et al. [112]

The physical method of power prediction is a method that expresses the contribution of various physical factors and wind–PV power through mathematical formulas. For example, the Prediktor system (widely used in Spain, Ireland, etc.), and the LocalPred-RegioPred tool (developed in Madrid, Spain) can predict wind power output.

Firstly, according to the wind speed, temperature, wind direction and pressure and other meteorological data obtained by Numerical Weather Prediction (NWP), a calculation method similar to the Wind Atlas Analysis and Application Program (WASP) is used to convert meteorological data into wind speed and wind direction at the height of the wind turbine hub, and then the corresponding output power is calculated according to the corresponding curve [88]. It is difficult to speculate the detailed weather conditions around the wind farm from the weather forecast, which makes the physical prediction method difficult to implement and causes large errors. Another common method is the statistical method, which runs calculations based on power and weather factors, finds the patterns and then obtains the regression model. In addition, there is a hybrid of the physical and statistical prediction methods. However, the physical method is not suitable for short-term prediction because it has the disadvantage of a large amount of calculation and a slow update speed [94].

The statistical method finds the functional relationship between weather factors and wind power through mathematical methods. It mainly inputs the historical data of wind power generation and its influencing factors, etc., and establishes the mapping model between the original data and the predicted data through the time series method, regression analysis, Bayesian statistics, Markov chain, etc., so as to obtain more accurate power prediction results [113]. Time series methods directly use historical data to establish black box models. The commonly used sequence methods include the following. (1) The ARMA prediction model, which uses the fitting of historical data to predict future data and is generally used in stationary series. Ref. [114] compared several forecasting methods for average hourly wind speed data using time-series analysis. (2) The wind power prediction model based on spatial correlation. Because the wind speed and wind power of the wind farm are not evenly distributed in space, but there is some correlation in space and time, the geographic and weather characteristics of the wind farm, wind speed and wind power can be predicted by modeling the spatial correlation at different known locations.

The common machine learning methods include the random forest algorithm, linear regression, naive Bayes algorithm, K-means algorithm, etc. Ref. [115] proposed a hybrid model for short-term prediction of wind and light consisting of variational modal decomposition (VMD), the K-means clustering algorithm, and the long-short-term memory (LSTM) network. The VMD was utilized to decompose the original wind power sequence into a certain number of sub-layers with different frequencies, and the K-means was utilized to decompose the data into approximate fluctuation layers according to the fluctuation level to achieve the power prediction. In order to evaluate the fitting ability of the proposed model, seven different models were compared on multiple scales in four wind turbine families, including the back-propagation neural network method, the Elman neural network method, the LSTM method, the VMD-BP method, the VMD-Elman method, the VMD-LSTM method, and the VMD-K-means-LSTM method. In order to solve the problem of inaccurate prediction due to the obvious fluctuation of wind power generation caused by weather changes, ref. [116] proposed a prediction model using an improved time series deep residual network, which combined the multilayer residual network and the tandem residual learning method of DenseNet and had high prediction accuracy and parameter efficiency. Ref. [117] proposed a combined forecasting method based on day-ahead numerical weather prediction (NWP) positioning technology for ultra-short-term wind power prediction (WPP). Firstly, the NWP information and time window were used to approximately locate the time point with low prediction accuracy of rolling WPP, and then a hybrid method combining the neural network and continuous method was proposed to predict future wind power output. In [118], a model combining VMD, a Convolutional Long Short-Term Memory network (ConvLSTM) and error analysis was used for short-term wind power prediction. The VMD algorithm was used to decompose the wind power signal into a set of different frequency components, and the convolution operation was embedded into the LSTM network to obtain the preliminary prediction results. The LSTM was used to model the error sequence trend of the preliminary prediction results. The prediction error sequence was integrated with the preliminary results to obtain the final prediction result. The results show that the proposed model has the highest prediction performance for hard-to-capture wind power generation sequences. In terms of power prediction, the machine learning method has a more obvious improvement in accuracy compared with the traditional statistical method. The common machine learning single model can complete the prediction of generation power well, but it requires a large amount of data compared with the traditional statistical method [119–121].

In recent years, deep learning has become more and more accurate in dealing with problems in various fields, and it also has relatively high accuracy in wind power prediction. With the development of deep learning, LSTMs, Bi-LSTMs, gated recurrent units (GRUs), artificial neural networks (ANNs) and other methods in deep learning algorithms have also been introduced into renewable energy generation power prediction [108]. These methods effectively mine spatiotemporal correlations between outputs and inputs and have advanced significantly for complex time series. However, in the process of model training, setting too long a time-series length may affect its stability [122–125].

Ref. [126] proposed two wind power prediction models, the K-means-LSTM network model for wind power spot prediction and a nonparametric kernel density estimation (KDE) model with bandwidth optimization for wind power probability interval prediction density estimation. The LSTM established correlations between the before and after data, and the K-Means clustering method composed different clusters of wind power impact factors to generate a new LSTM sub-prediction model. The bandwidth optimization of the nonparametric KDE was achieved using the mean integral squared error criterion. Ref. [127] introduced a deep learning approach to achieve a more accurate wind–solar towers (WST) power output prediction model. In order to predict the output power of the system, several machine learning models were evaluated based on quality metrics, a set of features and regression models were built using this data. Further, the prediction accuracy was improved by incorporating nonlinearity using second order polynomial regression. In addition, the data was trained and tested using deep neural networks to improve the performance of the power prediction. The fuzzy C-mean algorithm is often used to optimize other machine learning algorithm models for power prediction and the PSO is used to optimize the extreme learning machine. In view of the intermittency, randomness and fluctuation of PV power systems due to complex weather conditions, ref. [128] proposed short-term forecasting using an LSTM, which was based on the time scale of global horizontal irradiancy one hour in advance and one day in advance. In order to improve the prediction accuracy of cloudy weather, the clear sky index was introduced as input data in the LSTM model, and K-means was used to classify the weather types into cloudy weather and mixed (partly cloudy) weather during data processing. A neural network model was built to compare the accuracy of different methods. The traditional statistical method in power prediction can output more accurate prediction results to a certain extent, but it has a larger potential for error in power generation prediction results for the influence of stronger volatility of weather and other factors [129]. Ref. [130] proposed a PV power generation prediction algorithm based on an LSTM neural network. By combining statistical knowledge of historical solar irradiance data with publicly available weather forecast types for the city, a comprehensive weather forecast was created for the target PV plant location, enabling more reliable PV generation forecasts. The K-means algorithm was used to categorize the historical irradiance data into dynamic-type sky groups that change hourly during the same season. The performance limitations of using fixed-type sky categories were alleviated by converting them into dynamic and numerical irradiance predictions using historical irradiance data. The results showed that the proposed integrated weather prediction embedded statistical features from historical weather data, significantly improving accuracy. In addition, the superiority of the LSTM neural network with the proposed features was verified by comparing machine learning, including a recurrent neural network, generalized regression neural network, and extreme learning machine.

Ref. [131] proposed a hybrid deep learning model combining a time-series decomposition algorithm and a GRU network. The time-series decomposition algorithm consists of two parts: (1) the complete ensemble empirical mode decomposition with adaptive noise (CEEMDAN) and (2) Wavelet Packet Decomposition (WPD). The Normalized Wind Speed Time Series (WSTS) was processed by CEEMDAN to obtain pure fixed frequency components and residual signals. The WPD algorithm performed second-order decomposition on the first component of the original WSTS containing complex high-frequency signals. Finally, a GRU network was established for all relevant components of the signal, and the predicted wind speed was obtained by stacking the predictions of each component. An efficient deep learning-based wind power prediction model was proposed to address the uncertainty of wind power generation that makes it difficult to integrate into energy systems [132]. The proposed model was divided into two stages. In the first stage, the past wind power signal was decomposed using Wavelet Packet Transform (WPT). In addition to disaggregated signals and lagged wind power, multiple external inputs such as calendar variables and NWP were used as inputs to forecast wind power. In the second stage, a new prediction model, the Efficient Deep Convolutional Neural Network (EDCNN), was used to predict wind power. The demand-side management scheme was developed based on the predicted wind power, day-ahead demand and price, and the performance of the proposed forecasting model was evaluated on the big data of wind farms in the U.S. Figure 6 shows the structures of the GRU and BiGRU networks.

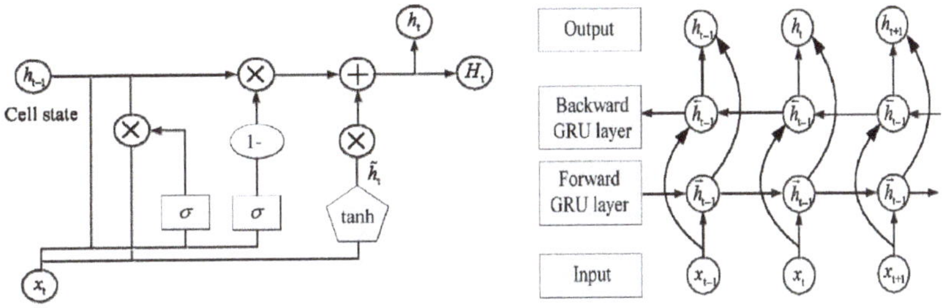

Figure 6. Structures of the GRU and BiGRU networks [132].

The deep learning model combining a nonlinear auto-regressive neural network, a convolutional neural network, and an LSTM recurrent network prediction was used to further improve the prediction accuracy [133–135]. Figure 7 shows the architecture of the GRU network for PV power forecasting.

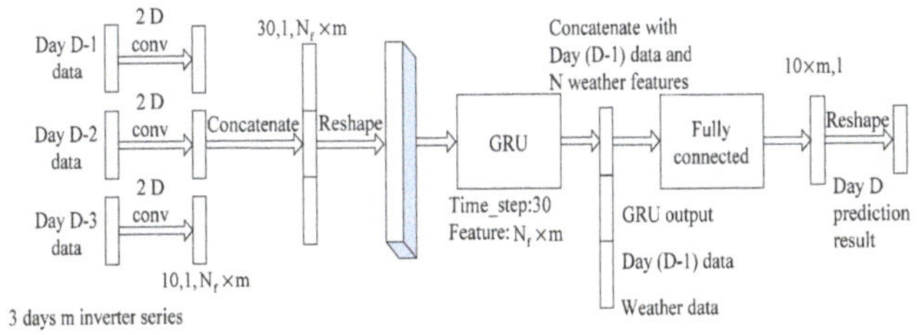

Figure 7. Architecture of the GRU network for PV power forecasting [134].

The artificial neural network approach combines artificial intelligence and neural network approaches for wind power prediction [136]. For the uncertainty and intermittency of wind power, a short-term wind power prediction model based on a Small-World BP Neural Network (SWBP) was constructed [137]. The input features of the SWBP were selected by the improved mutual information method. The selection criterion establishes the relationship between many input and output variables associated with wind power prediction by eliminating redundancy. Compared with the BP neural network and radial basis function neural network algorithm, this method effectively improves the prediction accuracy. Ref. [138] proposed a combined wind power prediction method based on Artificial Neural Networks (ANNs) and Support Vector Regression (SVR) algorithms. The wind power combination forecasting model was formed by the weighted average of each single forecasting algorithm, which improved the prediction accuracy of the model. However, the combined prediction algorithm based on the weighted average did not reflect the overall weighting effect of the prediction errors of individual samples in the sample set. Ref. [139] proposed a prediction method based on a sequential forward feature selection algorithm for wind–solar hybrid power generation, which combined two different objective functions with an artificial neural network method. For the diversity and dynamics of training samples, caused by the intermittency and volatility of wind and PV power, a variety of hybrid prediction models were proposed by using data mining methods composed of grey relational analysis, K-means clustering and a bagging neural network (NN) to solve complex control problems in wind and PV power systems and optimize the control of the microgrid [140–143].

According to the wind power ramp events in the economic operation and risk management of smart grids, ref. [144] proposed a hybrid prediction model based on a semi-supervised Generative Adversarial Network (GAN) to solve the problem of short-term wind power output and ramp events prediction. In the proposed model, the original time series of wind energy data was decomposed into several sub-sequences characterized by intrinsic mode functions (IMFs) with different frequencies, and the data was expanded using semi-supervised regression with label learning to extract nonlinear and dynamic behaviors from each IMF. The unlabeled virtual samples were obtained using the GAN to generate a model for capturing the data distribution characteristics of wind power outputs, while the semi-supervised regression layer was used to redesign the discriminant model to perform the point prediction of wind power. After two GAN models, a min–max game was formed to improve the quality of sample generation and reduce the prediction error. Ref. [145] quantified the prediction uncertainty caused by the intermittent and fluctuating characteristics of wind energy through wind power interval prediction. For the two most critical objectives in interval prediction, effective coverage and short interval length, a new interval prediction method was proposed, which combined the Conformal Quartile Regression algorithm with Time Convolution Networks (TCNs) without using any distributional assumptions. Compared to conventional RNN-based methods, the adopted TCN architecture avoided iterative propagation and gradient vanishing and could process very long sequences in a parallel fashion. According to the instability and volatility of load sequences, ref. [146] proposed a dynamic decomposition–reconstruction integration method by combining reconstruction and secondary decomposition techniques. The decomposition-integrated prediction framework was improved by introducing a dynamic classification and filtering-based decomposition reconstruction process and given criteria for determining the components that needed to be decomposed again. The model leveraged decomposition, complexity analysis, reconstruction, secondary decomposition and hyperparameter-optimized neural networks, demonstrating superiority in prediction accuracy, stability, correlation, overall performance and statistical tests. The short-term load prediction results provided effective support for the safe operation of the electric power system and rational scheduling.

In addition, extreme Gradient Boosting [147], Light Gradient Boosting Machine [148], Adaptive Boosting [149] and other tree integration algorithms, as representatives of artificial intelligence technology in wind and PV power prediction, have achieved good application results. The wind and PV power prediction technology has been widely studied throughout the world. The artificial neural network method, as an intelligent feature extraction method, uses a self-learning artificial intelligence method to help the management of mathematical models, and its research has practical application prospects [150–152].

3.2. *Research Status of Microgrid Load Forecasting Methods*

According to the scheduling needs of the power system dispatching department, the microgrid load forecasting is divided into short-term and medium-long-term, according to the forecasting time scale. According to the predictive modeling method, it is divided into a traditional statistical model and an AI-driven model.

Short-term load forecasting methods for microgrids mainly include traditional statistical methods and machine learning methods. Traditional statistical methods mainly include the Autoregressive Integrated Moving Average (ARIMA) model, Suppot Vector Machine (SVM), etc. [153]. These kinds of methods have simple principles and modeling and consider the time-series relationship of data but have limited prediction ability for nonlinear load data with relatively high complexity. Machine learning algorithms can effectively deal with nonlinear problems. Traditional machine learning methods mainly include Artificial Neural Network (ANN), Support Vector Regression (SVR) [154], random forest (RF), wavelet neural network [155], deep residual network [156], etc., but these methods are based on strongly correlated input. These methods can better reflect the nonlinear relationship between the data, but their common problem is the lack of consideration of the time correlation of time series data. Ref. [157] proposed a medium-short-term load forecasting, using a hybrid model of the Multilayer Feedforward Neural Network (MFFNN) and the Grasshopper Optimization Algorithm (GOA), which can be used to predict the load at different times and on different days of the month. The MFFNN processed the input layer and output layer, weather factors such as temperature were used as the input of MFFNN and, finally, the appropriate number of hidden layers was selected. The main steps included feeding data into the network, training the model and finally implementing the prediction process. The results show that the temperature has a clear influence on the predicted load in the proposed model. In addition, there are differences between the maximum and minimum loads in winter and summer. A regression model was introduced to determine the relationship between the dependent variable (load) and the independent variables affecting the load (such as temperature). Combined with load forecasting, ref. [158] developed a new power flow management algorithm for islanding power systems, which can simultaneously achieve more stable conventional unit operation and reduce the demand peak values. The forecasting module was based on the feedforward artificial neural network and can forecast short-time day-ahead load. Then, the predicted load curve of the previous day is used as the input of the pattern recognition algorithm to classify it according to the shape (pattern) of its load curve. Subsequently, if the classification result is a clear night peak pattern, the hourly trajectory of the diesel generator operation can be estimated, and the charging set point of the battery energy storage system can be derived. The SVM is a supervised learning method that always finds the global optimal solution and performs well on smaller datasets. However, when the data set becomes larger, it is prone to overfitting problems. Although the SVM can solve nonlinear problems well, the convergence speed of the SVM is slow due to the increase in computational complexity, and the relaxation variables and kernel parameters need to be set manually [159]. Ref. [160] proposed a multi-task learning model based on the Least Square Support Vector Machine (LSSVM), which realized the simultaneous output of electricity, heat, cold and gas load forecasting tasks.

The continuous development of deep learning has brought more choices for microgrid load forecasting. Different types of deep learning models are applied in the field of load forecasting [161,162]. It mainly includes convolutional neural networks (CNNs), recurrent neural networks (RNNs), LSTMs [163], deep belief networks (DBNs), etc.

Ref. [164] used CNNs to automatically extract features from input data and consider the correlation between different times of the day and different days of the week, which achieved good results compared to traditional machine learning methods, but the method only performs well on small data, and when the data fluctuates greatly, a single CNN model is unable to make good predictions. In ref. [165], the Multiple Time Series (MTS) was generated by combining macro and micro information of continuous time series and discrete time series to improve the performance of short-term load forecasting. The MTS included four information sequences: short-term, periodic, long-short-term, and trans-long-short-term. The MTS was used to build a short-term load forecasting system using a recurrent neural network (RNN) model that could learn sequential information between continuous and discrete sequences. Ref. [166] proposed a variant of the RNN based on the LSTM, which effectively solved the gradient explosion defect of recurrent neural networks by adding forgetting units. As an optimization of the LSTM network, GRU simplifies the internal unit structure of the LSTM, achieves similar prediction accuracy as the LSTM and has the advantages of fewer training parameters and high speed. The Bi-directional Long Short-Term Memory (Bi-LSTM) network can extract time series features from both the front and back directions and has better representation of continuous time series, and the multiplexing of weight parameters makes it less demanding on data [167]. The above network model can fully reflect the long-term historical process in the input time series data, but it cannot extract the effective information between the discontinuous data, so it cannot deeply explore the potential relationship between the data. Ref. [168] constructed an iterative multi-step load forecasting method by incorporating an extreme learning machine and dynamic neural network to mine features. The features of the two were learned and fused to output the predicted value for the current time step. Subsequently, the prediction was iteratively made into the future based on the predicted value and historical data. Ref. [169] proposed a hybrid attention LSTM network based on an encoder–decoder framework to better capture dynamic temporal features in an interpretable form.

In ref. [170], the DBN based on the restricted Boltzmann machine was used to extract the features of data, and multi-task learning was combined to realize the comprehensive energy load forecasting of electricity, heat and gas. Ref. [171] proposed a new hybrid ensemble deep learning method for low-voltage load side forecasting. In order to improve the regression ability of the DBN, a series of integrated learning methods, such as bagging and boosting variants, was introduced. In addition, the different transformation technique was used to ensure the stationarity of the load time series of the bagging and boosting methods. Based on the idea of integrated learning, a new hybrid integration algorithm was proposed by combining multiple independent integration methods. Considering the diversity of various ensemble algorithms, an effective k-nearest neighbor classification method was used to adaptively determine the weight of the sub-models. On this basis, using the inherent re-sample idea of bagging and boosting, a probabilistic prediction method based on hybrid ensemble deep learning (HEDL) was proposed. The above comprehensive energy multivariate load forecasting literature adopts different methods for forecasting, and most of them also consider the complex dynamic characteristics between energy sources, but do not fully mine the important information of more dimensions in the data.

With the diversification of training data types and the improvement of load forecasting accuracy requirements of the power grid, combined forecasting models have emerged to overcome the shortcomings of a single model in load forecasting accuracy [172]. Ref. [173] explored using a gated convolutional network and GRU network to solve a day-ahead multi-step load forecasting problem. By introducing a linear gated unit, the defect of convolutional neural networks having difficulty addressing time-series prediction was remedied. The recursive strategy was used for prediction, which greatly depended on the prediction results of the previous order. In the process of transmission, it is easy to cause the superposition of the prediction errors of the previous order, which leads to the decline in the prediction accuracy of the model. Ref. [174] proposed a hybrid network model by combining the attention mechanism. Compared with the unfused attention model, the proposed method is more accurate for real-time prediction and shows better adaptability in peak consumption situations. Ref. [175] adopted a Seq2Seq method in Encoder–Decoder architecture to enhance the load timing feature mining capability, and the CNN was added to extract the coupling feature information in the integrated energy system, which further helped Seq2Seq to achieve the accurate prediction of system load. Ref. [176] used the Copula method to analyze the nonlinear correlation between loads, screened the larger correlation factor features as the input features of the prediction model and constructed a bidirectional long- and short-term memory network to predict the electricity, heat and cold loads. Ref. [177] constructed three load CGRU-feature mining networks, established high-dimensional abstract features and cooperated with enhanced multi-task learning with homoskedastic uncertainty to output short-term prediction results for electricity, heat and cold loads. Ref. [178] proposed a load forecasting method based on nonlinear relationship extraction by training two convolutional neural networks individually to extract nonlinear load features and nonlinear load–temperature features, respectively, and the extracted features were fed to the SVR model for one hour ahead of load forecasting. Experimental results show that the method exhibits excellent prediction performance compared to LSTM, CNN and SVR methods.

With the continuous development of data decomposition algorithms, in order to reduce the influence of volatility and nonlinearity in the load sequence and further improve the accuracy of short-term load forecasting, a combined forecasting method that combines data decomposition algorithms with the existing forecasting models has been widely used in the field of load forecasting. The classical signal decomposition algorithms mainly include Wavelet Decomposition (WD), Empirical Mode Decomposition (EMD) and Variational Mode Decomposition (VMD). These combined prediction methods firstly decompose the original load sequence into multiple components with different characteristics by means of data decomposition algorithms and then build a prediction model for each component and reconstruct the prediction results of each component to obtain the final prediction value. Ref. [179] proposed a probabilistic load forecasting model based on wavelet transform. Before using quantile regression forest and random forest to establish a probabilistic forecasting model, the wavelet decomposition was used to preprocess a load time series, which effectively improved the forecasting accuracy. However, the decomposition effect of the WD is related to the selection of the wavelet basis and the amount of decomposition, lacking adaptability. The EMD method absorbs the advantage of multi-resolution of the wavelet transform and, at the same time, overcomes the difficulty of selecting the wavelet basis and determining the decomposition scale in the wavelet transform. So, it is more suitable for nonlinear and nonsmooth signal analysis and is an adaptive signal decomposition method. A hybrid method for STLF in the microgrid was proposed, which integrated the EMD, the PSO and an adaptive network based the fuzzy inference system [180]. Ref. [181] proposed an STLF method combining the EMD, the Bi-LSTM and the attentional mechanism, where the load sequence was decomposed into several Intrinsic Mode Functions (IMFs) by EMD, and, subsequently, an attentional mechanism-based Bi-LSTM neural network was applied to each of the extracted IMFs in order to predict the variation trends of these IMFs. However, the EMD needed to solve the problems of modal aliasing and endpoint effects. Ref. [182] used the VMD algorithm to decompose the original data and eliminate the noise present in the data, which helped the LSTM improve the accuracy of the STLF. Compared with the WD and EMD algorithms, the VMD is more adaptive and has the ability to overcome modal aliasing, so it has been widely used in the field of load forecasting. However, the VMD method lacks evaluation criteria to guide the parameter setting, and the parameters are often given empirically, which leads to unsatisfactory decomposition results.

Power prediction is a necessary precondition to ensure the balance of power supply and demand in the microgrid. Only with effective generation of power prediction and system consumption can load prediction be used as the condition of power dispatching at the same time. High-precision power prediction can ensure the correctness of power dispatch, reduce supply deviation and improve the operational stability and safety of the power system. In the prediction time, the current new energy power prediction main research focuses on ultra-short-term (same day) and short-term (next day) power prediction. And less research and results are oriented towards the future 3–10 days duration prediction. If the future is predicted for a longer time, the time resolution of the weather forecast is lower, the weather forecast model may have drift and the accuracy of the weather forecast will decrease day by day. Due to this constraint, when the traditional method is facing the future 3–10 days, the power prediction effect will be significantly reduced compared with the ultra-short-term and short-term. Therefore, medium-term power prediction is one of the urgent problems to be solved in microgrids, and its research has far-reaching significance and practical engineering value. Through the close integration of AI technology and the IoT, microgrid power prediction is shifting from a "passive response" to an "active sensing–decision-making" mode, which has become a key technology pivot for the intelligence of new power systems.

4. Research Status of Virtual Synchronous Active Support Control Technology

4.1. VSG Fundamentals and Control Strategies

When the power electronic interface circuit adopts droop control or other control methods, it has the characteristics of fast response and low inertia. With the emergence of a large number of distributed power supplies connected to the power system through the power electronic interface circuit, and the rotating reserve capacity in the system is reduced, which reduces the operation stability of the power system. In order to solve this problem, the inverter has the external characteristics of a synchronous generator by adopting the appropriate algorithm, which is the virtual synchronous control of the inverter. Virtual synchronous generator (VSG) technology facilitates the transition of renewable energy generation from passive regulation to active grid support. It not only enables inertia support, primary frequency regulation and active voltage regulation capabilities but also addresses challenges in power systems with high renewable penetration and power electronics ("double-high" systems).

VSG includes current mode and voltage mode. In the development of VSG, the early control scheme was current type. The concept of Static Synchronous Generator (SSC) was first proposed by the IEEE Task Force in 1997 [183]. On this basis, Johan Morren, Sjoerd W. H and other scholars from Delft University of Technology in the Netherlands proposed the idea of using a distributed power supply to participate in system voltage and frequency regulation and creatively proposed the principle of simulating synchronous generator primary frequency modulation characteristics and moment of inertia in 2005 [184], when, VSG's research was still in early stages. In 2007, Beck H. P of the Technical University of Lausktal in Germany took the lead in proposing the VISMA (Virtual Synchronous Machine) scheme, which enabled the inverter to have the same moment of inertia and damping characteristics as the synchronous generator [185]. The European VSYNC project was jointly launched by several scientific research institutions such as Devteco University of Technology and the Netherlands Energy Research Center, focusing on improving the overall stability of microgrid systems through inverter control and energy storage devices, and the concept of VSG was first proposed in the same year [186]. The above schemes are current-type VSG controls, and the whole system can be equivalent to the current source. Although these schemes can improve the dynamic performance of the system by simulating the characteristics of synchronous generators to a certain extent, due to the limitations of the current source itself, they cannot be applied to the off-grid mode of microgrids.

In order to solve the shortcomings of current-type VSG control and ensure that reliable voltage and frequency can be provided in off-grid mode, voltage-type VSG control has received extensive attention. In 2009, Professor Qingchang Zhong of Loughborough University proposed the concept of voltage VSG control technology with Synchronverter. By simulating the second-order electromagnetic model of synchronous generators, the distributed power script was equivalent to the electrical and mechanical characteristics, as well as the frequency and voltage regulation characteristics of SG [187]. In 2012, Salvatore D'Arco of the Norwegian Academy of Industrial Sciences and Jon Are Suul of the Norwegian University of Science and Technology proposed the VSM (Virtual Synchronous Machine) scheme. By adding an output voltage control loop (voltage and current double closed-loop control) to the power control loop in the VSG control, the output performance of the system was improved, and the design of various parameters of the control link was analyzed [188]. Since the European VSYNC Engineering Center first proposed virtual synchronous control technology in 2008, relevant technical theories have been continuously supplemented and developed [189,190].

4.2. Advanced VSG Inverter Control Strategies

The inverter control strategy based on virtual synchronous generator (VSG) technology can fully simulate the operating characteristics of a synchronous generator and provide inertial damping support for the system in the dynamic process while realizing the functions of primary frequency modulation and primary voltage regulation. Virtual synchronous generator (VSG) control is different from traditional microgrid inverter control. By simulating the mechanical and electrical characteristics of a synchronous generator, VSG can not only achieve a droop effect similar to droop control but also have further inertia and damping characteristics, so that the transient process of the microgrid inverter becomes slow, providing voltage and frequency support for the system. Finally, the anti-interference ability of the system is improved [191]. The circuit structure of the inverter based on VSG technology is shown in Figure 8.

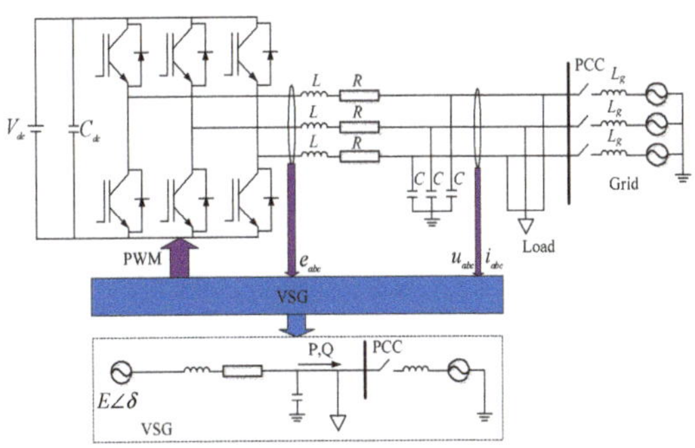

Figure 8. Circuit structure of the inverter based on VSG technology [191].

The traditional virtual synchronous generator adopts fixed parameters, which cannot reflect the advantages of flexible parameter regulation. Some scholars proposed a

VSG control method with adaptive parameter adjustment to reduce the duration of the transient process. When the system is disturbed, according to the rate of change of the rotor angular speed or the difference between the angular speed and the rated value of the virtual synchronous generator, the moment of inertia is changed adaptively. For large disturbances, greater inertia is used, and for small disturbances, less inertia is used, so that the system quickly enters a stable state. The concept of inertial midpoint was proposed, and according to this concept, the parameter design of the inverter in the multistage parallel system of virtual synchronous generators was guided, so as to optimize the frequency response of the entire isolated network system and improve the operation stability of the isolated network system [192]. An adaptive inertia control method was proposed to enable virtual synchronous generators to have different amounts of inertia under different working conditions, so as to improve the frequency regulation process of microgrids in reference [193]. The inertial response of the PMSM wind turbine control system and hybrid energy storage system controlled by a virtual synchronous generator were analyzed respectively, which expanded the application prospect of virtual synchronous generator control [194]. Ref. [195] derived the closed-loop characteristic equation of a VSG power loop, establishing relationships between dynamic performance and designing parameters (inertia coefficient and damping coefficient). Ref. [196] designed and proposed a rotational inertia adaptive control strategy based on VSG control and studied the dynamic response of VSG through parameters such as peak time and overshoot. By studying the change law of angular frequency, the adaptive adjustment scheme of moment of inertia is designed. Through the real-time change of moment of inertia, the control effect is obviously improved, the overshoot of the system is reduced and the adjustment time is shortened.

The real-time measured network frequency was used as the frequency reference value of VSG, and the constant power control of VSG was realized under a grid-connected steady state. A virtual synchronous generator control method for suppressing harmonic current and power frequency fault inrush current under non-ideal large power grid conditions was proposed, and the proposed method was verified [197]. Ref. [198] proposed an enhanced VSG control method to address the problem of active power oscillation and improper distribution of transient active power in VSG control. Based on state space analysis, it achieved oscillation damping and proper transient active power sharing by adjusting virtual stator reactance. Ref. [199] proposed a mode-switching control method that would change the VSG mode when the power grid failed. However, differential detection was used, which can lead to a lot of noise and additional system instability. At the same time, with the mode switching control method, if the VSG does not have an equilibrium point, it may oscillate at the boundary. Ref. [200] proposed an improved virtual synchro control method based on Lyapunov energy function for direct power control under the condition of power grid voltage imbalance to solve the power fluctuation problem when the power grid voltage was unbalanced. Ref. [201] explored the influence and configuration of inertia and damping on energy storage and used linear optimization control to adjust power frequency fluctuations, so as to enable the microgrid system to have secondary frequency modulation capability and optimize power output. The proposed control scheme for the cascaded H-bridge-based BESS in grid-connected mode is shown in Figure 9.

Ref. [202] studied the application of VSG technology in different scenarios, such as wind power, photovoltaic and energy storage, and each scenario had grid-connected and networking operating conditions. For VSGs with dual-mode operation, it is necessary to ensure that the charged state SOC of the energy storage unit is within the safe range during both charging and discharging. Once the SOC of the energy storage unit exceeds the safety range due to the power fluctuation of the distributed power supply on the DC side of the

VSG or the participation of the VSG in the primary frequency modulation, it is necessary to stop discharging or charging. Ref. [203] clarified the energy conversion method of inertia and energy storage and used equivalent calculation to configure energy storage but ignored the constraint of state of charge (SOC) of energy storage. When the battery has a low SOC, its output power changes greatly, and it cannot meet the adjustment requirements of the microgrid. The overall control diagram of the VSG is shown in Figure 10.

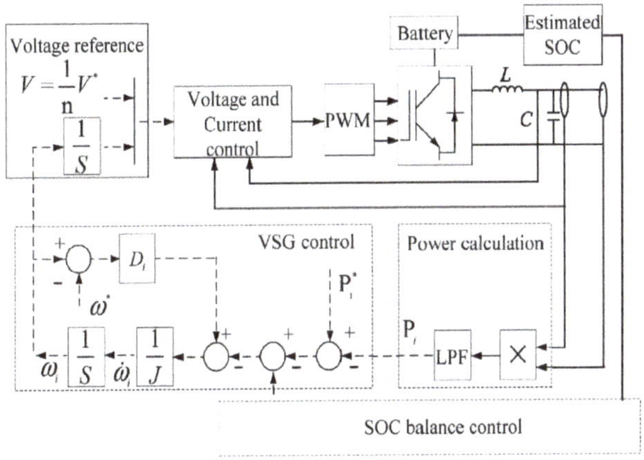

Figure 9. Proposed control scheme for the cascaded H-bridge-based BESS in grid-connected mode [201].

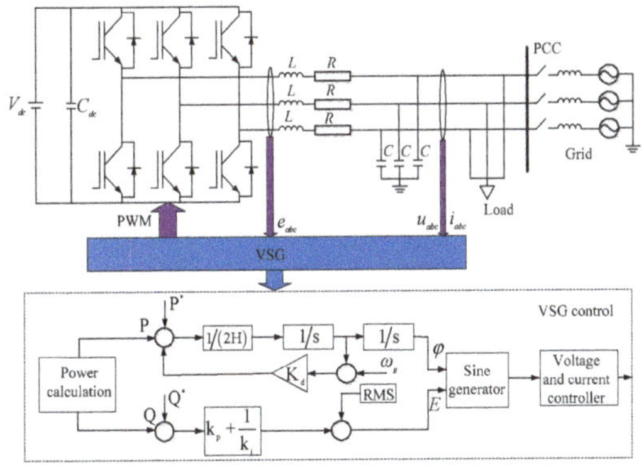

Figure 10. Overall control diagram of VSG [203].

The parallel operation of multiple VSGs is one of the hot topics in the research of high-proportion renewable energy power systems, and the core challenge is to solve the problems of stability reduction, model accuracy deficiency and multi-machine cooperative control [204]. In order to solve the stability problem caused by the high proportion of renewable energy connected to the power system, Wang X et al. studied the parallel system of a grid-constructing multi-machine VSG inverter and grid-following PQ inverter. By establishing Thevenin and Norton equivalent impedance models, and based on global admittance stability criteria, the influence of control parameters and line impedance on system stability was analyzed. They found parallel coupling could improve PQ inverter stability but potentially destabilize VSG inverters. The study provided a new theoretical support for the stability analysis of multi-inverter parallel systems [205]. Long proposed a power-frequency admittance model considering power coupling to solve the problem of insufficient model accuracy caused by power coupling in multi-VSG grid-connected systems. The model modified the second admittance by considering the power coupling, which significantly improved the accuracy of the model under the condition of high line impedance in the medium- and low-voltage microgrid. In addition, the conductivity matrix model of the multi-VSG system was extended to analyze the effects of line impedance and reactive power loop coefficients on system stability. Simulation results showed the proposed model's error was significantly lower than traditional models under strong power coupling [206]. Lin et al. proposed a multi-VSG mutual damping control strategy based on a model predictive control and consistency algorithm. The system power oscillation was effectively suppressed by distributed calculation of the power increment and information exchange. This study not only proved the stability of the controller through the Lyapunov theory but also introduced the gain adjustment function to improve the robustness in non-ideal communication environments. Simulations demonstrated significant advantages in dynamic response optimization and economic dispatch [207].

4.3. VSG Stability Mechanisms

The stability problem is one of the basic problems in power system analysis, which mainly includes power angle stability, voltage stability and frequency stability. The proposal and application of VSG technology is to solve the problem that large-scale new energy enters the power grid through power electronic equipment and affects the stability of the power grid. According to the magnitude of the disturbance, it can be divided into static stability and transient stability.

In terms of static stability, ref. [208] studied the stability of DC microgrids under the high permeability of high-voltage power electronic convertors and proposed three active damping solutions. The stability of the system under different compensation schemes is evaluated by small signal analysis. In addition, the reconstructed source impedance and modified voltage tracking dynamic performance of the voltage source inverter interface under different compensation schemes are derived. The dynamic coupling between active damping and voltage tracking controllers is evaluated by sensitivity and robustness analysis. The problem of secondary control of voltage and frequency in isolated island microgrids is transformed into the problem of first-order system tracking synchronization using feedback linearization technology [209]. The stability of VSGs was proved by using the eigenvalue method of small signal modeling, and key parameters such as inertia coefficient, excitation coefficient, sag coefficient and virtual impedance were analyzed by eigenvalue and sensitivity [210]. Ref. [211] used the impedance method to establish impedance models of the VSG, power grid and load, respectively, and gave the Nyquist stability criterion. The influence of each impedance on system stability was reflected from the separate frequency response of load and source impedance, and the cause of instability of a digitally controlled voltage source inverter operating as a microgrid was explained through experiments. The state space method was used to analyze the stability of small signals in power electronic systems. After linearizing the study variables at the operating point, the state space equation was formed, and the stability of the system was judged by the eigenvalue of the state matrix. Compared with the impedance method, the state–space method simplifies the system model by linearizing the system equation and can study the degree of influence of different parameters of the system on system stability and the interaction between different parts of the system [212]. Ref. [213] modeled each dynamic element of the VSG to form a state–space model, and then the eigenvalue and singular value decomposition (SVD) analysis were used to evaluate the impact of the network model and virtual impedance on the dynamics of the VSG.

Ref. [214] focused on the small-signal modeling and parameter design of the VSG power loop, pointing out that in order to avoid serious distortion of VSG output voltage, the bandwidth of the power loop should be much less than twice the bandwidth of the line frequency. On this basis, the linear frequency average small-signal model of the VSG was derived, which was used for system analysis and parameter design. The coupling effect between active loop and reactive loop was analyzed from the point of view of system stability. Ref. [215] pointed out that when dynamic changes of inverters and circuits are considered, system stability is constrained by virtual inertia and damping parameters. Ref. [216] studied a VSG implementation based on a voltage source converter (VSC), including a virtual impedance and an outer loop frequency droop controller. The linearized small-signal model, including the converter and its network-side filter, the control system and a network-side equivalent circuit, was established and verified by time-domain simulation of the nonlinear system model. However, there are 19 state variables in this model, and it is difficult to extend it to parallel or even multi-machine models because of too many variables considered. A full analytical model of the VSG's small-signal output impedance in the dq coordinate system was established, considering the dynamics of sag control and virtual inertia [217,218]. The small-signal models of power level, external power control loop and internal voltage and current controller were derived. The full analytical model's accuracy was then verified using SABER simulation measurements. Based on the generalized Nyquist criterion (GNC), the impedance model was used to analyze the stability of a VSG in grid-connected mode, and the information required for system stability was analyzed. However, this method can only be applied to systems with access to stable AC points, otherwise the operating points will no longer be constant. Ref. [219] proposed an additional damping strategy to suppress power oscillations. After analyzing the relationship between the output power and angular acceleration, an acceleration control damping strategy with disturbance compensation was designed. The block diagram of small-signal model for a paralleled VSG system is shown in Figure 11a. The stability of the method was studied by small-signal analysis. A small-signal model of an inverter based on sag control and VSG control was established, and the dynamic response of time was compared. An adaptive time constant control strategy based on improved sag control was proposed, which improved the response characteristics of output frequency and active power when the power command changes in grid-connected mode and can provide sufficient inertia characteristics [220]. Ref. [221] studied the synchronization stability of virtual synchronous generators under power grid faults. Qualitative analysis was carried out based on a linearization model, and quantitative analysis was combined with a nonlinear model to reveal the internal mechanism of a low-pass filter to improve synchronization stability in a reactive power control loop. Figure 11b shows the block diagram for enhancing VSG transient stability with a low-pass filter. Ref. [222] proposed a hybrid harmonic suppression scheme consisting of a local voltage harmonic control loop and an adaptive grid current control loop, aiming at the problem that local nonlinear load and network distortion will affect the power quality of virtual synchronous generators. Small-signal modeling is used to study the stability of the system and its robustness to parametric perturbations. Refs. [223,224] provided the small-signal models of grid-connected mode and island mode and analyzed the stability of the system by analyzing the eigenvalues of the small-signal models and their parameter sensitivity.

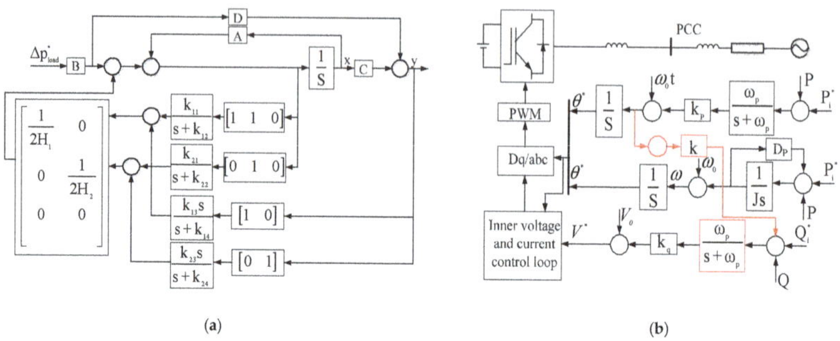

Figure 11. Block diagram of small-signal model for paralleled VSG system to enhance the VSG transient stability with low-pass filter (**a**) Block diagram of small-signal model (**b**) Configuration of a three-phase grid-connected VSG [219,221].

In terms of transient stability, ref. [225] used the phase portrait method to conduct an in-depth analysis of the transient stability of power synchronous controlled voltage source converters. The results showed that, following a transient disturbance, synchronization with the grid can be maintained as long as an equilibrium point exists. The critical clearance angle of a power synchronous control voltage source converter, that is, the power angle of unstable equilibrium point after fault, is identified when there is no equilibrium point in the power network fault. Ref. [226] proposed a method to improve the transient stability of the power grid of multiple virtual synchronous generators (VSGs) by controlling the oscillation of the electromagnet relative to the inertia center frequency of the power grid in the case of a short circuit. The proposed method can prevent the increase of the relative irreversible distance angle of the multi-virtual synchronous generator, which was the main factor preventing the multi-virtual synchronous generator from reaching the instability point, and its stability was verified by the Lyapunov function. The PSO algorithm was introduced to optimize and adjust the VSG unit parameters, and a scheme of alternating the moment of inertia value was applied to the dynamic voltage control system of a microgrid to rapidly suppress oscillations and improve transient stability after large disturbances [227]. In this scheme, the moment of inertia of the VSG is switched by the angular frequency of the VSG relative to the equilibrium point and its rate of change. Ref. [228] studied the transient rotor angle stability of parallel SG-VSG systems through comparative analysis. It showed that differences between parallel units affect transient stability, particularly during faults. A control method to improve the transient stability of shunt SG-VSG systems was proposed. On this basis, the nonlinear model of isolated island microgrids was established by using the Lyapunov method, and the attractive domain of parallel systems was quantified.

Ref. [229] designed a simple energy function to study the transient stability of VSGs and then proposed a Bang-Bang control strategy using the power angle relationship to improve the transient stability, but it is difficult to implement. Ref. [230] studied the transient stability of grid-connected voltage source converters under large signal interference. The dynamic performance and transient stability of four typical control schemes—power synchronization control, basic droop control, low-pass filter droop control and VSG control—were compared. The results show that transient stability and basic droop control can maintain stable operation as long as the equilibrium point exists, because they have a non-inertial transient response. However, the low-pass filter droop control and the virtual synchronous generator control are unstable even in the presence of an equilibrium point due to the lack of damping in their inertial transient responses. Through a phase portrait, the stability mechanism was clearly explained, and the quantitative effects of controller gain and virtual inertia were analyzed. The system stability of a static synchronous compensator (STATCOM) using a virtual synchronous generator was deeply analyzed [231]. The synchronizer and vector control performance were compared using different mathematical tools, such as eigenvalue analysis, numerical simulation and the Lyapunov theory. Finally, the feasibility of using synchronizer control strategy was verified on an IEEE 39-bus system. The results show that the synchronizer brings the advantage of artificially increasing system inertia, which is an important problem in modern power systems. The virtual synchronous generator was essentially a nonlinear system, and nonlinear control design was needed to maintain the stable operation of the synchronous generator when the power grid failed. Ref. [232] proposed an adaptive method to solve this problem, which can realize the full power injection of the power grid under unbalanced conditions, and analyzed the stability of the system through the Lyapunov function. In this method, the supplementary internal nonlinear controller was used as the external subsystem, and the power angle and inverter voltage were adjusted step by step using the on-line reverse updating law. The transient angular stability of VSGs was studied by the Lyapunov direct method. The influence of reactive control loop on transient angular stability was analyzed, and then the voltage change was considered using the Lyapunov direct method. The effects of different parameters on the transient angular stability were studied. Finally, a control method was proposed to improve the stability margin of VSGs by adjusting the reference active power.

Due to its excellent characteristics and the gradual maturity of distributed power technology, VSG technology will become the core technology of power structure transformation and smart microgrid construction in the future. Although there have been a large number of studies on the small-signal and large-signal stability of VSG, and there are a large number of studies on the parameter optimization design and parameter adaptive of VSG systems, most of the research results are based on the simplified model of a single machine connecting to an infinite power grid or island parallel, which can be regarded as a long-distance transmission model because it ignores the parameters of some transmission lines. This is inconsistent with the actual situation of the current user-side energy storage facilities connected to the grid, so it is of great significance to study the multi-machine VSG system with transmission line parameters. At the same time, because it is difficult to guide the system parameter design with the Lyapunov method, it is necessary to consider a new large-signal stability analysis method.

Although VSG technology can significantly improve the stability of the power grid and the ability to connect renewable energy to the grid, it still faces many economic and technical challenges in practical applications. VSGs have high initial investment costs and require high performance inverters and DSP controllers, which are more expensive than traditional synchronous machines [233]. The VSG itself does not store energy and requires a battery (BESS) or supercapacitor (SC) to provide actual power support. The VSG needs to simulate the inertia, damping and other characteristics of synchronous generators but under different grid conditions (such as microgrid, large grid, high proportion of new energy access), and parameters need to be dynamically adjusted, otherwise it may lead to oscillation or instability [234]. At the same time, it is necessary to realize the functions of frequency modulation (P-f), voltage regulation (Q-V) and harmonic suppression simultaneously, and the control strategy is complex. In the case of power grid faults (such as a short circuit or voltage drop), the VSG needs to quickly switch control mode to avoid going off grid [235]. At high grid impedance, such as remote microgrids, VSGs can cause resonance or voltage instability. When multiple VSG systems are connected in parallel, a mismatch of control parameters may lead to power oscillation or circulating current problems. A high proportion of new energy grids have low inertia, and the virtual inertia of the VSG may aggravate the frequency fluctuation if it is improperly designed. VSG technology has great potential in improving power grid stability, but it still needs to solve key problems such as high control complexity, energy storage dependence and high cost. In the future, through algorithm optimization, hybrid energy storage and policy support, VSGs are expected to become one of the core technologies of the smart grid.

5. Conclusions and Future Perspectives

5.1. Conclusions

Microgrids enhance grid reliability and flexibility while enabling effective renewable energy utilization. This paper summarizes recent microgrid research progress, focusing on operation optimization strategies, power prediction methods and the latest advances in VSG active support control technology. Key conclusions are drawn as follows:

(1) Operation Optimization Strategies: Energy Management Systems (EMSs) leveraging heuristic algorithms (e.g., PSO, GWO), multi-objective optimization and AI-driven strategies significantly enhance economic efficiency and renewable energy utilization. Challenges remain in handling high-dimensional computations and real-time adaptability in complex scenarios.

(2) Power Prediction Methods: Hybrid models integrating decomposition techniques (e.g., VMD, CEEMDAN) with deep learning (e.g., LSTM, GRU) improve the accuracy of wind/PV and load forecasting. Medium-term (3–10 days) prediction remains underdeveloped due to meteorological uncertainty and data resolution limitations.

(3) VSG Control Technology: Voltage-type VSG technologies emulate synchronous generator characteristics, providing inertia support and frequency regulation. Stability mechanisms (e.g., impedance modeling, Lyapunov-based analysis) are established for grid-connected/islanded modes, yet multi-VSG coordination and transient stability require further validation.

These findings underscore the maturity of core microgrid technologies while highlighting persistent gaps in scalability, real-time performance, and cross-domain integration. This review aims to provide insights for enhancing economic, reliable, safe and stable microgrid operation, promoting further research and application

5.2. Future Perspectives

To address unresolved challenges and propel microgrid technology toward industrial deployment, future research should prioritize the following:

(1) Quantum Optimization for EMSs: Address computational bottlenecks in large-scale, high-dimensional microgrid optimization using quantum algorithms (e.g., leveraging superposition and entanglement [236]). This can reduce optimization time from minutes to seconds, enabling real-time EMSs for 24 h scheduling. Cross-energy coupling optimization (electricity, hydrogen, heat) across multiple timescales is also crucial.

Advanced Power Prediction: Enhance prediction accuracy via multi-modal deep learning and spatiotemporal graph neural networks, integrating diverse data sources (satellite, radar, local sensors). Develop physical-information fusion systems using digital twins updated with real-time data [237]. Implement ultra-low-latency edge forecasting by integrating LSTM cores into dedicated weather chips (<50 ms) [238].

Next-Generation VSG Hardware: Utilize superconducting materials for near-zero-loss, ultra-fast response VSGs (100× speed increase) [239]. Employ wide-bandgap materials (SiC, GaN) for high-frequency switching (>10 kHz), reducing losses by 50% and enabling wider impedance bandwidth control.

Distributed VSG Coordination: Develop intelligent, self-organized VSG cluster control strategies for synchronous frequency regulation, reducing reliance on centralized systems and aligning with distributed microgrid trends.

(2) There is an urgent need to deepen research in the field of automation and intelligence of microgrid systems. Advanced technologies such as AI, IoT and big data analytics are deeply integrated with microgrid technology to build a comprehensive "source–grid–load–storage" intelligent synergistic system. Intelligent microgrid systems have the ability to automatically identify load demand, make accurate power predictions and optimize dispatch. Through independent management and automatic control, efficient and low-carbon energy utilization is achieved. Through the close integration of AI technology and the IoT, the microgrid is shifting from a "passive response" to an "active sensing-decision-making" mode.

(3) There is an urgent need to strengthen the in-depth research of multi-level hybrid microgrid technology, so as to promote the transformation process of regional multistage distribution systems of microgrids and realize the evolution of microgrids in the large-capacity and regional hybrid direction. At the same time, it is essential to deepen the research on application technology and integration technology, in order to promote the development of microgrids in the direction of integration. VSG technology faces a number of challenges, including high control complexity, dependence on energy storage systems and relatively high cost. These problems have become obstacles to the further development and wide application of VSG technology. However, with the continuous progress and innovation of technology, through the continuous optimization of algorithms, the in-depth application of hybrid energy storage technology and strong support from the policy level, VSG technology is expected to overcome these challenges in the future. Once these problems are properly solved, VSG technology will most likely become one of the core technologies in the field of smart grids, providing strong technical support for the stable operation of smart grids and the efficient use of energy.

(4) There is an urgent need to strengthen the research on energy storage technology. As a core component of microgrids, energy storage systems play a crucial role in optimizing energy structure, improving energy efficiency and ensuring the stability of the energy supply. With the continuous expansion of the scale of microgrids, the energy regulation function of energy storage systems in time series has become the key to solving the inherent randomness and volatility of renewable energy and maintaining the stable operation of microgrids. The VSG itself does not have the function of energy storage and must be combined with battery energy storage systems or energy storage devices such as supercapacitors to provide actual power support. By effectively integrating VSGs with these storage devices, power distribution can be optimized to improve the stability and efficiency of the entire system. Therefore, the combination of energy storage systems and VSG technology is the key to promoting the development of more intelligent and efficient microgrids.

In the process of rapid development of distributed generation technology, the microgrid has emerged to provide a boost to the operation of the power system. In the future, the microgrid will continue to face the challenges of technological innovation and marketing. The microgrid is a nonlinear complex system. In order to further improve the microgrid, it is necessary to flexibly apply more advanced intelligent energy management systems, power prediction technology and VSG active support control technology to further enhance the efficiency and flexibility of the microgrid, thereby promoting the development of the microgrid towards intelligence and holistic integration. Overall, with the continuous improvement of microgrid technology and the refinement of the system, the microgrid has broad prospects and will make significant contributions to sustainable energy supply and environmental protection.

Funding: This research was funded by Research and Application of Virtual Synchronous Support Control Technology for Village-level AC/DC Hybrid Microgrid OF State grid Henan Electric power company technology project, grant number 52170224001R.

Conflicts of Interest: The authors declare no conflicts of interest.

References

1. Srikanth, M.; Venkata Pavan Kumar, Y.; Pradeep Reddy, C.; Mallipeddi, R. Multivariable control-based dq decoupling in voltage and current control loops for enhanced transient response and power delivery in microgrids. *Energies* **2024**, *17*, 3689. [CrossRef]
2. Phani Raghav, L.; Seshu Kumar, R.; Koteswara Raju, D.; Singh, A.R. Analytic hierarchy process (AHP)-swarm intelligence based flexible demand response management of grid-connected microgrid. *Appl. Energy* **2022**, *306*, 118058. [CrossRef]
3. García-Trivio, P.; de Oliveira-Assís, L.; Soares-Ramos, E.; Sarrias-Mena, R.; García-Vázquez, C.; Fernández-Ramírez, L. Supervisory control System for a grid-connected MVDC microgrid based on Z-Source converters with PV, battery storage, green hydrogen system and charging station of electric vehicles. *IEEE Trans. Ind. Appl.* **2023**, *59*, 2650–2660. [CrossRef]
4. Zamanzad Ghavidel, B.; Liao, Y. A new power-sharing strategy with photovoltaic farms and concentrated diesel generators to increase power system resilience. *Energies* **2024**, *17*, 3742. [CrossRef]
5. Rouabhi, R.; Herizi, A.; Djerioui, A. Performance of robust type-2 fuzzy sliding mode control compared to various conventional controls of doubly-fed induction generator for wind power conversion systems. *Energies* **2024**, *17*, 3778. [CrossRef]
6. China Renewable Energy Development Report 2024: Wind Power Capacity Is Expected to Increase by Approximately 80 Million Kilowatts in 2025. Available online: https://baijiahao.baidu.com/s?id=1833417805276293481&wfr=spider&for=pc (accessed on 30 May 2025).
7. The National Energy Administration: In 2024, the Newly Installed Capacity of Photovoltaic Power Across the Country Will Reach 278 Million Kilowatts, Representing a Year-On-Year Growth of 28%. Available online: https://mp.weixin.qq.com/s?__biz=MzI2NjQ0NzUxNA==&mid=2247533597&idx=2&sn=f34602fc683f8ae1b02f72a43c7a5900&chksm=ebcc75d06373759e8511670b0fe780d94ad0349d39a3144b03ab12823c94d2f57919164748e2&scene=27 (accessed on 30 January 2025).
8. Alamir, N.; Kamel, S.; Hashim, F.A.; Bouaouda, A.; Safaraliev, M.; Abdelkader, S.M. Optimizing two-stage energy management in renewable-based multi-microgrid using a modified student psychology-based optimization with demand response and hydrogen storage. *Int. J. Hydrogen Energy* **2024**, *63*, 696–719. [CrossRef]
9. Wu, Y.; Liu, J.; Li, S.; Jin, M. Physical model and long short-term memory-based combined prediction of photovoltaic power generation. *J. Power Electron.* **2024**, *24*, 1118–1128. [CrossRef]
10. Wang, X.; Zhong, Z.; Jiang, W.; Zhao, X. Consensus-based secondary frequency control for parallel virtual synchronous generators. *Sustain. Energy Grids Netw.* **2024**, *38*, 101341. [CrossRef]
11. Brem, A.; Adrita, M.; O'Sullivan, D.; Bruton, K. Industrial smart and micro grid systems-A systematic mapping study. *J. Clean. Prod.* **2019**, *244*, 118828. [CrossRef]
12. Yolda, Y.; Nen, A.; Muyeen, S.; Vasilakos, A.; Alan, R. Enhancing smart grid with microgrids: Challenges and opportunities. *Renew. Sustain. Energy Rev.* **2017**, *72*, 205–214. [CrossRef]
13. Ardriani, T.; Dahono, P.; Rizqiawan, A.; Garnia, E.; Sastya, P.; Arofat, A.; Ridwan, M. A DC microgrid system for powering remote areas. *Energies* **2021**, *14*, 493. [CrossRef]
14. Pothireddy, K.; Vuddanti, S. Alternating direction method of multipliers based distributed energy scheduling of grid connected microgrid by considering the demand response. *Discov. Appl. Sci.* **2024**, *6*, 543–552. [CrossRef]
15. Cai, S.; Xie, Y.; Zhang, Y.; Zhang, M.; Wu, Q.; Guo, J. A simulation-assisted proactive scheduling method for secure microgrid formation under static and transient islanding constraints. *IEEE Trans. Smart Grid* **2024**, *15*, 272–285. [CrossRef]
16. Lou, H.; Fujimura, S. ADMM-based distributed algorithm for energy management in multi-microgrid system. *IEEJ Trans. Electr. Electron. Eng.* **2024**, *19*, 79–89. [CrossRef]
17. Wang, C.; Han, J.; Jiang, D.; Zhang, W.; Yang, W.; Song, H.; Deng, Y.; Qu, Y. Research on autonomous operation and mutual aid strategy of AC/DC hybrid microgrid cluster. *Electr. Power Syst. Res.* **2024**, *231*, 1103. [CrossRef]
18. Dou, C.; Liu, B. Multi-agent based hierarchical hybrid control for smart microgrid. *IEEE Trans. Smart Grid* **2013**, *4*, 771–778. [CrossRef]
19. Dolatabadi, S.; Soleimani, A.; Ebtia, A.; Shafie-Khah, M.; Bhuiyan, T. Enhancing voltage profile in islanded microgrids through hierarchical control strategies. *Electr. Power Syst. Res.* **2024**, *231*, 5434. [CrossRef]

20. Arou, A.; Tsuji, T. Coordinated Q-V/P-F control strategy using virtual power factor for autonomous AC microgrid with PV smart inverters. *IEEJ Trans. Electr. Electron. Eng.* **2024**, *19*, 327–341. [CrossRef]
21. Wang, T.; Lin, C.; Zheng, K.; Zhao, W.; Wang, X. Research on grid-connected control strategy of photovoltaic (pv) energy storage based on constant power operation. *Energies* **2023**, *16*, 8056. [CrossRef]
22. Pompodakis, E.; Orfanoudakis, G.; Yiannis, K.; Karapidakis, E. Robust secondary controller for islanded microgrids with unexpected electrical partitions under fault conditions. *Energies* **2024**, *17*, 3727. [CrossRef]
23. Xu, H.; Sun, J.; Huang, J.; Lin, X.; Ma, C. Distributed optimization of islanded microgrids integrating multi-type VSG frequency regulation and integrated economic dispatch. *Energies* **2024**, *17*, 1618. [CrossRef]
24. Orihara, D.; Taoka, H.; Otani, K. Influence of wind-turbine-generator power control on the performance of a virtual synchronous machine. *Energies* **2024**, *17*, 234. [CrossRef]
25. Ma, C.; Sun, J.; Huang, J.; Wang, K. Transient stability enhancement strategy for islanded microgrids based on energy storage–virtual synchronous machine control. *Energies* **2023**, *16*, 6390. [CrossRef]
26. Mrey, J.; Avera, G.; Acevedo-rueda, P.; Solano, J.; Amantilla, M.; Llanos, J.; Sáez, D. A review of microgrids in latin America: Laboratories and test systems. *IEEE Latin Am. Trans.* **2022**, *20*, 1000–1011.
27. Ali, A.; Li, W.H.; Hussain, R. Overview of current microgrid policies, incentives and barriers in the European Union, United States and China. *Sustainability* **2017**, *9*, 1146. [CrossRef]
28. Yokoyama, A. Toward deregulated, smart and resilient power systems with massive integration of renewable energy in Japan. *IEEJ Trans. Electr. Electron. Eng.* **2022**, *17*, 1242–1254. [CrossRef]
29. Yoshida, Y.; Farzaneh, H. Optimal design of a stand-alone residential hybrid microgrid system for enhancing renewable energy deployment in Japan. *Energies* **2020**, *13*, 1737. [CrossRef]
30. Zhang, L.; Yang, Y.; Li, Q.; Gao, W.; Qian, F.; Song, L. Economic optimization of microgrids based on peak shaving and CO_2 reduction effect: A case study in Japan. *J. Clean. Prod.* **2021**, *321*, 128973. [CrossRef]
31. Nosratabadi, S.; Hooshmand, R.; Gholipour, E. A comprehensive review on microgrid and virtual power plant concepts employed for distributed energy resources scheduling in power systems. *Renew. Sustain. Energy Rev.* **2017**, *67*, 341–363. [CrossRef]
32. Meng, L.; Guerrero, J.; Sanseverino, E.; Luna, A. Microgrid supervisory controllers and energy management systems: A literature review. *Renew. Sustain. Energy Rev.* **2016**, *60*, 1263–1273. [CrossRef]
33. Choudhury, S. Review of energy storage system technologies integration to microgrid: Types, control strategies, issues, and future prospects. *J. Energy Storage* **2022**, *48*, 103966. [CrossRef]
34. Rana, M.; Uddin, M.; Sarkar, M.; Shafiullah, G.; Mo, H.; Atef, M. A review on hybrid photovoltaic-Battery energy storage system: Current status, challenges, and future directions. *J. Energy Storage* **2022**, *51*, 104597. [CrossRef]
35. Kang, M.; Thanikanti, S.; Vigna, K.; Padmanathan, K.; Sunil, G.; Shangari, K. Empowering smart grid: A comprehensive review of energy storage technology and application with renewable energy integration. *J. Energy Storage* **2021**, *39*, 102591.
36. Li, J.; Liu, Y.; Wu, L. Optimal operation for community based multi-party microgrid in grid-connected and islanded modes. *IEEE Trans. Smart Grid* **2016**, *9*, 756–765. [CrossRef]
37. Belboul, Z.; Toual, B.; Kouzou, A.; Mokrani, L.; Bensalem, A.; Kennel, R.; Abdelrahem, M. Multiobjective optimization of a hybrid PV/Wind/Battery/Diesel generator system integrated in microgrid: A case study in Djelfa, Algeria. *Energies* **2022**, *15*, 3579. [CrossRef]
38. Zhang, G.; Wang, W.; Du, J.; Sheng, H. Multi-objective economic optimal dispatch for the island isolated microgrid under uncertainty based on interval optimization. *Math. Probl. Eng.* **2021**, *2021*, 9983104.
39. Hartmann, B.; Táczi, I.; Talamon, A.; Vokony, I. Island mode operation in intelligent microgrid-Extensive analysis of a case study. *Int. Trans. Electr. Energy Syst.* **2021**, *31*, e12950. [CrossRef]
40. Liu, W.L.; Liu, C.L.; Lin, Y.J.; Bai, K.; Ma, L.Y.; Chen, W.Y. Multi-objective optimal scheduling method for a grid-connected redundant residential microgrid. *Processes* **2019**, *7*, 296. [CrossRef]
41. Shreerajshakya, A.; Bajracharya, I.; Vaidya, R.; Prakashbhave, C.; Sharma, A.; Rupakheti, M.; Bajracharya, T. Estimation of air pollutant emissions from captive diesel generators and its mitigation potential through microgrid and solar energy. *Energy Rep.* **2022**, *8*, 3251–3262.
42. Zhao, X.-G.; Zhang, Z.-Q.; Xie, Y.-M.; Meng, J. Economic-environmental dispatch of microgrid based on improved quantum particle swarm optimization. *Energy* **2020**, *195*, 117011–117014.
43. Abhishek, A.; Ranjan, A.; Devassy, S.; Verma, B.; Dhakar, A. Review of hierarchical control strategies for DC microgrid. *IET Renew. Power Gener.* **2020**, *14*, 1631–1640. [CrossRef]
44. Zhang, J.; Qin, D.; Ye, Y.; He, Y.; Fu, X.; Yang, J.; Shi, G.; Zhang, H. Multi-time scale economic scheduling method based on day-ahead robust optimization and intraday MPC rolling optimization for microgrid. *IEEE Access* **2021**, *9*, 140315–140324. [CrossRef]

45. Li, D. Prediction of short-term load of microgrid based on multivariable and multistep long short-term memory. *Sens. Mater.* **2022**, *34*, 1275–1285. [CrossRef]
46. Yue, D.; Zhang, H.; Dou, C. MAS-based distributed cooperative control for DC microgrid through switching topology communication network with time-varying delays. *IEEE Syst. J.* **2017**, *13*, 615–624.
47. Khalid, M.; Ahmadi, A.; Savkin, A.; Agelidis, V. Minimizing the energy cost for microgrids integrated with renewable energy resources and conventional generation using controlled battery energy storage. *Renew. Energy* **2016**, *97*, 646–655. [CrossRef]
48. Kumar Venayagamoorthy, G.; Sharma, R.K.; Gautam, P.K.; Ahmadi, A. Dynamic energy management system for a smart microgrid. *IEEE Trans. Neural Netw. Learn. Syst.* **2016**, *27*, 1643–1656. [CrossRef]
49. Das, A.; Ni, Z. A computationally efficient optimization approach for battery Ssystems in islanded microgrid. *IEEE Trans. Smart Grid* **2017**, *9*, 6489–6499. [CrossRef]
50. Cheng, C.-H. Implementation of a small type DC microgrid based on fuzzy control and dynamic programming. *Energies* **2016**, *10*, 781. [CrossRef]
51. Ali, A.; Deldar, M.; Anwar, S. Optimal control of hydrostatic drive wind turbines for improved power output in low wind-speed regions. *Energies* **2021**, *14*, 5001. [CrossRef]
52. Deldar, M.; Anwar, S. Optimal control of a hydrostatic wind turbine drivetrain for efficiency improvements. In Proceedings of the ASME 2017 Dynamic Systems and Control Conference, Tysons, VI, USA, 11–13 October 2017.
53. Askarzadeh, A. A memory-based genetic algorithm for optimization of power generation in a microgrid. *IEEE Trans. Sustain. Energy* **2018**, *9*, 1081–1089. [CrossRef]
54. Zheng, Z.; Yang, S.; Guo, Y.; Jin, X.; Wang, R. Meta-heuristic techniques in microgrid management: A survey. *Swarm Evol. Comput.* **2023**, *78*, 101256. [CrossRef]
55. Torkan, R.; Ilinca, A.; Ghorbanzadeh, M. A genetic algorithm optimization approach for smart energy management of microgrid. *Renew. Energy* **2022**, *197*, 852–863. [CrossRef]
56. Li, B.; Roche, R.; Miraoui, A. Microgrid sizing with combined evolutionary algorithm and MILP unit commitment. *Appl. Energy* **2017**, *188*, 547–562. [CrossRef]
57. Hafez, A.; Abdelaziz, A.; Hendy, M.; Ali, A. Optimal sizing of off-line microgrid via hybrid multi-objective simulated annealing particle swarm optimizer. *Comput. Electr. Eng.* **2021**, *94*, 107294. [CrossRef]
58. He, Y. Research on the optimal economic power dispatching of a multi-microgrid cooperative operation. *Energies* **2022**, *15*, 8194. [CrossRef]
59. Sharma, S.; Bhattacharjee, S.; Bhattacharya, A. Grey wolf optimisation for optimal sizing of battery energy storage device to minimise operation cost of microgrid. *IET Gener. Transm. Distrib.* **2016**, *10*, 625–637. [CrossRef]
60. Wang, Y.; Li, F.; Yu, H.; Wang, Y.; Qi, C.; Yang, J.; Song, F. Optimal operation of microgrid with multi-energy complementary based on moth flame optimization algorithm. *Energy Sources* **2020**, *42*, 758–806. [CrossRef]
61. Pesaran, H.; Nazari-Heris, M.; Mohammadi-Ivatloo, B.; Seyedi, H. A hybrid genetic particle swarm optimization for distributed generation allocation in power distribution networks. *Energy* **2020**, *209*, 118218. [CrossRef]
62. Dey, B.; Bhattacharyya, B.; Srivastava, A.; Shivam, K. Solving energy management of renewable integrated microgrid systems using crow search algorithm. *Soft Comput.* **2020**, *24*, 10433–10454. [CrossRef]
63. Khalil, G.; Abbasi, M.; Ali, A.; Li, L. An efficient bi objective approach for dynamic economic emission dispatch of renewable integrated microgrids. *J. Ambient. Intell. Humaniz. Comput.* **2022**, *14*, 10695–10714.
64. Barua, S.; Merabet, A.; Al-Durra, A.; El-Fouly, T.; El-Saadany, E. Lévy arithmetic optimization for energy management of solar wind microgrid with multiple diesel generators for off-grid communities. *Appl. Energy* **2024**, *371*, 123736. [CrossRef]
65. Dey, B.; Bhattacharyya, B.; Marquez, F. A hybrid optimization-based approach to solve environment constrained economic dispatch problem on microgrid system. *J. Clean. Prod.* **2021**, *307*, 127196. [CrossRef]
66. Abualigah, L.; Diabat, A.; Mirjalili, S.; Abd Elaziz, M.; Gandomi, A. The arithmetic optimization algorithm. *Comput. Meth. Appl. Mech. Eng.* **2021**, *376*, 113609. [CrossRef]
67. Khadanga, R.; Das, D.; Kumar, A.; Panda, S. Sine augmented scaled arithmetic optimization algorithm for frequency regulation of a virtual inertia control based microgrid. *ISA Trans.* **2023**, *138*, 534–545. [CrossRef]
68. Kharrich, M.; Abualigah, L.; Kamel, S.; Abdel-Sattar, H.; Tostado-Veliz, M. An improved arithmetic optimization algorithm for design of a microgrid with energy storage system: Case study of El Kharga Oasis, Egypt. *J. Energy Storage* **2022**, *51*, 104343. [CrossRef]
69. Sujoy, B.; Adel, M. Lévy Arithmetic Algorithm: An enhanced metaheuristic algorithm and its application to engineering optimization. *Expert Syst.* **2024**, *241*, 122335.

70. Elkholy, M.; Yona, A.; Senjyu, T.; Elsayed, L.; Elymany, M.; Takahashi, H. Experimental validation of an AI-embedded FPGA-based Real-Time smart energy management system using Multi-Objective Reptile search algorithm and gorilla troops optimizer. *Energy Convers. Manag.* **2023**, *282*, 116860. [CrossRef]
71. Dey, B.; Bhattacharyya, B.; Basak, S. A comparative analysis between price-penalty factor method and fractional programming method for combined economic emission dispatch problem using novel hybrid CSA-JAYA algorithm. In Proceedings of the Michael Faraday IET International Summit 2020 (MFIIS 2020), Online Conference, 3–4 October 2021.
72. Lotfi, H.; Khodaei, A. Hybrid AC/DC microgrid planning. *Energy* **2017**, *118*, 37–46. [CrossRef]
73. Du, W.; Lasseter, R.; Khalsa, A. Survivability of autonomous microgrid during overload events. *IEEE Trans. Smart Grid* **2020**, *4*, 3515–3524.
74. Liu, W. Research on frequency control of islanded microgrid with multiple distributed power sources. *Processes* **2020**, *8*, 193. [CrossRef]
75. Prabaakaran, K.; Srividhya, R.; Senthil Kumar, R.; Hemanth Kumar, D.; Mohan Raj, D.; Sham Prabu, A. Energy management system for small-scale hybrid wind solar battery-based microgrid. *IEEE Access* **2022**, *8*, 8336–8345.
76. Dhifli, M.; Lashab, A.; Guerrero, J.; Abusorrah, A.; Al-Turki, Y.; Cherif, A. Enhanced intelligent energy management system for a renewable energy-based AC microgrid. *Energies* **2020**, *13*, 3268. [CrossRef]
77. Hossein, L.; Amin, K. AC versus DC microgrid planning. *IEEE Trans. Smart Grid* **2017**, *8*, 296–304.
78. Jung, S.; Yoon, Y. Optimal operating schedule for energy storage system: Focusing on efficient energy management for microgrid. *Processes* **2019**, *7*, 80. [CrossRef]
79. Vaka, S.; Matam, S. Optimal sizing and management of battery energy storage systems in microgrids for operating cost minimization. *Electr. Power Compon. Syst.* **2021**, *49*, 1319–1332. [CrossRef]
80. Fouladfar, M.; Saeed, N.; Marzband, M.; Franchini, G. Home-microgrid energy management strategy considering EV's participation in DR. *Energies* **2021**, *14*, 5971. [CrossRef]
81. Li, B.; Wang, H.; Tan, Z. Capacity optimization of hybrid energy storage system for flexible islanded microgrid based on real-time price-based demand response. *Int. J. Electr. Power Energy Syst.* **2022**, *136*, 107581. [CrossRef]
82. Yong, W.; Liew, P.; Woon, K.; Alwi, S.; Kleme, J. A pinch-based multi-energy targeting framework for combined chilling heating power microgrid of urban-industrial symbiosis. *Renew. Sustain. Energy Rev.* **2021**, *150*, 111482. [CrossRef]
83. Abdulazizalanazi, A.; Hosseinlotfi, B.; Aminkhodaei, B. Market clearing in microgrid-integrated active distribution networks. *Electr. Power Syst. Res.* **2020**, *183*, 106263.
84. Shuai, H.; He, H. Online scheduling of a residential microgrid via monte-carlo tree search and a learned model. *IEEE Trans. Smart Grid* **2021**, *12*, 1073–1087. [CrossRef]
85. Ali, K.; Abusara, M.; Tahir, A.; Das, S. Dual-layer Q-learning strategy for energy management of battery storage in grid-connected microgrids. *Energies* **2023**, *16*, 1334. [CrossRef]
86. Wang, Y.; Zhao, K.; Hao, Y.; Yao, Y. Short-term wind power prediction using a novel model based on butterfly optimization algorithm-variational mode decomposition-long short-term memory. *Appl. Energy* **2024**, *366*, 123313. [CrossRef]
87. Xue, Y.; Yin, J.; Hou, X. Short-term wind power prediction based on multi-feature domain learning. *Energies* **2024**, *17*, 3313. [CrossRef]
88. Louka, P.; Galanis, G.; Siebert, N.; Kariniotakis, G.; Katsafados, P.; Pytharoulis, I.; Kallos, G. Improvements in wind speed forecasts for wind power prediction purposes using Kalman filtering. *J. Wind. Eng. Ind. Aerodyn.* **2008**, *96*, 2348–2362. [CrossRef]
89. Cutler, N.; Kay, M.; Jacka, K.; Nielsen, T. Detecting, categorizing and forecasting large romps in wind farm power output using meteorological observations and WPPT. *Wind Energy* **2007**, *10*, 453–470. [CrossRef]
90. Smith, C.; Koracin, D.; Horvath, K. Day-Ahead Predictability of Complex Terrain Flows for Wind Resource Production: A Case Study of the Washoe Zephyr. *Weather Forecast.* **2014**, *29*, 1343–1355. [CrossRef]
91. Giebel, G.; Landberg, L.; Kariniotakis, G.; Brownsword, R. State-of-the-Art on Methods and Software Tools for Short-Term Prediction of Wind Energy Production. *Fidel. Fernández Bernal* **2006**, *23*, 81–86.
92. Lobo, M.; Sanchez, I. Aggregated wind power prediction methods based on distances between weather forecasting vectors. In Proceedings of the 2009 International Conference on Clean Electrical Power, Capri, Italy, 9–11 June 2009.
93. Wang, X.; Guo, P.; Huang, X. A review of wind power forecasting models. *Energy Procedia* **2011**, *12*, 770–778. [CrossRef]
94. Banakar, H.; Luo, C.; Ooi, B. Impacts of wind power minute-to-minute variations on power system operation. *IEEE Trans. Power Syst.* **2008**, *23*, 150–160. [CrossRef]
95. Pope, K.; Naterer, G.; Dincer, I.; Tsang, E. Power correlation for vertical axis wind turbines with varying geometries. *Int. J. Energy Res.* **2011**, *35*, 423–435. [CrossRef]

96. De Giorgi, M.; Campilongo, S.; Ficarella, A.; Congedo, P. Comparison between wind power prediction models based on wavelet decomposition with Least-Squares Support Vector Machine (LS-SVM) and Artificial Neural Network (ANN). *Energies* **2014**, *7*, 5251–5272. [CrossRef]
97. Pirjan, A.; Carutasu, G.; Petrosanu, D.-M. Designing, developing, and implementing a forecasting method for the produced and consumed electricity in the case of small wind farms situated on quite complex hilly Terrain. *Energies* **2018**, *11*, 2623. [CrossRef]
98. Kariniotakis, G.; Halliday, J.; Brownsword, R.; Marti, I.; Garrett, D. Next generation short-term forecasting of wind power-overview of the ANEMOS project. In Proceedings of the European Wind Energy Conference, EWEC 2006, Athens, Greece, 27 February–2 March 2006.
99. Zhang, J.; Chowdhury, S.; Messac, A. A comprehensive measure of the energy resource: Wind power potential (WPP). *Energy Conv. Manag.* **2014**, *86*, 388–398. [CrossRef]
100. Aggarwal, S.; Gupta, M. Wind Power Forecasting: A review of statistical models-wind power forecasting. *Int. J. Energy* **2013**, *3*, 1–10.
101. Wang, W.; Liu, J.; Wu, H.; Liang, T.; Zhang, J. Investigation of application and maintenance of wind power forecasting system. *Technol. Innov. Appl.* **2013**, *14*, 55–57.
102. Roberts, N. Assessing the spatial and temporal variation in the skill of precipitation forecasts from an NWP model. *Meteorol. Appl.* **2008**, *15*, 163–169. [CrossRef]
103. Shan, Y.; Fu, Q.; Geng, X. Combined forecasting of photovoltaic power generation in microgrid based on the improved BP-SVM-ELM and SOM-LSF with particlization. *Proc. CSEE* **2016**, *36*, 3334–3342.
104. Wai, R.-J.; Lai, P.-X. Design of Intelligent Solar PV power generation forecasting mechanism combined with weather information under lack of Real-Time power generation data. *Energies* **2022**, *15*, 3838. [CrossRef]
105. Voyant, C.; Muselli, M.; Paoli, C.; Nivet, M.-L. Numerical weather prediction (NWP) and hybrid ARMA/ANN model to predict global radiation. *Energy* **2012**, *39*, 341–355. [CrossRef]
106. Wang, C.; Li, M.; Cao, Y.; Lu, T. Gradient boosting dendritic network for ultra-short-term PV power prediction. *Front. Energy* **2024**, *18*, 785–798. [CrossRef]
107. Yao, Z.; Pan, F.; Shen, Y. Short-term prediction of photovoltaic power generation output based on GA-BP and POS-BP neural network. *Power Syst. Prot. Control* **2015**, *43*, 83–89.
108. Wang, H.; Zhang, N.; Du, E.; Yan, J.; Han, S.; Liu, Y. A comprehensive review for wind, solar, and electrical load forecasting methods. *Glob. Energy Interconnect.* **2022**, *5*, 9–30. [CrossRef]
109. Liu, Z.; Guo, J.; Wang, X.; Wang, Y.; Li, W.; Wang, X.; Fan, Y.; Wang, W. Prediction of long-term photovoltaic power generation in the context of climate change. *Renew. Energy* **2024**, *235*, 121263. [CrossRef]
110. Van der Meer, D.; Widen, J.; Munkhammar, J. Review on probabilistic forecasting of photovoltaic power production and electricity consumption. *Renew. Sustain. Energy Rev.* **2018**, *81*, 1484–1512. [CrossRef]
111. Jsanjari, M.; Bgooi, H. Probabilistic forecast of PV power generation based on higher order Markov Chain. *IEEE Trans. Power Syst.* **2017**, *32*, 2942–2952. [CrossRef]
112. Kodaira, D.; Tsukazaki, K.; Kure, T.; Kondoh, J. Improving forecast reliability for geographically distributed photovoltaic generations. *Energies* **2021**, *14*, 7340. [CrossRef]
113. Li, Y.; Su, Y.; Shu, L. An ARMAX model for forecasting the power output of a grid connected photovoltaic system. *Renew. Energy* **2014**, *66*, 78–89. [CrossRef]
114. Sfetsos, A. A comparison of various forecasting techniques applied to mean hourly wind speed time series. *Renew. Energy* **2000**, *21*, 23–35. [CrossRef]
115. Sun, Z.; Zhao, S.; Zhang, J. Short-term wind power forecasting on multiple scales using VMD decomposition, K-Means clustering and LSTM principal computing. *IEEE Access* **2019**, *7*, 166917–166929. [CrossRef]
116. Ko, M.; Lee, K.; Kim, J.; Hong, C.; Hur, K. Deep concatenated residual network with bidirectional LSTM for one-hour-ahead wind power forecasting. In Proceedings of the 2022 IEEE Power & Energy Society General Meeting (PESGM), Denver, CO, USA, 17–21 July 2022.
117. Sun, Y.; Li, Z.; Yu, X.; Li, B.; Yang, M. Research on ultra-short-term wind power prediction considering source relevance. *IEEE Access* **2020**, *8*, 147703–147710. [CrossRef]
118. Sun, Z.; Zhao, M. Short-term wind power forecasting based on VMD decomposition, ConvLSTM networks and error analysis. *IEEE Access* **2020**, *8*, 134422–134434. [CrossRef]
119. Carrera, B.; Kim, K. Comparative analysis of machine learning techniques in predicting wind power generation: A case study of 2018–2021 data from Guatemala. *Energies* **2024**, *17*, 3158. [CrossRef]

120. Sun, Y.; Zhang, S. A multiscale hybrid wind power prediction model based on least squares support vector regression-regularized extreme learning machine–multi-head attention–bidirectional gated recurrent unit and data decomposition. *Energies* **2024**, *17*, 2923. [CrossRef]
121. Lin, Y.; Dai, B.; Li, Z. An ensemble method for short-term wind power prediction considering error correction strategy. *Appl. Energy* **2022**, *322*, 119475.
122. Habib, M.A.; Hossain, M. Revolutionizing wind power prediction—The future of energy forecasting with advanced deep learning and strategic feature engineering. *Energies* **2024**, *17*, 1215. [CrossRef]
123. Nikulins, A.; Sudars, K.; Edelmers, E.; Namatevs, I.; Ozols, K.; Komasilovs, V.; Zacepins, A.; Kviesis, A.; Reinhardt, A. Deep learning for wind and solar energy forecasting in hydrogen production. *Energies* **2024**, *17*, 1053. [CrossRef]
124. Walczewski, M.; Wöhrle, H. Prediction of electricity generation using onshore wind and solar energy in Germany. *Energies* **2024**, *17*, 844. [CrossRef]
125. Habtemariam, E.; Kekeba, K.; Martínez-Ballesteros, M.; Martínez-Álvarez, F. A bayesian optimization-based LSTM model for wind power forecasting in the adama district, ethiopia. *Energies* **2023**, *16*, 2317. [CrossRef]
126. Zhou, B.; Ma, X.; Luo, Y.; Yang, D. Wind power prediction based on LSTM networks and nonparametric kernel density sstimation. *IEEE Access* **2019**, *7*, 165279–165292. [CrossRef]
127. Rushdi, M.; Yoshida, S.; Watanabe, K.; Ohya, Y.; Ismaiel, A. Deep learning approaches for power prediction in wind-solar tower systems. *Energies* **2024**, *17*, 3630. [CrossRef]
128. Yu, Y.; Cao, J.; Zhu, J. An LSTM short-term solar irradiance forecasting under complicated weather conditions. *IEEE Access* **2019**, *7*, 145651–145666. [CrossRef]
129. Tastu, J.; Pinson, P.; Kotwa, E.; Madsen, H.; Nielsen, H. Spatio-temporal analysis and modeling of short-term wind power forecast errors. *Wind Energy* **2011**, *14*, 43–60. [CrossRef]
130. Hossain, M.; Mahmood, H. Short-term photovoltaic power forecasting using an LSTM neural network and synthetic weather forecast. *IEEE Access* **2020**, *8*, 172524–172533. [CrossRef]
131. Wang, C.; Liu, Z.; Wei, H.; Chen, L.; Zhang, H. Hybrid deep learning model for short-term wind speed forecasting based on time series decomposition and gated recurrent unit. *Complex Syst. Model. Simul.* **2021**, *1*, 14. [CrossRef]
132. Mujeeb, S.; Alghamdi, T.; Ullah, S.; Fatima, A.; Saba, T. Exploiting deep learning for wind power forecasting based on big data analytics. *Appl. Sci.* **2019**, *9*, 4417. [CrossRef]
133. Massaoudi, M.; Chihi, I.; Sidhom, L.; Trabelsi, M.; Oueslati, F. An effective hybrid NARX-LSTM model for point and interval PV power forecasting. *IEEE Access* **2021**, *9*, 36571–36588. [CrossRef]
134. Cheng, H.-Y.; Yu, C.-C. Solar power generation forecast using multivariate convolution gated recurrent unit network. *Energies* **2024**, *17*, 3073. [CrossRef]
135. Aksan, F.; Suresh, V.; Janik, P. Optimal capacity and charging scheduling of battery storage through forecasting of photovoltaic power production and electric vehicle charging demand with deep learning models. *Energies* **2024**, *17*, 2718. [CrossRef]
136. Medina, S.; Ajenjo, U. Performance improvement of artificial neural network model in short-term forecasting of wind farm power output. *J. Mod. Power Syst. Clean Energy* **2020**, *8*, 484–490. [CrossRef]
137. Wang, S.; Li, M.; Zhao, L.; Jin, C. Short-term wind power prediction based on improved small-world neural network. *Neural Comput. Appl.* **2018**, *31*, 3173–3185. [CrossRef]
138. Buhan, S.; Cadirci, I. Multistage wind-electric power forecast by using a combination of advanced statistical methods. *IEEE Trans. Ind. Inform.* **2015**, *11*, 1231–1242. [CrossRef]
139. Couto, A.; Estanqueiro, A. Wind power plants hybridised with solar power: A generation forecast perspective. *J. Clean. Prod.* **2023**, *423*, 138791–138793. [CrossRef]
140. Shi, J.; Ding, Z.; Lee, W.; Yang, Y. Hybrid forecasting model for very-short term wind power forecasting based on grey relational analysis and wind speed distribution features. *IEEE Trans. Smart Grid* **2014**, *5*, 521–526. [CrossRef]
141. Zhang, C.; Wei, H.; Xie, L.; Shen, Y. Direct interval forecasting of wind speed using radial basis function neural networks in a multi-objective optimization framework. *Neuruting* **2016**, *205*, 53–63. [CrossRef]
142. Wu, W.; Peng, M. A data mining approach combining K-Means clustering with bagging neural network for short-term wind power forecasting. *IEEE Internet Things J.* **2017**, *4*, 979–986. [CrossRef]
143. Liang, G.; Su, Y.; Chen, F.; Long, H.; Gan, Y. Wind power curve data cleaning by image thresholding based on class uncertainty and shape dissimilarity. *IEEE Trans. Sustain. Energy* **2020**, *12*, 1383–1393. [CrossRef]
144. Zhou, B.; Duan, H.; Wu, Q.; Wang, H.; Meng, Y. Short-term prediction of wind power and its ramp events based on semi-supervised generative adversarial network. *Int. J. Electr. Power Energy Syst.* **2021**, *125*, 106411. [CrossRef]
145. Tang, J.; Heng, J.; Deng, Y.; Hu, J.; Luo, Q. Conformalized temporal convolutional quantile regression networks for wind power interval forecasting. *Energy* **2022**, *248*, 214.

146. Yang, D.; Guo, J.; Li, Y.; Sun, S.; Wang, S.; Lund, H.; Kaiser, M. Short-term load forecasting with an improved dynamic decomposition-reconstruction-ensemble approach. *Energy* **2023**, *263*, 125609. [CrossRef]
147. Si, Z.; Yang, M.; Yu, Y.; Ding, T. Photovoltaic power forecast based on satellite images considering effects of solar position. *Appl. Energy* **2021**, *302*, 117514. [CrossRef]
148. Liao, S.; Tian, X.; Liu, B.; Liu, T.; Su, H.; Zhou, B. Short-term wind power prediction based on LightGBM and meteorological reanalysis. *Energies* **2022**, *15*, 6287. [CrossRef]
149. Yang, X.; Wang, S.; Meng, L. Short-term photovoltaic power prediction with similar-day integrated by BP-AdaBoost based on the Grey-Markov model. *Electr. Power Syst. Res.* **2023**, *215 Pt A*, 108961–108966. [CrossRef]
150. Zhang, Z.; Deng, A.; Wang, Z.; Li, J.; Zhao, H.; Yang, X. Wind power prediction based on EMD-KPCA-BiLSTM-ATT model. *Energies* **2024**, *17*, 2568. [CrossRef]
151. Famoso, F.; Oliveri, L.; Brusca, S.; Chiacchio, F. A dependability neural network approach for short-term production estimation of a wind power plant. *Energies* **2024**, *17*, 1627. [CrossRef]
152. Zhou, W.; Xin, M.; Wang, Y.; Yang, C.; Liu, S.; Zhang, R.; Liu, X.; Zhou, L. An ultra-short-term wind power prediction method based on CNN-LSTM. In Proceedings of the 2024 IEEE 7th Advanced Information Technology, Electronic and Automation Control Conference (IAEAC), Chongqing, China, 15–17 March 2024.
153. Sharma, S.; Majumdar, A.; Arregui, V.; Chouzenoux, E. Blind kalman filtering for short-term load forecasting. *IEEE Trans. Power Syst.* **2020**, *35*, 4916–4919. [CrossRef]
154. Jiang, H.; Zhang, Y.; Muljadi, E.; Zhang, J.; Gao, W. A short-term and high-resolution distribution system load forecasting approach using support vector regression with hybrid parameters optimization. *IEEE Trans. Smart Grid* **2017**, *9*, 3341–3350. [CrossRef]
155. Chitsaz, H.; Shaker, H.; Zareipour, H.; Wood, D.; Amjady, N. Short-term electricity load forecasting of buildings in microgrids. *Energy Build.* **2015**, *99*, 56. [CrossRef]
156. Chen, K.; Chen, K.; Wang, Q.; He, Z.; Hu, J.; He, J. Short-term load forecasting with deep residual networks. *IEEE Trans. Smart Grid* **2019**, *10*, 3943–3952. [CrossRef]
157. Talaat, M.; Farahat, M.; Mansour, N.; Hatata, A. Load forecasting based on grasshopper optimization and a multilayer feed-forward neural network using regressive approach. *Energy* **2020**, *196*, 117081–117087. [CrossRef]
158. Spyridon, C.; Athanasios, N.; Petros, I.; Konstantinos, A.; Nikos, N.; Panagiotis, G.; Christos, Y.; Ioannis, A.; Emmanuel, K. Smart energy management algorithm for load smoothing and peak shaving based on load forecasting of an island's power system. *Appl. Energy* **2019**, *238*, 627–642.
159. Cai, Q.; Yan, B.; Su, B.; Liu, S.; Xiang, M.; Wen, Y.; Cheng, Y.; Feng, N. Short-term load forecasting method based on deep neural network with sample weights. *Int. Trans. Electr. Energy Syst.* **2020**, *30*, e12340. [CrossRef]
160. Tan, Z.; De, G.; Li, M.; Lin, H.; Tan, Q. Combined electricity-heat-cooling-gas load forecasting model for integrated energy system based on multi-task learning and least square support vector machine. *J. Clean. Prod.* **2019**, *248*, 119252. [CrossRef]
161. Wu, D.; Wang, B.; Precup, D.; Boulet, B. Multiple kernel learning-based transfer regression for electric load forecasting. *IEEE Trans. Smart Grid* **2019**, *11*, 1183–1192. [CrossRef]
162. Jian, L.; Tao, H.; Meng, Y. Real-time anomaly detection for very short-term load forecasting. *J. Mod. Power Syst. Clean Energy* **2018**, *6*, 235–243.
163. Lin, J.; Ma, J.; Zhu, J.; Cui, Y. Short-term load forecasting based on LSTM networks considering attention mechanism. *Int. J. Electr. Power Energy Syst.* **2022**, *137*, 107818. [CrossRef]
164. Wang, Y.; Chen, Q.; Gan, D.; Yang, J.; Kirschen, D.; Kang, C. Deep larning-based socio-demographic information identification from smart meter data. *IEEE Trans. Smart Grid* **2018**, *10*, 2593–2602. [CrossRef]
165. Zhang, B.; Wu, J.; Chang, P. A multiple time series-based recurrent neural network for short-term load forecasting. *Soft Comput.* **2017**, *22*, 4099–4112. [CrossRef]
166. Kwon, B.; Park, R.; Song, K. Short-term load forecasting based on deep neural networks using LSTM layer. *J. Electr. Eng. Technol.* **2020**, *15*, 1501–1509. [CrossRef]
167. Kiruthiga, D.; Manikandan, V. Levy flight-particle swarm optimization-assisted BiLSTM+dropout deep learning model for short-term load forecasting. *Neural Comput. Appl.* **2023**, *35*, 2679–2700. [CrossRef]
168. Xue, P.; Jiang, Y.; Zhou, Z.; Chen, X.; Fang, X.; Liu, J. Multi-step ahead forecasting of heat load in district heating systems using machine learning algorithms. *Energy* **2019**, *1*, 116085. [CrossRef]
169. Xu, C.; Li, C.; Zhou, X. Interpretable LSTM based on mixture attention mechanism for multi-step residential load forecasting. *Electronics* **2022**, *11*, 2189. [CrossRef]
170. Zhang, L.; Shi, J.; Wang, L.; Xu, C. Electricity, heat, and gas load forecasting based on deep multitask learning in industrial-park integrated energy system. *Entropy* **2020**, *22*, 1355. [CrossRef]

171. Cao, Z.; Wan, C.; Zhang, Z.; Li, F.; Song, Y. Hybrid ensemble deep learning for deterministic and probabilistic low-voltage load forecasting. *IEEE Trans. Power Syst.* **2019**, *35*, 1881–1897. [CrossRef]
172. Massaoudi, M.; Refaat, S.; Chihi, I.; Trabelsi, M.; Oueslati, F.; Aburub, H.; Lund, H.; Kaiser, M. A novel stacked generalization ensemble-based hybrid LGBM-XGB-MLP model for Short-Term Load Forecasting. *Energy* **2021**, *214*, 118874. [CrossRef]
173. Cai, M.; Pipattanasomporn, M.; Rahman, S. Day-ahead building-level load forecasts using deep learning vs. traditional time-series techniques. *Appl. Energy* **2019**, *236*, 1078–1088. [CrossRef]
174. Aouad, M.; Hajj, H.; Shaban, K.; Jabr, R.; El-Hajj, W. A CNN-Sequence-to-Sequence network with attention for residential short-term load forecasting. *Electr. Power Syst. Res.* **2022**, *211*, 108152. [CrossRef]
175. Zhang, G.; Bai, X.; Wang, Y. Short-time multi-energy load forecasting method based on CNN-Seq2Seq model with attention mechanism. *Mach. Learn. Appl.* **2021**, *5*, 100064. [CrossRef]
176. Zheng, J.; Zhang, L.; Chen, J.; Wu, G.; Ni, S.; Hu, Z.; Weng, C.; Chen, Z. Multiple-load forecasting for integrated energy system based on Copula-DBiLSTM. *Energies* **2021**, *14*, 2188. [CrossRef]
177. Wang, X.; Wang, S.; Zhao, Q.; Wang, S.; Fu, L. A multi-energy load prediction model based on deep multi-task learning and ensemble approach for regional integrated energy systems. *Int. J. Electr. Power Energy Syst.* **2021**, *126*, 106583.
178. Imani, M. Electrical load-temperature CNN for residential load forecasting. *Energy* **2021**, *227*, 120480. [CrossRef]
179. Alfieri, L.; Falco, P. Wavelet-based decompositions in probabilistic load forecasting. *IEEE Trans. Smart Grid* **2019**, *11*, 1367–1376. [CrossRef]
180. Kassa, Y.; Zhang, J.; Zheng, D. EMD-PSO-ANFIS based hybrid approach for short-term load forecasting in microgrids. *IET Gener. Transm. Distrib.* **2019**, *14*, 470–475.
181. Meng, Z.; Xie, Y.; Sun, J. Short-term load forecasting using neural attention model based on EMD. *Electr. Eng.* **2022**, *104*, 1857–1866. [CrossRef]
182. Jin, Y.; Guo, H.; Wang, J.; Song, A. A hybrid system based on LSTM for short-term power load forecasting. *Energies* **2020**, *13*, 6241. [CrossRef]
183. Edris, A.; Adapa, R.; Baker, M.; Bohmann, L.; Wood, R. Proposed terms and definitions for flexible AC transmission system (FACTS). *IEEE Trans. Power Deliv.* **1997**, *12*, 1848–1853.
184. Morren, J.; Haan, S.; Ferreira, J. Contribution of DG units to primary frequency control. *Eur. Trans. Electr. Power* **2006**, *16*, 507–521. [CrossRef]
185. Beck, H.; Hesse, R. Virtual synchronous machine. In Proceedings of the IEEE 2007 European Transactions on Electrical Power Electrical Power Quality and Utilisation, Barcelona, Spain, 9–11 October 2007.
186. Visscher, K.; Haan, S. Virtual synchronous machines (VSG'S) for frequency stabilisation in future grids with a significant share of decentralized generation. In Proceedings of the IET-CIRED, CIRED Seminar SmartGrids for Distribution, Frankfurt, Germany, 23–24 June 2008.
187. Zhong, Q.; Weiss, G. Synchronverters: Inverters that mimic synchronous generators. *IEEE Trans. Ind. Electron.* **2011**, *58*, 1259–1267. [CrossRef]
188. D'arco, S.; Suul, J.A.; Fosso, O.B. Control system tuning and stability analysis of Virtual Synchronous Machines. In Proceedings of the 2013 IEEE Energy Conversion Congress and Exposition, Denver, CO, USA, 15–19 September 2013.
189. Singh, S.; Singh, R.; Ashfaq, H.; Sharma, S. Virtual synchronous machine using ant colony optimization in inverter interfaced distributed generation (IIDG). *J. Electr. Eng. Technol.* **2023**, *18*, 167–179. [CrossRef]
190. Orihara, D.; Taoka, H.; Kikusato, H.; Hashimoto, J.; Otani, K.; Takamatsu, T.; Oozeki, T.; Matsuura, T.; Miyazaki, S.; Hamada, H. Internal induced voltage modification for current limitation in virtual synchronous machine. *Energies* **2022**, *15*, 901. [CrossRef]
191. Dai, Y.; Zhang, L.; Chen, Q.; Zhou, K.; Hua, T. Multi-VSG-based frequency regulation for uninterruptible power AC micro-grid with distributed electric vehicles. *Int. J. Electr. Power Energy Syst.* **2022**, *137*, 107785. [CrossRef]
192. Choopani, M.; Hosseinain, S.; Vahidi, B. A novel comprehensive method to enhance stability of multi-VSG grids. *Int. J. Electr. Power Energy Syst.* **2019**, *104*, 502–514. [CrossRef]
193. Hou, X.; Sun, Y.; Zhang, X.; Lu, J.; Wang, P.; Mguerrero, J. Improvement of frequency regulation in VSG-based AC microgrid via adaptive virtual inertia. *IEEE Trans. Power Electron.* **2020**, *35*, 1589–1602. [CrossRef]
194. Xi, J.; Geng, H.; Yang, G.; Ma, S. Inertial response analysis of PMSG-based WECS with VSG control. *J. Eng.* **2017**, *2017*, 897–901. [CrossRef]
195. Shintai, T.; Miura, Y.; Ise, T. Oscillation damping of a distributed generator using a virtual synchronous generator. *IEEE Trans. Power Deliv.* **2014**, *29*, 668–676. [CrossRef]
196. Wang, F.; Zhang, L.; Feng, X.; Guo, H. An adaptive control strategy for virtual synchronous generator. *IEEE Trans. Ind. Appl.* **2018**, *54*, 5124–5133. [CrossRef]

197. Zhou, L.; Liu, S.; Chen, Y.; Yi, W.; Wang, S.; Zhou, X.; Wu, W.; Zhou, J.; Xiao, C.; Liu, A. Harmonic current and inrush fault current coordinated suppression method for VSG under non-ideal grid condition. *IEEE Trans. Power Electron.* **2021**, *36*, 1030–1042. [CrossRef]
198. Liu, J.; Miura, Y.; Bevrani, H.; Ise, T. Enhanced virtual synchronous generator control for parallel inverters in microgrids. *IEEE Trans. Smart Grid* **2017**, *8*, 2268–2277. [CrossRef]
199. Wu, H.; Wang, X. A Mode-Adaptive Power-angle control method for transient stability enhancement of virtual synchronous generators. *IEEE J. Emerg. Sel. Top. Power Electron.* **2020**, *8*, 1034–1049. [CrossRef]
200. Rchowdhury, V.; Divan, D. Lyapunov energy function based direct power control of synchronverters under unbalanced grid voltage conditions. In Proceedings of the 2021 IEEE Energy Conversion Congress and Exposition (ECCE), Vancouver, BC, Canada, 10–14 October 2021.
201. Cai, Y.; Qi, D. A distributed VSG control method for a battery energy storage system with a cascaded H-bridge in a grid-connected mode. *Glob. Energy Interconnect.* **2022**, *5*, 343–352. [CrossRef]
202. Wang, S.; Hu, J.; Yuan, X. Virtual synchronous control for grid-connected DFIG-based wind turbines. *IEEE J. Emerg. Sel. Top. Power Electron.* **2015**, *3*, 932–944. [CrossRef]
203. Meng, J.; Sun, Y.; Wang, B. Cooperative adaptive inertial control for PV and energy storage units with multiple constraints. *IET Renew. Power Gener.* **2022**, *16*, 2076–2087. [CrossRef]
204. Liu, K.; Qian, T.; Zhang, W.; Wu, S.; Han, R.; Hou, K.; Wu, Z.; Hu, Q. Review on small-signal stability of multiple virtual synchronous generators. *Renew. Sustain. Energy Rev.* **2025**, *215*, 115543. [CrossRef]
205. Wang, X.; Wen, X.; Wang, S.; Zhao, X.; Niu, C. Stability analysis of multi-parallel inverters with different control strategies based on global admittance. *Electr. Power Syst. Res.* **2025**, *241*, 111373. [CrossRef]
206. Long, B.; Yang, W.; Zhu, S.; Cao, T.; Guerrero, J.; Rodriguez, J.; Zang, X. Power-frequency admittance model of multi-VSGs grid-connected system considering power coupling. *Int. J. Electr. Power Energy Syst.* **2024**, *155*, 109513. [CrossRef]
207. Lin, J.; Liu, S.; Tian, M.; Huang, M.; Wang, G. Power oscillation suppression of multi-VSG based on both consensus and model predictive control. *Int. J. Electr. Power Energy Syst.* **2024**, *155*, 109459. [CrossRef]
208. Aaradwan, A.; Arimohamed, Y. Linear active stabilization of converter-dominated DC microgrids. *IEEE Trans. Smart Grid* **2012**, *1*, 203–216.
209. Ahmadi, S.; Sadeghi, D.; Marzband, M.; Abusorrah, A.; Sedraoui, K. Decentralized bi-level stochastic optimization approach for multi-agent multi-energy networked micro-grids with multi-energy storage technologies. *Energy* **2022**, *245*, 123223. [CrossRef]
210. Hu, C.; Chen, K.; Luo, S.; Zhou, B.; Ding, L. Small signal modeling and stability analysis of virtual synchronous generators. In Proceedings of the 2017 20th International Conference on Electrical Machines and Systems (ICEMS), Sydney, Australia, 11–14 August 2017.
211. Turner, R.; Walton, S.; Duke, R. A Case Study on the Application of the nyquist stability criterion as applied to interconnected loads and sources on grids. *IEEE Trans. Ind. Electron.* **2013**, *60*, 2740–2749. [CrossRef]
212. Wang, W.; Zhang, X.; Xiao, F. Effect of grid-connected solid oxide fuel cell power generation on power systems small-signal stability. *IET Renew. Power Gener.* **2012**, *6*, 24–37.
213. Rodríguez-cabero, A.; Roldán-pérez, J.; Prodanovic, M. Virtual impedance design considerations for virtual synchronous machines in weak grids. *IEEE J. Emerg. Sel. Top. Power Electron.* **2020**, *8*, 1477–1489. [CrossRef]
214. Wu, H.; Ruan, X.; Yang, D.; Chen, X.; Zhao, W.; Lv, Z.; Zhong, Q. Small-signal modeling and parameters design for virtual synchronous generators. *IEEE Trans. Ind. Electron.* **2016**, *63*, 4292–4303. [CrossRef]
215. Chen, J.; O'donnell, T. Parameter constraints for virtual synchronous generator considering stability. *IEEE Trans. Power Syst.* **2019**, *34*, 2479–2481. [CrossRef]
216. D'arco, S.; Suul, J.A.; Fosso, O.B. Small-signal modelling and parametric sensitivity of a Virtual Synchronous Machine. In Proceedings of the 2014 Power Systems Computation Conference, Wroclaw, Poland, 18–22 August 2014.
217. Wang, S.; Liu, Z.; Liu, J. Modeling of D-Q small-signal impedance of Virtual Synchronous Generator. In Proceedings of the 2018 IEEE International Power Electronics and Application Conference and Exposition (PEAC), Shenzhen, China, 4–7 November 2018.
218. Li, W.; Wei, J.; Shi, J.; Li, F. Full-band output impedance model of Virtual Synchronous Generator in dq framework. In Proceedings of the 2018 International Power Electronics Conference (IPEC-Niigata 2018-ECCE Asia), Niigata, Japan, 20–24 May 2018.
219. Chen, M.; Zhou, D.; Blaabjerg, F. Active power oscillation damping based on acceleration control in paralleled Virtual Synchronous Generators system. *IEEE Trans. Power Electron.* **2021**, *36*, 9501–9510. [CrossRef]
220. Jia, Y.; Xiao, J.; Jia, B.; Wang, Y.; Li, Z.; Pan, Y. Adaptive time constant control strategy for Micro-Grid based on improved droop control. In Proceedings of the 2020 IEEE/IAS Industrial and Commercial Power System Asia (I&CPS Asia), Weihai, China, 13–15 July 2020.
221. Xiong, X.; Wu, C.; Blaabjerg, F. An improved synchronization stability method of Virtual Synchronous Generators based on frequency feedforward on reactive power control loop. *IEEE Trans. Power Electron.* **2021**, *36*, 9136–9148. [CrossRef]

222. Lou, G.; Yang, Q.; Gu, W.; Quan, X.; Mguerrero, J.; Li, S. Analysis and design of hybrid harmonic suppression scheme for VSG considering nonlinear loads and distorted grid. *IEEE Trans. Energy Convers.* **2021**, *36*, 3096–3107. [CrossRef]
223. D'Arco, S.; Suul, J.; Fosso, O. Small-signal modeling and parametric sensitivity of a virtual synchronous machine in islanded operation. *Int. J. Electr. Power Energy Syst.* **2015**, *72*, 3–15. [CrossRef]
224. D'Arco, S.; Suul, J.; Fosso, O. A virtual synchronous machine implementation for distributed control of power converters in Smart Grids. *Electr. Power Syst. Res.* **2015**, *122*, 180–197. [CrossRef]
225. Wu, H.; Wang, X. Design-oriented transient stability analysis of Grid-Connected converters with power synchronization control. *IEEE Trans. Ind. Electron.* **2019**, *66*, 6473–6482. [CrossRef]
226. Choopani, M.; Hhosseinian, S.; Vahidi, B. New transient stability and LVRT improvement of Multi-VSG grids using the frequency of the center of inertia. *IEEE Trans. Power Syst.* **2020**, *35*, 527–538. [CrossRef]
227. Alipoor, J.; Miura, Y.; Ise, T. Stability Assessment and optimization methods for microgrid with multiple VSG units. *IEEE Trans. Smart Grid* **2018**, *9*, 1462–1471. [CrossRef]
228. Cheng, H.; Shuai, Z.; Shen, C.; Liu, X.; Li, Z.; Jshen, Z. Transient angle stability of paralleled synchronous and virtual synchronous generators in islanded microgrids. *IEEE Trans. Power Electron.* **2020**, *35*, 8751–8765. [CrossRef]
229. Alipoor, J.; Miura, Y.; Ise, T. Power system stabilization using Virtual Synchronous Generator with alternating moment of inertia. *IEEE J. Emerg. Sel. Top. Power Electron.* **2015**, *3*, 451–458. [CrossRef]
230. Pan, D.; Wang, X.; Liu, F.; Shi, R. Transient stability of voltage-source converters with Grid-Forming control: A design-oriented study. *IEEE J. Emerg. Sel. Top. Power Electron.* **2020**, *8*, 1019–1033. [CrossRef]
231. Vetoshkin, L.; Müller, Z. Dynamic Stability improvement of power system by means of STATCOM with virtual inertia. *IEEE Access* **2021**, *9*, 116105–116114. [CrossRef]
232. Yazdani, S.; Ferdowsi, M. Robust backstepping control of synchronverters under unbalanced grid condition. In Proceedings of the 2019 IEEE Power and Energy Conference at Illinois (PECI), Champaign, IL, USA, 28 February–1 March 2019.
233. Ding, X.; Cao, J. Deep and reinforcement learning in Virtual Synchronous Generator: A Comprehensive Review. *Energies* **2024**, *17*, 2620. [CrossRef]
234. Zhang, X.; Liu, H.; Cao, Y.; Fu, Y.; Huang, Y.; Li, Y. Virtual shaft control of Virtual Synchronous Generator with nonlinear stiffness for power oscillation suppression. *IEEE Trans. Power Deliv.* **2025**, *40*, 497–508. [CrossRef]
235. Citro, C.; Al-Numay, M.; Siano, P. Extensive assessment of virtual synchronous generators in intentional island mode. *Int. J. Electr. Power Energy Syst.* **2024**, *157*, 109853. [CrossRef]
236. Blekos, K.; Brand, D.; Ceschini, A.; Chou, C.-H.; Li, R.-H.; Pandya, K.; Summer, A. A review on Quantum approximate optimization algorithm and its variants. *Phys. Rep.* **2024**, *1068*, 1–66. [CrossRef]
237. Dui, H.; Zhang, S.; Dong, X.; Wu, S. Digital twin-enhanced opportunistic maintenance of smart microgrids based on the risk importance measure. *Reliab. Eng. Syst. Saf.* **2025**, *253*, 110548. [CrossRef]
238. Shi, J.; Liu, N.; Huang, Y.; Ma, L. An Edge Computing-oriented net power forecasting for PV-assisted charging station: Model complexity and forecasting accuracy trade-off. *Appl. Energy* **2022**, *310*, 118456. [CrossRef]
239. Rajaguru, V.; Annapoorani, K. Virtual synchronous generator based superconducting magnetic energy storage unit for load frequency control of micro-grid using African vulture optimization algorithm. *J. Energy Storage* **2023**, *65*, 107343. [CrossRef]

energies

MDPI

Article

Multi-Level Asynchronous Robust State Estimation for Distribution Networks Considering Communication Delays

Xianglong Zhang *, Ying Liu, Songlin Gu, Yuzhou Tian and Yifan Gao

State Grid Economic and Technological Research Institute Co., Ltd., Beijing 102209, China
* Correspondence: xianglong1983@126.com

Abstract: With the hierarchical evolution of distribution network control architectures, distributed state estimation has become a focal point of research. To address communication delays arising from inter-level data exchanges, this paper proposes a multi-level, asynchronous, robust state estimation algorithm that accounts for such delays. First, a multi-level state estimation model is formulated based on the concept of a maximum normal measurement rate, and a hierarchical decoupling modeling approach is developed. Then, an event-driven broadcast transmission strategy is designed to unify boundary information exchanged between levels during iteration. A multi-threaded parallel framework is constructed to decouple receiving, computation, and transmission tasks, thereby enhancing asynchronous scheduling capabilities across threads. Additionally, a round-based synchronization mechanism is proposed to enforce fully synchronized iterations in the initial stages, thereby improving the overall process of asynchronous state estimation. Case study results demonstrate that the proposed algorithm achieves high estimation accuracy and strong robustness, while reducing the average number of iterations by nearly 40% and shortening the runtime by approximately 35% compared to conventional asynchronous methods, exhibiting superior estimation performance and computational efficiency under communication delay conditions.

Keywords: communication delay; multi-level decoupling in distribution networks; distributed state estimation; asynchronous communication

1. Introduction

State estimation, as a critical supporting technology for monitoring and control in distribution network operations, analyzes system states by integrating various types of measurement data [1–5], thereby providing auxiliary support for core functions such as fault diagnosis [6], dispatch decision making [7], and energy management [8]. With the increasing integration of distributed energy resources and intelligent edge devices, the operational scale of distribution networks continues to expand, and the control architecture is evolving toward a hierarchical and collaborative framework [9,10]. In this context, traditional centralized state estimation methods are gradually revealing limitations in terms of data exchange, communication overhead, and computational efficiency [11,12].

In this context, state estimation algorithms are progressively evolving toward distributed architectures. Distributed state estimation features structural decoupling, parallel computation, and hierarchical coordination, making it well aligned with the needs of distribution networks characterized by wide geographical coverage and multi-level control structures. It has thus emerged as a key research focus and development direction in the field [13–16]. Reference [17] proposes a multi-area state estimation model using a generalized loss function, where matrix splitting is applied to achieve regional decomposition. Reference [18] constructs a multi-level estimation framework at the area level based on asynchronous measurements, enabling distributed state estimation via encrypted communication. These distributed estimation approaches are essentially built upon same-layer partitioned architectures, where global state estimation is achieved by parallel estimation of subregions. However, they do not address the issue of coordinated estimation in multi-level hierarchical structures.

To address the hierarchical structure and multi-level control characteristics of power system operations, several studies have proposed hierarchical state estimation frameworks featuring top-down coordination mechanisms to enable cross-level state fusion and coordinated estimation [19,20]. Existing efforts have primarily focused on inter-level modeling and coordination mechanisms between transmission and distribution systems. Reference [21] proposes a master–slave state estimation model that partitions the transmission and distribution systems into master and slave regions, where cross-level state fusion is achieved through coordinated variables. Reference [22] develops a multi-level estimation procedure based on hierarchical topology, wherein state estimation tasks are executed in parallel across different control levels. At the distribution network level, some preliminary investigations into multi-level state estimation strategies have also emerged. Reference [23] designs a hierarchical state estimation method for radial distribution networks, integrating SCADA and AMI data via regional decoupling. Reference [24] proposes a two-layer distributed state estimation approach that combines local estimation with boundary coordination and fusion. However, these methods essentially belong to intra-layer multi-stage estimation schemes, in which state estimation is first performed within individual subregions, followed by secondary estimation on tie-lines based on boundary states to complete the overall coordination [25,26]. A truly hierarchical state estimation framework, comparable to those used in transmission–distribution systems and capable of representing multi-level structures and enabling coordinated solving, has yet to be established.

On the other hand, communication conditions are a critical factor influencing the performance of state estimation, particularly in multi-level distributed architectures of distribution networks. Variations in communication paths and data update cycles across different levels often lead to asymmetric delays and response misalignments, which in turn affect the convergence speed and stability of state estimation algorithms [27,28]. To mitigate the dependence on synchronous communication mechanisms, some studies have introduced asynchronous estimation algorithms. Reference [29] proposes an online asynchronous state estimation algorithm based on the proximal point method. Reference [30] designed a local asynchronous convergence detection mechanism. In addition, several studies have attempted to reduce the impact of communication constraints by simplifying the treatment of communication delays. Reference [31] utilizes the correlations between measurement data to infer and compensate for missing measurements caused by communication interruptions. Reference [32] adopts a hybrid strategy that combines consensus and incremental updates to ensure convergence under heterogeneous communication scenarios. Although these approaches alleviate the adverse effects of communication delays to some extent, they fall short in fully addressing the challenges posed by the accumulated delays and inter-level response mismatches inherent in multi-level distribution network structures, which can ultimately compromise the convergence and stability of state estimation.

Building upon existing research, this paper further conducts a comparative analysis of representative state estimation methods for distribution networks developed in recent years, with a focus on differences in control architecture, communication mechanisms, hierarchical adaptability, and robustness, as summarized in Table 1.

Table 1. Comparative analysis of different state estimation methods.

Method	Control Architecture	Communication Mode	Delay Adaptability	Hierarchical Adaptability
Centralized estimation (e.g., WLS)	Centralized control	Synchronous communication	Weak	Weak
Distributed synchronous estimation (e.g., synchronous ADMM)	Distributed control	Synchronous communication	Weak	Moderate
Distributed asynchronous estimation (e.g., asynchronous ADMM)	Distributed control	Asynchronous communication	Strong	Moderate
Graph neural network-based methods (e.g., GNN)	Data-driven control	Model training	Data-dependent	Weak
Filtering-based methods (e.g., UKF)	Recursive structure	Synchronous communication	Weak	Weak

In summary, although existing state estimation methods have achieved a certain degree of communication adaptability and distributed solving capability, they are mostly limited to same-layer architectures and lack systematic investigation into global state estimation under multi-level frameworks. To address this gap, this paper proposes an asynchronous robust state estimation algorithm that considers communication delays. The main contributions of this paper are as follows:

1. A multi-level state estimation modeling method for distribution networks is proposed based on the concept of a maximum normal measurement rate, along with a decoupling analysis tailored to the structural characteristics of distribution systems.
2. An event-driven broadcast transmission strategy and a synchronization initiation mechanism are designed to enhance the consistency of boundary information and the convergence stability during the initial estimation stage.
3. A multi-threaded parallel mechanism is introduced, decoupling the receive–compute–transmit process, which structurally optimizes the state estimation algorithm under asynchronous conditions and significantly improves the computational efficiency and concurrent processing capability.

The remainder of this paper is organized as follows: Chapter 2 develops a decoupled model for multi-level state estimation in distribution networks and introduces the synchronous solution algorithm; Chapter 3 presents an enhanced asynchronous state estimation algorithm considering communication delays; Chapter 4 conducts comparative case studies to validate the effectiveness and advantages of the proposed method; Chapter 5 concludes the paper.

2. Modeling and Synchronous Solution Algorithm for Multi-Level State Estimation in Distribution Networks

2.1. Maximum Normal Measurement Rate Mathematical Model

The measurement normal rate can be used to reflect the reliability of measurement data, and its value is closely related to the uncertainty parameters of the measured results [33]. Based on the probabilistic characteristics of measurement values, the expanded uncertainty can be further derived to construct a confidence interval model. Specifically, assuming that the true value of a measurement Z_i falls within the interval $[Z_i - V_i, Z_i + V_i]$ with a probability of pp, this confidence interval can statistically represent the credibility of the measurement result. The corresponding expression is:

$$P\left(\left|h_i(\bar{x}) - Z_i\right| \leqslant V_i\right) = p \tag{1}$$

where $\bar{x}$ represents the theoretical system state; $h_i(\bar{x})$ represents the measurement function of point i under state $\bar{x}$; and V_i is the expanded uncertainty of point i corresponding to probability p.

For measurement points i with normal measurements during actual operation, the probability p is typically close to 1. Therefore, measurement points that satisfy Equation (2) can be regarded as normal points:

$$|h_i(\boldsymbol{x}) - Z_i| \leqslant V_i \tag{2}$$

where x represents the system state obtained from state estimation.

To better evaluate the state estimation results at each point, the relative estimation deviation d_i of point i is defined based on Equation (2) as follows:

$$d_i = \frac{h_i(\boldsymbol{x}) - Z_i}{V_i} \tag{3}$$

According to Equation (3), a normal point satisfies $|d_i| \leq 1$, while a point with abnormal measurement satisfies $|d_i| > 1$. Based on this, a point evaluation function is constructed as follows:

$$f(d_i) = \begin{vmatrix} 0 & |d_i| \leqslant 1 \\ 1 & |d_i| > 1 \end{vmatrix} \tag{4}$$

At this point, maximizing the measurement normal rate can be transformed into minimizing the sum of the evaluation functions of all points, thereby formulating a state estimation model with MNMR as the objective function:

$$\begin{cases} \min\limits_{x} \sum\limits_{i=1}^{n} f(d_i) \\ \text{s.t. } d_i = \dfrac{h_i(\boldsymbol{x}) - Z_i}{U_i} \\ g(\boldsymbol{x}) = 0 \\ l(\boldsymbol{x}) \leq 0 \end{cases} \tag{5}$$

where n represents the number of measurement points; $g(x)$ represents the power flow constraints of the system; and $l(x)$ represents the upper and lower bounds of system components, such as the transmission power limits of lines and output limits of generators.

The MNMR model proposed in Equation (5) exhibits strong robustness; regardless of the magnitude of deviation between the measurement data and the state estimation result, its weight in the objective function remains constant at 1. This reduces sensitivity to large deviations and significantly decreases the model's responsiveness to abnormal measurement data, thereby enabling more stable identification of problematic measurements. However, since the point evaluation function is not continuously differentiable, the optimization problem becomes significantly more difficult to solve.

To address this issue, the point evaluation function is approximated using a sigmoid function, which preserves its evaluation capability while ensuring continuous differentiability. Specifically, it is defined as:

$$\begin{cases} f(d_i) = \delta(d_i) + \delta(-d_i) \\ \delta(d_i) = \frac{1}{1+e^{\frac{2\eta}{\lambda-1}[-d_i+(1+\frac{\lambda-1}{2})]}} \end{cases} \tag{6}$$

where λ represents a positive constant and η represents an integer greater than 2.

2.2. Decoupled Model for Multi-Level State Estimation in Distribution Networks

In response to the distinctive operational characteristics of modern distribution networks, characterized by clear multi-level structures and refined control granularity, this paper proposes a generalized decoupled modeling framework for multi-level state estimation. The framework is organized into three representative hierarchical levels: the primary distribution network layer, the substation transformer layer, and the feeder layer, featuring a well-defined hierarchical structure and clear boundary coupling logic. Taking a two-level decoupling model as an example, as illustrated in Figure 1, the upper-level region HD (shown in blue) is connected to multiple lower-level region LDs (shown in yellow), which differ significantly in structure and operational attributes. Since different LD regions may fall under separate management entities, their topological structures, communication protocols, and data standards often vary considerably, making unified global modeling and data sharing difficult to implement directly. To address this, a distributed solution mechanism is introduced, allowing each LD region to perform independent modeling based on its own structural characteristics and measurement configuration, while interacting with the HD layer solely through boundary parameters.

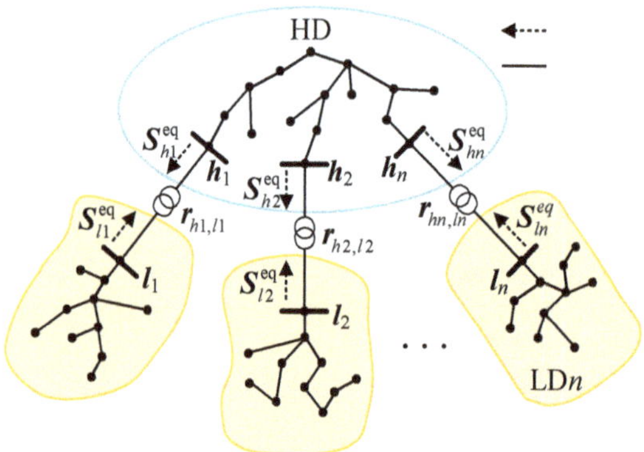

Figure 1. Two-level decoupled model for distribution networks with equivalent measurements.

The tie line between the HD layer and each LD is represented by $r_{hi,li}$, where h_i represents the node on the HD side and l_i represents the root node of LD_i. The voltage magnitude measurement at the root node is denoted as V_{li}. Considering the diverse topological configurations of distribution networks, such as radial and meshed structures, as well as nonlinear operational characteristics like three-phase imbalance, variables related to the tie line, including $r_{hi,li}$, and V_{li}, may be either scalars or vectors. When LD_i is connected to the HD layer via a single tie line, these variables are scalars; when LD_i has multiple tie lines or operates under three-phase unbalanced conditions, they are represented as vectors. In the subsequent analysis, no distinction is made between scalar and vector forms.

By incorporating the root node l_i of each LD into the HD layer, a complete HD region is formed. The state variables of this region include not only the original state variables of HD nodes, but also the state variables of each LD root node. To support this, equivalent pseudo-measurements are introduced, specifically the power flow measurements S_{hi}^{eq} and S_{li}^{eq} on both sides of the tie line connected to the root node of LD$_i$. Similarly, the complete set of state variables for each LD subregion can be obtained. Based on this, the measurements of the complete upper HD region and the complete lower LD subregions can be expressed as:

$$z_{EH} = \begin{bmatrix} z_{HD} & S_l^{eq} & V_l \end{bmatrix}^T \tag{7}$$

$$z_{EL} = \begin{bmatrix} z_{LD} & S_h^{eq} & V_l \end{bmatrix}^T \tag{8}$$

where EH represents the complete upper HD region and EL represents the complete lower LD subregions.

After decoupling, the global state estimation problem (5) can be decomposed into an EH subproblem and multiple EL$_i$ subproblems:

$$\min_{x_{HD}} \sum_{i=1}^{n} f(d_i) = \sum_{i=1}^{n} (\delta(d_i) + \delta(-d_i)) \tag{9}$$

$$\min_{x_{EL,i}} \sum_{k=1}^{n} f(d_k) = \sum_{k=1}^{n} (\delta(d_k) + \delta(-d_k)) \tag{10}$$

It can be observed that the state variables of node l_i, along with the line power flow measurements S_{hi}^{eq} and S_{li}^{eq}, are involved in both subproblems, which naturally leads to an iterative process between the subproblems:

$$\left[S_{h1}^{eq}, \ldots, S_{hi}^{eq}, \ldots, S_{hn}^{eq}; V_{l1}, \ldots, V_{li}, \ldots V_{ln}\right] = \Theta_{EH}\left[S_{l1}^{eq}, \ldots, S_{li}^{eq}, \ldots, S_{ln}^{eq}\right] \tag{11}$$

$$\left[S_{li}^{eq}\right] = \Theta_{EL,i}\left[S_{hi}^{eq}; V_{li}\right] \tag{12}$$

Equation (11) describes the iterative process of the EH subproblem, where Θ_{EH} represents its iterative mapping; Equation (12) describes the iterative process of the EL$_i$ subproblem, with $\Theta_{EL,i}$ denoting its iterative mapping. After each iteration, all state variables and equivalent pseudo-measurements are updated. By combining Equations (11) and (12), we obtain:

$$\left[S_{h,k+1}^{eq}; V_{l,k+1}\right] = \Theta_B\left[S_{h,k}^{eq}; V_{l,k}\right] \tag{13}$$

where $\Theta_B = \Theta_{EH}(\Theta_{EL,1}(\cdot), \Theta_{EL,2}(\cdot), \ldots, \Theta_{EL,n}(\cdot))$.

Equation (13) represents the fixed-point iterative form of multi-level state estimation in distribution networks. Compared with previous studies, the proposed model and iterative formulation are more aligned with engineering practice. In the solution of the two subproblems, the line power flow measurements on both sides of node l_i can be treated as node injection measurements and incorporated into the computation. This establishes a fixed-point iterative form for multi-level state estimation, enabling decoupled computation between the two parts. Each part can be solved using different estimation methods to accommodate the specific characteristics of measurement distribution, network topology, and other regional features.

2.3. Synchronous Algorithm for Multi-Level State Estimation in Distribution Networks

Equation (13) essentially represents the iterative form of the synchronous algorithm. Under the synchronous state estimation framework, the HD layer and each subregion LD_i are required to follow a centralized and strictly time-aligned synchronization mechanism. As shown in Figure 2, during each round of state estimation, the HD layer must synchronously receive boundary parameters from all LD_i subregions and can only proceed with its own state estimation after all data have been received. This synchronization strategy has the following inherent drawbacks: First, due to communication delays, the convergence process of the HD layer is constrained by the data arrival time of the slowest subregion. Second, for LD_i regions with complex topologies or limited computational capacity, the global synchronization cycle is further prolonged, leading to system-level asymmetric delay accumulation. Third, in the presence of data loss or timeout caused by communication delays, the HD layer may face the risk of divergence in state estimation due to incomplete parameters. As a result, traditional synchronous mechanisms severely limit the real-time performance and robustness of the system in multi-level state estimation scenarios.

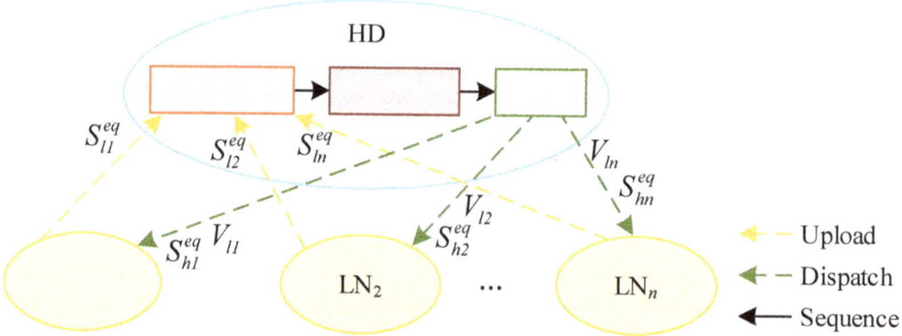

Figure 2. Data flow of the synchronous state estimation algorithm.

3. Asynchronous Algorithm for Multi-Level State Estimation Considering Communication Delays

3.1. Asynchronous Algorithm for Multi-Level State Estimation

The multi-level state estimation problem in distribution networks can be abstracted into the fixed-point iterative form described by Equation (13), which can be solved using an asynchronous iterative approach to improve computational efficiency [34,35]. Given an initial value $\left[S_{h,0}^{\mathrm{eq}}; V_{l,0}\right]$ for asynchronous iteration, the iterative sequence $\left[S_{h,k}^{\mathrm{eq}}; V_{l,k}\right]$ for $k = 0, 1\ldots$ is generated based on the mapping Θ_B, and can be expressed as:

$$\left[S_{hi,k+1}^{\mathrm{eq}}; V_{li,k+1}\right] = \begin{cases} \left[S_{hi,k}^{\mathrm{eq}}; V_{li,k}\right] & i \notin \Phi_k \\ \Theta_B\left[S_{hi,k}^{\mathrm{eq}}; V_{li,k}\right] & i \in \Phi_k \end{cases} \tag{14}$$

where Φ_k represents the set of LD regions whose updated values are successfully received by the HD layer without communication timeout during the (k+1)th iteration.

Equation (14) represents the basic iterative form of the asynchronous algorithm, in which the boundary state variables obtained in the (k+1)th iteration may vary depending on the communication conditions. By decoupling Equation (14), an asynchronous iterative form of the EH subproblem and the EL_i subproblems can be derived:

$$\left[S_{h,k+1}^{\mathrm{eq}}; V_{l,k+1}\right] = \Theta_{\mathrm{EH}}\left[S_{l1,s_1(k)}^{\mathrm{eq}}, \ldots, S_{li,s_i(k)}^{\mathrm{eq}}, \ldots, S_{ln,s_n(k)}^{\mathrm{eq}}\right] \tag{15}$$

$$\left[S^{eq}_{li,s_i(k)}\right] = \Theta_{EL,i}\left[S^{eq}_{hi,s_i(k)}; V_{li,s_i(k)}\right] \tag{16}$$

where $s_i(k)$ represents the iteration index of the return value from the ith LD used by the HD layer in the (k+1)th iteration of the EH subproblem. This index is identical to that of the HD-issued value used in the corresponding computation of the EL_i subproblem.

There are two possible values for $s_i(k)$: In the first case, $s_i(k) = k$, indicating that the LD received the value issued by the EH subproblem in the kth iteration, computed its EL_i subproblem accordingly, and the HD layer successfully received the return value in time. In the second case, $s_i(k) < k$, indicating that during the (k+1)th iteration of the EH subproblem, the HD layer either failed to receive the return value from the LD, or received outdated data.

Compared to Equation (13), the asynchronous iterative form primarily differs in the iteration index of the state variables used in each update, denoted as $s_i(k)$. To improve the convergence speed, it is necessary to incorporate asynchronous-related constraints into the above asynchronous algorithm. When performing the next iteration of the EH subproblem, the HD layer must receive a sufficient number of updated return values from the LD regions. The minimum required number of such updates is denoted by the parameter S. To ensure convergence, the HD layer should also use return values from the LD regions in a timely manner, with the maximum allowable staleness denoted by the parameter τ. Based on these considerations, the iteration of the EH subproblem is subject to the following constraints:

$$\begin{cases} s_i(k) - s_i(k-\tau) > 0 \\ \sum\limits_{i=1}^{n} \operatorname{sgn}(s_i(k) - s_i(k-1)) \geq S \\ s_i(k) \leq k \end{cases} \tag{17}$$

When $S = n$ or $\tau = 1$, the first two constraints in Equation (17) reduce to $s_i(k) - s_i(k-1) > 0$, meaning that updated data from every LD must be received in each successive iteration. In this case, the asynchronous state estimation degenerates into synchronous state estimation.

3.2. Broadcast Transmission Strategy

To address the issue of information synchronization lag inherent in synchronous mechanisms, traditional asynchronous state estimation relaxes the synchronization constraints by allowing the HD layer to initiate an iteration as soon as it receives updated data from at least S lower-level LD_i regions. This approach reduces the dependency of the upper-layer HD estimation process on the slowest LD region. However, in practical scenarios where communication delays and asynchronous triggers coexist, the data reporting frequencies across different LD_i regions often vary significantly. Some regions may have already completed multiple updates and transmitted new data, while others remain stalled at earlier iterations. As a result, the HD layer operates with input data that exhibit pronounced temporal heterogeneity. This heterogeneity introduces an accumulation of outdated information, which degrades the effectiveness of the estimation results and may even compromise the convergence performance of the algorithm.

To overcome this issue, this paper proposes an event-triggered broadcast transmission strategy. After each completion of the EH subproblem, the HD layer actively broadcasts the latest iteration index $\left[S^{eq}_{h,k}; V_{l,k}\right]$ to all LD_i regions. Regardless of whether a subregion has completed its current computation, each LD_i adopts the broadcasted result as the new starting point for subsequent estimation. This approach enables the dynamic calibration of the local state in each subregion.

Accordingly, after applying the broadcast transmission strategy, the second constraint in Equation (17) needs to be modified as follows:

$$\sum_{i=1}^{n} \mathrm{sgn}(s_i(k) - k) \geq n - S \tag{18}$$

The remaining two constraint conditions remain the same as those in Equation (17). The modified constraint ensures that, during each iteration, the HD layer uses the latest S updated values from LD regions with iteration index k. For any LD subregion that has not provided an updated value for τ consecutive iterations, the HD layer must wait for its current update before proceeding to the next computation.

By actively issuing the latest iteration index after each upper-layer EH subproblem is solved, the proposed strategy effectively unifies the boundary state information between the HD layer and all LD_i regions. This ensures that each region performs subsequent estimation based on the most up-to-date boundary inputs, thereby avoiding error accumulation caused by version inconsistencies. As a result, the strategy significantly enhances data consistency and convergence stability in multi-level state estimation. Compared with traditional asynchronous strategies that rely on passive progression triggered by subregion updates, this mechanism introduces an explicit synchronization anchor, which effectively suppresses the risk of divergence caused by asynchronous communication and uneven feedback. Furthermore, the broadcast strategy imposes no strict dependency on the local estimation progress of lower-layer regions, demonstrating strong adaptability to communication constraints and excellent scalability. It is particularly well suited for large-scale distributed systems with complex structures and non-ideal communication conditions.

3.3. Multithreaded Parallel Mechanism

Under the broadcast transmission strategy, the upper-layer region is required to push the latest boundary state to all lower-layer regions during each iteration. This may lead to a surge in communication traffic within a short period, resulting in significant concurrent transmission pressure. Limited by the system thread architecture and processing capacity, such communication peaks may cause issues such as data reception delays and resource scheduling bottlenecks, thereby reducing the overall responsiveness of the estimation process. To mitigate these impacts, a multithreaded parallel mechanism can be employed to achieve structural decoupling between computation and communication tasks [36]. Alternatively, a predictor–corrector mechanism can be introduced to enhance the data update speed and parallel fault tolerance by performing feedforward prediction and corrective iteration on key variables [37]. This paper introduces a multithreaded parallel mechanism within both the HD layer and each LD_i region. By separating the execution channels for reception, computation, and transmission tasks, this mechanism decouples communication and computation processes, thereby enhancing the system's processing capability and execution efficiency under high-concurrency conditions.

As illustrated in Figure 3a, the system deploys independent threads for reception, computation, and transmission. The reception thread continuously monitors and dynamically receives iteration data from other regions. The computation thread extracts the latest cached measurement information in real time and solves the corresponding EH or EL subproblem. Upon completion of the computation, the transmission thread promptly distributes the updated iteration sequence to the target regions. Dashed arrows represent inter-regional communication interactions, while solid arrows indicate the execution sequence of internal processes. These three types of threads operate collaboratively and are mutually decoupled, effectively alleviating resource idleness and execution blockages caused by concurrent broadcasting. This design significantly enhances the stability and execution efficiency of the asynchronous estimation algorithm under complex communication conditions.

Figure 3b illustrates the interaction process between the HD and each LD under asynchronous execution conditions, before and after the introduction of the parallel mechanism. Each rectangular block in the figure represents the execution time of a specific thread during the current iteration, with colors corresponding to the thread types shown in Figure 3a. For example, the yellow block for the HD indicates the operation time of its transmission thread. In the second iteration, due to reception delays at LD_3, the iteration sequence generated by the HD in the first round has not yet been delivered. As a result, the HD's transmission thread continues occupying resources, and under the serial mechanism, conflicts occur with the reception thread in the current round, delaying the start of the HD's next iteration, as marked by "x" in the figure. After introducing the parallel mechanism, the HD's transmission from the previous round no longer blocks the reception in the new round, effectively avoiding resource conflicts. At $k = 2$, since LD_3 experiences delayed reception in the previous round, its new reception thread must wait for the previous transmission thread to finish before starting, which extends the HD's broadcasting time. With the parallel mechanism, however, LD_3 can start its reception thread earlier, reducing the HD's broadcast latency. Moreover, LD_3 completes its transmission earlier than LD_1 in the current round, allowing the HD to enter the next computation cycle earlier under the asynchronous trigger condition $S = 2$, thereby accelerating the overall scheduling progress and improving execution efficiency.

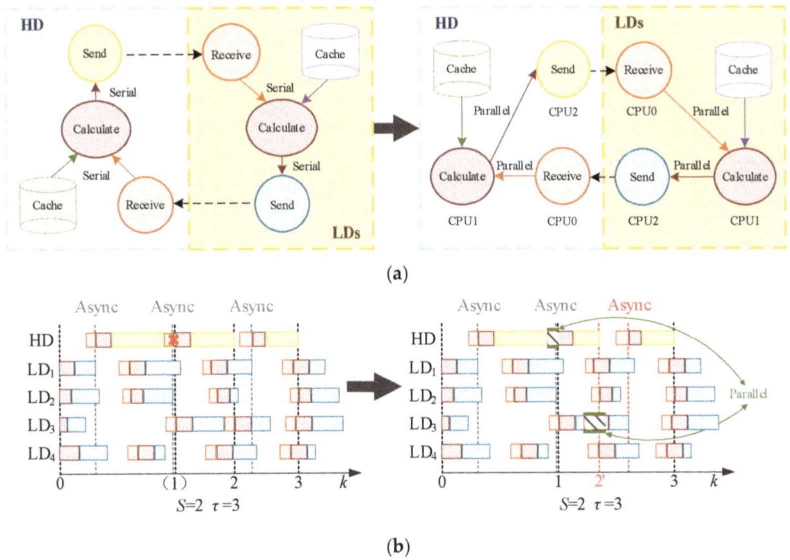

Figure 3. Structural and behavioral evolution of sequential–parallel mechanism. (**a**) Schematic of sequential-to-parallel structural transition. (**b**) Comparison of sequential and parallel execution timelines.

3.4. Synchronous Startup Mechanism

Meanwhile, to address the issue of significant disparities in response speeds among subregions during the initial phase of asynchronous estimation, where synchronizing the timing of boundary information feedback is difficult, the traditional asynchronous mechanism provides improved operational flexibility, but often causes premature computation to be triggered at the HD layer. Due to communication delays or computational lag in certain subregions during the initial iterations, the HD may begin calculations using incomplete or temporally inconsistent data. This can lead to the accumulation of estimation errors and adversely affect the overall convergence stability of the algorithm.

To mitigate the problem of error propagation during initial iterations, this paper proposes a synchronous bootstrapping mechanism. During the first k0k_0 iterations, the HD layer is required to execute the EH subproblem only after receiving the latest boundary states from all LD_i regions. This ensures that the system performs computations based on complete and synchronized data during the initialization phase, thereby enhancing the temporal consistency of the state estimation process.

Accordingly, after adopting the synchronous bootstrapping mechanism, an additional constraint must be introduced to enforce synchronous startup:

$$s_i(k) = \max(k, k_0), \quad i = 1, 2, \ldots, n \tag{19}$$

When $k < k_0$, the HD layer must wait for updated results from all LD_i regions before proceeding to the next iteration. Once the initial synchronous window is exceeded, the system automatically switches to the asynchronous mode, where the EH subproblem can be triggered as long as responses from at least S regions are received.

This mechanism addresses the issue of insufficient data coverage in the early stages of the asynchronous mode by enforcing synchronous execution during the initial iterations. It effectively enhances the stability and reliability of early-stage state estimation, thereby providing favorable initial conditions for the subsequent asynchronous phase.

3.5. Improved Asynchronous State Estimation Algorithm Flow

The multi-level asynchronous state estimation algorithm proposed in this section is illustrated in Figure 4. The flow on the left side of the figure represents the process executed on the HD layer, while the right side shows the process running on all LD_i regions. At the start of the algorithm, each LD_i independently solves its EL subproblem based on local measurement data and sends the result to the HD layer. The HD continuously monitors the iteration data received from the LD_i regions. During the initial k_0 iterations, the HD must wait until all LD_i regions have returned their iteration values $\boldsymbol{S}^{\text{eq}}_{li,k}$ before triggering the EH subproblem computation, after which, the result $\left[\boldsymbol{S}^{\text{eq}}_{h,k+1}; \boldsymbol{V}_{l,k+1}\right]$ is broadcast to all LD_i regions. Once the iteration count exceeds k_0, the HD immediately solves the EH subproblem upon receiving iteration values $\boldsymbol{S}^{\text{eq}}_{li,k}$ from no fewer than S regions and confirming that the staleness of all LD_i values does not exceed the threshold τ; the resulting $\left[\boldsymbol{S}^{\text{eq}}_{h,k+1}; \boldsymbol{V}_{l,k+1}\right]$ is then broadcast to all LD_i regions. If these conditions are not satisfied, the HD continues polling for data until the triggering criteria are met. Each LD_i then proceeds to a new round of EL subproblem computation based on the received results. This process iterates until the voltage change at the boundary nodes between two consecutive iterations remains below a predefined convergence threshold, at which point, the algorithm terminates and outputs the final state estimation results.

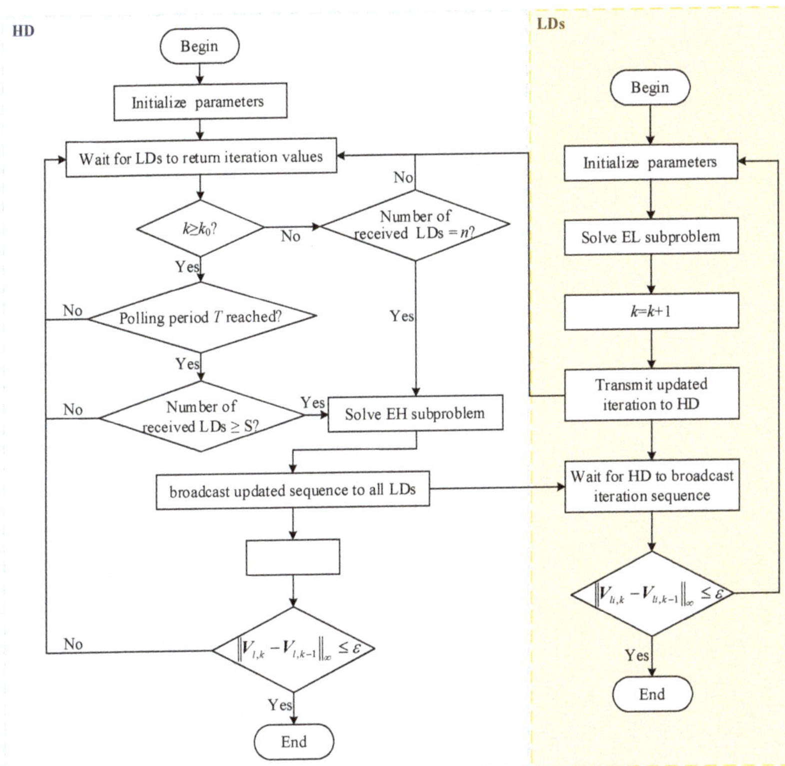

Figure 4. Flowchart of the multi-level asynchronous state estimation algorithm.

4. Case Analysis

4.1. Case Setup

This chapter verifies the effectiveness of the proposed algorithm. The simulations were conducted on a computer equipped with an Intel Core i9-14900 CPU and 32 GB of available memory. The communication interactions were simulated programmatically in a local environment. The HD layer adopts the IEEE Case33 system with a nominal voltage of 12.66 kV, where nodes 4, 19, 24, and 25 are connected to the LD layers, with the original loads at these nodes set to zero. Each LD layer is modeled using a modified 18-bus system [38], with a nominal voltage of 0.4 kV and cable-type lines. The HD and each LD are connected via a distribution transformer with a rated capacity of 0.63 MVA. The transmission line impedance is set to 0.002 + j0.01 (in per unit), and the overall system configuration follows that shown in Figure 1. Communication delays are simulated using a normal distribution $N(\mu,\sigma^2)$. The parameters for the MNMR model are set as $\lambda = 2$ and $\eta = 3$.

The measurement configuration is as follows: In the HD layer, measurements include the active power, reactive power, and voltage magnitude at the nodes, as well as the active and reactive power at injection points such as feeder inlets. In each LD layer, measurements include voltage magnitude at the nodes and active and reactive power at injection points such as feeder inlets.

To reflect the error characteristics of actual measurement devices, simulated measurement datasets are constructed by superimposing zero-mean Gaussian noise onto the true values obtained from power flow calculations in the HD and each LD region. For active and reactive power measurements at nodes, the standard deviation is set to $\sigma = 0.01$; for voltage magnitude measurements at nodes, $\sigma = 0.004$; for active and reactive power measurements at injection points, $\sigma = 0.01$; and for measurements related to boundary nodes, $\sigma = 0.005$. To generate bad data, a portion of the measurements is artificially perturbed using methods such as sign reversal, zeroing, or applying ±20% deviations from the original values. The proportion of bad data is set at 2%. The convergence error threshold is set to 10^{-6}, and the error tolerance for boundary state variables is set to 10^{-4}.

In this study, the voltage magnitudes, phase angles, and active and reactive power at the nodes are selected as state variables. The synchronous state estimation (S-SE) algorithm described in Section 2.3 and the traditional asynchronous state estimation (A-SE) algorithm described in Section 3.1 are used as benchmark methods to evaluate the estimation accuracy and convergence performance of the proposed improved asynchronous state estimation (HA-SE) algorithm. In addition, the runtime of each algorithm is adopted as an efficiency metric to assess the computational performance of the proposed method.

4.2. Analysis of State Estimation Results

To verify the estimation accuracy of the HA-SE algorithm, a comparison was conducted against the traditional WLS algorithm and the true values obtained from power flow calculations, with the asynchronous parameter set to $S = 1$ and $\tau = 3$. The results are presented in Table 2. As shown, the HA-SE algorithm yields estimation results for the voltage magnitude, phase angle, active power, and reactive power, which are highly consistent with both the WLS algorithm and the true values. Most deviations are controlled within 0.001, demonstrating excellent accuracy. With the introduction of boundary variable interaction, the voltage magnitudes at the boundary nodes of the upper HD and lower LD layers are nearly identical, indicating strong consistency between the hierarchical estimation results. In addition, the active power injected at the HD boundary nodes is negative, while the corresponding LD boundary injections are positive, with their sum approaching zero, which conforms to power flow conservation. These findings confirm that the proposed algorithm possesses high estimation accuracy and strong engineering applicability.

Table 2. Boundary node state estimation results.

Boundary Node		Parameter	True Value	WLS Algorithm	HA-SE Algorithm	Parameter	True Value	WLS Algorithm	HA-SE Algorithm
HD	4	V/p.u.	0.95291	0.95295	0.95294	P/MW	−0.40726	−0.41542	−0.41542
		θ/°	−1.72831	−1.76822	−1.76823	Q/MW	−0.13663	−0.13822	−0.13822
	19	V/p.u.	0.97574	0.97615	0.97614	P/MW	−0.40601	−0.40442	−0.40438
		θ/°	−1.77014	−1.76026	−1.76011	Q/MW	−0.13607	−0.13688	−0.13687
	24	V/p.u.	0.95164	0.951852	0.95184	P/MW	−0.40734	−0.40083	−0.40080
		θ/°	−1.94983	−1.90388	−1.90375	Q/MW	−0.13666	−0.14047	−0.14048
	25	V/p.u.	0.94849	0.94889	0.94887	P/MW	−0.40752	−0.40510	−0.40509
		θ/°	−2.02008	−1.99429	−1.99429	Q/MW	−0.13674	−0.13920	−0.13918
LD	4	V/p.u.	0.95291	0.95295	0.95295	P/MW	0.40726	0.41487	0.41486
		θ/°	−1.72831	−1.76822	−1.76823	Q/MW	0.13663	0.13839	0.13839
	19	V/p.u.	0.97574	0.97615	0.97614	P/MW	0.40601	0.40316	0.40311
		θ/°	−1.77014	−1.76026	−1.76011	Q/MW	0.13607	0.13820	0.13819
	24	V/p.u.	0.95164	0.95185	0.95184	P/MW	0.40734	0.40158	0.40155
		θ/°	−1.94983	−1.90388	−1.90375	Q/MW	0.13666	0.13896	0.13896
	25	V/p.u.	0.94849	0.94889	0.94888	P/MW	0.40752	0.40617	0.40616
		θ/°	−2.02008	−1.99429	−1.99429	Q/MW	0.13674	0.13744	0.13742

To further evaluate the robustness of the proposed algorithm under abnormal measurement scenarios, varying proportions of outlier data were introduced into the test cases, and the robust estimation capabilities of the WLS and HA-SE algorithms were compared. The maximum voltage deviation and overall deviation results for both algorithms are presented in Table 3. Here, the overall voltage deviation refers to the total deviation between the estimated and true voltage values across all nodes, reflecting the algorithm's global estimation error level. As the proportion of abnormal data increases from 0% to 14%, the WLS algorithm demonstrates greater sensitivity to outliers, with the overall deviation rising rapidly from 0.2397 to 0.3603 and the maximum deviation increasing significantly, indicating a notable decline in estimation stability. In contrast, the HA-SE algorithm exhibits much smaller variation in the overall deviation, increasing only from 0.2323 to 0.2684 under the same conditions, and maintains more stable estimation performance. These results indicate that the HA-SE algorithm possesses stronger robustness in the presence of abnormal measurements. The proposed multi-level decoupled structure and asynchronous coordinated iteration mechanism effectively suppress the propagation of local anomalies, thereby enhancing the stability and robustness of state estimation and ensuring the safe operation of the system in complex environments.

Table 3. Robust state estimation comparison.

Bad Data Ratio/%	WLS Algorithm		HA-SE Algorithm	
	Maximum Voltage. Deviation/p.u.	Overall Voltage Deviation/p.u.	Maximum Voltage Deviation/p.u.	Overall Voltage Deviation/p.u.
0	0.0125	0.2397	0.0122	0.2323
3	0.0144	0.2697	0.0125	0.2482
6	0.0163	0.2921	0.0131	0.2523
10	0.0192	0.3246	0.0139	0.2571
14	0.0224	0.3603	0.0144	0.2684

Furthermore, this section compares the boundary consistency of state estimation results obtained using the S-SE, A-SE, and HA-SE algorithms. With asynchronous parameters fixed, the boundary node state estimation results of each algorithm are tested and recorded, as shown in Table 4. The mismatch value refers to the deviation between the HD and LD layer estimation results for the same boundary node; smaller values indicate better coordination achieved by the algorithm under the multi-layer decoupling framework. As shown in Table 4, when each layer performs fully independent computation without any information exchange, the mismatch at boundary nodes increases significantly, and the overall error level is much higher than that of the other three algorithms. This indicates that such an approach is not suitable for coordinated operation in multi-layer distribution networks. The S-SE algorithm, leveraging a globally synchronized triggering mechanism and consistently utilizing the latest exchanged results at each iteration, significantly reduces mismatch values, and achieves strong boundary consistency. The A-SE algorithm, based on a traditional asynchronous triggering strategy, improves iterative efficiency but suffers from notable boundary inaccuracies due to the lag in lower-layer iteration sequences. In contrast, the proposed HA-SE algorithm optimizes the traditional asynchronous triggering strategy and achieves superior performance in terms of the voltage magnitude, phase angle, and power metrics. The mismatch values are significantly lower than those of the A-SE algorithm and nearly identical to those of the S-SE algorithm, confirming the effectiveness of the HA-SE method in enhancing coordination consistency under asynchronous estimation.

Table 4. State estimation results of different algorithms.

Algorithm	Maximum Mismatch of Boundary Voltage Magnitude/p.u.	Maximum Mismatch of Boundary Voltage Phase Angle/°	Maximum Mismatch of Boundary Active Power/MW	Maximum Mismatch of Boundary Reactive Power/MW	Maximum Deviation of Node Voltage/p.u.
Independent computation	0.0011	0.0844	0.0269	0.0341	0.0125
S-SE	1.2161×10^{-9}	9.1948×10^{-8}	0.0018	0.0013	0.0122
A-SE	6.4501×10^{-9}	9.1948×10^{-8}	0.0018	0.0018	0.0122
HA-SE	2.1709×10^{-9}	9.1319×10^{-8}	0.0013	0.0013	0.0122

4.3. Algorithm Performance Under Communication Delays

Based on the consideration of random communication delays, this section presents a comparative analysis of the runtime and parameter sensitivity of the three algorithms. Figure 5 shows the runtime comparison under different asynchronous parameter settings. It can be observed that the runtime of the S-SE algorithm remains steady at approximately 1.16 s, unaffected by changes in asynchronous parameters, which reflects the operational stability of its synchronous mechanism. In contrast, the runtime of the A-SE algorithm is more sensitive to parameter variations, fluctuating between 0.95 and 1.45 s. This indicates that although the A-SE algorithm adopts an asynchronous mechanism that can reduce waiting time to some extent, it may also suffer from resource contention or computational disorder, ultimately resulting in lower overall efficiency compared to the S-SE algorithm. By comparison, the HA-SE algorithm consistently achieves runtimes between 0.70 and 1.05 s across various parameter settings. It not only demonstrates lower computational costs, but also frequently achieves optimal or near-optimal runtimes close to 0.7 s under multiple parameter combinations, highlighting its robustness and stability. This suggests that the HA-SE algorithm is less prone to being trapped in local optima and is more capable of finding global or near-global solutions, thereby enabling efficient and stable state estimation under diverse and heterogeneous communication conditions. Overall, the proposed HA-SE algorithm significantly improves operational efficiency and exhibits superior and more stable performance.

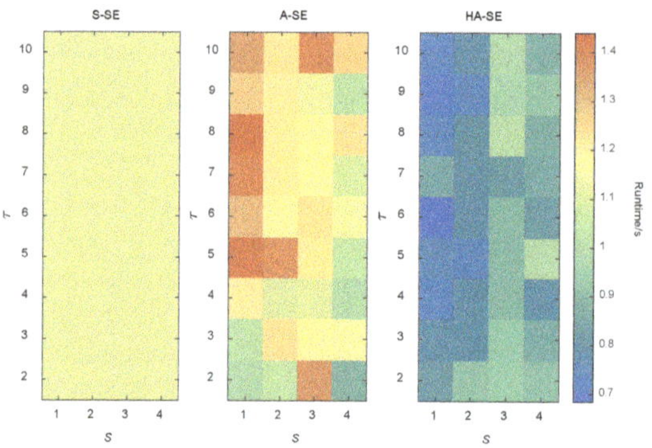

Figure 5. Comparison of runtime between three algorithms.

Furthermore, Table 5 presents the average number of iterations, average runtime, and minimum runtime of the three algorithms under various asynchronous parameter settings. The comparison shows that the A-SE algorithm exhibits significantly higher iteration counts and runtime in most parameter combinations compared to both the S-SE and the proposed HA-SE algorithm, highlighting its drawbacks in terms of slow convergence and low computational efficiency. However, under certain specific parameter settings, the minimum runtime of the A-SE algorithm is slightly better than that of the S-SE algorithm, indicating that it possesses some efficiency potential under favorable conditions. Nevertheless, this potential is highly dependent on the precise tuning of the asynchronous parameters, making it less stable and controllable in practical applications.

Table 5. Performance comparison of the three algorithms.

Algorithm	Average Iteration Count	Average Runtime/s	Minimum Runtime/s
S-SE	12.22	1.137	1.137
A-SE	23.86	1.230	0.997
HA-SE	14.37	0.833	0.685

In contrast, the HA-SE algorithm demonstrates a significantly lower average iteration count than the A-SE algorithm and is only slightly higher than that of the S-SE algorithm. This indicates that the HA-SE algorithm effectively mitigates the accumulation of asynchronous communication errors and enhances convergence stability in the early stages through its broadcast transmission strategy and synchronous bootstrapping mechanism. As a result, it significantly improves the convergence performance and overall stability of asynchronous state updates. Regarding runtime, the HA-SE algorithm outperforms both the S-SE and A-SE algorithms in terms of both average and minimum values, demonstrating a clear advantage in execution efficiency. This improvement is primarily attributed to the multithreaded parallel mechanism, which enables the decoupled and concurrent execution of receiving, computation, and transmission tasks, greatly enhancing thread resource utilization and system computational throughput.

To further account for the diversity of topologies in practical distribution networks, the possibility of non-radial configurations is considered. In addition to the commonly observed radial structure, the connections between LD and HD layers may exhibit ring or meshed characteristics, where a single LD region can be connected to the HD system through multiple tie lines. To reflect such cases, a more representative system model is constructed. The HD layer adopts the IEEE 69-bus distribution system from MATPOWER [39], while the LD layer remains composed of four regions: Two of these are based on the previously modified 18-bus systems, and the other two are derived from a modified case16ci system [40], with the voltage level set to 0.4 kV and a base power of 1 MVA. This paper uses MATPOWER 8.0, which was developed by the Power Systems Engineering Research Center (PSERC) at Cornell University, Ithaca, NY, USA.

In terms of connection configuration, the two 18-bus LDs are connected to nodes 28 and 46 of the HD system, respectively. The two 16-bus LDs are connected to the HD system through multiple feeders: one group is connected to nodes 52, 53, and 59, while the other group is connected to nodes 13, 14, and 68.

Figure 6 illustrates the runtime performance of the three algorithms under different asynchronous parameter configurations. It can be observed that the A-SE algorithm is highly sensitive to asynchronous parameters, resulting in significant fluctuations in runtime. In contrast, the proposed HA-SE algorithm exhibits relatively stable runtime across various parameter settings, with limited sensitivity to parameter perturbations, further validating the stability of its performance.

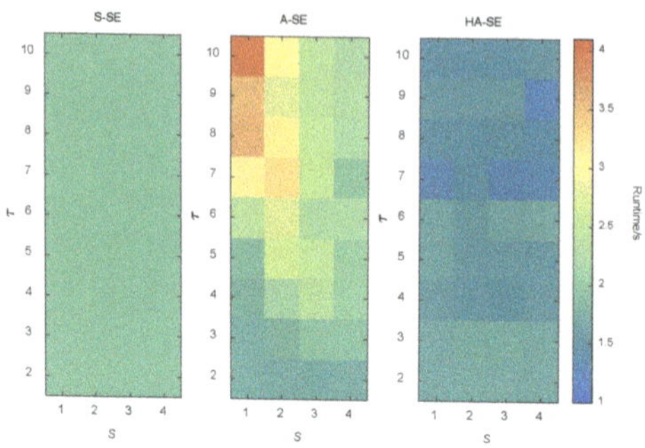

Figure 6. Runtime comparison of the three algorithms under ring topology configuration.

A performance comparison of the HA-SE, S-SE, and A-SE algorithms was also conducted, with Table 6 presenting the average number of iterations and runtime results under different parameter settings. Although the system includes LD subsystems with ring structures, the observed trends remain consistent with those in the earlier radial LD scenarios. The A-SE algorithm still exhibits a relatively high average iteration count, whereas the proposed HA-SE algorithm, benefiting from asynchronous scheduling and boundary state broadcasting, achieves iteration counts comparable to the S-SE algorithm. Moreover, HA-SE outperforms both benchmark methods in terms of average and minimum runtime, demonstrating superior computational efficiency and concurrent control capability.

Table 6. Performance comparison of the three algorithms under ring topology configuration.

Algorithm	Average Iteration Count	Average Runtime/s	Minimum Runtime/s
S-SE	14.94	1.955	1.955
A-SE	31.46	2.408	1.717
HA-SE	16.21	1.575	1.189

Based on the above results, it can be concluded that the proposed HA-SE algorithm not only achieves a good balance between estimation accuracy and convergence performance, but also demonstrates excellent computational efficiency. It effectively overcomes the efficiency bottlenecks and stability issues associated with traditional A-SE algorithms in complex communication environments, thereby exhibiting superior practicality and significant potential for engineering applications and deployment.

5. Conclusions

This paper proposes an asynchronous robust state estimation algorithm tailored for the multi-level structure of distribution networks. The main findings are as follows: First, in terms of estimation accuracy and robustness, the proposed algorithm achieves high estimation precision under multi-layer asynchronous conditions. It demonstrates excellent boundary consistency and strong robustness against outliers, reflecting superior accuracy and reliability in state estimation. Second, under the influence of communication delays and uneven asynchronous update frequencies, the algorithm incorporates a broadcast transmission strategy and a synchronous bootstrapping mechanism to address the issue of insufficient data coverage in the early estimation stages. This enhances the consistency of boundary information and stabilizes the convergence process, significantly reducing the number of iterations compared to traditional asynchronous methods. Finally, regarding computational efficiency, the algorithm adopts a multithreaded parallel mechanism that decouples the tasks of receiving, computation, and transmission. Across various asynchronous parameter configurations, it reduces the average runtime by approximately 30–50% compared to synchronous and conventional asynchronous methods, thereby validating its superior execution efficiency while maintaining estimation accuracy.

Future research can be pursued in several directions: First, the communication modeling capabilities can be extended by incorporating non-Gaussian delay characteristics to improve the algorithm's adaptability in complex network environments. Second, the proposed method can be validated in practical engineering scenarios by integrating real-world distribution network topologies and measurement data. Third, for real-time applications, the asynchronous scheduling and iterative mechanisms can be further optimized to enhance the algorithm's online operation and rapid response performance.

Author Contributions: Methodology: X.Z., Y.L., S.G., Y.T. and Y.G.; writing—original draft preparation: X.Z. and Y.G.; writing—review and editing: X.Z., Y.L., S.G., Y.T. and Y.G. All authors have read and agreed to the published version of the manuscript.

Funding: This research was supported by the Science & Technology Project of the State Grid Corporation of China, "Research on Key Technologies of Distributed Collaborative Computing and Control for New-Type Distribution Networks" (Project Code: 5400-202356680A-3-3-JC).

Data Availability Statement: The original contributions presented in the study are included in the article; further inquiries can be directed to the corresponding author/s.

Conflicts of Interest: The authors declare that the research was conducted in the absence of any commercial or financial relationships that could be construed as a potential conflict of interest.

Abbreviations

The following abbreviations are used in this manuscript:

SE	State Estimation
MNMR	Maximum Normal Measurement Rate State estimation
WLS	Weighted Least Squares

References

1. Radhoush, S.; Bahramipanah, M.; Nehrir, H.; Shahooei, Z. A Review on State Estimation Techniques in Active Distribution Networks: Existing Practices and Their Challenges. *Sustainability* **2022**, *14*, 2520. [CrossRef]
2. Monticelli, A. Electric Power System State Estimation. *Proc. IEEE* **2000**, *88*, 262–282. [CrossRef]
3. Song, W.; He, J.; Lin, J.; Ye, H.; Ling, X.; Lu, C. Bias Analysis of PMU-Based State Estimation and Its Linear Bayesian Improvement. *IEEE Trans. Ind. Inform.* **2024**, *20*, 1607–1617. [CrossRef]
4. Zhang, J.; Zhao, J.; Cheng, G.; Rouhani, A.; Chen, X. Explainable Multi-Fidelity Bayesian Neural Network for Distribution System State Estimation. *Appl. Energy* **2025**, *392*, 125972.
5. Primadianto, A.; Lu, C.-N. A Review on Distribution System State Estimation. *IEEE Trans. Power Syst.* **2017**, *32*, 3875–3883. [CrossRef]
6. Qiu, H.; Yu, J.; Chyad, M.H.; Singh, N.S.S.; Hussein, Z.A.; Jasim, D.J.; Khosravi, M. Comparative Analysis of Kalman Filters, Gaussian Sum Filters, and Artificial Neural Networks for State Estimation in Energy Management. *Energy Rep.* **2025**, *13*, 4417–4440. [CrossRef]
7. Pagani, G.A.; Aiello, M. Power Grid Complex Network Evolutions for the Smart Grid. *Phys. A* **2014**, *396*, 248–266. [CrossRef]
8. Pignati, M.; Zanni, L.; Romano, P.; Cherkaoui, R.; Paolone, M. Fault Detection and Faulted Line Identification in Active Distribution Networks Using Synchrophasors-Based Real-Time State Estimation. *IEEE Trans. Power Deliv.* **2017**, *32*, 381–392. [CrossRef]
9. Hao, W.; Xu, M.; Lin, J.; Fu, L.; Cao, X.; Jia, Q. Multi-Time Scale Cloud-Edge Collaborative Scheduling Strategy for Distribution Network Considering Spatiotemporal Characteristics of Demand Response. *Energies* **2024**, *17*, 1933. [CrossRef]

10. Loji, K.; Sharma, S.; Loji, N.; Sharma, G.; Bokoro, P.N. Operational Issues of Contemporary Distribution Systems: A Review on Recent and Emerging Concerns. *Energies* **2023**, *16*, 1732. [CrossRef]
11. Wang, Y.; Gu, J.; Yuan, L. Distribution Network State Estimation Based on Attention-Enhanced Recurrent Neural Network Pseudo-Measurement Modeling. *Prot. Control Mod. Power Syst.* **2023**, *8*, 31. [CrossRef]
12. Zhou, X.; Liu, Z.; Guo, Y.; Zhao, C.; Huang, J.; Chen, L. Gradient-Based Multi-Area Distribution System State Estimation. *IEEE Trans. Smart Grid* **2020**, *11*, 5325–5338. [CrossRef]
13. Jiang, S.; Yang, S. Distributed State Estimation Method of Distribution Networks Based on LSTM. *J. Phys. Conf. Ser.* **2023**, *2418*, 12089. [CrossRef]
14. Liu, Z.; Gao, S.; Li, P.; Ji, H.; Xi, W.; Yu, H.; Wu, J.; Wang, C. Distributed State Estimation in Digital Distribution Networks Based on Proximal Atomic Coordination. *IEEE Trans. Instrum. Meas.* **2022**, *71*, 1–11. [CrossRef]
15. Jiang, S.; Li, S.; Wu, H.; Hua, Y.; Xu, B.; Ding, M. Distributed State Estimation Method Based on WLS-AKF Hybrid Algorithm for Active Distribution Networks. *Int. J. Electr. Power Energy Syst.* **2023**, *145*, 108732. [CrossRef]
16. Wang, W.; Liu, X.; Wu, Z.; Li, Y.; Guo, Z.; Sun, Q. Distributed State Estimation of Interconnected Power Systems with Time-varying Disturbances and Random Communication Link Failures. *Energy Convers. Econ.* **2024**, *5*, 382–395. [CrossRef]
17. Chen, T.; Liu, F.; Li, P.; Sun, L.; Amaratunga, G.A.J. A Distributed Multi-Area Power System State Estimation Method Based on Generalized Loss Function. *Meas. Sci. Technol.* **2023**, *34*, 115010. [CrossRef]
18. Tian, S.; Zhu, F.; Shen, J.; Yang, X.; Fu, Y.; Mi, Y.; Ling, P. Distributed State Estimation of Active Distribution Network Considering Mixed-Frequency Measurement Data Hierarchical Encryption. *Appl. Energy* **2025**, *388*, 125661. [CrossRef]
19. Abbasi, A.R.; Seifi, A.R. A New Coordinated Approach to State Estimation in Integrated Power Systems. *Int. J. Electr. Power Energy Syst.* **2013**, *45*, 152–158. [CrossRef]
20. Tang, Y.; Tang, K.; Dong, S. A Parallel and Asynchronous State Estimation for Coupled Transmission-Distribution Networks. *Int. J. Electr. Power Energy Syst.* **2022**, *141*, 108163. [CrossRef]
21. Sun, H.B.; Zhang, B.M. Global State Estimation for Whole Transmission and Distribution Networks. *Electr. Power Syst. Res.* **2005**, *74*, 187–195. [CrossRef]
22. Madurasinghe, D.; Venayagamoorthy, G.K. Multilevel Distributed Linear State Estimation Integrated with Transmission Network Topology Processing. *Appl. Sci.* **2024**, *14*, 3422. [CrossRef]
23. Jin, Y.; Wang, T.; Liu, Y. Hierarchical State Estimation Method for Tree Structure Distribution Network. In Proceedings of the 2021 IEEE 5th Conference on Energy Internet and Energy System Integration (EI2), Taiyuan, China, 22–24 October 2021; IEEE: Piscataway, NJ, USA, 2021; pp. 1117–1122.
24. Gholami, M.; Tehrani-Fard, A.A.; Lehtonen, M.; Moeini-Aghtaie, M.; Fotuhi-Firuzabad, M. A Novel Multi-Area Distribution State Estimation Approach for Active Networks. *Energies* **2021**, *14*, 1772. [CrossRef]
25. Gomez-Exposito, A.; De La Villa Jaen, A. Two-Level State Estimation with Local Measurement Pre-Processing. *IEEE Trans. Power Syst.* **2009**, *24*, 676–684. [CrossRef]
26. Korres, G.N.; Contaxis, G.C. Application of a Reduced Model to a Distributed State Estimator. In Proceedings of the 2000 IEEE Power Engineering Society Winter Meeting, Singapore, 23–27 January 2000; Conference Proceedings (cat. No.00ch37077). IEEE: Piscataway, NJ, USA, 2000; Volume 2, pp. 999–1004.
27. Xie, L.; Choi, D.-H.; Kar, S. Cooperative Distributed State Estimation: Local Observability Relaxed. In Proceedings of the 2011 IEEE Power and Energy Society General Meeting, San Diego, CA, USA, 24–29 July 2011; IEEE: Piscataway, NJ, USA, 2011; pp. 1–11.
28. Lakshminarasimhan, S.; Girgis, A.A. Hierarchical State Estimation Applied to Wide-Area Power Systems. In Proceedings of the 2007 IEEE Power Engineering Society General Meeting, Tampa, FL, USA, 24–28 June 2007; IEEE: Piscataway, NJ, USA, 2007; pp. 1–6.
29. Cavraro, G.; Comden, J.; Dall'Anese, E.; Bernstein, A. Real-Time Distribution System State Estimation with Asynchronous Measurements. *IEEE Trans. Smart Grid* **2022**, *13*, 3813–3822. [CrossRef]
30. Ren, Z.; Chen, Y.; Huang, S.; Heleno, M.; Xia, Y. A Fully Distributed Coordination Method for Fast Decoupled Multi-Region State Estimation. *IEEE Access* **2019**, *7*, 132859–132870. [CrossRef]
31. Anandini, G.N.; Gupta, I. Estimating the Lost Real-Time Measurements under Communication Failure for Distribution System State Estimation. In Proceedings of the 2016 National Power Systems Conference (NPSC), Bhubaneswar, India, 19–21 December 2016; IEEE: Piscataway, NJ, USA, 2016; pp. 1–5.
32. Wang, J.; Li, T. Distributed Multi-Area State Estimation for Power Systems with Switching Communication Graphs. *IEEE Trans. Smart Grid* **2021**, *12*, 787–797. [CrossRef]
33. He, G.; Dong, S.; Qi, J.; Wang, Y. Robust State Estimator Based on Maximum Normal Measurement Rate. *IEEE Trans. Power Syst.* **2011**, *26*, 2058–2065. [CrossRef]

34. Stathopoulos, G.; Jones, C.N. An Inertial Parallel and Asynchronous Forward–Backward Iteration for Distributed Convex Optimization. *J. Optim. Theory Appl.* **2019**, *182*, 1088–1119. [CrossRef]
35. Baudet, G.M. Asynchronous Iterative Methods for Multiprocessors. *J. ACM* **1978**, *25*, 226–244. [CrossRef]
36. Ding, B.; Li, Z.; Li, Z.; Xue, Y.; Chang, X.; Su, J.; Sun, H. Cooperative Operation for Multiagent Energy Systems Integrated with Wind, Hydrogen, and Buildings: An Asymmetric Nash Bargaining Approach. *IEEE Trans. Ind. Inform.* **2025**, 1–12. [CrossRef]
37. Li, Z.; Sun, H.; Xue, Y.; Li, Z.; Jin, X.; Wang, P. Resilience-Oriented Asynchronous Decentralized Restoration Considering Building and E-Bus Co-Response in Electricity-Transportation Networks. *IEEE Trans. Transp. Electrif.* **2025**, 1–13. [CrossRef]
38. CIGRE Networks—Pandapower 3.1.0.Dev0 Documentation. Available online: https://pandapower.readthedocs.io/en/latest/networks/cigre.html (accessed on 11 May 2025).
39. Das, D. Optimal Placement of Capacitors in Radial Distribution System Using a Fuzzy-GA Method. *Int. J. Electr. Power Energy Syst.* **2008**, *30*, 361–367. [CrossRef]
40. Zhu, J.Z. Optimal Reconfiguration of Electrical Distribution Network Using the Refined Genetic Algorithm. *Electr. Power Syst. Res.* **2002**, *62*, 37–42. [CrossRef]

energies

MDPI

Article

Maximized Autonomous Economic Operation and Aggregated Equivalence for Microgrids with PVs and ESSs

Zhiwei Wang [1], Shumin Sun [1,*], Yan Cheng [1], Peng Yu [1], Jiawei Xing [1], Wanting Shen [2] and Jinquan Zhao [2]

1 State Grid Shandong Electric Power Research Institute, Jinan 250003, China; wyysws126@126.com (Z.W.); 15822923227@163.com (Y.C.); 19811716052@163.com (P.Y.); miaoxiaeryi@163.com (J.X.)

2 College of Electrical Engineering, Shanghai University of Electric Power, Yangpu District, Shanghai 200090, China; shenwt2001@163.com (W.S.); zhaojinquan@shiep.edu.cn (J.Z.)

* Correspondence: 19301393701@163.com

Abstract: With the rapid development of renewable energy, more microgrids integrating photovoltaics and energy storage systems (MGPEs) have been deployed. Frequent grid faults have consequently increased the likelihood of MGPEs operating in off-grid mode, highlighting the need to maximize their autonomous operation. This paper proposes an optimal day-ahead scheduling method, which aims to maximize autonomous economic operation and minimize dependence on the main grid. Based on the autonomous results, the internal resources of MGPEs are aggregated into an equivalent power unit (EPU) and an aggregated energy storage system (AESS). The aggregated demands and residual capabilities of these aggregated models are then quantified, thereby facilitating coordinated scheduling by the main grid. Compared with traditional cost-optimization approaches, this case study shows that the proposed method reduces the daily outage probability of MGPEs by at least 24.9%.

Keywords: microgrid with photovoltaic and energy storage systems (MGPE); autonomous economic operation (AEO); aggregated equivalence

1. Introduction

With the increasing depletion of fossil resources and the growing emphasis on environmental sustainability, the transition of the energy structure toward low carbon has become inevitable [1,2]. Microgrids (MGs) are regarded as one of the essential technologies for the future development of the energy sector [3]. MGs can efficiently utilize distributed generators (DGs), energy storage systems (ESSs), and flexible loads to achieve localized power supply [4,5]. Microgrids with photovoltaics (PVs) and ESSs (MGPEs) have rapidly developed due to the increasing penetration of renewable energy.

An MGPE operates in two modes: on-grid and off-grid. In the event of a fault, an on-grid MGPE may be disconnected from the main grid and forced to operate in off-grid mode. Therefore, it is crucial for MGPEs to enhance their autonomous capabilities even during on-grid operation to reduce reliance on the main grid. This implies that MGPEs must be capable of maintaining power supply during off-grid operation. Consequently, optimal scheduling strategies that maximize autonomous operation are essential for MGPEs.

The optimal scheduling of MGs has been extensively studied in the current works. Most of these studies focus on proposing optimization models aimed at minimizing operational costs, primarily emphasizing the economic performance of MGs [6–14]. For example, Silva et al. [6] analyzed MG operating costs under time-of-use electricity pricing

and considered intentional islanding with potential load shedding. Both Gomes et al. [7] and Li and Xu [8] establish cost-optimal economic models for MG operation in both on-grid and off-grid modes. Both Nwulu and Xia [9] and Witharama et al. [10] incorporate demand response mechanisms into their optimal-dispatch model to further reduce MG operating costs. Sun et al. [11] propose a day-ahead economic dispatch strategy that maximizes the benefits for utilities and users. Silvente et al. [12] evaluate the economic advantages of selling surplus photovoltaic energy to the main grid and include the resulting revenue in the cost–benefit analysis. Hou et al. [13] propose a multi-objective economic dispatch approach that co-optimizes the operation of electric vehicles and transferable loads within the microgrid. Moreover, Aghajani and Ghadimi [14] present a multi-objective energy management framework that simultaneously minimizes electricity and environmental costs. However, these studies often overlook the priority of utilizing internal resources to enhance the self-sufficiency of MGs. This limitation increases the risks associated with unplanned islanding and undermines the security and reliability of MGs operation.

To address the risk of energy shortfall without relying on full-scale resilience modeling, several studies have incorporated uncertainty- or risk-based metrics into cost-oriented MG-scheduling frameworks. Cao et al. [15] propose a distributionally robust optimization model for MG energy management, which explicitly bounds the probability of energy shortfall under uncertain PV generation and load forecasts. Hou et al. [16] introduce a distributed model-predictive scheme for islanded MG that jointly accounts for battery degradation and cycle-life uncertainty, with aims to reduce the risk of premature capacity loss during extended islanding periods. Tan et al. [17] develop a two-stage robust optimization approach that exploits a data-driven uncertainty set to hedge against worst-case PV and electric-vehicle load deviations. Consequently, the above studies demonstrate that the cost-oriented MGs can incorporate probabilistic or risk-based metrics to assess internal resource adequacy. The present work addresses this gap while adhering to a primarily economic optimization framework.

MGs are increasingly required to aggregate internal resources to support flexible scheduling by the main grid. To address this requirement, models and methods for resource aggregation have been extensively studied. Zhao et al. [18] and Wu and Wang [19] aggregate diverse resources within MG to develop models of models for flexible resource clusters composed of equivalent generators and equivalent ESSs. Utkarsh et al. [20] aggregate DGs into a network-aware framework to achieve flexibility and optimal cost. Wen et al. [21] and Yi et al. [22] derive the feasible aggregation region of MG and equivalently aggregate flexible resources into a unified ESS representation. Moreover, He et al. [23] construct the flexibility model of MG, characterizing the flexibility boundaries that MG can provide to the main grid. However, these studies have not quantified the external demand and residual capacity of MGs following autonomous economic operation.

Given the above insights, the aim of this paper is primarily to maximize the utilization of internal resources to reduce external dependence in the event of faults and to aggregate resources to quantify the residual capabilities for the main grid. Compared to previous studies, the main contributions of this paper are summarized as follows:

1. Develop a sensitivity-based autonomous economic optimization for an MGPE to minimize electricity exchange while covering internal costs, concurrently lowering outage probability to secure supply reliability.
2. Establish aggregated equivalent models of the MGPE to reduce the scheduling complexity of scheduling and enhance the operational efficiency of the main grid.

3. Quantify both the physical and economic parameters of the aggregated models to provide valuable information for the main grid, enabling better resource management and improved energy utilization.

The paper is organized as follows. In Section 2, the autonomous economic operation of an MGPE is described. In Section 3, the proposed method is presented. In Section 4, a comprehensive case study analysis is provided. In Section 5, the case study results are discussed and interpreted. Finally, the paper is concluded in Section 6.

2. Autonomous Economic Operation (AEO)

The configuration of the MGPE is illustrated in Figure 1. The PVs and ESSs are connected to the AC bus via their respective power conversion devices. The MGPE is interconnected with the main grid via a Point of Common Coupling (PCC) [24].

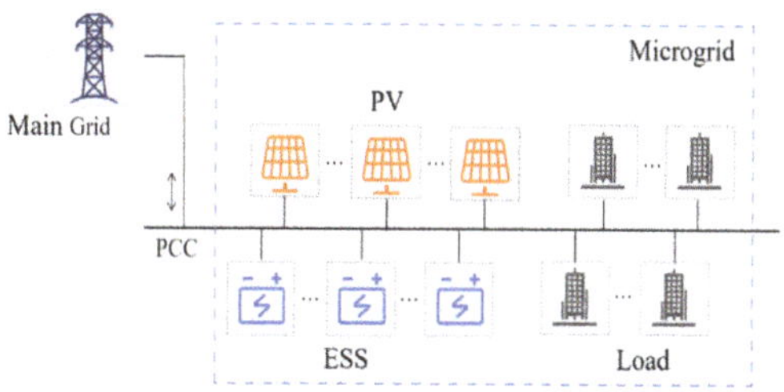

Figure 1. Schematic of the MGPE.

In autonomous economic operation, internal resources of the MGPE, such as PVs, ESSs, and flexible loads, are prioritized. The load demand is primarily met through self-consumption to reduce dependence on external resources, while surplus power is fed into the main grid. This approach aims to enhance the independence and reliability of the MGPEhile the potential for economic operation of internal resources is explored to reduce operational costs.

Overall, the optimized operation provides a favorable guarantee for the security, reliability, and economic viability of the MGPE.

3. Methodology

3.1. Optimization Model

3.1.1. Objective Function

The optimization objective of the MGPE is formulated to prioritize the utilization of internal resources, aiming to achieve self-consumption, surplus power export, and economic operation. Dependence on the main grid is reduced and the economic operation of the MGPE is further optimized. Thus, the objective function in this study is established as follows:

Objective Function 1: Minimization of Electricity Deficit and Surplus

$$\min F_1 = \sum_{t=1}^{T} \left(\left| P_{buy,t} \right| + \left| P_{sell,t} \right| \right) \Delta T \tag{1}$$

where $P_{buy,t}$ and $P_{sell,t}$ is the power purchased from and sold to the main grid at t. T is the total number of time periods and ΔT is the duration of each time period.

Objective Function 2: Minimization of Internal Resource Cost

$$\min F_2 = \sum_{t=1}^{T} (\sum_{i=1}^{N_V} c_{v,i} P_{v,i,t} + \sum_{i=1}^{N_{ess}} c_{ess,i}^{dch} P_{ess,i,t}^{dch} + \sum_{i=1}^{N_{tl}} c_{tl,i,t} |\Delta P_{tl,i,t}|) \Delta T \tag{2}$$

where $c_{ess,i}^{dch}$, $c_{v,i}$, and $c_{tl,i,t}$ denote the discharge cost of the ith ESS, the generation cost of the ith PV, and the adjustment cost of the ith flexible load at t. $P_{v,i,t}$, $P_{ess,i,t}^{dch}$ and $\Delta P_{tl,i,t}$ represents the generation power of the ith PV, the discharge power of the ith ESS, and the active power change in the ith flexible load at t. N_v, N_{ess}, and N_{tl} correspond to the total number of PVs, ESSs, and flexible loads.

To ensure the autonomy and minimize the costs of the MGPE, the aforementioned multiple objective functions are converted into a single objective function:

$$\min F = \omega F_1 + F_2 \tag{3}$$

where ω is the penalty factor, incorporated to ensure that the MGPE prioritizes the use of its internal resources for AEO.

3.1.2. Constraints

(a). Power Balance Constraint

$$P_{buy,t} + P_{sell,t} + \sum_{i=1}^{N_v} P_{v,i,t} + \sum_{i=1}^{N_{ess}} (P_{ess,i,t}^{dch} - P_{ess,i,t}^{ch}) = \sum_{i=1}^{N_{bl}} P_{bl,i,t} + \sum_{i=1}^{N_{tl}} P_{tl,i,t} \tag{4}$$

where $P_{ess,i,t}^{ch}$ represents the charging power of the ith ESS at t. $P_{bl,i,t}$ represents the active power of the ith fixed load at t. N_{bl} corresponds to the total number of fixed loads.

(b). ESS Operation Constraints

$$0 \leq P_{ess,i,t}^{ch} \leq z_{i,t} P_{ess,i,\max}^{ch} \tag{5}$$

$$0 \leq P_{ess,i,t}^{dch} \leq (1 - z_{i,t}) P_{ess,i,\max}^{dch} \tag{6}$$

where $P_{ess,i,\max}^{ch}$ and $P_{ess,i,\max}^{dch}$ represent the maximum charging and discharging power of the ith ESS, respectively. $z_{i,t}$ is a 0–1 binary variable; a value of 1 indicates that the ith ESS is charging at t, and a value of 0 indicates it is discharging.

The constraints for the state of charge (SOC) of the ESS are given by Equations (7) and (8).

$$S_{ess,i,t} = S_{ess,i,t-1} + \gamma_{ess,i}^{ch} \Delta T P_{ess,i,t}^{ch} / E_{ess,i} - \Delta T P_{ess,i,t}^{dch} / \gamma_{ess,i}^{dch} E_{ess,i} \tag{7}$$

$$S_{ess,i,\min} \leq S_{ess,i,t} \leq S_{ess,i,\max} \tag{8}$$

Within the scheduling period, the ESSs must satisfy the charge/discharge energy balance to ensure the sustainability of the dispatch strategy, as given by Equation (9).

$$S_{ess,i,0} = S_{ess,i,T} \tag{9}$$

where $S_{ess,i,t}$ denotes the SOC of the ith ESS at t. $S_{ess,i,\min}$ and $S_{ess,i,\max}$ is the minimum and maximum SOC limits of the ith ESS, respectively. $E_{ess,i}$ indicates the rated capacity of the ith ESS at t. $\gamma_{ess,i}^{ch}$ and $\gamma_{ess,i}^{dch}$ is the charging and discharging efficiencies of the ith ESS, respectively. $S_{ess,i,0}$ and $S_{ess,i,T}$ denote the initial SOC and end-of-period SOC of the ith ESS.

(c). Flexible Load Constraints

The flexible loads are subject to constraints including the upper and lower limits of load shifting, the ramp rate constraint, and the requirement that total energy consumption remains invariant over a scheduling period.

$$P_{tl,i,t} \triangleq P_{tl,i,t,base} + \Delta P_{tl,i,t} \tag{10}$$

$$\theta_{tl,i,t} P_{tl,i,t,\min} \le P_{tl,i,t} \le \theta_{tl,i,t} P_{tl,i,t,\max} \tag{11}$$

$$-\Delta P_{tl,i,t,d} \le \Delta P_{tl,i,t} - \Delta P_{tl,i,t-1} \le \Delta P_{tl,i,t,u} \tag{12}$$

$$\sum_{t=1}^{T} \Delta P_{tl,i,t} \Delta T = 0 \tag{13}$$

where $P_{tl,i,t,base}$ and $P_{tl,i,t}$ represent the ith baseline load and the ith load after shifting at t. $P_{tl,i,t,\min}$ and $P_{tl,i,t,\max}$ represent the minimum and maximum active power bounds of the ith load i after shifting at t, respectively. $\Delta P_{tl,i,t,u}$ and $\Delta P_{tl,i,t,d}$ represent the maximum upward and downward active power adjustments of the ith flexible load at t. $\theta_{tl,i,t}$ is a binary variable representing the shifting status of the ith flexible load at t, where 1 indicates that the load is being shifted.

3.2. Aggregation and Equivalence

Power flow constraints are typically neglected. Based on AEO dispatch results, the overall MGPE is characterized through net power output by an aggregated equivalent model consisting of an equivalent power unit (EPU) and an aggregated ESS (AESS). The aggregation of the MGPE is illustrated in Figure 2. On this basis, the residual regulation capability and surplus capability available for external interaction are further quantified, and the corresponding price coefficients are also calculated.

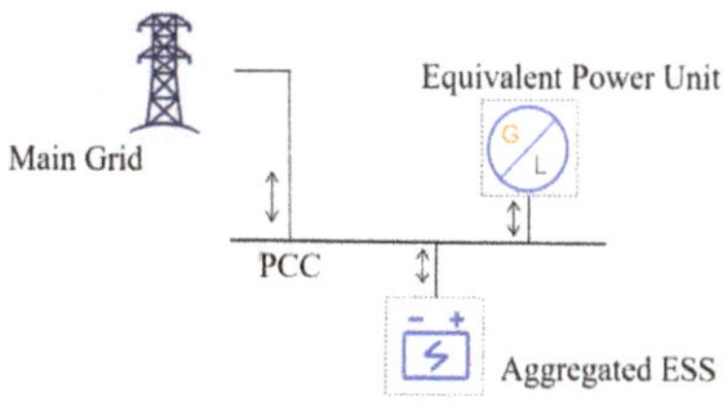

Figure 2. Aggregated Equivalence of the MGPE.

3.2.1. Residual Regulation Capability

When power deficits occur in the main grid, additional regulation power can be supplied by the MGPE, primarily through the discharge of the AESS. Thus, the residual regulation capability of the MGPE is quantified from the remaining capacity and discharge power of the AESS after AEO optimization.

The parameters of the AESS model are formulated as Equations (14)–(18).

$$P_{ess,\max}^{B,ch} = \sum_{i=1}^{N_{ess}} P_{ess,i,\max}^{ch} \tag{14}$$

$$P^{B,dch}_{ess,\max} = \sum_{i=1}^{N_{ess}} P^{dch}_{ess,i,\max} \tag{15}$$

$$E^{B}_{ess} = \sum_{i=1}^{N_{ess}} E_{ess,i} \tag{16}$$

$$S^{B}_{ess,\min} = \frac{\sum_{i=1}^{N_{ess}} S_{ess,i,\min} E_{ess,i}}{E^{B}_{ess}} \tag{17}$$

$$S^{B}_{ess,\max} = \frac{\sum_{i=1}^{N_{ess}} S_{ess,i,\max} E_{ess,i}}{E^{B}_{ess}} \tag{18}$$

where $P^{B,ch}_{ess,\max}$ and $P^{B,dch}_{ess,\max}$ represent the maximum charging and discharging power of the AESS. $S^{B}_{ess,\min}$ and $S^{B}_{ess,\max}$ denote the minimum and maximum SOC of the AESS. E^{B}_{ess} indicates the rated capacity of the AESS.

The residual regulation capability of the AESS is quantified by Equations (19)–(21).

$$P^{res,ch}_{ess,t} = \sum_{i=1}^{N_{ess}} (P^{ch}_{ess,i,\max} - P^{ch}_{ess,i,t}) \tag{19}$$

$$P^{res,dch}_{ess,t} = \sum_{i=1}^{N_{ess}} (P^{dch}_{ess,i,\max} - P^{dch}_{ess,i,t}) \tag{20}$$

$$S^{res}_{ess,t} = \frac{E^{B}_{ess,t}}{E^{B}_{ess}} = \frac{\sum_{i=1}^{N_{ess}} S_{ess,i,t} E_{ess,i}}{E^{B}_{ess}} \tag{21}$$

where $P^{res,ch}_{ess,t}$ and $P^{res,dch}_{ess,t}$ represent the residual regulation capability of the AESS in charging and discharging states at t. $E^{res}_{ess,t}$ indicates the remaining capacity of the AESS at t. $S^{res}_{ess,t}$ is the SOC of the AESS at t.

When the main grid requires regulation power from the MGPE, it prioritizes discharging the less-costly ESSs within the AESS to ensure economic efficiency. Thus, the segmented pricing of the residual regulation capability of the AESS ($c^{dch}_{ess,1} \leq c^{dch}_{ess,2} \leq \ldots \leq c^{dch}_{ess,N_{ess}}$) is then determined.

$$c^{tl}_{MG,t} = \begin{cases} c^{dch}_{ess,1}, & 0 < E^{tl}_{UG,t} \leq E_{1,t} \\ c^{dch}_{ess,2}, & E_{1,t} < E^{tl}_{UG,t} \leq E_{2,t} \\ \vdots & \\ c^{dch}_{ess,i}, & E_{i-1,t} < E^{tl}_{UG,t} \leq E_{i,t} \\ \vdots & \\ c^{dch}_{ess,N_{ess}}, & E_{N_{ess}-1,t} < E^{tl}_{UG,t} \leq E_{N_{ess},t} \end{cases} \tag{22}$$

$$E_{i,t} = \sum_{j=1}^{i} S_{ess,j,t} E_{ess,j} \quad i = 1, 2, \ldots, N_{ess} \tag{23}$$

where $c^{tl}_{MG,t}$ indicates the segmented regulation price of the MGPE to the main grid at t. $E^{tl}_{UG,t}$ represents the required regulation electricity of the main grid at t. $E_{i,t}$ indicates the cumulative remaining capacity of the first i ESS at t. Moreover, the constraints on maximum discharge power and minimum SOC must still be satisfied for the external regulation of each ESS.

Thus, the regulation revenue of the MGPE at t is calculated as the sum of the electricity prices of each segmented power.

$$R_t^{tl} = \sum_{j=1}^{i-1} c_{ess,j}^{dch} S_{ess,j,t} E_{ess,j} + c_{ess,i}^{dch} \left(E_{UG,t}^{tl} - E_{i-1,t} \right) \tag{24}$$

where R_t^{tl} represents the total regulation revenue of the MGPE to the main grid at t.

3.2.2. Residual Regulation Capability Residual Surplus Capability

After the MGPE maximizes its AEO, surplus power can be exchanged with the main grid. This surplus is mainly derived from the residual capacity of the EPU, particularly through excess PV generation. Thus, the residual surplus capability of the MGPE is quantified based on the remaining power of the EPU after AEO.

The EPU is composed of PVs and the total load, and its behavior is characterized by the net generation power and net load power.

$$P_{epu,t}^{B,G} = \sum_{i=1}^{N_v} P_{v,i,t} - \sum_{i=1}^{N_{bl}} P_{bl,i,t} - \sum_{i=1}^{N_{tl}} P_{tl,i,t} \geq 0 \tag{25}$$

$$P_{epu,t}^{B,G} = \sum_{i=1}^{N_{bl}} P_{bl,i,t} + \sum_{i=1}^{N_{tl}} P_{tl,i,t} - \sum_{i=1}^{N_v} P_{v,i,t} \geq 0 \tag{26}$$

where $P_{eq,t}^{B,G}$ and $P_{eq,t}^{B,L}$ represent the net generation power and net load power of the EPU at t.

The residual deficit or surplus electricity exchanged between the MGPE and the main grid is quantified by the net power measured at the PCC. Surplus electricity that arises from PV generation within the EPU after its autonomous demands has been fully satisfied.

$$S_{MG,t}^{nd} = P_{buy,t} \Delta T \tag{27}$$

$$S_{MG,t}^{sp} = P_{sell,t} \Delta T = \sum_{j=1}^{N_V} P_{v,j,t}^{res} \Delta T \tag{28}$$

where $S_{MG,t}^{nd}$ and $S_{MG,t}^{sp}$ denote the remaining demands and surplus electricity of the MGPE after AEO at t. $P_{v,j,t}^{res}$ indicates the surplus power of PV j at t.

When surplus power is drawn from the MGPE, priority is given to the less-costly PV generation within the EPU to achieve cost savings. Thus, the segmented pricing of the surplus power of the EPU ($c_{v,1} \leq c_{v,2} \leq \ldots \leq c_{v,N_V}$) is then determined.

$$c_{MG,t}^{sp} = \begin{cases} c_{v,1}, \ 0 < S_{UG,t}^{nd} \leq S_{1,t} \\ c_{v,2}, \ S_{1,t} < S_{UG,t}^{nd} \leq S_{2,t} \\ \vdots \\ c_{v,i}, \ S_{i-1,t} < S_{UG,t}^{nd} \leq S_{i,t} \\ \vdots \\ c_{v,N_V}, \ S_{N_V-1,t} < S_{UG,t}^{nd} \leq S_{N_V,t} \end{cases} \tag{29}$$

$$S_{i,t} = \sum_{j=1}^{i} P^{res}_{v,j,t}\Delta T \qquad i = 1, 2, \cdots, N_V \tag{30}$$

where $c^{sp}_{MG,t}$ indicates the segmented surplus electricity price of the MGPE to the main grid at t. $S^{nd}_{UG,t}$ represents the electricity demand of the main grid at t. $S_{i,t}$ indicates the cumulative surplus electricity of the first i PV at t.

Thus, the surplus electricity revenue of the MGPE at t is calculated as the sum of the electricity prices of each segmented power.

$$R^{sp}_t = \sum_{j=1}^{i-1} c_{v,j} P^{res}_{v,j,t}\Delta T + c_{v,i}\left(S^{nd}_{UG,t} - S_{i-1,t}\right) \tag{31}$$

where R^{sp}_t represents the total surplus electricity revenue of the MGPE to the main grid at t. Equations (29) and (30) partition the main grid demand into successive segments, with the marginal cost of each segment determined by the ascending-ordered cost of surplus PV generation. Equation (31) calculates the total revenue received by the EPU from the main grid as the cumulative product of each segment quantity and its corresponding price.

3.3. Procedure

The procedure for Maximized Autonomous Economic Operation and Equivalent Aggregation of the MGPE is illustrated in Figure 3, and the pseudocode of the optimization scheduling is provided in Appendix A Algorithm A1.

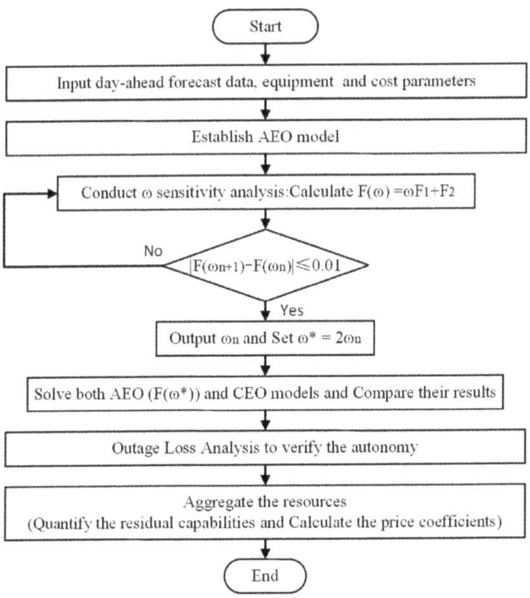

Figure 3. Flowchart of the AEO and aggregated equivalence.

1. Data Input: The input set is constructed by reading data of ESSs, PVs, loads, flexible loads, and cost parameters.
2. Establish Modeling: The AEO model is constructed using the optimization objective defined by Equations (1)–(3) together with the relevant constraints.

3. ω Sensitivity Analysis: the parameter ω is varied from 0 to 1 with a step size of 0.04. At each point, the MIQP is solved using CPLEX and YALMIP, and the F–ω curve is plotted. To maintain an adequate margin, the optimal weight ω^* is set to twice the knee-point value.
4. Optimization: Optimization with Fixed ω^* and Comparison between AEO and the minimized cost economic operation (CEO).
    a. Substituting ω^* the MIQP is solved again to obtain the AEO scheduling result.
    b. Using the same constraints but minimizing only the cost function ($F = \sum_{t=1}^{T}(c_{buy}P_{buy,t} + c_{sell}P_{sell,t} + \sum_{i=1}^{N_V} c_{v,i}P_{v,i,t} + \sum_{i=1}^{N_{ess}} c_{ess,i}^{dch}P_{ess,i,t}^{dch} + \sum_{i=1}^{N_{tl}} c_{tl,i,t}|\Delta P_{tl,i,t}|)\Delta T$), the CEO scheduling scheme is derived as a benchmark.
5. Outage Loss Analysis: the outage probability and outage losses under AEO and CEO scheduling are compared to quantify the autonomy benefits.
6. Aggregated equivalence: according to the AEO results, the internal resources are aggregated; the residual external capability and its corresponding price coefficient are then quantified.

4. Case Study Analysis

4.1. Model Verification

The proposed model is verified before the case-specific optimization results are presented.

(1) Internal consistency

All instances were solved using CPLEX and YALMIP, achieving an optimality gap of $\leq$1% and an average runtime of 2.9 s. This confirms the robustness and applicability of the approach.

(2) Scalability and sensitivity

The MGPE model is tested by varying irradiance profiles and ESS capacities. The resulting grid-import energy decreases monotonically and converges asymptotically, demonstrating robustness to parameter perturbations and scalability.

(3) Data availability statement

24 h operational data from a fully instrumented microgrid and access to a real-time hardware-in-the-loop platform, are currently unavailable. Consequently, experimental validation on a physical system has not yet been conducted; such validation will be carried out once field data or platform access becomes available.

4.2. Basic Data

The experimental data in this section correspond to the standard input set read in Step 1 of Section 3.3. MGPE comprises PVs, ESSs, and daily loads, with flexible loads also considered. Specifically, it includes three PVs, three ESS units, and two types of flexible loads. The generation costs of the PVs are set at 0.16, 0.20, and 0.32 CNY/kWh. The adjustment costs of the flexible loads in different time periods are presented in Appendix A Table A1. The rated capacities of the ESS units are 600, 800, and 1000 kWh, and their discharge costs are set at 0.35, 0.65, and 0.75 CNY/kWh [25]. The charging and discharging efficiencies of each ESS are both specified at 95%. Other specific equipment parameters are detailed in Table 1. The cost of purchasing electricity from the main grid is determined to be 0.60 CNY/kWh, while the cost of selling electricity to the main grid is determined to be 0.20 CNY/kWh. The cost of power outage due to fault islanding is substantial [26]. The cost of power outage due to fault islanding is calculated as 3.00 CNY/kWh [27].

Table 1. Equipment Parameters of Each ESS.

ID	E_{ess}	$S_{ess,0}$	$S_{ess,\min}$	$S_{ess,\max}$	$P^{ch}_{ess,\max}$	$P^{dch}_{ess,\max}$
1	600	0.5	0.3	0.7	300	300
2	800	0.5	0.2	0.8	450	450
3	1000	0.5	0.1	0.9	500	500

Forecast data for total PV generation and load demand are used as inputs for optimization, as illustrated in Appendix A Figures A1 and A2. The predicted PV generation is obtained under three typical weather conditions: sunny, cloudy, and rainy, with corresponding total outputs of 6399 kWh, 5434 kWh, and 4758 kWh, respectively. The total load demand is 5700 kWh. Given that the forecasts span 24 h, the day-ahead optimization horizon is defined as $T = 24$ h with a temporal resolution of $\Delta t = 1$ h.

4.3. Dispatch Analysis

4.3.1. Sensitivity Analysis

A sensitivity analysis is performed under different weather conditions, as illustrated in Figure 4, whereby the impact of the penalty factor ω on Objective Functions 1 and 2 is evaluated. The values of both objectives are observed to stabilize once ω, representing the maximum marginal cost of internal resources, exceeds 0.75.

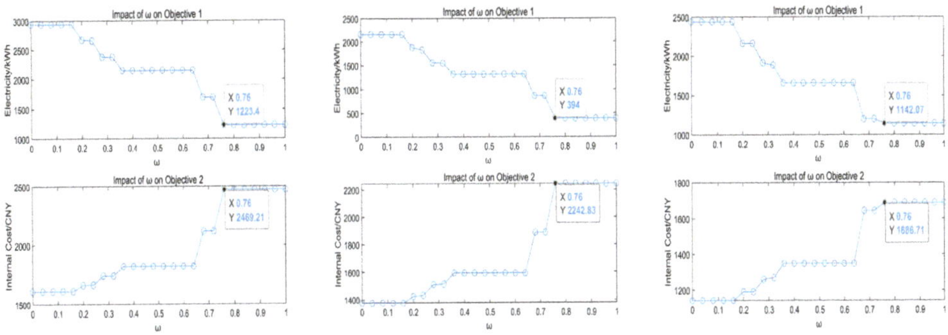

(a) (b) (c)

Figure 4. Sensitivity analysis of w under different weather conditions: (**a**) sunny, (**b**) cloudy, and (**c**) rainy. **Note:** The data in this figure are generated from Steps 2 and 3 in Section 3.3.

As a result, the penalty factor w should be set higher than this internal cost cap. In this study, ω is defined as twice the maximum internal resource cost in order to ensure a sufficient safety margin while maintaining computational simplicity. Therefore, $\omega^* = 1.50$ (i.e., 2×0.75) is selected as the penalty factor for the re-optimization in Step 4, and the corresponding results are shown in Figures 5 and 6.

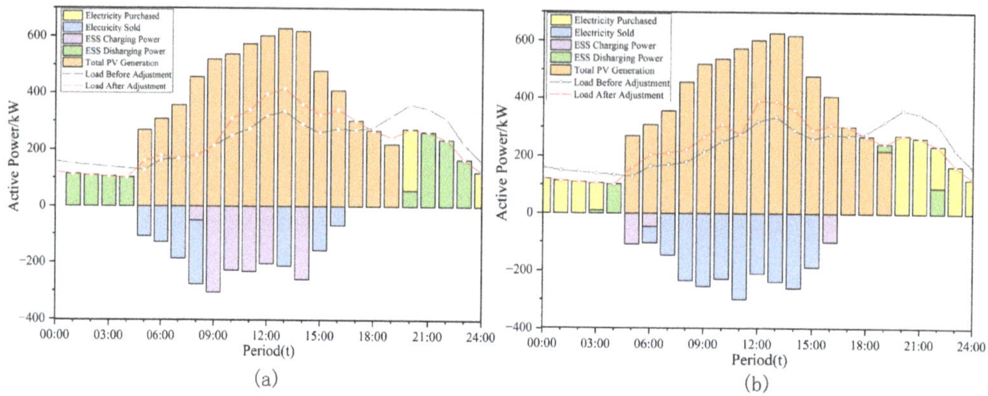

Figure 5. Day-ahead Scheduling Results of the MGPE under the Sunny Condition. (**a**) Results of AEO. (**b**) Results of CEO.

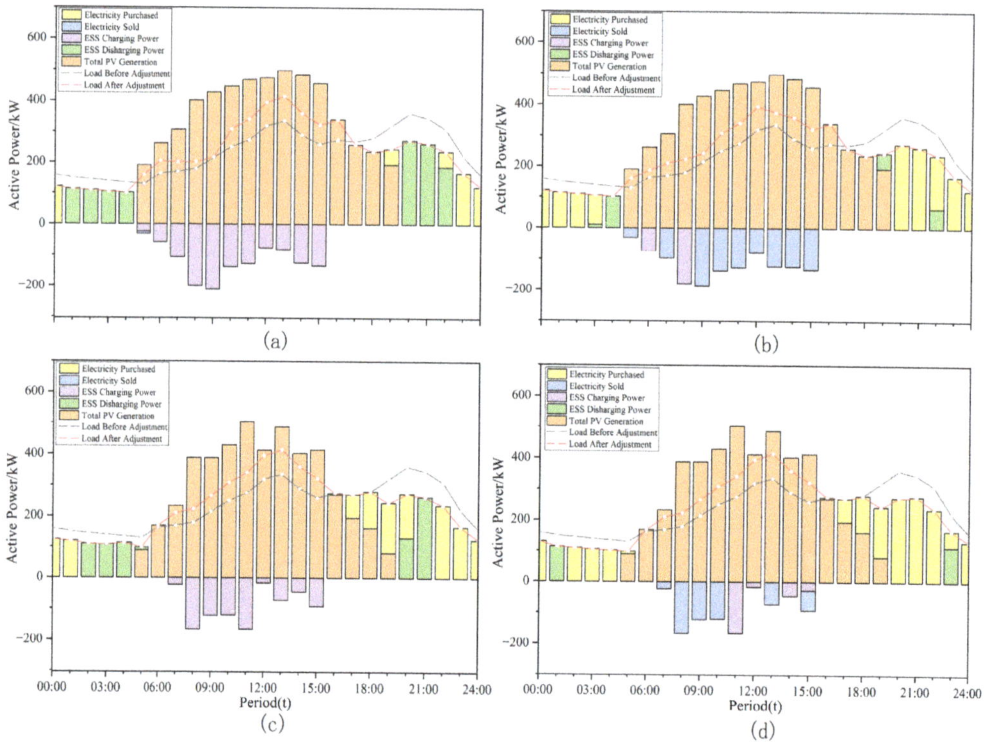

Figure 6. Day-ahead Scheduling Results of the MGPE. (**a**) Results of AEO under the Cloudy Condition. (**b**) Results of CEO under the Cloudy Condition. (**c**) Results of AEO under the Rainy Condition. (**d**) Results of CEO under the Cloudy Condition.

4.3.2. Optimized Scheduling

This section follows Step 4 in Section 3.3: the AEO scheduling scheme is re-solved with the selected ω^* obtained in Step 3, and the CEO scheduling scheme is also derived.

These two schemes are then employed for subsequent comparisons of cost and regulation capability.

Case 1: Scheduling of the Sunny Day

Under the sunny condition, the total PV generation significantly exceeds the load demand. The day-ahead scheduling results obtained from the AEO and the CEO of the MGPE are presented in Figure 5.

The optimal results of the AEO under the Sunny Condition are depicted in Figure 5a. Between 01:00 and 04:00, PV generation is absent, and the internal ESSs are utilized to discharge preferentially to supply demand. By 05:00, the lower-cost ESSs (ESS 1 and ESS 2) are depleted to their minimum SOC. The power generation is concentrated between 05:00 and 19:00, when output is abundant and generation cost is low. During this period, demand is primarily satisfied by PVs, and the surplus is used to charge the ESSs. By 18:00, the ESSs are fully charged and charging is halted. Flexible loads tend to be shifted to this period to reduce reliance on grid exchange. Between 20:00 and 24:00, PV generation is again absent, and demand is met through the ESSs' discharge. However, since the ESSs are required to return to their initial SOC by the end of the 24 h horizon, part of the demand is supplied by electricity purchased from the external grid.

The optimal results of the CEO under the Sunny Condition are depicted in Figure 5b. Between 01:00 and 04:00, PV generation is absent. The purchase price of grid electricity is lower than the discharge costs of ESS 2 and ESS 3. Therefore, electricity is purchased from the grid, and ESS 1 is discharged to meet the load demand. In contrast, ESS 2 and ESS 3 remain inactive due to their higher discharge costs. By 05:00, only ESS 1 is depleted to its minimum SOC. From 05:00 to 19:00, abundant PVs are available. The surplus is utilized both to sell electricity to the grid and to charge ESS 1 to its maximum SOC. This strategy prepares the system for potential load deficits in subsequent hours. Between 20:00 and 23:00, no power is generated by the PVs. During this period, the load demand is supplied by ESS 1 and by electricity purchased from the main grid. During this period, ESS 1 is also discharged to return to its initial SOC.

Case 2–3: Scheduling of the Cloudy/Rainy Day

Compared with the sunny condition, the total PV generation is insufficient to meet the load demand under both cloudy and rainy conditions. The PV output under rainy conditions is significantly lower than that under cloudy conditions. The day-ahead scheduling results of the MGPE based on AEO and CEO for these two conditions are presented in Figure 6.

The optimal results of the AEO under cloudy and rainy conditions are illustrated in Figure 6a,c. Between 01:00 and 04:00, PV generation is absent. During this interval, demand is supplied by preferential ESS discharge, and flexible loads are predominantly shifted to this period. By 05:00, PV output remains low under rainy conditions, necessitating partial discharge from the ESSs. From 05:00 to 19:00, after meeting demand, a limited PV surplus is used to charge the ESSs. However, due to the overall shortage of PVs, power is no longer sold to the main grid. Since PV generation under the rainy condition is lower than that under the cloudy conditions, the charging level is correspondingly reduced. Between 20:00 and 24:00, demand continues to be met through ESSs' discharge. Nevertheless, constrained by SOC requirements and the PV generation deficit, additional electricity must be purchased from the grid, particularly under rainy conditions. Due to the lower charging amount in the earlier period, the available power from the ESSs is more limited under rainy conditions, leading to higher grid purchases compared with cloudy conditions.

The CEO results under cloudy and rainy conditions, presented in Figure 6b,d, confirm that the CEO closely follows the pattern observed under sunny conditions. When PV generation is absent, demand is supplied by electricity purchased from the grid and by discharge from the lower-cost ESSs. During PV generation periods, the surplus after meeting demand is utilized to charge the ESSs.

Motivated by time-of-use tariffs, the lower-cost type-1 loads are dispatched first until their common ceiling of 600 kWh is reached in all three weather scenarios, with any residual imbalance covered by type-2 loads. Consequently, the total adjustable energy amounts to 1120.8 kWh (type-2: 520.8 kWh) on sunny days, 1210.8 kWh (type-2: 610.8 kWh) on cloudy days, and 1111.6 kWh (type-2: 511.6 kWh) on rainy days. On sunny days, abundant PV surplus enables substantial battery charging between 09:00 and 15:00 followed by high-rate discharge during the evening peak (19:00–24:00), yielding a modest system imbalance. Under cloudy conditions, reduced PV surplus curtails battery charging and enlarges the evening discharge deficit, forcing a significant upward adjustment of type-2 loads. During rainy days, PV surplus is confined to 05:00–15:00, the overall imbalance is limited, and the requirement imposed by Equation (14) that upward and downward energy must balance within each cycle further constrains type-2 participation, resulting in a total adjustment slightly lower than that observed on cloudy days.

5. Case Study Improvements

5.1. Comparison of Outage Loss Performance

This section performs the 3 h islanding fault experiment in Step 5 using the AEO and CEO scheduling schemes obtained in Step 4. The outage cost and outage probability are compared to verify the autonomy benefits.

If a 3 h islanding fault occurs on the following day, the outage cost associated with CEO is significantly higher than that of AEO, as shown in Table 2, and the gap further widens with longer islanding durations. Although the day-ahead costs of AEO (CNY 2719.16, 2703.12, and 2637.18 under sunny, cloudy, and rainy conditions, respectively) are slightly higher than those of CEO (CNY 2420.21, 2404.17, and 2487.81), the fault scenarios reveal substantial differences. Specifically, under sunny conditions, the outage cost decreases to the maximum extent from CNY 2041.8 (CEO) to CNY 217.8 (AEO); under cloudy conditions, it decreases to the maximum extent from CNY 2119.8 to CNY 148.2; and under rainy conditions, it decreases to the maximum extent from CNY 957.6 to CNY 0.

Table 2. Costs of a Three-Hour Power outage (/CNY) under the Cloudy and Rainy Conditions.

Sunny			Cloudy			Rainy		
Time	AEO	CEO	Time	AEO	CEO	Time	AEO	CEO
1:00–3:00	0	957.6	1:00–3:00	0	957.6	1:00–3:00	0	649.8
2:00–4:00	0	615.6	2:00–4:00	0	615.6	2:00–4:00	0	957.6
3:00–5:00	0	285	3:00–5:00	0	285	3:00–5:00	0	653.4
						4:00–06:00	0	334.2
4:00–6:00 to 17:00–19:00	0	0	4:00–6:00 to 16:00–18:00	0	0	5:00–7:00 to 14:00–16:00	0	0
						15:00–17:00	222	240
						16:00–18:00	573	591
			17:00–19:00	147.6	0	17:00–19:00	573	1053.6
18:00–20:00	217.8	820.8	18:00–20:00	147.6	820.8	18:00–20:00	1171.8	1652.4
19:00–21:00	217.8	1607.4	19:00–21:00	147.6	1607.4	19:00–21:00	1607.4	2131.2
20:00–22:00	217.8	2041.8	20:00–22:00	148.2	2119.8	20:00–22:00	2314.2	2357.4
21:00–23:00	501.6	1722.6	21:00–23:00	649.8	1800.6	21:00–23:00	1667.6	1696.2
22:00–24:00	866.4	1300.8	22:00–24:00	1014.6	1378.8	22:00–24:00	1245.8	1259.4

The outage probabilities of AEO and CEO can be analyzed from Figures 5 and 6, as summarized in Table 3. The results further show that AEO also lowers the outage probability: from 33.3% to 8.33% under sunny conditions, from 33.3% to 12.5% under cloudy conditions, and from 41.6% to 33.3% under rainy conditions.

Table 3. Performance comparison of AEO vs. CEO under three weather conditions.

Conditions	Method	SSR	OP	RIP
sunny	CEO	66.7%	33.3%	-
	AEO	91.7%	8.3%	75.1%
cloudy	CEO	66.7%	33.3%	-
	AEO	87.5%	12.5%	62.5%
rainy	CEO	58.4%	41.6%	-
	AEO	66.7%	33.3%	24.9%

Note: The Supply Security Rate (SSR) is defined as SSR = 100% − OP, where OP denotes the Outage Probability. The Relative Improvement Rate (RIP) is calculated as $(OP_{CEO} - OP_{AEO})/OP_{CEO} \times 100\%$, and it indicates the relative reduction in outage probability achieved by AEO compared with CEO.

In summary, regardless of whether under sunny, cloudy, or rainy condition, the cost of AEO is always lower than that of CEO after a fault occurs. Moreover, the AEO can reduce the probability of power outages, ensure the continuous supply of electricity, and enhance the power supply reliability of the MGPE.

5.2. Aggregation and Quantification

Once the autonomy has been verified in Steps 1–5, resources of the MGPE are aggregated, and its overall demands and residual capability are quantified. This enables flexible dispatch of the main grid. However, due to the unknown demand from the main grid, the relevant price coefficients cannot be calculated in this case study.

The demand/surplus capacity curve at the PCC is derived from Equations (27) and (28), as illustrated in Figure 7. Positive values indicate surplus exports from the MGPE to the main grid, whereas negative values represent the demand for grid imports on an hourly basis.

The charging and discharging power of the ESSs and their SOC are aggregated over a 24 h period. It aims to facilitate the utilization of the residual regulation capability of the MGPE by the main grid. This aggregation aims to quantify the residual charging and discharging power and the SOC of the AESS, as shown in Figure 8. Due to the different capacities of the internal ESSs, the SOC of the AESS cannot be simply summed. Instead, the SOC of each ESS must be calculated for each time period according to Equation (22).

The AESS is prioritized for discharging by the MGPE until renewable PV generation becomes available as depicted in Figure 8. Furthermore, PVs with lower generation costs are utilized by the MGPE for charging after meeting the load demands. It ensures that the AESS can discharge preferentially during subsequent PV shortages. The operational strategy reduces dependence on the main grid, thereby enhancing the autonomy and reliability of the MGPE. It also optimizes AEO by scheduling the charging and discharging of the AESS in a reasonable manner. If the residual regulation capability of the AESS after AEO is utilized by the main grid, the exchanged power must satisfy both the residual input/output limits and the SOC constraints of the AESS at each time step. By quantifying and reporting, flexible dispatch can be achieved by the main grid to effectively improve the resource consumption rate. Similarly, the results under cloudy and rainy conditions are aggregated and quantified, as presented in Appendix A Figure A3.

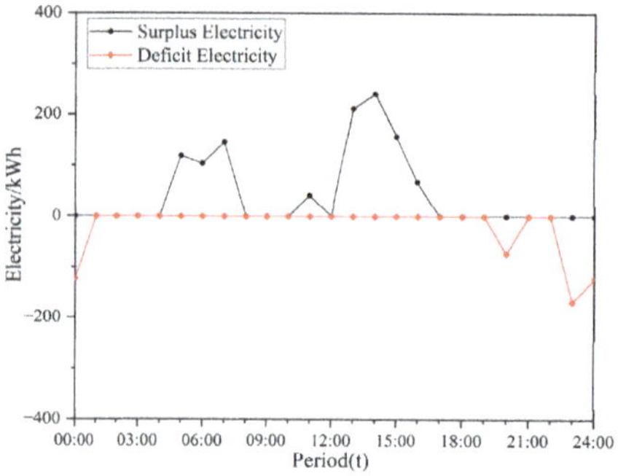

Figure 7. Demand and Residual Surplus Capability (under the sunny condition).

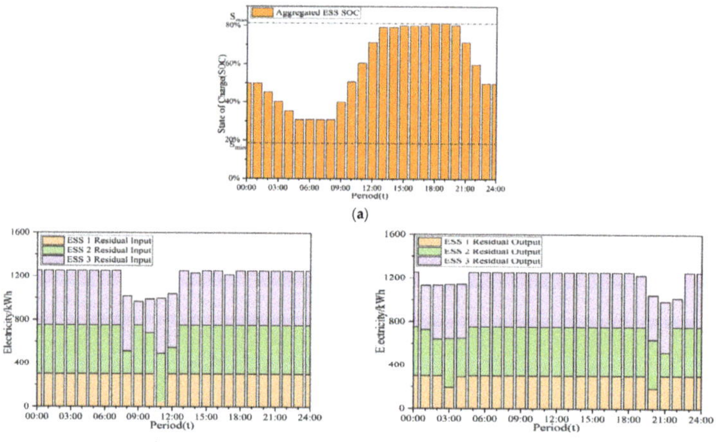

Figure 8. Residual Regulation Capability. (**a**) SOC of the AESS. (**b**) Residual output power of the AESS. (**c**) Residual input power of the AESS.

5.3. *Further Experiments*

5.3.1. Scalability

Scalability is examined by expanding the MGPE with two, six, and ten additional ESSs, each rated at 400 kWh and 200 kW charging/discharging limits. The discharge cost (0.30 CNY/kWh), SOC bounds, and round-trip efficiency of the baseline ESS 1 are applied to all supplementary units.

The impact of varying ESS capacities is illustrated in Figure 9. Under sunny conditions, abundant PV surplus is available. The additional ESS capacity stores this surplus and later discharges it during non-PV periods, thereby considerably reducing external energy imports compared with the baseline. However, once ESS capacities exceeds the PV-chargeable threshold, external demand stabilizes. The PV surplus cannot fully charge the enlarged fleet, and terminal SOC constraints force a net purchase from the grid. Under the cloudy condition, the reduced PV surplus limits additional charging. As a result, only a modest reduction in external demand is achieved, which saturates at the same threshold. During the rainy condition, no PV surplus is available even without additional ESSs.

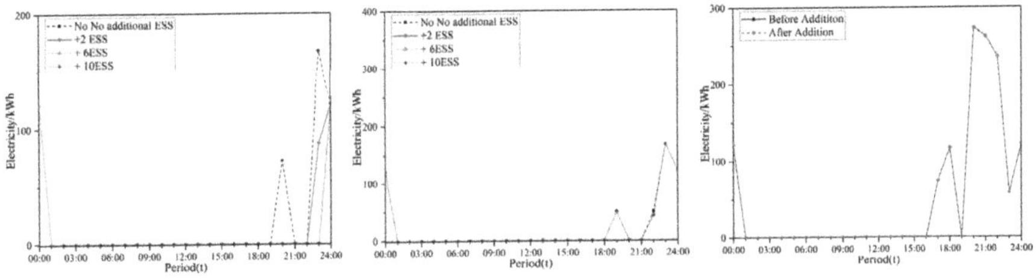

(a) (b) (c)

Figure 9. Impact of Varying ESS capacities on External Electricity Dependence under Different Weather Conditions: (**a**) sunny, (**b**) cloudy, and (**c**) rainy.

Consequently, further expansion of ESS capacity has little effect on external dependence.

5.3.2. Sensitivity Analysis of PV and ESS Capacity

Sensitivity analyses were conducted to examine the impact of varying PV conditions and ESS configurations on the penalty factor w. These analyses considered different weather scenarios and the addition of two, six, and ten ESSs, as described in Section 5.2 (see Appendix A Figure A4). Due to the limited PV surplus and the discharge cost hierarchy, PV generation first charges, and later discharges, the lowest-cost units. Because the added ESSs have lower discharge costs than the original units, w decreases as PV surplus shrinks and as the number of ESSs increases. Nevertheless, w always remains above the maximum discharge cost observed in the dispatch results.

This finding confirms that w must be set above the highest marginal cost of internal resources. For practical implementation, w is set to twice the observed maximum internal cost. This guarantees a sufficient safety margin without adding engineering complexity.

6. Conclusions

This study proposes a day-ahead scheduling framework for an MGPE that prioritizes autonomous operation while achieving internal economic optimality. Compared with traditional cost-oriented optimization, the framework reduces the daily outage probability by at least 24.9% while simultaneously lowering outage-related losses. A novel EPU–AESS dual aggregation structure is introduced to quantify the residual flexibility as closed-form cost coefficients of the MGPE, enabling direct dispatch by the main grid. Additionally, it provides a systematic method for setting the penalty factor threshold applicable to various weather conditions and ESS capacities. Due to the deterministic nature of the current model, forecast uncertainties are not yet considered. Future work will integrate stochastic optimization, conduct field data validation, and extend the approach to the cooperative operation of networked microgrids.

Author Contributions: Conceptualization, Z.W.; methodology, S.S. and J.Z.; software, J.X.; validation, J.X.; formal analysis, Y.C.; investigation, Z.W.; resources, P.Y.; data curation, Y.C. and P.Y.; writing—original draft preparation, Z.W., S.S. and W.S.; writing—review and editing, S.S., W.S. and J.Z.; visualization, Y.C.; supervision, P.Y.; project administration, S.S.; funding acquisition, Z.W. All authors have read and agreed to the published version of the manuscript.

Funding: This research was funded by the State Grid Shandong Electric Power Company Technology Project, "Research on Coordinated Control Technology of Distribution Network-Microgrid Group Supporting the Safe and Stable Operation of Power Grid with High Proportion of Renewable Energy" (520626230115).

Data Availability Statement: The original contributions presented in this study are included in the article. Further inquiries can be directed to the corresponding author(s).

Conflicts of Interest: The authors declare no conflict of interest.

Appendix A

Algorithm A1: CPLEX-Based Maximized AEO for the MGPE with ω-Sensitivity

Input:
-Time horizon T, step ΔT
-Fixed load: $P_{bl,i,t}, i \in 1, \ldots, N_{bl}$; flexible loads: $\Delta P_{tl,i,t,base}$, $P_{tl,i,t,max/min}$, $\Delta P_{tl,i,t,u/d}$, $i \in 1, \ldots, N_{tl}$
-PV generation: $P_{v,i,t}, i \in 1, \ldots, N_v$
-ESS parameters: $S_{ess,i,max/min}$, $S_{ess,i,0}$, $P^{ch}_{ess,i,max/min}$, $P^{dch}_{ess,i,max/min}$, $\eta^{ch}_{ess,i}$, $\eta^{dch}_{ess,i}$, $i \in 1, \ldots, N_{ess}$
-Cost coefficients: $c_{v,i}$, $c^{dch}_{ess,i}$, $c_{l,i,t}$
Hyperparameters:
-Penalty factor set Ω (e.g., 0:0.04:1)
-Solver: IBM CPLEX via YALMIP (MIQP model, mipgap = 0.01, timelimit = 60 s)
Output:
-Optimal schedules: $P_{buy,t}$, $P_{sell,t}$, $P^{ch}_{ess,i,t}$, $P^{dch}_{ess,i,t}$, $S_{ess,i,t}$, $\Delta P_{tl,i,t}$
-F_1, F_2, selected ω*
1. Define Objective
  1.1 Objective 1:
  $F_1 = \sum_{t=1}^{T}\left(\left|P_{buy,t}\right|+\left|P_{sell,t}\right|\right)\Delta T$
  1.2 Objective 2:
  $F_2 = \sum_{t=1}^{T}\left(\sum_{i=1}^{N_v} c_{v,i}P_{v,i,t} + \sum_{i=1}^{N_{ess}} c^{dch}_{ess,i}P^{dch}_{ess,i,t} + \sum_{i=1}^{N_{tl}} c_{l,i,t}\left|\Delta P_{tl,i,t}\right|\right)\Delta T$
  1.3 Total objective:
  $F(\omega) = \omega \cdot F_1 + F_2$
2. Constraints
  2.1 Power balance: grid exchange, PV generation, ESS charge/discharge must meet total load demand (fixed + flexible)
  2.2 ESS operation: SOC dynamics, capacity limits, and mutually exclusive charge/discharge
  2.3 Flexible load: shifting and ramp limits, total energy conservation $\sum \Delta P_{l,i,t}\Delta T = 0$
3. ω-Sensitivity Analysis
  For each ω in Ω:
  -Update objective F(ω)
  -Solve MIQP model with CPLEX
  -CPLEX internally applies Branch-and-Bound for integer vars, Barrier/Simplex for QP relaxation
  -Presolve, cut generation, heuristics are activated to improve efficiency
  -Record F(ω), F_1, and F_2
4. Select ω
  4.1 Choose ω' based on trade-off between grid exchange and internal cost
  4.2 Typically select ω' yielding minimal F(ω) under constraints
5. Final Scheduling
  5.1 Fix ω = 2ω', re-solve MIQP via CPLEX
  5.2 Output dispatch

Table A1. Adjustment Cost of Flexible Loads in Different Time Periods.

Time Period	Flexible Load 1 Cost	Flexible Load 2 Cost
01:00–08:00; 19:00–24:00	0.18	0.28
12:00–14:00	0.25	0.35
09:00–11:00; 15:00–18:00	0.20	0.30

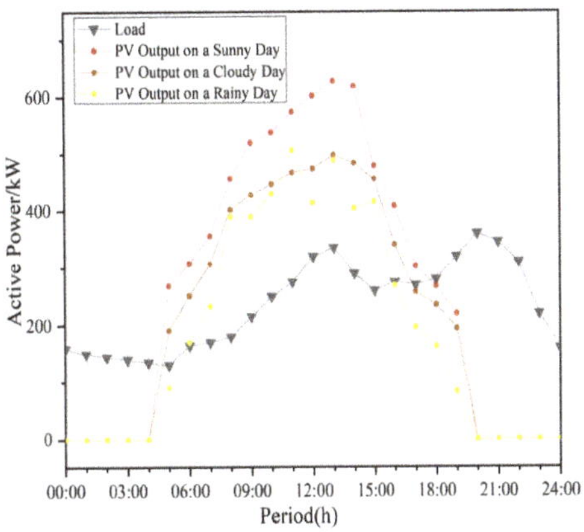

Figure A1. Total PV Generation and Load Forecasting Curve.

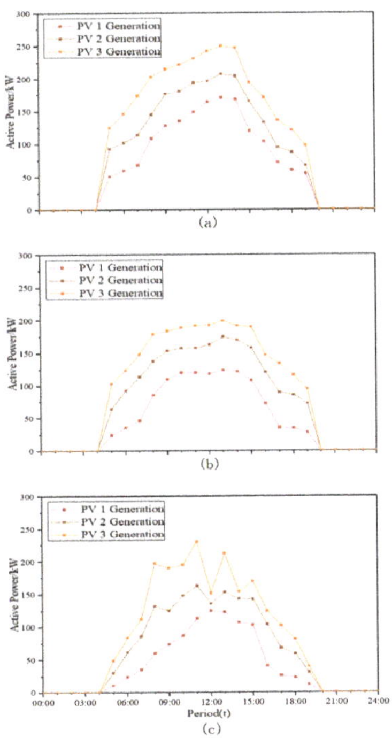

Figure A2. PV Forecast under Three Weather Conditions: (**a**) sunny; (**b**) cloudy; (**c**) rainy.

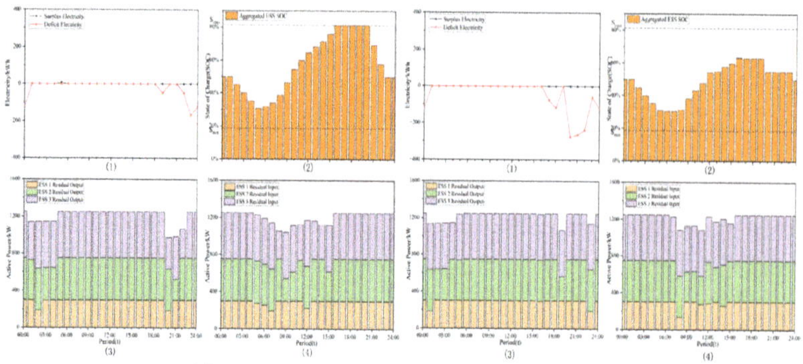

(**a**) (**b**)

Figure A3. Aggregation and Quantification Results. (**a**) Results for Cloudy. (**b**) Results for Rainy.

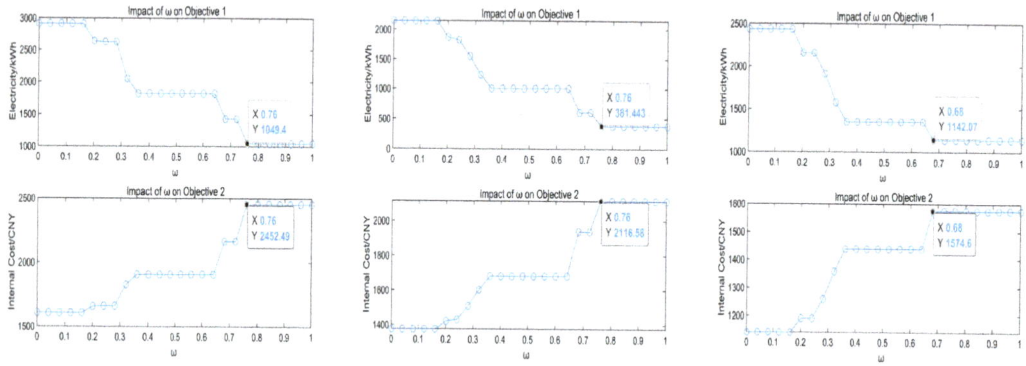

(**a**) (**b**) (**c**)

A4.1 w-Sensitivity with +2 ESSs: (**a**) sunny, (**b**) cloudy, (**c**) rainy.

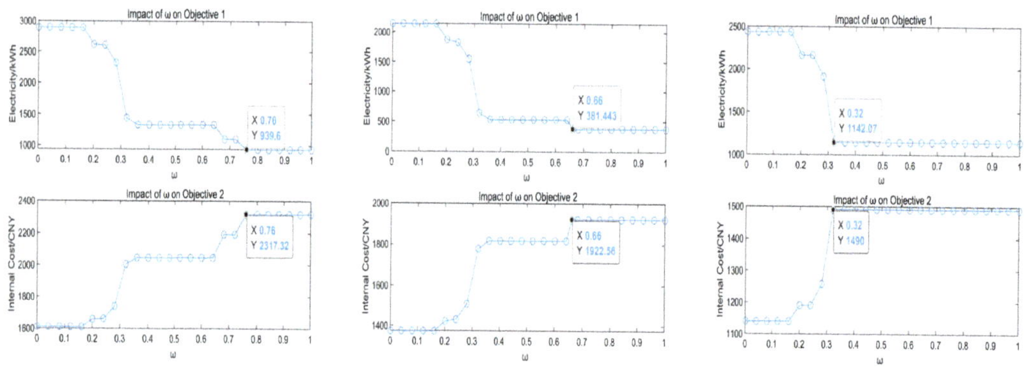

(**a**) (**b**) (**c**)

A4.2 w-Sensitivity with +6 ESSs: (**a**) sunny, (**b**) cloudy, (**c**) rainy.

Figure A4. *Cont.*

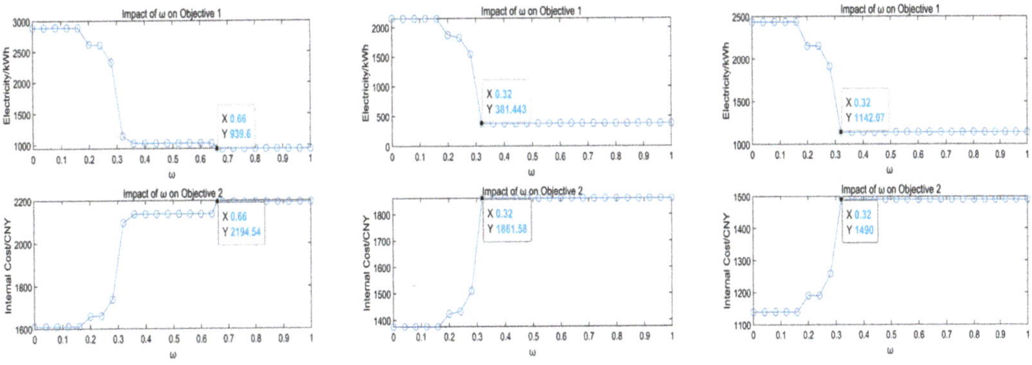

(a) (b) (c)

A4.3 w-Sensitivity with +10 ESSs: (**a**) sunny, (**b**) cloudy, (**c**) rainy.

Figure A4. Sensitivity of w to Multi-Step ESS Additions (two, six, ten units) under Different Weather Conditions.

References

1. Gielen, D.; Boshell, F.; Saygin, D.; Bazilian, M.D.; Wagner, N.; Gorini, R. The Role of Renewable Energy in the Global Energy Transformation. *Energy Strategy Rev.* **2019**, *24*, 38–50. [CrossRef]
2. Kroposki, B.; Johnson, B.; Zhang, Y.C.; Gevorgian, V.; Denholm, P.; Hodge, B.-M.; Hannegan, B. Achieving a 100% Renewable Grid: Operating Electric Power Systems with Extremely High Levels of Variable Renewable Energy. *IEEE Power Energy Mag.* **2017**, *15*, 61–73. [CrossRef]
3. Faisal, M.; Hannan, M.A.; Ker, P.J.; Begum, R.A.; Dong, Z.Y. Review of Energy Storage System Technologies in Microgrid Applications: Issues and Challenges. *IEEE Access* **2018**, *6*, 35143–35164. [CrossRef]
4. Dawoud, S.M.; Lin, X.; Okba, M.I. Hybrid Renewable Microgrid Optimization Techniques A Review. *Renew. Sustain. Energy Rev.* **2018**, *82*, 2039–2052. [CrossRef]
5. Muhtadi, A.; Pandit, D.; Nguyen, N.; Yu, D.C. Distributed Energy Resources Based Microgrid Review of Architecture, Control, and Reliability. *IEEE Trans. Ind. Appl.* **2021**, *57*, 2223–2235. [CrossRef]
6. Silva, V.A.; Aoki, A.R.; Lambert-Torres, G. Optimal Day-Ahead Scheduling of Microgrids with Battery Energy Storage System. *Energies* **2020**, *13*, 5188. [CrossRef]
7. Gomes, J.G.; Xu, H.J.; Yang, Q.; Zhao, C.Y. An Optimization Study on a Typical Renewable Microgrid Energy System with Energy Storage. *Energy* **2021**, *234*, 121210. [CrossRef]
8. Li, Z.; Xu, Y. Optimal Coordinated Energy Dispatch of a Multi-Energy Microgrid in Grid-Connected and Islanded Modes. *Appl. Energy* **2018**, *210*, 974–986. [CrossRef]
9. Nwulu, N.I.; Xia, X. Optimal Dispatch for a Microgrid Incorporating Renewables and demands Response. *Renew. Energy* **2017**, *101*, 16–28. [CrossRef]
10. Witharama, W.M.N.; Bandara, K.M.D.P.; Azeez, M.I.; Logeeshan, V.; Wanigasekara, C. Advanced Genetic Algorithm for Optimal Microgrid Scheduling Considering Solar and Load Forecasting, Battery Degradation and demands Response Dynamics. *IEEE Access* **2024**, *12*, 123456–123470. [CrossRef]
11. Sun, L.; Xu, Q.; Chen, X.; Fan, Y. Day-Ahead Economic Dispatch of Microgrid Based on Game Theory. *Energy Rep.* **2020**, *6*, 633–638. [CrossRef]
12. Silvente, J.; Kopanos, G.M.; Pistikopoulos, E.N. A Rolling Horizon Optimization Framework for the Simultaneous Energy Supply and Demand Planning in Microgrids. *Appl. Energy* **2015**, *155*, 485–501. [CrossRef]
13. Hou, H.; Xue, M.; Xu, Y.; Xiao, Z.; Deng, X.; Xu, T.; Liu, P.; Cui, R. Multi-Objective Economic Dispatch of a Microgrid Considering Electric Vehicle and Transferable Load. *Appl. Energy* **2020**, *262*, 114489. [CrossRef]
14. Aghajani, G.; Ghadimi, N. Multi-Objective Energy Management in a Micro-Grid. *Energy Rep.* **2018**, *4*, 218–225. [CrossRef]
15. Cao, Y.; Lu, D.; Zhang, Y.; Tang, Q.; Khodaei, A.; Zhang, H.; Han, Z. Optimal Energy Management for Multi-Microgrid under a Transactive Energy Framework with Distributionally Robust Optimization. *IEEE Trans. Smart Grid* **2022**, *13*, 599–612. [CrossRef]
16. Nawaz, A.; Wu, J.; Ye, J.; Dong, Y.; Long, C. Distributed MPC-Based Energy Scheduling for Islanded Multi-Microgrid Considering Battery Degradation and Cyclic Life Deterioration. *Appl. Energy* **2023**, *329*, 120168. [CrossRef]

17. Tan, B.; Chen, H.; Zheng, X. Two-Stage Robust Optimization Dispatch for Multiple Microgrids with Electric Vehicle Loads Based on a Novel Data-Driven Uncertainty Set. *Int. J. Electr. Power Energy Syst.* **2022**, *134*, 107359. [CrossRef]
18. Zhao, H.; Wang, B.; Wang, X.; Pan, Z.; Sun, H.; Liu, Z.; Guo, Q. Active Dynamic Aggregation Model for Distributed Integrated Energy System as Virtual Power Plant. *J. Mod. Power Syst. Clean Energy* **2020**, *8*, 831–840. [CrossRef]
19. Wang, S.; Wu, W. Aggregation Reference Model and Quantitative Metric System of Flexible Energy Resources. *Autom. Electr. Power Syst.* **2024**, *48*, 1–9. (In Chinese)
20. Utkarsh, K.; Ding, F.; Jin, X.; Blonsky, M.; Padullaparti, H.; Balamurugan, S.P. A Network-Aware Distributed Energy Resource Aggregation Framework for Flexible, Cost-Optimal, and Resilient Operation. *IEEE Trans. Smart Grid* **2021**, *13*, 1213–1224. [CrossRef]
21. Wen, Y.; Hu, Z.; You, S.; Duan, X. Aggregate Feasible Region of DERs: Exact Formulation and Approximate Models. *IEEE Trans. Smart Grid* **2022**, *13*, 4405–4423. [CrossRef]
22. Yi, Z.; Xu, Y.; Gu, W.; Yang, L.; Sun, H. Aggregate Operation Model for Numerous Small-Capacity Distributed Energy Resources Considering Uncertainty. *IEEE Trans. Smart Grid* **2021**, *12*, 4208–4224. [CrossRef]
23. He, G.; Huang, Y.; Huang, G.; Liu, X.; Li, P.; Zhang, Y. Assessment of Low-Carbon Flexibility in Self-Organized Virtual Power Plants Using Multi-Agent Reinforcement Learning. *Energies* **2024**, *17*, 3688. [CrossRef]
24. Yang, Z.; Wu, R.; Yang, J.; Long, K.; You, P. Economical Operation of Microgrid with Various Devices via Distributed Optimization. *IEEE Trans. Smart Grid* **2016**, *7*, 857–867. [CrossRef]
25. He, Y.; Chen, Y.; Liu, Y.; Liu, H.; Liu, D.; Sun, C. Analysis on Cost per kWh and Cost per mileage for energy storage technologies. *Adv. Technol. Electr. Eng. Energy* **2019**, *38*, 1–10. (In Chinese) [CrossRef]
26. Hamidieh, M.; Ghassemi, M. Microgrids and Resilience: A Review. *IEEE Access* **2022**, *10*, 106059–106080. [CrossRef]
27. Guangdong Provincial Development and Reform Commission. *Implementation Plan for High-Quality Development of Guangzhou Virtual Power Plant (2025)*; Guangdong Provincial Development and Reform Commission: Guangzhou, China, 2023.

energies

MDPI

Article

Planning Scheme for Optimal PMU Location Considering Power System Expansion

Gandhi Carvajal [1,*], Diego Carrión [2] and Manuel Jaramillo [2]

1 Master's Program in Electricity, Salesian Polytechnic University, Quito EC170702, Ecuador
2 Smart Grid Research Group—GIREI (Spanish Acronym), Electrical Engineering Department, Salesian Polytechnic University, Quito EC170702, Ecuador; dcarrion@ups.edu.ec (D.C.); mjaramillo@ups.edu.ec (M.J.)
* Correspondence: gcarvajalm@est.ups.edu.ec

Abstract: In the present research, a methodology was developed for the deployment and optimal location of PMUs with redundancy and system observability constraints at 100%, considering the topological changes generated by transmission expansion planning (TEP). To determine the expansion scenarios and topology changes of the power system, the original IEEE test systems of 14, 30, and 118 buses were used as a starting point, in which a 20-year horizon was considered and an increase in load nodes was proposed, with which the new lines needed were determined by applying TEP. Under these temporal and expansion scenarios, the optimal location of PMUs was determined based on mixed-integer linear optimization, with which the order of location of PMUs was determined. The results of the methodology allow defining the order of PMU implementation in power systems that expand over time.

Keywords: phasor measurement units; transmission expansion planning; power system measurements; mixed-integer linear programming; optimal PMU location

1. Introduction

In recent years, the development of measurement systems has evolved, resulting in improved performance in obtaining and recording system parameters in real time [1]. These systems have an advanced monitoring and control infrastructure called WAMS (Wide-Area Measurement System) [2], which is based on the implementation of Phasor Measurement Units (PMUs) and fast communication systems [3,4].

PMUs are devices composed of microprocessors; the current and voltage signals coming from the measurement equipment are collected and sent in the form of pulses per second (PPS) via global positioning system (GPS) satellites and recorded in the control center. The quality of the information sent by these synchrophasors enables the real-time estimation of the electric power system (EPS). It makes direct and accurate measurements of the phase angles possible [5]. These attributes make PMUs currently the most effective option for obtaining state estimations based on complete topological observations of the system.

One of the main drawbacks to the deployment of PMUs is the limited storage capacity of communication systems [2], which does not allow the sending and storage of the large number of records generated by PMUs. These records are used to derive the optimal PMU placement (OPP) problem. Based on the OPP problem, several strategies have been developed to minimize the number of PMUs, considering various constraints and operating conditions, including total observability [6], N-1 contingencies [7,8], N-2 contingencies, measurement redundancy [9], existing conventional meters [10], PMU channel capacity [10–12], and the PMU loss [12], among others [13]. Some researchers have investigated transmission system expansion [2,8,14].

The OPP problem is an NP-complete binary optimization problem with 2^n possible solutions in an n-bus electric power system (EPS). To solve the optimization problem, several numerical techniques [15] and heuristic models [10] have been developed, which can be categorized into conventional, modern, and miscellaneous techniques [4]. In recent years, several modern methods using heuristics [16] and metaheuristics [12,17–20] have been developed. However, conventional methods are the most widely used, with mixed-integer linear programming (MILP) standing out among them [2,7,9].

On the other hand, the growth of demand in electric power systems poses a challenge for those in charge of planning the optimal expansion of electric transmission systems [21], considering factors and restrictions such as reliability, maximum loadability, angular deviation, voltage profile, resolution method, contingencies, and system loss [22]. In many cases, optimal TEP seeks to strengthen the network to make it stable, safe, and reliable [23], thus supplying demand when a single element fails based on the criterion (N-1) established by the regulations [24].

To address optimal TEP, mathematical [22] and heuristic optimization models can be found, as well as metaheuristic models that result from their combination. Mathematical models provide practical results; however, by not considering all factors or restrictions, an optimal TEP is obtained that omits important technical, environmental, and financial details.

From the perspective of mathematical modeling, TEP is a mixed-integer nonlinear optimization problem [22,25], and the flexibility of the TEP problem allows evaluating it considering different objective functions and constraints that take into account, for example, the cost of a new line, reliability cost, congestion cost, electricity market cost [26], etc. The model constraints can be categorized as either mandatory or optional. Mandatory constraints include generator output power limits, voltage level limits, and transmission power limits on lines. Optional constraints, on the other hand, include investment limits, reliability and safety limits, and environmental impact limits. Taking these into account, the TEP problem is analyzed as a minimization objective function that must consider the mandatory constraints and, depending on the object of study, it can also consider the optional constraints [24].

There are two methods and mathematical models used to solve the TEP problem. The AC model is comprehensive and practical; it considers reactive power and power losses, allows for the inclusion of other devices such as FACTS, and facilitates the analysis of other studies, including voltage stability. However, its resolution becomes lengthy and complex [25,27,28]. In contrast, the DC model is simpler but involves simplifying the constraints; reactive power is not considered, and losses cannot be estimated [29]. However, the agility of its resolution allows for obtaining a good solution to the TEP problem, optimizing time and resources [22].

Heuristic models (HMs) allow the development of skills, procedures, and processes to obtain a reasonable possibility for a specific problem [30]. The results obtained from MHs, due to the lack of a solid foundation, do not offer reliable results for complex networks, which has motivated researchers to use metaheuristic models (MHMs). These models, by using common sense rules, enable an intelligent search for the optimum [31]. Based on the research conducted, MHMs are divided into three categories: evolutionary algorithms, swarm intelligence algorithms, and a combination that merges the principles of the first two. These methods search for an optimum under certain considerations and assumptions; they do not seek to approach the global optimum, so their efficiency is determined by the conditions initially considered by the researcher [24,32].

Considering the studies carried out on the solution to the optimal PMU location and transmission expansion, the present research proposes an optimization model to reduce PMU deployment in the EPS, taking into account the redundancy and observability restrictions, based on mixed-integer programming, within a scheme of optimal planning of the transmission system by stages using the mathematical model from the DC power flow. For this, it is considered that the communications network operates perfectly and is correctly deployed. To test the efficiency of the model, the IEEE 14-, 30-, and 118-bus test systems are utilized to verify the algorithm's functionality by comparing the number and location of PMUs in two scenarios. The first scenario considers the optimal location of PMUs, disregarding investment costs. In contrast, the second scenario considers a limited number of PMUs and their staged deployment, taking into account implementation costs, to evaluate the levels of observability achieved and thus determine the order of PMU deployment.

The main scientific contribution of this work is the integration of optimal PMU placement with dynamic transmission expansion planning over a multi-year horizon. Unlike previous studies, which typically consider static topologies or fixed planning windows, the proposed methodology adapts PMU deployment to topological changes, thereby avoiding relocation and supporting observability throughout the expansion process. This approach reflects the evolving nature of real power systems, offering planners a scalable and cost-conscious framework for developing long-term measurement infrastructure.

The article is organized as follows. Section 2 describes the optimal PMU location and optimal transmission expansion planning. Section 3 describes the problem formulation. Section 4 analyzes the proposed algorithm's performance. Conclusions are shown in Section 5.

2. Power System Expansion Planning

The characteristics of the EPS must be considered when analyzing the optimal location of the PMU. Establishing the connectivity matrix (CM) is the first step. The CM is a binary matrix of dimensions $n_x n$, where n is the number of buses in the system. The topology of the SEP is presented with the CM and is determined by (1) [7].

$$CM_{n_x n} = \begin{cases} 1 & If\ bus_i\ and\ bus_j\ are\ connected \\ 0 & otherwise \end{cases} \quad (1)$$

Taking the topological conditions and characteristics of the SEP, the following rules can be applied to quantify observability:

1. If the voltage and current on one side of the branch are known, the values on the other can be determined by applying Ohm's Law.
2. ZIB and its repercussions: If only one bus is not observable, applying Kirchhoff's Law of Currents makes the bus observable.
3. If the voltage phasors on both sides of a branch are known, the corresponding current phasors can be obtained.

2.1. Optimal PMU Placement Considering Redundancy and Observability Restrictions

To reduce the number of PMUs that must be deployed in the SEP, considering observability and redundancy, (2) is applied; restricted by (3), which assigns at least one PMU to the system, redundancy is maximized with (4), and observability at 100% with (5).

O.F.:

$$\min\ \#_{PMU} = \sum_{i=1}^{n} w_i * X_i \quad (2)$$

s.t.:

$$f(X) \geq 1 \tag{3}$$

$$\max R = (N_{PMU}, X_i) \tag{4}$$

$$X_i + \sum_{\substack{j=1 \\ \forall j \in buses}}^{n} X_j \geq \alpha_i \tag{5}$$

Consider:

$$X_i = \begin{cases} 1 & if\ PMU\ is\ required\ at\ node\ i \\ 0 & otherwise \end{cases} \tag{6}$$

While the individual cost of a PMU may appear relatively low, especially when weighed against the benefits it offers in terms of system observability and control, the reality of deploying these devices is more complex. Beyond the hardware, there are significant associated costs such as communication infrastructure, system integration, security protections, and ongoing maintenance that can quickly add up, particularly in large or evolving power systems. For this reason, minimizing the number of PMUs remains a practical and meaningful objective. Rather than undermining system performance, this approach helps ensure a cost-effective deployment strategy, allowing planners to implement PMUs in the most critical locations first and gradually expand coverage as needed. This is especially valuable in scenarios where budgets are limited or where phased implementation aligns better with long-term expansion plans.

2.2. Location of the Minimum Number of PMUs to Maximize Observability

Considering that the cost of implementing PMUs is usually a limiting factor, the question arises of how many PMUs should be implemented and where they should be located to maximize the system's observability when there is a limited number of PMUs available.

This problem can be solved through the following expression:

Objective function:

$$\max_{X_i, \alpha_i} OF = \sum_{i \in \Omega_B} \alpha_i \tag{7}$$

Subject to:

$$\sum_{i \in \Omega_B} X_i \leq N_{PMU} \tag{8}$$

$$X_i + \sum_{j \in \Omega_{ij}^l} X_j \geq \alpha_i; \forall i \in \Omega_B \tag{9}$$

$$1 \leq \alpha_i \tag{10}$$

2.3. Transmission Expansion Planning

The planning of a power system must be implemented before the design and construction activities. It must consider defined processes in the short, medium, and long term, taking into account that short-term planning is related to limitations in the system's operation, which must be addressed to avoid inconveniences with meeting demand. In contrast, medium-term planning is related to the repowering of the WSS while maintaining a backup or redundancy criterion that ensures system reliability. Long-term planning is oriented towards the construction of new facilities based on projected demand growth [33,34].

The objective function for the TEP is defined in Equation (11), subject to the restrictions expressed in Equations (12) and (13), which relate to operating costs and initial investment.

In addition, the criteria of maximum loadability of transmission lines expressed in (14)–(17) are applied, the power limits of the generators in (18) are also considered and the balance between demand and generation expressed in (19).

F.O:

$$minOF = T \times OC \ + IC \tag{11}$$

s.t.:

$$OC = \sum_{i=1}^{n_{bus}} (C_g \times P_g);\ \forall\, g \in n_g \tag{12}$$

$$IC = (\eta^0_{i,j} + \sum \beta^k_{i,j}) \times CL_{i,j} \tag{13}$$

$$P^k_{i,j} - B_{i,j}(\delta_i - \delta_j) \leqslant (1 - \beta^k_{i,j}) \times M \tag{14}$$

$$P^k_{i,j} - B_{i,j}(\delta_i - \delta_j) \geq -(1 - \beta^k_{i,j}) \times M \tag{15}$$

$$M = max\big(B_{i,j} \times (\delta_i - \delta_j)\big) \tag{16}$$

$$-P^{max}_{i,j} \times \beta^k_{i,j} \leqslant\ P^k_{i,j} \leqslant\ P^{max}_{i,j} \times \beta^k_{i,j};\ \forall\, i,j\ \in n_L \tag{17}$$

$$Pg^{min}_i \leqslant\ Pg_i \leqslant\ Pg^{max}_i;\ \forall\, g_i\ \in n_g \tag{18}$$

$$\sum P_{i,j} = \sum P_g + LS_i - L_i;\ \forall\, i,j\ \in n_{bus} \tag{19}$$

To provide a broader perspective on existing approaches to PMU placement and transmission expansion planning, we have included a comparative summary of representative studies in Table 1. This table highlights the main characteristics, advantages, and limitations of each methodology, with an emphasis on their optimization techniques, observability criteria, and treatment of system growth or contingencies. By reviewing these contributions side by side, the distinctiveness of our proposed method, particularly its integration of stepwise transmission expansion and observability-aware PMU deployment, becomes more evident. This contextual analysis also helps underline the practical need for a planning framework that adapts dynamically as the power system evolves.

Table 1. Comparative summary of representative PMU placement and expansion planning methodologies.

Reference	Methodology	Optimization Approach	Key Features	Advantages	Limitations
[1]	PMU placement taxonomy	Various (ILP, metaheuristics)	Comprehensive classification of methods	Broad overview, useful for method selection	Does not propose a new optimization model
[7]	PMU placement with N-1 contingencies	MILP	Based on Ecuadorian EPS	Realistic case constraints	No cost or staged planning
[8]	PMU placement with N-1 contingencies	MILP	Integrates reliability constraints	Practical for contingency analysis	No temporal expansion modeled

Table 1. *Cont.*

Reference	Methodology	Optimization Approach	Key Features	Advantages	Limitations
[9]	Probabilistic PMU placement	MILP (multi-objective)	Considers uncertainty in observability	Robust under system variability	Focused on static systems
[14]	Multi-stage PMU placement with expansion	Hybrid (mathematical + heuristic)	Accounts for network expansion	Introduces staging	Complexity increases with scale
[This paper]	PMU placement with expansion over time	MILP (stepwise planning)	Observability + redundancy + expansion	Adaptable, cost-aware, supports scheduling	Currently excludes contingencies and cyber-resilience

3. Problem Formulation and Methodology

The optimal location of PMUs has been studied extensively. Various methods exist for determining the optimal number of devices to be deployed; similarly, the TEP problem can be analyzed from multiple perspectives to inform the expansion planning model of the EPS.

The present research seeks to establish a methodology to determine the minimum number of PMUs that should be deployed in each stage of a transmission expansion model to obtain 100% observability of the new system topology considering the location of existing PMUs and then evaluate the levels of observability obtained by having limited numbers of PMUs for deployment; in this way, the order and number of PMUs to be deployed is obtained under the observability criterion, algorithm 1.

Algorithm 1 describes the location of PMUs considering TEP, for which in step 1 the EPS to be studied is initialized, in step two the PMU location is carried out considering observability and redundancy restrictions, starting from a previous location from a previous year, and of course for year 1 there will be no PMUs in the EPS. In step 3, the planning of the EPS expansion is carried out every T years. For each analysis, it returns to step 2 to identify if a new PMU location is required. Finally, in step 4, the results of the expansion and the PMU location are shown for each year T.

Figure 1 illustrates the overall workflow of the proposed methodology, combining PMU placement with transmission expansion planning (TEP) across multiple stages. The iterative nature of the algorithm allows the system to adapt its measurement infrastructure as the network grows over time.

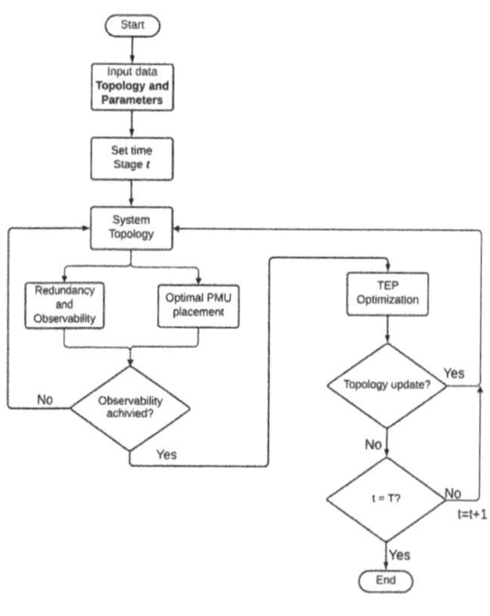

Figure 1. Flowchart of the proposed methodology.

Algorithm 1 Optimal PMU location considering TEP

Step: 1 **Input data**
EPS Parameterization
Gen_data, line_data, load_data
$CM_{nxn}, W_i, year_t$
Step: 2 **PMU Minimization**
Only consider nodes with $X_i = 0$
O.F.:
Equation (2)
s.t.:
Equations (3)–(5), (8) and (9)
Save X_{i_new}
for i=1 to n
If $X_{i\_new} = 1$
$X_i = 1$
Save X_i
Else
$X_i = 0$
Save X_i
endfor
Step: 3 **TEP**
for $year_t$ = 1 to T
O.F.:
Equation (11)
s.t.:
Equations (12)–(19)
Save results in $Data_{new}(:, year_t)$
Go to Step 2
endfor
Step: 4 **Show results for each year**
$X_i, NSW, v_{i,j}, \alpha^k_{i,j}, P^k_{i,j}, \delta_i$

4. Results Analysis

The IEEE 14-bus model is used to explain the results obtained; two transmission expansion scenarios are proposed to demonstrate the effectiveness and benefits of the proposed model. In the first scenario, previous stages of expansion are proposed, starting from systems of 10 buses, 12 buses, and 14 buses. It is proposed that the system expansion be carried out every 5 years; the topology changes are generated by incorporating the lines that connect the new buses. In each stage, algorithms 1 and 3 are applied to determine the order of location of the minimum number of PMUs; in the first system, the minimum number of PMUs is obtained that guarantees observability equal to or greater than 85%, and after this it is proposed that the investment can be made to install a PMU every two years.

Figure 2 shows how the location of PMUs in the EPS is planned, taking into account the expansion works in the power system. In year 0 an initial location is planned, which is the basis for future PMU locations; from there, in year 5, where two new nodes are implemented, the proposed methodology suggests the location of a new PMU, and finally, in year 10, where two nodes are added, a new PMU is required, giving a total of six PMUs in the EPS.

Table 2 shows the results obtained, and a chronological analysis is conducted of the number of PMUs and the observability achieved considering the PMU expansion and deployment scenarios; at the end of the expansion and deployment stages, six PMUs are needed on buses 2, 5, 6, 8, 10 and 14 to guarantee 100% observability of the system (Figure 2); when applying traditional PMU optimal location algorithms to the

IEEE model of 14 buses, it is obtained that four PMUs are required on buses 2, 6, 7 and 9, contrasting the results obtained; the number of PMUs under an expansion scheme without considering relocation is greater to guarantee total observability of the system, which implies a greater investment of resources.

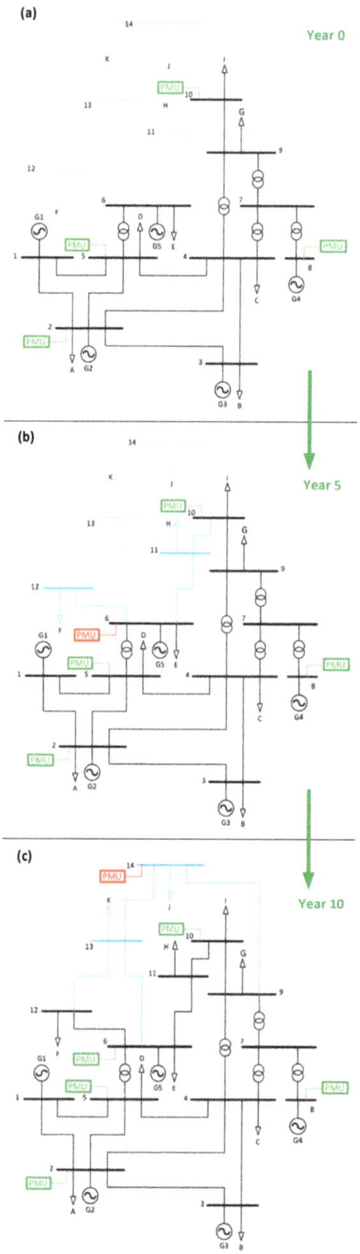

Figure 2. PMU optimal location: (**a**) year 0, (**b**) year 5, (**c**) year 10.

Table 2. IEEE 14-bus PMU location, every 2 years.

Year	# Nodes	# PMUs	Nodes with PMU	# Buses Observed
0	10	4	2, 5, 8, 10	9 (90%)
2	10	4	2, 5, 8, 10	10 (100%)
4	10	4	2, 5, 8, 10	10 (100%)
6	12	5	2, 5, 6, 8, 10	12 (100%)
8	12	5	2, 5, 6, 8, 10	12 (100%)
10	14	6	2, 5, 6, 8, 10, 14	14 (100%)

To provide a clearer perspective on the effectiveness of our approach, we compared it with a conventional PMU placement strategy that assumes a static system topology. For the IEEE 14-bus test case, traditional models generally identify four PMUs (e.g., at buses 2, 6, 7, and 9) as sufficient to achieve full observability. In contrast, our methodology, designed to accommodate system expansion, proposes a total of six PMUs, placed progressively at buses 2, 5, 6, 8, 10, and 14. While the number may seem higher at first glance, it is essential to note that these PMUs are deployed incrementally, and none require relocation as the network expands. This characteristic is particularly valuable in real-world settings, where infrastructure changes are implemented over time and budgets are typically allocated in stages. By ensuring observability at every step without redundancy losses or reinstallation costs, the proposed model offers a practical and forward-looking alternative to static placement strategies.

Figure 3 shows the evolution of PMU deployment and observability for the IEEE 14-bus system over six expansion stages. The left axis indicates the number of PMUs installed, while the right axis reflects the percentage of observability achieved.

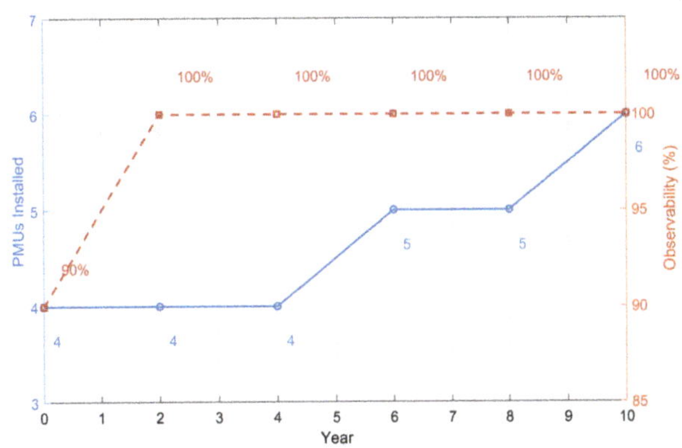

Figure 3. Evolution of PMUs installed (left axis) and system observability (right axis) under the proposed deployment strategy for the IEEE 14-bus system.

Initially, four PMUs yield 90% observability, and full coverage is reached by year 2 without additional devices. As new buses are added in years 6 and 10, the method adds only the necessary PMUs to maintain 100% observability, demonstrating a deployment strategy that adapts efficiently to system growth.

By anticipating future expansions, the method minimizes reconfiguration and ensures long-term monitoring continuity with minimal investment.

Figure 4 shows the observability coverage of the EPS nodes for each stage considered. In year 0, there is 100% coverage with redundancy in nodes 1 and 5 in only four zones. in year 5, there are five zones to achieve 100% observability and redundancy in nodes 1,

5, and 6; finally, in year 10, there is 100% observability with six zones and redundancy in nodes 1, 5, 6, and 9.

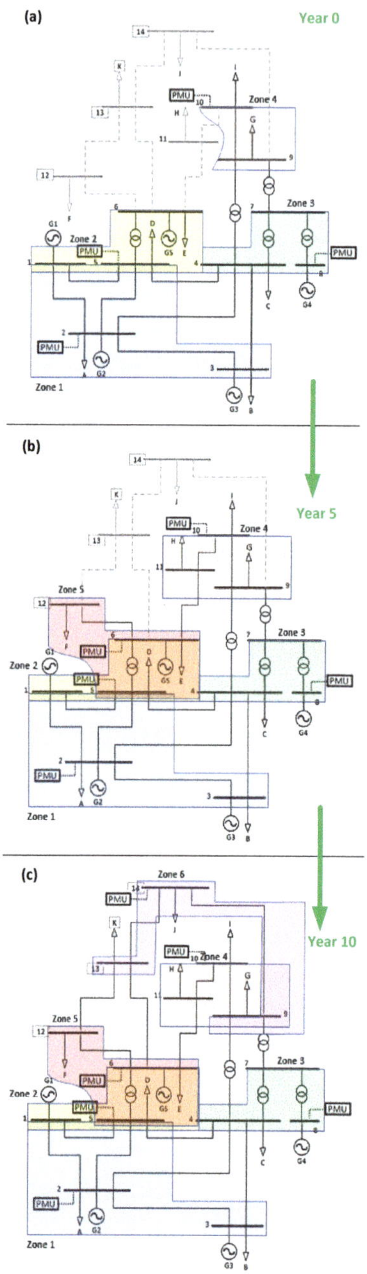

Figure 4. IEEE 14 bus observability coverage: (**a**) year 0, (**b**) year 5, (**c**) year 10.

The second scenario considers future stages of transmission expansion based on assumptions, and the minimum number of PMUs is deployed considering redundancy and observability restrictions; taking into consideration the locations found in the first scenario, the proposed algorithm is then applied to determine the order of implementation; similarly, it is considered that the system is expanded every 5 years and that the implementation of a new PMUs can be carried out within a period of two years; the results are shown in Table 3.

Table 3. PMU location IEEE 14 buses, future stages.

Year	# Nodes	# PMUs	Nodes with PMU	# Buses Observed
12	14	6	2, 5, 6, 8, 10, 14	14 (100%)
14	14	6	2, 5, 6, 8, 10, 14	14 (100%)
16	15	7	2, 5, 6, 8, 10, 14, 15	15 (100%)
18	15	7	2, 5, 6, 8, 10, 14, 15	15 (100%)
20	17	8	2, 5, 6, 8, 10, 14, 15, 16	16 (100%)

The results presented in Tables 2 and 3 show the PMU placement order obtained to maximize observability with a restricted number of PMUs per expansion stage, which depends on the system topology. The generation of deployment scenarios provides the planner with a clearer perspective on the levels of observability achieved after a stage of transmission system expansion and the levels that can be attained by installing new PMUs.

Table 4 presents three possible assumptions for increasing the number of load nodes in the IEEE 14-bus model, specifying their power requirements and the potential transmission lines that must be constructed to meet these new demands. It is expected that the new node 15 will come into operation by 2015, and nodes 16 and 17 by 2020. Figure 5 shows the increases in lines due to the expansion of the system and the increase in PMUs while maintaining what was already planned in the original system.

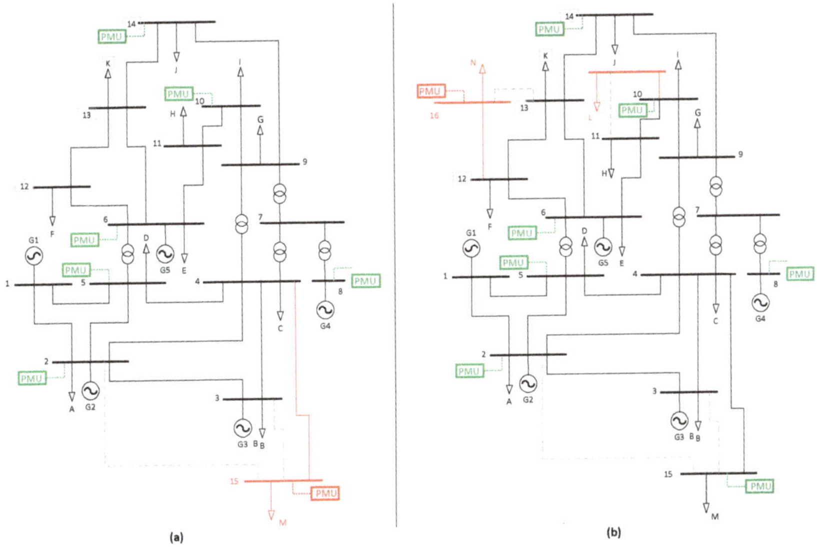

Figure 5. PMU optimal location: (**a**) year 15, (**b**) year 20.

To reflect real-world planning conditions, the cost of PMU implementation is addressed indirectly in our methodology, not through a direct monetary value in the objective function, but by limiting the number of devices installed at each stage. This staged approach helps keep the model computationally efficient while still accounting for typical investment constraints. In practice, utilities rarely install all PMUs at once; budgets, logistics, and system growth often require a gradual rollout. Our model captures that reality by ensuring observability improves with each expansion step, without the need to relocate already installed units. While no explicit financial figures are provided, the strategy aligns with how deployment decisions are made in the field, helping to minimize unnecessary costs and support long-term, resource-conscious implementation, particularly in systems with limited budgets or in developing regions.

Table 4. Expansion scenarios.

Total Nodes	New Loads	Candidates Lines	New Lines
14	bus 15 45 kW	2–15, 3–15, 4–15	4–15
15	bus 16 8 kW	12–16, 13–16	12–16
16	bus 17 5 kW	10–17, 11–17	10–17

Figure 6 shows the results of the modified IEEE 14-bar system coverage, based on what was planned in year 10 and considering the new nodes and lines. In year 15, the system has 100% observability with seven zones and redundancy in nodes 1, 4, 5, 6, and 9. On the other hand, in year 20, the observability coverage zone 4 grows, with direct visualization of nodes 9, 10, and 11, plus the new node 17, for a total of eight zones to achieve 100% observability with redundancy in nodes 1, 4, 5, 6, 9, and 12.

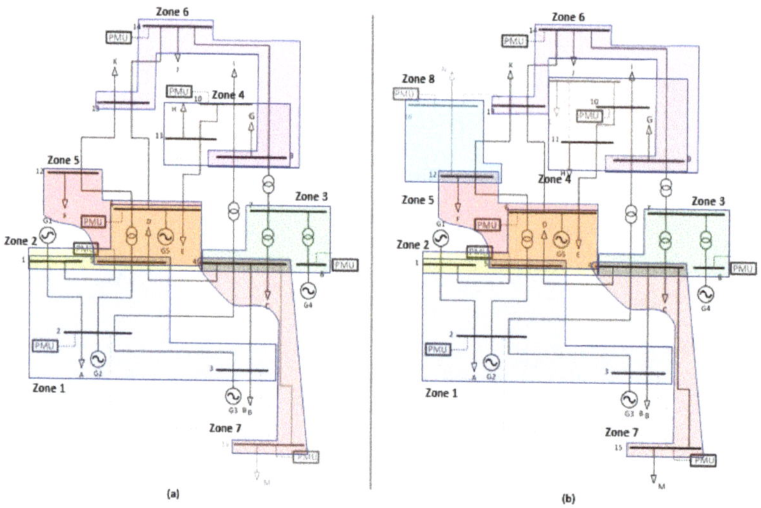

Figure 6. IEEE 14 modified bus observability coverage: (**a**) year 15, (**b**) year 20.

The chronology of system expansion and PMU deployment for the 30 and 118 busbar systems is shown in Tables 5 and 6; for the 30-bus system, a system expansion is considered every five years and a PMU is deployed every two years; for the 118-bus system, a system expansion is regarded as every five years and two PMUs are deployed every year.

Figure 7 presents the staged deployment of PMUs and the evolution of system observability for the IEEE 30-bus network. The left axis shows the number of PMUs installed over time, while the right axis reflects the percentage of observed buses at each expansion stage.

Table 5. PMU location IEEE 30 busbars, without PMU relocation.

Year	# Nodes	# PMUs	Nodes with PMU	# Buses Observed
0	20	4	1, 6, 12, 19	17 (85%)
2	20	5	1, 2, 6, 9, 12, 19	18 (90%)
4	20	6	1, 2, 6, 9, 12, 19	19 (95%)
5	24	6	1, 2, 6, 9, 12, 19	20 (83%)
6	24	7	1, 2, 6, 9, 10, 12, 19	22 (92%)
8	24	8	1, 2, 6, 9, 10, 12, 19, 23	24 (100%)
10	28	9	1, 2, 6, 9, 10, 12, 19, 23, 25	28 (100%)
15	30	9	1, 2, 6, 9, 10, 12, 19, 23, 25	28 (93%)
16	30	10	1, 2, 6, 9, 10, 12, 19, 23, 25, 27	30 (100%)

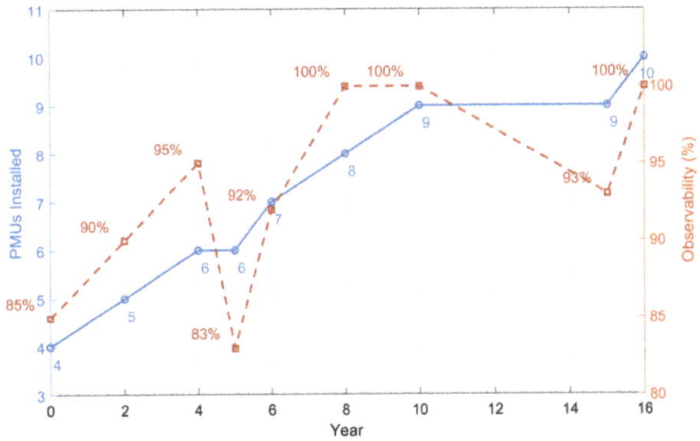

Figure 7. Evolution of PMUs installed (left axis) and system observability (right axis) under the proposed deployment strategy for the IEEE 30-bus system.

Table 6. PMU location IEEE 118 busbars, without PMU relocation.

Year	# Nodes	# PMUs	Nodes with PMU	# Buses Observed
0	100	19	1, 7, 9, 11, 17, 21, 25, 28, 34, 37, 49 56, 59, 65, 70, 77, 80, 85, 94	86 (86%)
1	100	21	1, 7, 9, 11, 17, 21, 25, 28, 34, 37, 49, 52 56, 59, 62, 65, 70, 71, 77, 80, 85, 94	90 (90%)
2	100	23	1, 7, 9, 11, 17, 21, 25, 28, 34, 37, 45, 49 52, 56, 59, 62, 65, 70, 71, 77, 80, 85, 94	94 (94%)
3	100	25	1, 7, 9, 11, 15, 17, 21, 25, 28, 34, 37, 45, 49 52, 56, 59, 62, 65, 70, 71, 77, 80, 85, 91, 94	97 (97%)
4	100	27	1, 7, 9, 11, 15, 17, 21, 25, 28, 31, 34, 37, 40, 45 49, 52, 56, 59, 62, 65, 70, 71, 77, 80, 85, 91, 94	99 (99%)
6	100	29	1, 7, 9, 11, 15, 17, 21, 25, 28, 31, 34, 37, 40, 45, 49, 52 56, 59, 62, 65, 70, 71, 77, 80, 85, 91, 94, 100, 101	101 (100%)
8	100	30	1, 7, 9, 11, 15, 17, 21, 25, 28, 31, 34, 37, 40, 45, 49, 52 56, 59, 62, 65, 70, 71, 77, 80, 85, 87, 91, 94, 100, 101	105 (97%)
8	100	31	1, 7, 9, 11, 15, 17, 21, 25, 28, 31, 34, 37, 40, 45, 49, 52, 56 59, 62, 65, 70, 71, 77, 80, 85, 87, 91, 94, 100, 101, 105	118 (100%)

Initially, four PMUs provide 85% coverage. As the network grows and more nodes are integrated, observability improves gradually through targeted additions of PMUs. Notably, observability briefly drops in year 5 due to system expansion outpacing measurement capacity, but is quickly recovered and optimized by year 8. From that point on, full observability is maintained, even as the system expands to 30 buses.

This progression highlights the methodology's ability to anticipate structural changes and adjust placements accordingly, minimizing redundancy while ensuring consistent monitoring. The result is a robust and forward-looking deployment strategy that aligns closely with real-world system growth patterns.

Figure 8 shows the progression of PMU deployment and system observability for the IEEE 118-bus network. The left axis represents the number of PMUs installed over time, while the right axis displays the percentage of observability achieved at each stage.

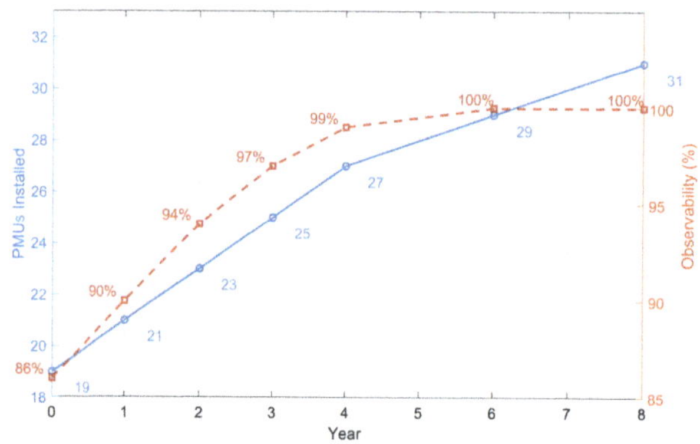

Figure 8. Evolution of PMUs installed (left axis) and system observability (right axis) under the proposed deployment strategy for the IEEE 118-bus system.

Starting with 19 PMUs in year 0, the system achieves 86% observability. As the method adds PMUs in anticipation of future expansion, observability improves gradually, reaching 99% by year 4. Full observability (100%) is attained by year 6 with 29 devices and remains stable even as the network continues to evolve.

This behavior demonstrates how the method strategically distributes PMUs to adapt to a large-scale system while minimizing unnecessary relocations. Each device contributes to both present and future coverage, ensuring scalability and measurement continuity as the system grows in complexity.

Although emergency operation scenarios were not the primary focus of this study, the proposed methodology can support emergency management by ensuring full observability throughout the system at every stage of expansion. With PMUs strategically deployed over time, system operators gain access to real-time, synchronized data across the grid, which enhances situational awareness, accelerates fault detection, and improves decision making during abnormal operating conditions. In this way, the planning approach contributes not only to the efficient deployment of infrastructure but also to strengthening the system's readiness to respond to critical events.

In all test cases, the expansion of the power system directly influences the number and positioning of new PMUs. The proposed algorithm integrates topological forecasts to ensure that each PMU installed during earlier stages remains effective throughout the entire expansion process. It is important to note that none of the PMUs deployed in early stages required relocation, demonstrating the model's robustness in anticipating long-term observability needs.

5. Conclusions

In contrast to traditional static approaches, this method ensures full observability at every step of network growth while maintaining deployment continuity, which is essential for utility planning under evolving grid topologies and constrained investment cycles. The dual-axis visualizations introduced in this work (Figures 3, 7 and 8) effectively illustrate the temporal evolution of PMU deployment alongside system observability, allowing intuitive evaluation of coverage efficiency across staged expansions.

The developed research presents a methodology to determine the minimum number of PMUs to be deployed in power systems, considering the topology changes determined by the transmission expansion, where the main contribution is focused on determining the timing of implementation of the measurement devices to maximize the observability of the EPS during the time interval considered in the EPS expansion.

Expansion planning typically focuses on the implementation or reinforcement of transmission lines, power plants, substations, and other key elements of the power system. However, it must also consider aspects that involve monitoring, such as PMUs. The location of PMUs in power systems depends on the SEP topology, and if this changes in the future, the present location may become unnecessary. This consideration was taken into account in the proposed methodology.

Planning for PMU expansion, considering the optimal location, maximizing observability, and minimizing the number of PMUs can be regarded as a multi-objective problem. Still, to avoid this in the proposed methodology, the optimization problem is relaxed by converting the observability maximization into a constraint, which gives reliable results and allows planning the PMU implementation time in the candidate nodes.

This is because the methodology considers the planning period, which for this study was 10 years. Therefore, not having the EPS in its final state does not achieve full observability of the system. Still, as the years pass and the topology changes, PMUs are placed in such a way that it is not necessary to remove the existing ones; instead, they serve as the basis for the new PMUs.

From a practical standpoint, the proposed methodology offers a flexible and cost-aware approach to PMU deployment. Instead of assuming full installation from the outset, it reflects the way many utilities operate, implementing gradually as budgets and network expansions progress. By limiting the number of PMUs per stage and avoiding device relocation, the model helps reduce unnecessary expenditures and aligns with the realities of long-term planning. This makes it especially valuable for developing systems or scenarios where financial and logistical constraints are key decision factors. While the model does not include explicit monetary terms, its structure effectively supports economically sustainable implementation.

While the present study focuses on transmission systems, the proposed methodology is flexible and can be adapted to distribution networks. By redefining the connectivity matrix and adjusting observability constraints to match the characteristics of radial or weakly meshed systems, the same optimization framework can support staged PMU deployment in distribution environments. This opens the possibility of extending the model to a broader range of practical applications.

The current version of the model does not explicitly incorporate N-1 or N-2 contingency scenarios. However, the optimization framework can be extended to include such reliability constraints by adding additional observability conditions under failure modes. This represents a promising direction for future work, particularly in applications related to critical infrastructure planning.

Author Contributions: Conceptualization, D.C.; Methodology, G.C.; Formal analysis, G.C.; Data curation, M.J.; Writing original draft, G.C.; Writing review & editing, D.C. and M.J.; Supervision, D.C.; Project administration, D.C.; Funding acquisition, D.C. All authors have read and agreed to the published version of the manuscript.

Funding: This work was supported by Salesian Polytechnic University and GIREI-Smart Grid Research Group under the project "Integration of Artificial Intelligence in the Planning, Operation, and Protection of Electrical Power Systems", Resolution No.: 083-02-2024-04-12.

Data Availability Statement: The original contributions presented in this study are included in the article. Further inquiries can be directed to the corresponding author.

Conflicts of Interest: The authors declare no conflicts of interest.

Abbreviations

The abbreviations used in this article are as follows:

CM_{nxn}	EPS connectivity matrix
w_i	PMU implementation cost at node i
X_i	PMU location binary variable at node i
α_i	EPS observability percentage
N_{PMU}	PMU quantity
T	Planning period
OC	Operation costs
IC	Investment costs
n_g	generators set
n_{bus}	Nodes set
C_g	Generation production costs
P_g	Generator real power
$\eta^0_{i,j}$	Initial state of the line between nodes i, j
$\beta^k_{i,j}$	Binary variable representing the status of the line
$CL_{i,j}$	Cost of the candidate line between nodes i, j
$P_{i,j}$	Power flow limit per line
$P^{max}_{i,j}$	Maximum power flow limit per line
$B_{i,j}$	Susceptance of the line between nodes $i - j$
δ_i	Voltage angle at node i
M	maximum line load capacity
P_{g_i}	Real power generated by the generator i
$P^{max}_{g_i}$	Maximum active power limit of generators
$P^{min}_{g_i}$	Minimum active power limit for generator
LS_i	Load shedding at node i
L_i	Load at node i

References

1. Manousakis, N.M.; Korres, G.N.; Georgilakis, P.S. Taxonomy of PMU placement methodologies. *IEEE Trans. Power Syst.* **2012**, *27*, 1070–1077. [CrossRef]
2. Ghamsari-Yazdel, M.; Esmaili, M.; Amjady, N. Optimal substation-based joint allocation of PMUs and measuring channels considering network expansion planning. *Int. J. Electr. Power Energy Syst.* **2019**, *106*, 274–287. [CrossRef]
3. Aminifar, F.; Fotuhi-Firuzabad, M.; Safdarian, A.; Davoudi, A.; Shahidehpour, M. Synchrophasor Measurement Technology in Power Systems: Panorama and State-of-the-Art. *IEEE Access* **2014**, *2*, 1607–1628. [CrossRef]
4. Mohanta, D.K.; Murthy, C.; Sinha Roy, D. A Brief Review of Phasor Measurement Units as Sensors for Smart Grid. *Electr. Power Compon. Syst.* **2016**, *44*, 411–425. [CrossRef]
5. Johnson, T.; Moger, T. A critical review of methods for optimal placement of phasor measurement units. *Int. Trans. Electr. Energy Syst.* **2021**, *31*, 1–25. [CrossRef]

6. Babu, R.; Raj, S.; Bhattacharyya, B. Weak bus-constrained PMU placement for complete observability of a connected power network considering voltage stability indices. *Prot. Control Mod. Power Syst.* **2020**, *5*, 28. [CrossRef]
7. Carrión Galarza, D.F.; González Sánchez, J.W. Ubicación óptima de PMU considerando restricciones de contingencias N-1 en sistemas eléctricos de potencia. *Enfoque UTE* **2019**, *10*, 1–12. [CrossRef]
8. Asgari, A.; Firouzjah, K.G. Optimal PMU placement for power system observability considering network expansion and N-1 contingencies. *IET Gener. Transm. Distrib.* **2018**, *12*, 4216–4224. [CrossRef]
9. Aghaei, J.; Baharvandi, A.; Rabiee, A.; Akbari, M.A. Probabilistic PMU Placement in Electric Power Networks: An MILP-Based Multiobjective Model. *IEEE Trans. Ind. Inform.* **2015**, *11*, 332–341. [CrossRef]
10. Manousakis, N.M.; Korres, G.N. Optimal PMU arrangement considering limited channel capacity and transformer tap settings. *IET Gener. Transm. Distrib.* **2020**, *14*, 5816–5822. [CrossRef]
11. Manousakis, N.M.; Korres, G.N. Optimal PMU Placement for Numerical Observability Considering Fixed Channel Capacity-A Semidefinite Programming Approach. *IEEE Trans. Power Syst.* **2016**, *31*, 3328–3329. [CrossRef]
12. Hanis, N.; Rahman, A.; Zobaa, A.F. Binary PSO Algorithm for Optimal PMUs Placement. *IEEE Trans. Ind. Inform.* **2017**, *13*, 3124–3133.
13. Bae, J. Cost-Effective placement of phasor measurement units to defend against false data injection attacks on power grid. *Energies* **2020**, *13*, 3862. [CrossRef]
14. Ali, Z.M.; Razavi, S.E.; Javadi, M.S.; Gandoman, F.H.; Abdel Aleem, S.H. Dual enhancement of power system monitoring: Improved probabilistic multi-stage PMU placement with an increased search space & mathematical linear expansion to consider zero-injection bus. *Energies* **2018**, *11*, 1429. [CrossRef]
15. Bečejac, V.; Stefanov, P. Groebner bases algorithm for optimal PMU placement. *Int. J. Electr. Power Energy Syst.* **2020**, *115*, 105427. [CrossRef]
16. Abd el Ghany, H.A. Optimal PMU allocation for high-sensitivity wide-area backup protection scheme of transmission lines. *Electr. Power Syst. Res.* **2020**, *187*, 106485. [CrossRef]
17. Mohammadi, P.; Mehraeen, S.; Nazaripouya, H. Sensitivity analysis-based optimal PMU placement for fault observability. *IET Gener. Transm. Distrib.* **2021**, *15*, 737–750. [CrossRef]
18. Ghosh, P.K.; Chatterjee, S.; Roy, B.K.S. Optimal PMU placement solution: Graph theory and MCDM-based approach. *IET Gener. Transm. Distrib.* **2017**, *11*, 3371–3380. [CrossRef]
19. Razavi, S.E.; Falaghi, H.; Esmaeel Nezhad, A.; Hadidian Moghaddam, M.J.; Gandoman, F.H. An Effective Approach for the Probabilistic and Deterministic Multistage PMU Placement Using Cuckoo Search: Iran's National Power System. *Iran. J. Sci. Technol.-Trans. Electr. Eng.* **2020**, *44*, 237–252. [CrossRef]
20. Saleh, A.A.; Adail, A.S.; Wadoud, A.A. Optimal phasor measurement units placement for full observability of power system using improved particle swarm optimisation. *IET Gener. Transm. Distrib.* **2017**, *11*, 1794–1800. [CrossRef]
21. Gacitua, L.; Gallegos, P.; Henriquez-Auba, R.; Lorca.; Negrete-Pincetic, M.; Olivares, D.; Valenzuela, A.; Wenzel, G. A comprehensive review on expansion planning: Models and tools for energy policy analysis. *Renew. Sustain. Energy Rev.* **2018**, *98*, 346–360. [CrossRef]
22. Giral Ramírez, D.A.; Montoya Giraldo, O.D.; Vargas Robayo, C.Y.; Blanco Valbuena, D.F. Evaluación de modelos de programación lineal y no lineal para la planeación de sistemas de transmisión en el software GAMS. *Tecnura* **2021**, *25*, 16–50. [CrossRef]
23. Bolaños Ocampo, R.A.; Correa Flórez, C.A. Planeamiento de la transmisión considerando seguridad e incertidumbre en la demanda empleando programación no lineal y técnicas evolutivas. *Rev. Tecnura* **2013**, *18*, 62. [CrossRef]
24. Palacios, J. Estado del arte de la planeación de expansión de sistemas de transmisión. *I+D Tecnol.* **2020**, *16*, 78–85.
25. Hamidpour, H.; Aghaei, J.; Pirouzi, S.; Niknam, T.; Nikoobakht, A.; Lehtonen, M.; Shafie-khah, M.; Catalão, J.P. Coordinated expansion planning problem considering wind farms, energy storage systems and demand response. *Energy* **2022**, *239*, 122321. [CrossRef]
26. Zhao, J.H.; Dong, Z.Y.; Lindsay, P.; Wong, K.P. Flexible transmission expansion planning with uncertainties in an electricity market. *IEEE Trans. Power Syst.* **2009**, *24*, 479–488. [CrossRef]
27. Vilaça, P.; Street, A.; Colmenar, J.M. A MILP-based heuristic algorithm for transmission expansion planning problems. *Electr. Power Syst. Res.* **2022**, *208*, 107882. [CrossRef]
28. Escudero, P.; Carrión, D. Expansión de un sistema de transmisión mediante LOPF-AC Expansion of a transmission system using LOPF-AC. *INGE CUC* **2018**, *14*, 116–125. [CrossRef]
29. Tejada-Arango, D.; López-Lezama, J.M.; Rider, M.J. Planeamiento de la expansión de la red de transmisión en sistemas eléctricos de potencia considerando reconfiguraciones de red. *Inf. Tecnol.* **2014**, *25*, 85–96. [CrossRef]
30. Li, C.; Conejo, A.J.; Liu, P.; Omell, B.P.; Siirola, J.D.; Grossmann, I.E. Mixed-integer linear programming models and algorithms for generation and transmission expansion planning of power systems. *Eur. J. Oper. Res.* **2022**, *297*, 1071–1082. [CrossRef]

31. Esmaili, M.; Ghamsari-Yazdel, M.; Amjady, N.; Chung, C.Y.; Conejo, A.J. Transmission Expansion Planning including TCSCs and SFCLs: A MINLP Approach. *IEEE Trans. Power Syst.* **2020**, *35*, 4396–4407. [CrossRef]
32. Hemmati, R.; Hooshmand, R.A.; Khodabakhshian, A. State-of-the-art of transmission expansion planning: Comprehensive review. *Renew. Sustain. Energy Rev.* **2013**, *23*, 312–319. [CrossRef]
33. Alhamrouni, I.; Khairuddin, A.; Ferdavani, A.; Salem, M. Transmission expansion planning using AC-based differential evolution algorithm. *IET Gener. Transm. Distrib.* **2014**, *8*, 1637–1644. [CrossRef]
34. Carrión, D.; Francisco, J.; Paul, M. Revisión para la restauración optima de la operación del sistema eléctrico basado en criterios de calidad de energía y estabilidad. *Rev. I+D Tecnol.* **2021**, *17*, 9. [CrossRef]

energies

MDPI

Article

Event-Driven Edge Agent Framework for Distributed Control in Distribution Networks

Xianglong Zhang *, Ying Liu, Songlin Gu, Yuzhou Tian and Yifan Gao

State Grid Economic and Technological Research Institute Co., Ltd., Beijing 102209, China
* Correspondence: xianglong1983@126.com

Abstract: With the large-scale integration of heterogeneous energy resources and the increasing demand for flexible control, centralized control is facing challenges in terms of operational efficiency and system responsiveness when handling high-precision regulation tasks. To address this issue, this paper proposes an event-driven edge agent framework for distributed control in power distribution networks. First, based on the diverse requirements of distributed control in distribution networks, an edge agent architecture is constructed with modular components such as configuration management at its core. Second, considering the hybrid system characteristics of distribution networks, a control configuration technique based on activity-on-edge is designed, which decouples and discretizes continuous control processes through event-driven mechanisms. Furthermore, an edge-oriented automatic differentiation solver and a lightweight web application framework are developed to address the challenges of real-time optimization under resource-constrained environments. Finally, a semi-physical simulation is conducted using station-level economic dispatch as a case study to verify the effectiveness of the proposed technology. The results demonstrate that, compared to centralized control, the designed distributed agent maintains optimization accuracy while reducing event-triggering frequency by 40% and improving communication response speed by 70%, showing strong performance in operational efficiency at the edge.

Keywords: edge agent; distribution network; control configuration; distributed control; event-driven; automatic differentiation

1. Introduction

With the large-scale integration of distributed generation, energy storage systems, and various controllable loads [1–4], power distribution systems exhibit significant differences in dynamic response characteristics, regulation capabilities, and control interaction mechanisms [5–7], thereby increasing the complexity of system operation. This high degree of heterogeneity substantially raises the difficulty of unified scheduling and global coordination, posing severe challenges to traditional centralized control approaches in managing complex control tasks [8,9]. Meanwhile, the growing demand for flexible control continues to push the system toward higher requirements in regulation frequency, response latency, and edge-side execution capability [10–12]. In particular, as emerging applications such as electric vehicle clusters [13] and virtual power plant aggregation [14] are rapidly deployed, distribution network control tasks are becoming increasingly decentralized and dynamic, further intensifying the communication and computational burdens of centralized control architectures.

To address the above challenges, edge agents have been introduced as novel solutions for distributed control in power distribution networks [15–17]. By offloading computationally lightweight tasks from the central controller to devices closer to the field, the system enables localized real-time data perception and coordinated regulation with the control center, thereby improving response speed and operational flexibility [18]. Researchers have explored the engineering pathway of deploying edge agents in power systems. Reference [19] systematically elaborates on the integrated architecture of edge agents and artificial intelligence technologies in the context of the power Internet of Things. Reference [20] constructs a single-phase ground fault section location model by integrating deep learning algorithms into the edge agent framework. References [21,22] propose optimized deployment strategies for edge agent clusters in distribution networks from the perspectives of software-defined network architecture and fault handling requirements, respectively. Existing studies mainly focus on how edge agents overcome the computational bottlenecks of centralized control [23,24]. However, achieving optimal control performance under resource-constrained conditions remains a key challenge between academic research and practical deployment. Current approaches to address this issue include lightweight deployment [25] and rational allocation of computing resources [26–28].

Lightweight deployment encompasses both algorithm-level and model-level strategies. Current algorithm deployment often relies on pre-debugged embedding [29], which is not suitable for large-scale, dynamic deployment and online updating across massive terminal devices. At the model level, existing continuous system operation models for edge agents typically rely on gradient-based iterative mechanisms, which lead to continuous occupation of communication bandwidth. In recent years, the hybrid characteristics of power systems have been recognized [30], giving rise to hybrid system modeling approaches that abstract high-dimensional nonlinear systems as discrete event-triggered state transition systems. These methods achieve model order reduction by decoupling dynamic systems from discrete events [31]. Moreover, modeling based on discrete event-driven mechanisms initiates state evolution computation only when events are triggered, significantly reducing the computational resource consumption of edge agents. Related technologies have already been validated in microgrid energy management systems [32,33].

In addition, in multi-objective optimization control of distribution networks [34,35], time-iterative continuous control often results in high redundancy of control commands, which severely limits the real-time responsiveness of control actions and increases the computational burden on edge agents. Current research primarily focuses on resource allocation algorithms to improve resource utilization efficiency [36,37]. However, systematic solutions for optimizing continuous control mechanisms under time-step constraints remain lacking. By adopting event-driven control, the system's dynamic evolution can be transformed into a state transition trigger mechanism, thereby avoiding repetitive time-step iterations and improving computational resource efficiency [38].

Based on the above background and challenges, this paper conducts research on distributed edge agent technology for distribution networks. First, an edge agent control architecture is designed in response to the diverse requirements of distributed control in distribution networks, enabling flexible deployment of control strategies through modules such as control configuration and intelligent computation. Second, a hybrid system model for distribution networks is established based on hybrid control theory, and an event-driven control configuration technique is proposed to discretize complex continuous control processes. The control strategy is formally modeled using AOE, where a joint dynamic mechanism of node events and edge actions is employed to achieve on-demand control. Furthermore, an automatic differentiation-based solver and a lightweight web application framework tailored for edge environments are developed to address the challenge of real-time optimization in resource-constrained settings. Finally, using station-level optimal dispatch as an application scenario, the proposed control configuration technique is applied to model and configure a distributed optimization strategy based on consensus algorithms, and the applicability of the proposed architecture and control method is validated through semi-physical simulation.

2. Edge Agent Architecture for Distributed Control

Edge agents are typically located close to distributed and flexible resources such as photovoltaics and energy storage systems, enabling localized computation and fast control near the source side by leveraging their edge computing capabilities. Through low-latency communication mechanisms, edge agents can not only execute localized control strategies but also respond to centralized optimization and dispatch, serving as a key enabler of local autonomy and global coordination in modern distribution networks. Based on this, the edge agent architecture constructed in this paper is shown in Figure 1. The main functional modules are described as follows.

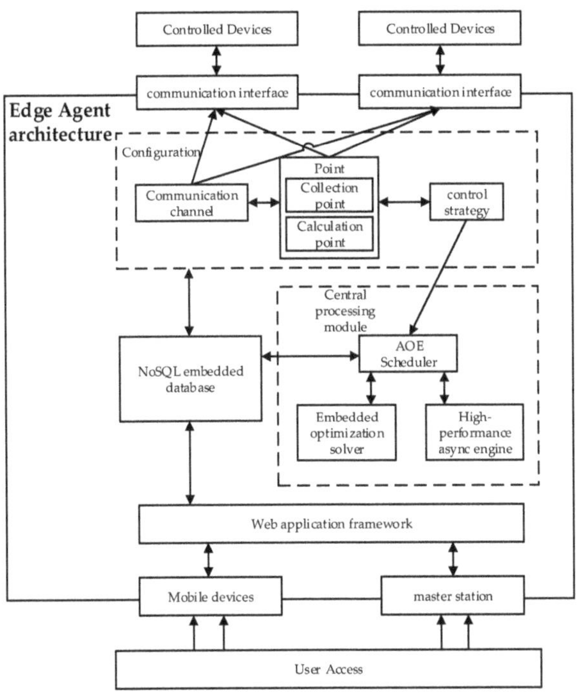

Figure 1. Edge agent architecture diagram.

1. Configuration Module

The edge agent provides a secondary development interface to users in the form of configuration, including control strategies, communication channels, and measurement points. The control strategy adopts a control configuration technique based on hybrid control theory, enabling custom editing of edge computing strategies. Channels are used to establish communication between the edge agent and controlled devices, supporting configuration of various communication protocols. Measurement points describe the state of the controlled objects and serve as the data foundation for control strategies or communication. They are categorized into acquisition points and calculation points.

2. Central Processing Module

The AOE scheduler verifies, parses, and executes the configuration to enable optimized operation for the designed control scenario. The edge agent performs mathematical computations such as solving optimization models and equation systems within the control strategy using an automatic differentiation-based optimization solver. This paper uses the IPOPT optimization solver, which was developed by the Computational Infrastructure for Operations Research (COIN-OR) Foundation, located in Chicago, IL, USA. A coroutine-based high-performance asynchronous execution engine supports high-concurrency task scheduling, and an embedded database is integrated to handle data storage and management under resource-constrained conditions. The asynchronous execution engine in this paper is designed based on the Tokio asynchronous runtime framework, which is developed by contributors to the Tokio project and maintained by an organization based in the USA. The database is built on RocksDB, which was developed by Facebook Inc., headquartered in Menlo Park, CA, USA.

3. Web Application Interface Module

The edge agent employs a browser-based web application as the user interface, providing an interactive platform for user operations and offering intuitive visualization of various data, reports, and the execution statuses of control strategies. The web application interface in this paper is designed using WebAssembly, which is led by the World Wide Web Consortium (W3C) and jointly promoted by major technology companies including Mozilla, Google, Microsoft, and Apple. The standardization and development efforts are primarily carried out in the USA.

3. Edge Agent Control Configuration Technology Based on Hybrid Control

3.1. Hybrid Control Theory

As a hybrid dynamic system, the distribution network requires coordinated operation between continuous variable control and discrete logical constraints to achieve multi-objective optimization [32]. Its dynamic behavior is typically represented mathematically by high-dimensional nonlinear differential–algebraic equations, expressed as

$$\begin{aligned} \min J(\boldsymbol{x},\boldsymbol{y}) &= \int_0^{\infty} \sum_{i=1}^{n} q_i F_i(\boldsymbol{x},\boldsymbol{y}), \\ s.t.\ \dot{\boldsymbol{x}} &= f(\boldsymbol{x},\boldsymbol{u}), \\ 0 &= \varphi(\boldsymbol{x},\boldsymbol{y}), \end{aligned} \tag{1}$$

where $\boldsymbol{x} \in R^n$, $\boldsymbol{y} \in R^m$, and $\boldsymbol{u} \in R^r$ represent the system state vector, output vector, and control vector, respectively. Traditional methods struggle to obtain satisfactory solutions within a limited time frame [39]. To address this high-dimensional constrained variational problem, edge agents adopt a hybrid control approach. The full state space of the system is partitioned based on whether the multi-objective optimal operating conditions are satisfied, resulting in a satisfactory state set and an unsatisfactory state set, defined as

$$X(T) = X_S(T) \cup X_{\overline{S}}(T) \tag{2}$$

where $X_s(T)$ represents the satisfactory state set of the system at time T, that is, the state set where multi-objective optimal conditions are met; $X_{\bar{S}}(T)$ represents the unsatisfactory state set at time T; and $X(T)$ represents the full state space of the system at time T.

When the unsatisfactory state set $X_{\bar{S}}(T)$ is an empty set, the system is considered to have achieved the desired operational objectives. At any time T, when the system operating state enters $X_{\bar{S}}(T)$, an event E is said to be triggered:

$$E = f_E(D) \tag{3}$$

where E represents the set of events, D represents the system operating data, and f_E represents the logical evaluation function that determines whether an event in E is generated based on D.

When an event E triggers the corresponding control strategy that drives the system operating state back to $X_s(T)$, the system's multi-objective optimal operation is achieved to a certain extent:

$$C = f_C(E) \tag{4}$$

where C represents the set of action commands executed by the edge agent, and f_C represents the transition function that generates control strategies from the event set E.

The edge agent issues control commands to operate devices within the system, leading to a change in the system state that brings it back to the satisfactory state space. At this point, the system is in a no-event operational state, expressed as

$$\dot{x} = f_x(x, C) \tag{5}$$

$$A(C) \rightarrow E \subseteq 0 \tag{6}$$

where x represents the state variable of the system, and f_x is the state equation that describes how the system state is changed through control. Equation (6) indicates that, through control actions, the event set E becomes an empty set.

When Equation (6) holds, it indicates that the system is operating in a satisfactory state. In other words, satisfying Equation (6) also signifies that the system has achieved multi-objective optimal operation. In summary, the system is continuously in the process of event detection, handling, and elimination. The control method based on this theory is referred to as event-driven control.

3.2. *Control Configuration Technology*

3.2.1. Control Configuration

As shown in Section 3.1, the hybrid control approach can be modeled using two key elements—events and action commands—and their interactions. Based on this, a control configuration technology is proposed in this paper. Control configuration is similar to "building blocks", where control strategies are developed by selecting basic components, setting parameters, and combining them through a "configuration" process rather than traditional programming [40]. Compared with conventional programming-based development, this approach offers advantages such as lower development barriers, inherent correctness, and greater generality.

3.2.2. Control Process Model Based on AOE

In distributed scenarios, the design and execution of control strategies are essential to ensuring efficient system operation. The control process can essentially be abstracted as an orderly transition of system states across multiple state sets. During the transition from an initial satisfactory state to the next, the system typically passes through several intermediate unsatisfactory states, which require corresponding control actions for correction. As this process cannot be handled by a single rule, a clear and structured modeling method for control logic is required. To this end, this paper adopts an AOE structure to intuitively describe the control process, with the corresponding control module workflow illustrated in Figure 2. The AOE structure is fundamentally a directed acyclic graph with a single input and a single output, composed of nodes and directed edges. Its implementation details are presented as follows.

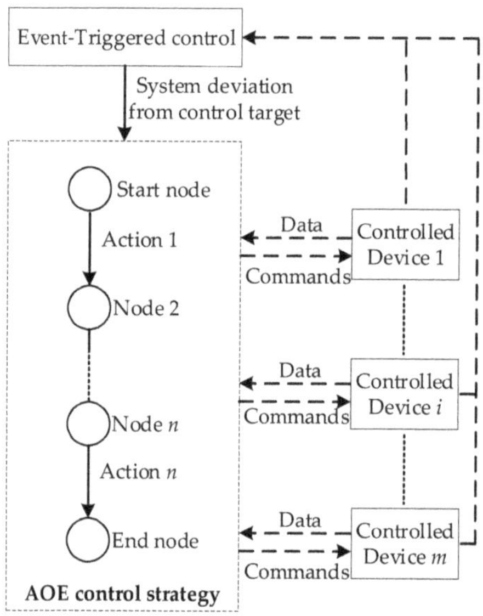

Figure 2. Operation process of the AOE-based control module.

In terms of static modeling, the node set in the AOE structure corresponds to a subset of the event set E defined in Equation (3), with each node expression formulated according to the rules of f_E. Each directed edge represents a control action, i.e., a subset of the instruction set C defined in Equation (4), and its action logic and execution expression are specified based on the rules of f_C. The connectivity between nodes and edges reflects the logical dependencies and execution sequences between events and actions, and the resulting topological structure is consistent with the causal relationships among the elements in sets E and C. The terminal node serves as a validity check for the AOE-based control strategy, corresponding to Equations (5) and (6), to determine whether the current strategy can guide the system back to a satisfactory operational state, i.e., whether the intended control objective has been achieved.

In terms of dynamic execution, the control process starts from the initial node. The system continuously evaluates the event trigger expression defined at this node based on real-time measurement data. Once the expression is satisfied, the node is considered triggered, corresponding to the event generation described in Section 3.1. The system then follows the conditional logic specified at the node to select the corresponding outgoing edge and executes the control action defined on that edge. The execution of the edge action is typically accompanied by updates to system state variables or the issuance of control commands, thereby driving the system toward the satisfactory state space. Upon completion of the action, the process transitions to the next node pointed to by the edge, where the evaluation cycle continues until the next triggering condition is met. Starting from the AOE initial node, this process continues to drive the system until the end node is reached, thus completing the entire execution of the control strategy. The migration at each step is determined by the triggering of node events, thereby implementing event-driven control strategy execution.

Taking the AOE-based control flow illustrated in Figure 2 as an example, the edge agent first collects real-time operating data from nearby devices. Once the system is determined to have deviated from its target and entered an unsatisfactory state, the corresponding AOE control strategy is immediately activated to guide the system back to a satisfactory condition. The control process begins at the initial node, where the predefined logical expression is evaluated to determine whether the triggering condition is satisfied. If the condition holds true, the system proceeds along edge 1–2 to execute the corresponding control action and then transitions to the next node, node 2. This intermediate node typically represents a temporary state, which may be used to assess control effects or evaluate downstream branch logic. If the path condition is met, the system continues along edge 2–3 to execute the next control action and proceeds accordingly until the terminal node is reached. The terminal node is associated with a termination event, which is used to verify whether the control task has been successfully completed—for example, whether the command was issued correctly, key actions were executed as expected, or optimization tasks were fulfilled—thus completing the full closed-loop execution of the control strategy.

3.3. Control Configuration Implementation

When editing control strategies using edge agents, the AOE nodes, edges, and their attributes are configured according to the defined standard format. The AOE scheduler will automatically validate, parse, solve, and activate the user-defined AOE for execution.

First, based on the fundamental concepts and properties of the AOE data structure, the following rules and constraints are applied:

1. A node with an in-degree of zero is called the AOE start node, and a node with an out-degree of zero is called the end node. There is exactly one start node in each AOE, and there are one or more end nodes.
2. When a node has multiple out-degrees, the multiple outgoing edges are numbered in the order of configuration, such as edge 1, edge 2, and so on.
3. Each node has an in-degree greater than or equal to zero. When a node has multiple incoming edges, the node can be reached when any of the incoming edges is triggered and its associated action is completed. At this point, the event represented by the node is in a waiting-trigger state.
4. Multiple expressions can be listed on a single edge. Once an edge is entered, all expressions on that edge are executed simultaneously.

Next, in terms of the specific configuration format for the basic unit nodes and edges of the AOE, their definitions are as described in Section 3.2.2. The specific expressions for configuration vary depending on the type and strategy design, and due to space limitations, only the basic forms of both are presented here without further elaboration.

In terms of node types, the behavior of each node in the AOE control flow determines how its outgoing edges are handled. Based on different control logic requirements, two basic execution modes are commonly defined: the first is sequential execution, in which all outgoing edges are executed in order once the node is triggered; the second is conditional branching, where one edge is selected from multiple candidates based on the current system state or a specific logical expression, enabling decision-making, switching, and conditional transitions within the control strategy.

Accordingly, the event types of nodes in this paper are categorized into two major types: condition-execution nodes, which trigger all subsequent actions and emphasize execution completeness; and branch-selection nodes, which determine the next control path based on logical conditions and emphasize control divergence. Complex control processes can be constructed by combining these two types of nodes, providing strong generality and formal consistency in the expression of structured control logic.

In terms of edge action types, for various computational requirements in distribution networks, such as solving nonlinear equation systems in distributed power flow calculations or solving optimal economic dispatch models formulated as MILP or NLP, the control configuration provides directly callable properties for computing edge actions. Meanwhile, for control commands involving remote switching and setpoint adjustment by the agent, the control configuration also supports assignment edge actions to set measurement point values. Based on this, complex control strategies can be developed and implemented to meet the control requirements of distributed scenarios.

It should be noted that the solver invoked in the control process is an open-source implementation, which is solely responsible for executing mathematical computations and cannot directly determine the feasibility of the constructed optimization model. When the optimization problem is mathematically solvable and successfully converges, the solver returns a valid optimal solution. Conversely, if the model is infeasible or fails to converge during computation, the solving process is considered unsuccessful. Therefore, it is typically necessary to define branching logic at the node following the corresponding computation edge. If the solution is successful, the system retrieves the optimization results and issues control commands accordingly. If the solution fails, the edge agent automatically enters an exception handling process, triggers a failure alert, and executes a predefined backup control strategy to ensure the continuity and safety of device operation.

Finally, in the overall configuration of the control strategy, the activation of the AOE and the execution of the entire network's behavior from the start node are defined as the AOE triggers. Based on the requirements of practical engineering applications, the AOE trigger is divided into three types, as shown in Table 1.

Table 1. Trigger types of AOE network.

Trigger Type	Name	Definition
EventDriven	Event-driven	Enable the AOE network when the first node event condition is met
TimeDriven	Time-driven	Enable the AOE network at the set time
SimpleRepeat	Simple repetition drive	Enable and loop the AOE network at the set time

4. Automatic Differentiation-Based Optimization Solving Technology for the Edge Side

As described in Section 3.3, the actions in the control configuration involve computational requirements such as NLP and MILP, which often require large-scale derivative calculations including Jacobian and Hessian matrices of multivariable functions. To address this, this paper applies automatic differentiation algorithms to optimize the solving process within edge agents. By decoupling control modeling from the solving engine through control configuration and automatic differentiation, the approach enables solver reuse, thereby enhancing the operational efficiency and reliability of edge agents. At the same time, it allows users to focus solely on the design of control strategies, simplifying mathematical computation and reducing implementation complexity.

4.1. Model Solving Based on Automatic Differentiation

Automatic differentiation technology is divided into forward mode and reverse mode [41]. By employing a hierarchical composite function and layer-by-layer derivative solving strategy, the computational complexity is effectively reduced and the final expressions are simplified. As a result, the overall storage space required for computation is significantly reduced compared to symbolic differentiation methods [42].

For the function $y = f(x)$, it can be represented as

$$\begin{cases} y = f_i(w_i) \\ w_i = f_{i-1}(w_{i-1}) \\ \cdots\cdots \\ w_1 = f_0(w_0) \\ w_0 = x \end{cases} \tag{7}$$

In the forward mode, the computation proceeds by first evaluating $\mathrm{d}w_0/\mathrm{d}x$ using Equation (7),and then recursively deriving $\mathrm{d}w_k/\mathrm{d}x$ up to $\mathrm{d}w_i/\mathrm{d}x$, from which $\mathrm{d}y/\mathrm{d}x$ can be obtained.

$$\frac{\mathrm{d}w_1}{\mathrm{d}x} = \frac{\mathrm{d}w_1}{\mathrm{d}w_0}\frac{\mathrm{d}w_0}{\mathrm{d}x}, \frac{\mathrm{d}w_2}{\mathrm{d}x} = \frac{\mathrm{d}w_2}{\mathrm{d}w_1}\frac{\mathrm{d}w_1}{\mathrm{d}x}, \ldots, \frac{\mathrm{d}w_i}{\mathrm{d}x} = \frac{\mathrm{d}w_i}{\mathrm{d}w_{i-1}}\frac{\mathrm{d}w_{i-1}}{\mathrm{d}x}, \frac{\mathrm{d}y}{\mathrm{d}x} = \frac{\mathrm{d}y}{\mathrm{d}w_i}\frac{\mathrm{d}w_i}{\mathrm{d}x} \tag{8}$$

The computation process of the reverse mode is opposite to that of the forward mode. It begins with calculating $\mathrm{d}y/\mathrm{d}w_i$ using Equation (7) and then proceeds downward to derive $\mathrm{d}y/\mathrm{d}w_k$ and eventually $\mathrm{d}y/\mathrm{d}w_0$, from which $\mathrm{d}y/\mathrm{d}x$ can be obtained.

$$\frac{\mathrm{d}y}{\mathrm{d}w_{i-1}} = \frac{\mathrm{d}y}{\mathrm{d}w_i}\frac{\mathrm{d}w_i}{\mathrm{d}w_{i-1}}, \frac{\mathrm{d}y}{\mathrm{d}w_{i-2}} = \frac{\mathrm{d}y}{\mathrm{d}w_{i-1}}\frac{\mathrm{d}w_{i-1}}{\mathrm{d}w_{i-2}}, \ldots, \frac{\mathrm{d}y}{\mathrm{d}w_0} = \frac{\mathrm{d}y}{\mathrm{d}w_1}\frac{\mathrm{d}w_1}{\mathrm{d}w_0}, \frac{\mathrm{d}y}{\mathrm{d}x} = \frac{\mathrm{d}y}{\mathrm{d}w_0}\frac{\mathrm{d}w_0}{\mathrm{d}x} \tag{9}$$

From a mathematical perspective, both the forward and reverse modes of automatic differentiation are derived from the chain rule and thus provide theoretically equivalent accuracy in gradient computation. Meanwhile, by comparing Equations (8) and (9), it can be observed that in the forward mode, each input variable corresponds to a distinct intermediate derivative path, requiring multiple propagation steps to obtain the complete gradient. In contrast, the reverse mode shares a unified propagation path for all intermediate derivatives, allowing all partial derivatives with respect to input variables to be obtained after computing a single backward pass [43]. In high-dimensional gradient computation tasks, the reverse mode can significantly reduce overall computational load, making it more suitable for edge agents that demand fast response and operate under constrained resources. Therefore, this study adopts the reverse mode to improve computational speed.

4.2. Case Study Validation

Taking the optimal power flow problem in a distribution network as an example, the validation of the edge agent's model-solving capability based on automatic differentiation is performed. Mathematically, the optimal power flow calculation is a typical nonlinear optimization problem, and the mathematical model is as follows:

$$\begin{cases} \min F(\boldsymbol{x}, \boldsymbol{u}) \\ s.t.\ G(\boldsymbol{x}, \boldsymbol{u}) = 0, H(\boldsymbol{x}, \boldsymbol{u}) \leq 0 \end{cases} \tag{10}$$

where F represents the objective function; G represents the equality constraints; H represents the inequality constraints; $\boldsymbol{x}$ is the state variable; and $\boldsymbol{u}$ is the control variable.

The standard case 9 test case from MATPOWER 8.0 is selected, with the objective function set to minimizing system operating cost, i.e., minimizing the active power output of generating units under fixed fuel prices. This paper uses MATPOWER 8.0, which was developed by the Power Systems Engineering Research Center (PSERC) at Cornell University, Ithaca, NY, USA.

The AOE control strategy network for optimal power flow calculation consists of four nodes: start node 1, computation node 2, assignment node 3, and end node 4. The action executed on edge 2–3 is a nonlinear programming task for solving the optimal power flow, while edge 3–4 performs an assignment operation used to display the optimal power flow calculation results.

The results obtained by executing the AOE network are shown in Appendix A, Table A1. In this table, voltage angles are expressed in radians, and power values are in per unit (p.u.), which must be multiplied by the base value of 100 MW to obtain physical units. The results computed by MATPOWER 8.0 are shown in Appendix A, Figure A1, where the voltage angles are expressed in degrees. A comparison shows that the two sets of results are consistent, thereby verifying the computational accuracy of the edge agent based on automatic differentiation.

5. Web Application Technology for Edge Agents

The deployment of edge agents requires deep coupling with the underlying hardware architecture, and their design must follow the principles of high performance and lightweight implementation to adapt to heterogeneous edge computing resources and the constraints of real-time response and power sensitivity. To this end, the edge agent adopts a microservice architecture for backend development and uses a web application frontend developed in WebAssembly. Both frontend and backend are implemented in Rust to enable unified full-stack programming and improve code reuse, thereby enhancing system operational efficiency.

5.1. Database Technology for Edge Agents

Embedded database technology is used for data storage and management, with access control handled by the application itself, eliminating the need for a separately deployed database server. Embedded databases are customizable, lightweight, and integrated within the process, making them suitable for embedded systems. In this paper, the open-source arbitrary binary K-V (key-value, key-value pair) data storage engine RocksDB is adopted as the embedded database implementation for the edge agent.

The data structure and management method of RocksDB are introduced as follows. RocksDB has three types of data structures: (1) mentable, a memory data structure, where all write requests enter; (2) logfile, an ordered write storage structure, where write requests enter selectively; and (3) sstfile, when the mentable is full, data enter the sstfile and are stored, and the related logfile is safely deleted. The data in the sstfile are sorted, so that they can be quickly searched based on the key. In terms of data management, RocksDB stores data as binary key-value pairs and supports sharding of a database instance by column families. RocksDB organizes all data in sequence and can specify compression algorithms to sort the key-value pairs. Common operations include retrieving a value by key (Get), inserting a key-value pair (Put), deleting a key-value pair (Delete), and iteration (Iter).

Compared with traditional storage solutions that use high-capacity servers to store raw text and other files, embedded databases offer higher reliability and stronger low-level control capabilities. They can operate stably and reliably over long periods without human intervention and are capable of directly interacting with underlying hardware devices. The use of embedded databases eliminates the overhead of client-server architecture, resulting in more compact code, faster execution speed, and improved performance when running in embedded environments.

5.2. *Asynchronous Non-Blocking Communication Technology*

In various edge computing scenarios, reliable acquisition of measurement data from devices is a prerequisite for controlling them. Distribution networks contain a wide range of complex device types, requiring the edge agent to support parallel data acquisition from different devices using multiple communication protocols and to standardize and store the collected data.

The edge agent application adopts asynchronous non-blocking technology to implement communication gateways for the power Internet of Things, supporting multiple protocols including IEC 104, Modbus, DL/T645, and MQTT. Taking the non-blocking implementation of the IEC 104 protocol as an example, the protocol is designed with a three-layer structure: TCP/IP layer, application layer, and user process.

Based on the Tokio asynchronous framework, a non-blocking I/O network communication framework is developed, covering client connection, server listening and connection, and read/write method design, enabling data exchange over lower-layer TCP/IP connections. Asynchronous channels are used to support information sharing and command transmission between the application layer and the user process. In the application layer, asynchronous tasks are used to implement IEC 104 protocol functions such as link transmission rules, message parsing and assembly, and application functionalities. The overall structure and data flow of IEC 104 are shown in Figure 3.

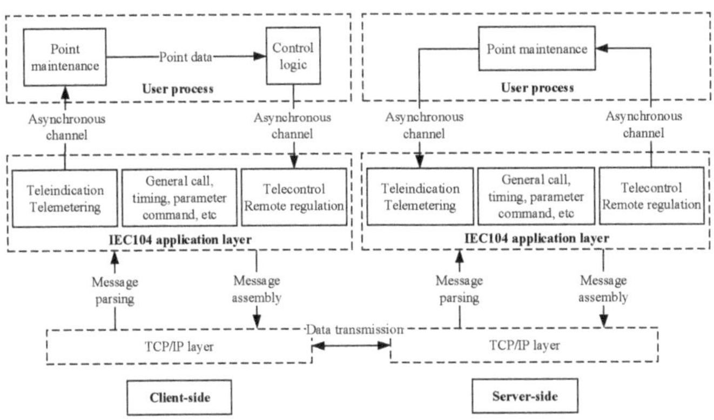

Figure 3. IEC 104 general structure and data flow.

In the asynchronous non-blocking communication framework described above, the underlying non-blocking network communication effectively conserves hardware resources and enhances the parallel processing capability of multi-channel communication. The user process standardizes the collected data from each channel, converting them into a unified measurement point format, which enables consistent computation and logical analysis across data from different devices using a generalized approach. By decoupling data acquisition from data processing, each stage operates in parallel, improving overall data flow and processing speed.

Consistency and performance tests were conducted on the communication system developed using this framework. As shown in Figure 4, under high communication load, the delay rate of the proposed method is significantly lower than that of traditional solutions. This verifies that the developed communication application can complete data transmission in accordance with protocol specifications. Based on asynchronous non-blocking technology, the edge agent can still ensure the accuracy and real-time performance of communication data on resource-constrained embedded devices.

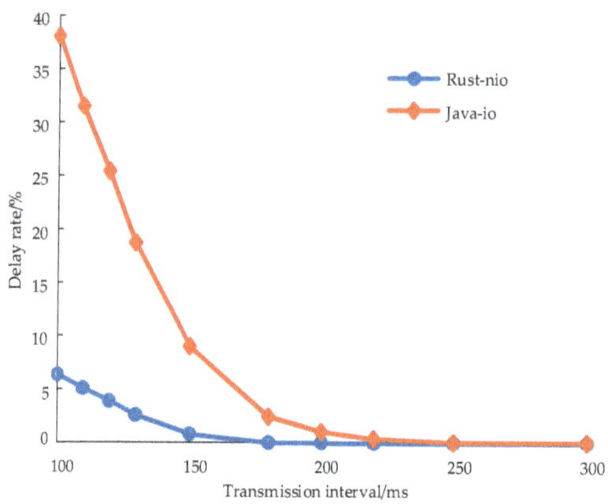

Figure 4. Delay rate comparison with traditional scheme.

6. Station-Level Optimal Dispatch in Distribution Networks Based on Distributed Edge Agents

This section verifies the distributed control technology of edge agents using the economic dispatch of distribution network stations. First, a station-level economic optimization dispatch model based on a discrete consensus algorithm is designed. Then, a digital twin of the actual system is simulated, and the distributed control performance of the edge agents is analyzed.

6.1. Economic Dispatch Optimization Model for Stations

To achieve economic optimality in a station, the total generation cost within the station's jurisdiction should be minimized. The generation cost C_i of the distributed generation unit at bus i in the station can be approximated by a quadratic function [44], as shown below.

$$C_i = \frac{1}{2}a_i P_i^2 + b_i P_i + c_i \tag{11}$$

where a_i, b_i, and c_i represent the cost coefficients of the distributed generation unit, and P_i represents the actual active power output of the unit.

Given that the bus voltage remains stable under actual station operating conditions and the deviation is negligible, the actual active power output P_i (in p.u.) of the distributed generation unit is approximately equivalent to its output current I_i (in p.u.) [45]. Accordingly, Equation (11) can be rewritten as

$$C_i = \frac{1}{2}a_i I_i^2 + b_i I_i + c_i \tag{12}$$

Based on Equation (12), an economic dispatch optimization model for the station is constructed. Considering the power transfer requirements among the buses within the station, in addition to satisfying the station power balance equation and the output current constraints of the distributed generation units, corresponding voltage constraints are formulated. The arithmetic mean of bus voltages is used as the reference value. The objective function and constraints are defined as follows:

$$\begin{aligned} & f = \min \sum_{i=1}^{n} C_i \\ & \text{s.t.} \begin{cases} \sum_{i=1}^{n} I_i = \sum_{i=1}^{n} L_i \\ 0 \le I_i \le I_i^{\max} \\ \dfrac{\sum_{i=1}^{n} U_i}{n} = U_{\text{mean}} \end{cases} \end{aligned} \tag{13}$$

where L_i represents the load current injected at bus I; $I_i^{\max}$ represents the maximum output current of the distributed generation unit at bus I; U_i represents the measured voltage at bus i; and U_{mean} represents the reference value of the bus voltage.

6.2. Mathematical Design of Control Strategy

Based on Lagrangian duality theory, the optimization functional of the station economic dispatch model is constructed. Through gradient conditions and the first-order optimality conditions of the Karush–Kuhn–Tucker (KKT) framework, it is analytically derived that the total generation cost of the station reaches its minimum when the marginal costs at all buses are equal. The corresponding expression is

$$\frac{\partial C_1}{\partial I_1} = \frac{\partial C_2}{\partial I_2} = \ldots = \frac{\partial C_n}{\partial I_n} \tag{14}$$

Define $\lambda_i = \partial C_i / \partial I_i$ as the marginal cost at bus i. Considering the property of multi-agent systems to achieve state consensus through distributed information exchange, a distributed consensus computation framework is constructed using marginal cost as the consensus state variable. Under this framework, each agent iteratively updates its marginal cost until convergence is achieved. For bus i, the control steps are as follows.

Step 1: Initialization. When the economic dispatch task is initiated, the edge agent acquires bus information and calculates the marginal cost $\lambda_i[k]$ of bus i at time step k as well as the marginal cost deviation $\mu_i[k]$ with respect to its topological neighboring buses. The calculations are as follows:

$$\lambda_i[k] = a_i I_i[k] + b_i \tag{15}$$

$$\mu_i[k] = \sum_{j=1, j \in L_i}^{n} g_{ij}(\lambda_i[k] - \lambda_j[k]) \tag{16}$$

where g_{ij} represents the element in the i-th row and j-th column of the nodal conductance matrix G. L_i represents the set of topological neighboring buses of bus i.

Step 2: Determine iteration status. If the event trigger condition defined in Equation (17) is satisfied [46], proceed to Step 3; otherwise, maintain the current state until the next time step and terminate the task.

$$\begin{cases} error = e_i[k] - \lambda_i[k] \\ error^2 > \frac{\sigma\varepsilon[1-2\varepsilon G(i,i)]}{2G(i,i)} \mu_i[k]^2 \end{cases} \tag{17}$$

where $e_i[k]$ represents the event-triggered marginal cost stored at bus i at time step k; and σ and ε represent error coefficients, where $0 < \sigma < 1$. To ensure that Equation (17) remains non-negative, it must also satisfy $0 < \varepsilon < 1/\boldsymbol{G}(i,i)$.

Step 3: Update the event-triggered marginal cost, calculate the voltage adjustment value for bus i at time step $k + 1$, and issue the control command to complete the task.

$$e_i[k+1] = \lambda_i[k] \tag{18}$$

$$V_i[k+1] = V_i[k] - q\mu_i[k] \tag{19}$$

where q represents the step-size coefficient. To ensure control stability, it must satisfy $0 < q < 1/\lambda_{An}\lambda^2{}_{Gn}$ [47], where λ_{An} is the maximum value among all cost coefficients a_i, and λ_{Gn} represents the largest eigenvalue of the conductance matrix $\boldsymbol{G}$.

6.3. Strategy Description Based on Control Configuration Technology

The operational states and dynamic responses in the control steps of Section 6.2 are defined as events and actions, respectively. Combined with control configuration technology, the control steps can be described using an AOE structure, as shown in Figure 5. Taking bus 1 as an example, the configuration expressions for each event and action are provided in Appendix B, thereby completing the control strategy design.

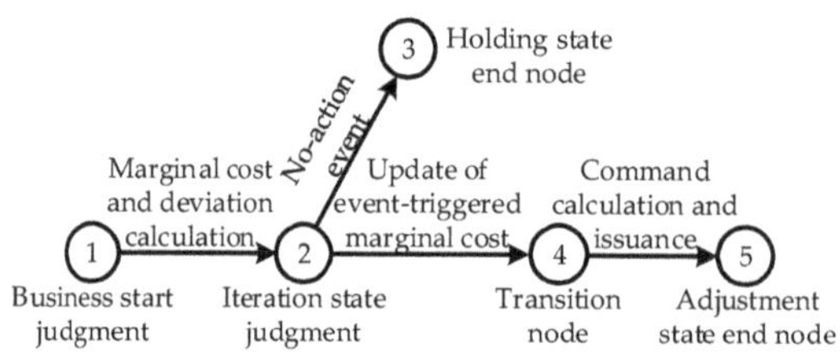

Figure 5. AOE description based on control configuration technology.

Specifically, node 1 is the start node with an EventDriven trigger type. When the agent receives the start command, Start_command1, the strategy is initiated, and edge 1–2 is executed to calculate the marginal cost and deviation based on Equations (15) and (16). Node 2 is a switch-type event node; if the condition is satisfied, the process proceeds to node 4; otherwise, it jumps to termination node 3 to exit the process. Then, actions on edges 2–4 and 4–5 are executed, corresponding to Step 3 in Section 6.2. After completion, the process reaches end node 5, closing the AOE-based control loop.

6.4. Simulation Analysis

6.4.1. Station Simulation Environment Construction

A digital twin model of a five-bus station system in a distribution network is constructed in Simulink to simulate real-time economic dispatch at the station level. The topology of the five-bus station is shown in Figure 6, and the corresponding parameters are listed in Appendix C. The simulation model is built using Simulink 10.2 in MATLAB 2020b, and the simulation is run on a computer equipped with an Intel(R) i7-8700 processor and 16 GB of RAM.

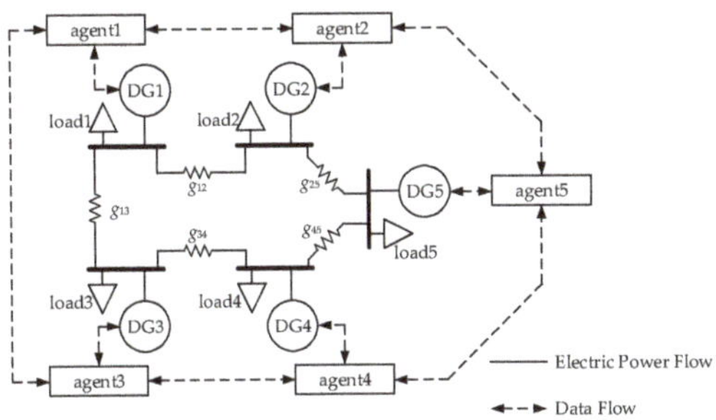

Figure 6. Topology of the five-bus station area.

Figure 7 shows the hardware-in-the-loop simulation architecture between the edge agent and Simulink. The edge agent is deployed on hardware and is responsible for collecting model data, executing the AOE control strategy, and sending control commands to the Simulink model. In Simulink, Modbus communication with each edge agent is implemented using the S-Function module, which serves as the control module in the model, linking the input signals from data acquisition and the output signals for control commands. The hardware device used in this study is equipped with an ARMV7 CPU, 2 GB of RAM, and a 4 GB hard drive.

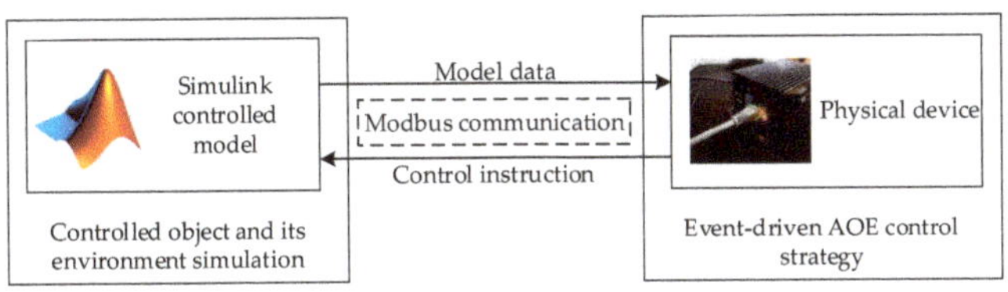

Figure 7. Semi-physical simulation architecture.

6.4.2. Economic Dispatch Results and Analysis

Figure 8 shows the simulation results of station-level economic dispatch, where the dispatch task is initiated at 1 s. It can be observed that after each load fluctuation, the edge agents dynamically adjust the output power of the distributed generation units by modifying the marginal cost. The marginal costs quickly converge to a common value, thereby achieving economic optimality. Additionally, when the load at bus 1 exceeds 1 p.u., the edge agents are able to control other buses to support bus 1, ensuring that the DG output at bus 1 does not exceed its upper limit. Throughout the control process, the average voltage across all five buses remains at the rated level, demonstrating the effectiveness of the proposed distributed control algorithm.

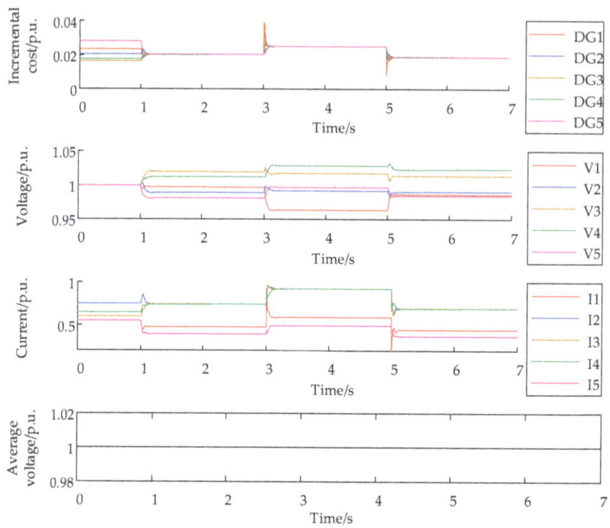

Figure 8. Simulation effect of station economic scheduling.

6.4.3. Comparison of Dispatch Trigger Frequency

During the operation of control systems, frequent communication and repetitive computation often lead to waste of computational resources and device energy, which has become a major bottleneck for the large-scale deployment of edge agents. In the context of distribution networks, if edge agents communicate too frequently, resource exhaustion may occur, posing a serious threat to the stable operation of the system. Figure 9 shows the number of event triggers for each edge agent during the distributed control process at the edge.

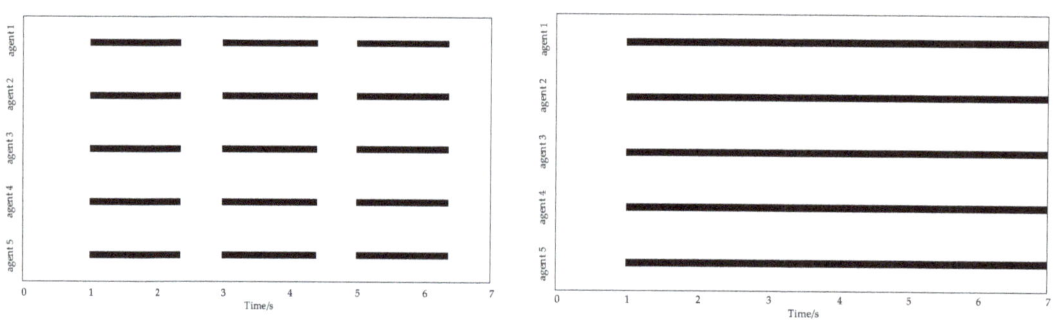

(**a**) With event-driven control (**b**) Without event-driven control

Figure 9. Edge agent event triggering in distributed system.

When event-driven control is adopted, the number of triggers for the five edge agents is 2119, 2147, 1907, 1942, and 2006, respectively. Without event-driven control, each edge agent is triggered 3251 times. The comparison shows that the trigger counts at each node are reduced by 34.82%, 33.96%, 41.34%, 40.26%, and 40.26%, respectively. When Equation (17) is not satisfied, the edge agent does not trigger the task, significantly reducing the number of activations. This demonstrates that edge-based distributed control effectively reduces unnecessary task triggering, improves the efficiency of concurrent task execution, and achieves better real-time control performance.

6.4.4. Dispatch Efficiency Comparison

Using traditional centralized control as a baseline, the efficiency of the proposed edge-based distributed control is compared, as shown in Table 2. It can be observed that, compared to centralized control, the edge-based distributed approach significantly reduces communication latency by leveraging the geographically proximal deployment of edge agents, thereby shortening the data upload and command transmission paths. Although distributed control introduces additional parallel processing overhead due to multi-node coordination, its localized decision-making mechanism avoids the data aggregation bottleneck inherent in centralized systems. Overall, the edge-based distributed control method responds more quickly to changes in distribution network conditions and demonstrates superior global responsiveness in dynamic scenarios.

Table 2. Comparison of time consumption across control method stages.

Control Method	Consumed Time/ms		
	Data Upload	Control Computation	Command Issuance
Edge-based distributed control	6.1	5.0	6.3
Centralized control	25.4	3.0	24.9

Furthermore, an economic dispatch control strategy for a distribution network station with a larger number of nodes is constructed and computed. As the number of nodes increases, the computation times required for strategy execution using the traditional centralized control method and the proposed edge-based distributed control method are shown in Table 3.

Table 3. Comparison of calculation times of two method strategies.

Control Method	Computation Time/ms		
	5 Nodes	14 Nodes	39 Nodes
Edge-based distributed control	5.0	5.0	5.0
Centralized control	3.0	12.0	34.0

The computation time increases when using the centralized control method, while it remains nearly constant when using the edge-based distributed control method. This is because, in edge-based distributed control, each edge agent only communicates with directly connected agents, resulting in smaller computation scales and immunity to increasing network complexity. In contrast, with the centralized control method, as the number of nodes increases, both the communication network and the computation scale expand, leading to longer strategy computation times.

7. Discussion

In recent years, edge agent technology has been gradually applied in distributed control of power distribution networks, but several key challenges remain. First, most existing edge control schemes are designed for specific scenarios and thus struggle to accommodate the broad adaptability required for managing heterogeneous multi-source resources. Second, many methods lack a structured representation of control logic, making it difficult to support modular configuration and reuse. Moreover, in typical emerging scenarios, edge nodes must respond to frequent state changes, which can lead to communication congestion and compromise system stability under resource-constrained conditions.

To address these challenges, this paper proposes a control configuration method oriented toward modular deployment. Specifically, an activity-on-edge model is employed to abstract control logic, explicitly representing the execution sequence and triggering relationships of control flows. By incorporating hybrid control theory, discrete events and continuous dynamics are unified within a structured framework. This approach improves the reusability and adaptability of control logic across diverse application scenarios.

At the execution level, a web-based edge agent architecture is adopted, integrating event-driven control with automatic differentiation-based optimization. The event-driven mechanism activates control only when predefined conditions are met, effectively improving responsiveness and resource efficiency in dynamic environments. Automatic differentiation avoids symbolic expression expansion and enables fast computation of high-dimensional gradients. By introducing web application technology, the frontend and backend of the system are decoupled to improve module reuse and execution efficiency. In addition, asynchronous communication enables non-blocking data exchange between agents, ensuring timely information delivery and supporting the real-time responsiveness required in distributed control systems.

Despite these advantages, the proposed method still has some limitations. The current framework assumes relatively stable edge hardware and communication conditions, while in real-world deployments, device heterogeneity and unstable communication links may affect coordination among agents. Furthermore, the event-driven mechanism is sensitive to parameter settings, and inappropriate parameter selection may lead to delayed responses or frequent triggering, negatively affecting overall system performance.

Looking ahead, the proposed method offers strong scalability and integration potential. The edge agent architecture can be further integrated with IoT-based sensing platforms or extended to coordinate flexible loads such as electric vehicles with dynamic charging behavior. Future work will focus on validating the proposed method in more complex and diverse distribution network scenarios to further improve and refine its performance.

8. Conclusions

This paper proposes a distributed edge agent technology for distribution networks with several key contributions. First, a highly adaptable edge agent control architecture is designed based on hybrid control theory, considering the diverse requirements of distributed control in distribution networks. Furthermore, a control configuration technology suitable for edge agents is proposed, enabling standardized configuration of distribution network tasks. This facilitates fast and lightweight deployment, as well as flexible updates, of event-driven control strategies. In addition, by taking economic optimization dispatch at the station level as a representative edge computing application, the proposed edge-based distributed control method is compared with traditional centralized control. The results validate the effectiveness of the proposed approach, which demonstrates advantages such as low communication latency and reduced computation scale, making it suitable for large-scale distribution network control scenarios

Author Contributions: Methodology: X.Z., Y.L., S.G., Y.T. and Y.G.; writing—original draft preparation: X.Z. and Y.G.; writing—review and editing: X.Z., Y.L., S.G., Y.T. and Y.G. All authors have read and agreed to the published version of the manuscript.

Funding: This research was supported by the Science & Technology Project of State Grid Corporation of China, "Research on Key Technologies of Distributed Collaborative Computing and Control for New-Type Distribution Networks" (Project Code: 5400-202356680A-3-3-JC).

Data Availability Statement: The original contributions presented in the study are included in the article, and further inquiries can be directed to the corresponding author.

Acknowledgments: The authors sincerely acknowledge the contribution of all individuals, reviewers, and editors for their contribution toward the production of this manuscript.

Conflicts of Interest: Authors Xianglong Zhang, Ying Liu, Songlin Gu, Yuzhou Tian, and Yifan Gao are affiliated with State Grid Economic and Technological Research Institute Co., Ltd.. The authors declare that the research was conducted in the absence of any commercial or financial relationships that could be construed as potential conflicts of interest.

Abbreviations

The following abbreviations are used in this manuscript:

AOE	Activity-on-edge network
MILP	Mixed-Integer Linear Programming
NLP	Nonlinear programming

Appendix A

Table A1. Operation results of the optimal power flow calculation case.

Serial Number	Point Number	Name	Another Name	Current Value
1	1001	Point1	V1	1.099999920539086
2	1002	Point2	V2	1.0973546514586485
3	1003	Point3	V3	1.0866203282032385
4	1004	Point4	V4	1.094221462595771
5	1005	Point5	V5	1.084448462899668
6	1006	Point6	V6	1.1000000109999306
7	1007	Point7	V7	1.0894894953643341
8	1008	Point8	V8	1.1000000109967762
9	1009	Point9	V9	1.0717554151378195
10	1010	Point10	THETA1	0.0000000000000000
11	1011	Point11	THETA2	0.0854101800932149
12	1012	Point12	THETA3	0.05671518263242919
13	1013	Point13	THETA4	−0.04298613947999122
14	1014	Point14	THETA5	−0.06949871348651646
15	1015	Point15	THETA6	0.010522383653749938
16	1016	Point16	THETA7	−0.020878902993962594
17	1017	Point17	THETA8	0.015806264325692185
18	1018	Point18	THETA9	−0.0805511204216152
19	1019	Point19	PG1	0.8979870754479929
20	1020	Point20	PG2	1.3432060080841293
21	1021	Point21	PG3	0.9418738048899336
22	1022	Point22	QG1	0.1296559578027687
23	1023	Point23	QG2	0.0003187190823897824
24	1024	Point24	QG3	−0.22634181614061613
25	1025	Point25	Docal_POINT	0

Bus Data								
Bus	Voltage		Generation		Load		Lambda($/MVA-hr)	
#	Mag(pu)	Ang(deg)	P (MW)	Q (MVAr)	P (MW)	Q (MVAr)	P	Q
1	1.100	0.000*	89.80	12.94	-	-	24.756	-
2	1.097	4.893	134.32	0.05	-	-	24.035	-
3	1.087	3.249	94.19	-22.62	-	-	24.076	-
4	1.094	-2.463	-	-	-	-	24.756	0.004
5	1.084	-3.982	-	-	90.00	30.00	24.998	0.027
6	1.100	0.602	-	-	-	-	24.076	-
7	1.089	-1.197	-	-	100.00	35.00	24.254	0.036
8	1.100	0.905	-	-	-	-	24.035	-
9	1.072	-4.616	-	-	125.00	50.00	24.999	0.112
		Total:	318.31	-9.63	315.00	115.00		

Figure A1. Operation results of MATPOWER 8.0.(* indicates that the bus is the slack bus).

Appendix B

Table A2. Node event configuration representation.

Node Name	Event Type	Expression
Business start judgment	Condition	Start_command1 > 0
Iteration state judgment	Switch	(triggerE1 − lambd1)^2 − (sigma * epsilon * miu1^2 * (1 − 2 * epsilon * G1[1])/2 * G1[1]) > 0
Transition node	Condition	1
Adjustment state end node	Condition	1
Holding state end node	Condition	1

Table A3. Action configuration representation.

Start and End Nodes	Action Definition	Expression
1:2	Marginal cost and deviation calculation	lambd1:a1 * I1 + b1; miu1:G1[2] * (labmd1 − lambd2) + G1[3] * (labmd1 − lambd3) + G1[4] * (labmd1 − lambd4);
2:3	No-action event	\
2:4	Update of event-triggered marginal cost	triggerE1:lambd1;
4:5	Command calculation and issuance	V1:V1 − q * miu1;

Appendix C

Table A4. Station parameters.

Parameters	Value
$a_1, a_2, a_3, a_4, a_5/10^{-2}$ p.u.	4.175, 2.675, 2.675, 2.675, 5.025
$b_1, b_2, b_3, b_4, b_5/10^{-4}$ p.u.	5.889, 5.841, 5.841, 5.841, 6.003
$c_1, c_2, c_3, c_4, c_5/10^{-3}$ p.u.	8.060, 7.576, 7.576, 7.576, 8.560
$g_{12}, g_{13}, g_{25}, g_{34}, g_{45}/\text{p.u.}$	4, 5, 4, 5, 4
$U_{\text{mean}}/\text{V}, I_{\text{max},i}/\text{p.u.}$	400, 1

Table A5. Node load parameters.

Node ID	Load Value/p.u.		
	0–3 s	3–5 s	5–7 s
1	0.55	1.00	0.60
2	0.75	0.85	0.60
3	0.60	0.70	0.60
4	0.65	0.70	0.55
5	0.55	0.60	0.55

Table A6. Consensus algorithm parameters.

Parameters	Value
σ	10^{-6}
ε	0.03
q	0.02

References

1. Lu, P.; Lan, H.; Yuan, Q.; Jiang, Z.; Cao, S.; Ding, J.; Wei, Q.; Fan, J.; Cai, Q.; Zhang, N.; et al. A Bi-Level Solution Strategy Based on Distributed Proximal Policy Optimization for Transmission and Distribution Network Dispatch with EVs and Variable Energy. *Appl. Energy* **2025**, *384*, 125405. [CrossRef]
2. Huang, Q.; Cheng, H.; Zhuang, Z.; Duan, M.; Fang, K.; Huang, Y.; Wang, L. Distributed Dispatch of Distribution Network Operators, Distributed Energy Resource Aggregators, and Distributed Energy Resources: A Three-Level Conditional Value-at-Risk Optimization Model. *Inventions* **2024**, *9*, 117. [CrossRef]
3. Iftikhar, M.Z.; Imran, K. Network Reconfiguration and Integration of Distributed Energy Resources in Distribution Network by Novel Optimization Techniques. *Energy Rep.* **2024**, *12*, 3155–3179. [CrossRef]
4. Gao, X.; Xu, H.; Zhao, X.; Gao, S.; Xie, H.; Zhang, N.; Li, Q.; Yang, Y. Dispatching of Distribution Networks with High Permeability Distributed New Energy Based on Phase Shifting Transformer Loop-Closing Device. *Front. Energy Res.* **2025**, *13*, 1516459. [CrossRef]
5. Caballero-Peña, J.; Cadena-Zarate, C.; Parrado-Duque, A.; Osma-Pinto, G. Distributed Energy Resources on Distribution Networks: A Systematic Review of Modelling, Simulation, Metrics, and Impacts. *Int. J. Electr. Power Energy Syst.* **2022**, *138*, 107900. [CrossRef]
6. Xu, X.; Zhao, N.; Wang, L.; Yao, X.; Zhou, L. Research on Time-Varying Dynamic Response Aggregation Model of Distributed Generator Participating in Active Distribution Network. *Energy Rep.* **2023**, *9*, 1546–1556. [CrossRef]
7. Varathan, G.; Belwin Edward, J. A Review of Uncertainty Management Approaches for Active Distribution System Planning. *Renew. Sustain. Energy Rev.* **2024**, *205*, 114808. [CrossRef]
8. Chen, Y.; Hayawi, K.; Fan, M.; Chang, S.Y.; Tang, J.; Yang, L.; Zhao, R.; Mao, Z.; Wen, H. A Bilevel Optimization Model Based on Edge Computing for Microgrid. *Sensors* **2022**, *22*, 7710. [CrossRef]
9. Wenzhi, S.; Zhang, H.; Tseng, M.-L.; Weipeng, Z.; Xinyang, L. Hierarchical Energy Optimization Management of Active Distribution Network with Multi-Microgrid System. *J. Ind. Prod. Eng.* **2022**, *39*, 210–229. [CrossRef]

10. Ahrens, M.; Kern, F.; Schmeck, H. Strategies for an Adaptive Control System to Improve Power Grid Resilience with Smart Buildings. *Energies* **2021**, *14*, 4472. [CrossRef]
11. Thenmozhi, T.; Suganyadevi, M.V.; Ramya, R.; Muthukumar, P. Hybrid Energy Management on Electric Vehicles for Power Grids with Renewables System. *Environ. Chall.* **2022**, *9*, 100647. [CrossRef]
12. Aslam, M.M.; Lateef, U.; Khalid, Z.; Saleem, U.; Liu, W.; Li, W. Energy Management in Smart Grids: Innovative Solutions for Demand-Side Optimization Considering Control Constraints. In Proceedings of the 2024 3rd International Conference on Emerging Trends in Electrical, Control, and Telecommunication Engineering (ETECTE), Lahore, Pakistan, 26–27 November 2024; pp. 1–6.
13. Rubino, L.; Rubino, G.; Esempio, R. Linear Programming-Based Power Management for a Multi-Feeder Ultra-Fast DC Charging Station. *Energies* **2023**, *16*, 1213. [CrossRef]
14. Feng, J.; Ran, L.; Wang, Z.; Zhang, M. Optimal Bidding Strategy for Virtual Power Plant in Multiple Markets Considering Integrated Demand Response and Energy Storage. *J. Energy Storage* **2025**, *124*, 116706. [CrossRef]
15. Chen, Y.; Liu, Y.; Shen, X.; Xu, L.; Liu, J.; Li, L. Review of Edge Intelligence Technology for Distributed Energy Resources in Urban Energy Systems. *Autom. Electr. Power Syst.* **2022**, *46*, 142–152. [CrossRef]
16. Shen, S.; Ren, Y.; Ju, Y.; Wang, X.; Wang, W.; Leung, V.C.M. *EdgeMatrix*: A Resource-Redefined Scheduling Framework for SLA-Guaranteed Multi-Tier Edge-Cloud Computing Systems. *IEEE J. Sel. Areas Commun.* **2023**, *41*, 820–834. [CrossRef]
17. Liu, Z.; Song, J.; Qiu, C.; Wang, X.; Chen, X.; He, Q.; Sheng, H. Hastening Stream Offloading of Inference via Multi-Exit DNNs in Mobile Edge Computing. *IEEE Trans. Mob. Comput.* **2024**, *23*, 535–548. [CrossRef]
18. Nie, Y.; Peng, C.; Hu, Y.; He, Y.; Ma, G.; Huang, C. Parallel distributed optimal economic dispatch of High penetration microgrid based on edge computing. *South. Power Syst. Technol.* **2023**, *17*, 114–124. [CrossRef]
19. Jie, T.; Qi, Z.; Pu, T.; Song, R.; Zhang, J.; Tan, Y.; Wang, X. Edge Intelligence in Power Internet of Things: Concept, Architecture, Technology and Application. *Proc. CSEE* **2024**, *44*, 5473–5496. [CrossRef]
20. Zhang, D.; Li, X.; Tao, W. Location of single-phase grounding fault segment in limited information distribution network based on edge computing and deep learning. *Power Syst. Prot. Control* **2023**, *51*, 22–32. [CrossRef]
21. Qin, Q.; Liu, W.; Tan, W.; Cai, Z.; Cen, B.; Kuang, P. Optimal deployment method of distribution edge computing terminal for software defined network. *Electr. Power Constr.* **2023**, *44*, 82–90. [CrossRef]
22. Pan, S.; Liu, B. Optimal allocation method of edge computing unit for fast Fault processing of distribution network. *Electr. Power Constr.* **2022**, *43*, 31–41. [CrossRef]
23. Liu, D.; Zeng, X.; Wang, Y. Control Strategy of virtual power station in Distribution area based on Edge Computing Architecture. *Trans. China Electrotech. Soc.* **2021**, *36*, 2852–2860. [CrossRef]
24. Yuan, L.; Gu, J.; Jin, Z. User-side data application framework of Power Internet of Things based on Cloud-edge-end collaboration. *Electr. Power Constr.* **2020**, *41*, 1–8. [CrossRef]
25. Hinton, G.; Vinyals, O.; Dean, J. Distilling the Knowledge in a Neural Network. *Comput. Sci.* **2015**, *14*, 38–39. [CrossRef]
26. Mao, Y.; Zhang, J.; Letaief, K.B. Dynamic Computation Offloading for Mobile-Edge Computing with Energy Harvesting Devices. *IEEE J. Sel. Areas Commun.* **2016**, *34*, 3590–3605. [CrossRef]
27. You, C.; Huang, K.; Chae, H.; Kim, B.-H. Energy-Efficient Resource Allocation for Mobile-Edge Computation Offloading. *IEEE Trans. Wirel. Commun.* **2017**, *16*, 1397–1411. [CrossRef]
28. Zhang, K.; Mao, Y.; Leng, S.; Zhao, Q.; Li, L.; Peng, X.; Pan, L.; Maharjan, S.; Zhang, Y. Energy-Efficient Offloading for Mobile Edge Computing in 5G Heterogeneous Networks. *IEEE Access* **2016**, *4*, 5896–5907. [CrossRef]
29. Krizhevsky, A.; Sutskever, I.; Hinton, G.E. ImageNet Classification with Deep Convolutional Neural Networks. *Commun. ACM* **2017**, *60*, 84–90. [CrossRef]
30. Hu, W.; Lu, Q. Hybrid Power Control System and Its Application. *Trans. China Electrotech. Soc.* **2005**, *20*, 11–16. [CrossRef]
31. Liu, G.; Park, J.H.; Hua, C.; Li, Y. Hybrid Dynamic Event-Triggered Load Frequency Control for Power Systems with Unreliable Transmission Networks. *IEEE Trans. Cybern.* **2023**, *53*, 806–817. [CrossRef]
32. Dou, C.; Teng, S.; Zhang, T.; Zhang, B.; Ma, K. Layered Management and Hybrid Control Strategy Based on Hybrid Automata and Random Forest for Microgrid. *IET Renew. Power Gener.* **2019**, *13*, 3113–3123. [CrossRef]
33. Liu, F.; Zhang, Y.; Meng, D.; Zhu, P. Voltage hybrid control of microgrid. *J. Henan Univ. Sci. Technol.* **2017**, *38*, 49–53, 59, 118. [CrossRef]
34. Zhang, C.; Xu, Y.; Dong, Z.Y.; Zhang, R. Multi-Objective Adaptive Robust Voltage/VAR Control for High-PV Penetrated Distribution Networks. *IEEE Trans. Smart Grid* **2020**, *11*, 5288–5300. [CrossRef]
35. Li, Y.; Li, K.; Fan, R.; Chen, J.; Zhao, Y. Multi-Objective Planning of Distribution Network Based on Distributionally Robust Model Predictive Control. *Front. Energy Res.* **2024**, *12*, 1478040. [CrossRef]

36. Zhang, J.; Tang, Y.; Wang, J.; Lv, D.; Zhu, L.; Dong, F. Simulation-Based Joint User Assignment and Edge Resource Allocation Optimization for Hybrid Tasks in Vehicular Edge Computing. *Simul. Model. Pract. Theory* **2023**, *128*, 102810. [CrossRef]
37. Lan, Y.; Wang, X.; Wang, C.; Wang, D.; Li, Q. Collaborative Computation Offloading and Resource Allocation in Cache-Aided Hierarchical Edge-Cloud Systems. *Electronics* **2019**, *8*, 1430. [CrossRef]
38. Shi, B.; Zhao, M.; Yu, Z.; Yuan, L.; Chen, K.; Ji, S. Development status and prospect of power electronics Simulation technology and simulation software based on discrete state events. *Proc. CSEE* **2022**, *42*, 6005–6016. [CrossRef]
39. Zhang, M.; Lu, Q. Hybrid Control System for Super Energy Management System of China Southern Power Grid. *Power Grid Technol.* **2007**, *31*, 5–9. [CrossRef]
40. Ahmad, Z.; Abbasi, M.H.; Khan, A.; Mall, I.S.; Khan, M.F.N.; Sajjad, I.A. Design of IoT Embedded Smart Energy Management System. In Proceedings of the 2020 International Conference on Engineering and Emerging Technologies (ICEET), Lahore, Pakistan, 22–23 February 2020; pp. 1–5. [CrossRef]
41. Li, C. Research on Automatic Differentiation Implementation and Optimization Algorithm Based on Source Code Conversion. Master's Thesis, Beijing Jiaotong University, Beijing, China, 2022.
42. Margossian, C.C. A Review of Automatic Differentiation and Its Efficient Implementation. *WIREs Data Min. Knowl. Discov.* **2019**, *9*, e1305. [CrossRef]
43. Coleman, T.F.; Verma, A. The Efficient Computation of Sparse Jacobian Matrices Using Automatic Differentiation. *SIAM J. Sci. Comput.* **1998**, *19*, 1210–1233. [CrossRef]
44. Yu, Z. Research on Source-Load Distributed Cooperative Optimized Operation of Active Distribution Network. Master's Thesis, Northeastern University, Shenyang, China, 2024.
45. Peng, J.; Fan, B.; Liu, W. Voltage-Based Distributed Optimal Control for Generation Cost Minimization and Bounded Bus Voltage Regulation in DC Microgrids. *IEEE Trans. Smart Grid* **2021**, *12*, 106–116. [CrossRef]
46. Dimarogonas, D.V.; Frazzoli, E.; Johansson, K.H. Distributed Event-Triggered Control for Multi-Agent Systems. *IEEE Trans. Autom. Control* **2012**, *57*, 1291–1297. [CrossRef]
47. Xu, Y.; Liu, W. Novel Multiagent Based Load Restoration Algorithm for Microgrids. *IEEE Trans. Smart Grid* **2011**, *2*, 152–161. [CrossRef]

energies

MDPI

Article

A Coordinated Emergency Frequency Control Strategy Based on Output Regulation Approach for an Isolated Industrial Microgrid

Xin Ding [1,*] and Sujie Zhang [2,3]

[1] School of Mechanical Engineering, University of Shanghai for Science and Technology, Shanghai 200093, China
[2] State Grid Shanghai Municipal Electric Power Company, Shanghai 200122, China; zhangsujiesgcc@163.com
[3] College of Electrical Engineering, Shanghai University of Electric Power, Shanghai 200090, China
* Correspondence: dingxin2016@whu.edu.cn

Abstract: Constructing isolated industrial microgrids with wind power is beneficial for improving the economic benefits of high-energy-consuming production, such as the electrolytic aluminum industry. Due to the specialized structure of industrial microgrids and the unique characteristics of the electrolytic aluminum load (EAL), the common emergency frequency control methods do not apply to the specific operational requirements of isolated industrial microgrids. Since EALs have huge regulating capacities and fast responses, this paper proposes a coordinated emergency frequency control scheme to deal with power disturbances in isolated industrial microgrids. The coordinated frequency control model of an industrial microgrid considering demand-side participation is derived. With the help of output regulation theory, a practical, feasible coordinated frequency controller is designed by introducing frequency deviation and power disturbance as feedback control signals. The proposed control scheme achieves reserve power distribution between the generation and demand sides. The microgrid frequency can be maintained within a permitted range in the presence of large power imbalances. The simulation results conducted in an actual isolated industrial microgrid validate the effectiveness and dynamic performance of the proposed control scheme.

Keywords: frequency control; isolated industrial microgrid; electrolytic aluminum; output regulation; wind power

1. Introduction

1.1. Background

In order to deal with environmental pollution and energy shortages, the Chinese government issued carbon peaking and carbon neutrality targets in 2020 [1]. Toward this end, carbon dioxide emissions in China will peak and stabilize by 2030, and carbon neutrality will be achieved by 2060 [2]. As a type of renewable energy, wind power generation plays an important role in Chinese low-carbon energy system construction, which has been developed rapidly in recent years [3]. According to the statistical reports issued by the National Energy Administration, by the end of 2023, the total wind power installed capacity in China had already reached to 441.3 GW, representing a 20.7% increase over the previous year [4]. Therefore, in order to obtain cheaper electricity, high-energy-consuming enterprises in China, such as the electrolytic aluminum industry, are gradually transiting production to regions with abundant thermal and wind power [5]. To facilitate wind power utilization and improve the economic profit of industries, isolated industrial microgrid construction is encouraged in China [6,7].

However, the isolated industrial microgrids are normally designed with simple electrical structures and disconnected from the utility grid, leading to smaller system inertia than with large power grids. In addition, when the isolated industrial microgrid is integrated with a high proportion of wind power, the power system stability may be further deteriorated in terms of the large power disturbances and insufficient reserve capacity [8]. Therefore, emergency frequency control is one of the major technical issues for the operation of the isolated industrial microgrid [9].

1.2. Literature Review

The traditional emergency frequency control methods include the active power regulation of generators (primary and secondary frequency control) and load-shedding control strategies. In [10], an optimal primary frequency controller based on an adaptive optimal output–feedback approach was proposed to deal with uncertain disturbances in islanded microgrids. The authors of [11] presented a distributed robust secondary control approach to achieve frequency restoration and accurate active power sharing. The authors of [12] investigated a coordinated load-shedding control scheme to eliminate the power deficit for an islanded microgrid with distributed energy resources. In [13], the authors studied a deep learning-based load-shedding control strategy to determine the shedding action selection accurately. In addition, constructing an energy storage system (ESS) is a feasible solution to improve the stability of an isolated power grid with renewable energy. A joint frequency modulation control strategy combining wind farms and ESSs was employed to enhance the dynamic response characteristics under different wind speeds in [14]. The authors of [15] studied a variable droop strategy for an actual power grid with a battery ESS and evaluated its impact on the dynamic response.

Although the above literature is well studied, the common methods may not be particularly practical to guarantee the specific production requirements of the isolated industrial microgrid. The electrolytic aluminum load (EAL), as a type of energy-intensive industrial load, usually works as a whole part with an installed megawatt-class capacity, which means that the normal load-shedding strategy is not applicable for EALs.

As the time derivative of the power system frequency (df/dt), the rate of change of frequency (ROCOF) is important for qualifying the robustness of a power grid. Due to the small inertial of isolated industrial microgrids, extreme working scenarios, such as short faults on generators, may lead to the rapid decrease in the rate of change of frequency (ROCOF). Large-scale ESS construction will bring higher production costs for the high-energy-consuming enterprises, which is neither practical nor economical for isolated industrial microgrids.

Developing the regulating potential of EALs can be another feasible method, since the load control and demand response of EALs have been investigated by many research works. The authors of [16] studied the load control methods of the EAL based on its electrical model, such as controlling the transformer ratio and equivalent inductance of saturable reactors. By regulating the active power consumption of the EAL, the authors of [17] developed a frequency control method to eliminate the power imbalance in an isolated microgrid. In [18,19], the fast demand response provided by EALs was shown to smooth or track wind power fluctuations effectively by using tie-line power deviation as a feedback control signal. A hierarchical dispatch architecture of EALs and thermal generators in [20] was shown to provide secondary frequency auxiliary service for industrial microgrids. The aforementioned research works indicate that EALs are able to provide huge demand response in terms of emergency frequency control in isolated industrial microgrids.

Therefore, this paper aims to coordinate the generation and demand sides to achieve the emergency frequency control of isolated industrial microgrids with a high proportion of wind power. The traditional control strategies mainly focus on PI or PID controllers, which are incapable of achieving satisfactory effects. The PI control methods in [21,22] were unable to achieve zero-steady-error tracking in the presence of wind power fluctuations. Several research works applied model predictive control (MPC) methods to improve the dynamic performance. In [23], an MPC-based frequency controller was formulated to regulate the demand side and address the large power disturbance in microgrids. The authors of [24] proposed an MPC-based demand-side management strategy to smooth distributed photovoltaic power fluctuations. Although MPC controllers are capable of achieving the required performance, the complexity of online optimization algorithms brings significant difficulties for practical applications.

On the other hand, frequency control issues have been formulated as output regulation problems by several works. The authors of [25] investigated a frequency and excitation voltage control scheme by applying regulator theory to achieve demand-side regulation. In [26], the authors proposed a voltage and frequency regulation strategy based on the output regulation approach to track voltage and frequency references accurately in terms of the occurrence of disturbance events. Motivated by the positive error tracking ability and the practical feasibility, this paper adopts output regulation theory to develop a coordinated emergency frequency control for isolated industrial microgrids.

1.3. Contributions

In order to address the frequency control issues in isolated industrial microgrids with a high penetration of wind power, this paper aims to investigate the regulation potential of EALs by developing a detailed control model. For the purpose of restoring frequency in the presence of active power disturbances caused by emergencies or wind power fluctuations, a coordinated frequency control scheme based on the output regulation approach is proposed to combine the thermal generators and EALs.

In summary, the main contributions of this work are summarized as follows:

1. The emergency frequency control issues in isolated aluminum production industrial microgrids are discussed in detail. The control model of EALs and the advantages of demand-side control in isolated industrial microgrids are also illustrated.
2. A coordinated frequency control model combining thermal generators and EALs is developed for emergency frequency regulation in the isolated industrial microgrid in the presence of low system inertia and insufficient reserve capacity.
3. The problem of emergency frequency control is formulated as an output regulation problem. With the help of the output regulation approach, a practically feasible frequency controller is investigated with the feedback control signals of frequency deviation and power disturbances, verifying that it increases the rapid frequency regulating capacity of isolated industrial microgrids.

1.4. Organization

The remainder of this paper is organized as follows. Section 2 describes the emergency frequency control issues in an actual isolated industrial microgrid with a high proportion of wind power. In Section 3, the coordinated frequency control model is developed, including the detailed control model of EALs. In Section 4, the output regulation theory is applied to derive solutions for coordinated frequency control problems and to implement the frequency control scheme. The simulation results and analyses are presented in Section 5. The conclusions are given in Section 6.

2. Emergency Frequency Control Issues in Isolated Industrial Microgrids

As a northern province in China, Inner Mongolia is abundant with coal and wind energy resources, which is attractive for high-energy-consuming industries due to the lower electricity price. Electrolytic aluminum enterprises in China are gradually transiting production to the northern areas with abundant renewable energy. The technical and economic assessment in [27] has revealed the economic efficiency and engineering feasibility of constructing isolated industrial aluminum production microgrids. Thus, an actual industrial microgrid located in Inner Mongolia for aluminum production is selected as the study case, with the grid configuration shown in Figure 1. The main components of the industrial microgrid include generation sources, transformers, transmission lines, and loads. The generation sources consist of eight self-owned thermal generators (G1–G8) with a total installed capacity of 1800 MW and two 400-MW wind farms (WF1–WF2). Since the isolated industrial microgrid is constructed specifically for aluminum production, the main loads in the microgrid include three aluminum production lines (EAL1–EAL3) and regular heating loads. The loads are supplied by the thermal generators and wind farms directly via transformers and transmission lines. The detailed installed capacity of the thermal generators and wind farms, and the power demand of the loads, are listed in Tables 1 and 2, respectively.

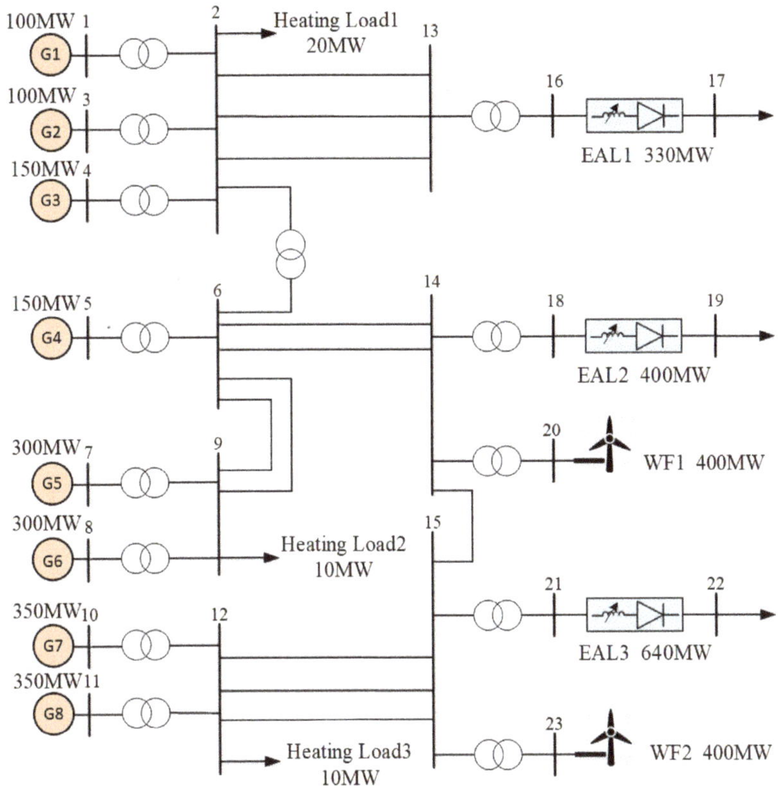

Figure 1. The structure of the actual isolated industrial microgrid for aluminum production.

Table 1. The installed capacity of thermal generators and wind farms.

Generators/Wind Farms	G1/G2	G3/G4	G5/G6	G7/G8	WF1/WF2
Installed Capacity (MW)	100	150	300	350	400

Table 2. The power demand of the EALs and heating loads.

EALs/Heating Loads	EAL1	EAL2	EAL3	Heating Load1	Heating Load2	Heating Load3
Power Demand (MW)	330	400	640	20	10	10

The isolated industrial microgrid operation is supposed to meet the security requirements of the power system, such as the "N-1" principle. The microgrid is supposed to have a sufficient reserve capacity to cope with active shortages and maintain a stable frequency when any one of the generation units is out of service. Generally, the primary frequency control reserve from thermal generators is set as 5% of the installed capacity, which means 90 MW in total for the power reserve. Table 3 lists a typical working situation for thermal generators and wind farms. For instance, when the thermal generator G5 is out of service because of short-circuit faults, the huge active power shortage of 170.2 MW in the industrial microgrid is much higher than the primary frequency reserve from the rest of the generators at 75 MW. The frequency nadir, representing the lowest point of the frequency when a power disturbance occurs, is an essential critical indicator for evaluating the power system stability.

Table 3. The output power of thermal generators and wind farms in a typical working situation.

Generation Units	G1/G2	G3/G4	G5/G6	G7/G8	WF1/WF2
Output Power (MW)	60.4	85.4	170.2	191.5	212.8

The generator G5 is tripped off from the microgrid at $t = 2$ s, leading to the frequency dropping down to the nadir of 48.48 Hz within 3 s (with the ROCOF of 4.45 Hz/s at the beginning of the emergency) and remaining steady at 49.5 Hz in the end, as shown in Figure 2.

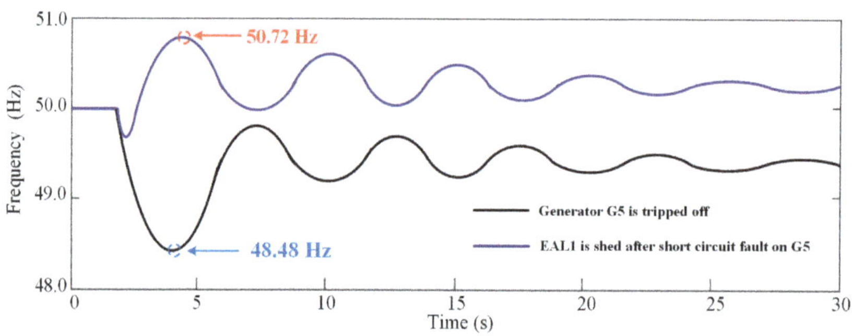

Figure 2. The frequency curves of different situations in the isolated industrial microgrid.

However, the traditional load-shedding strategy is not practical in this industrial microgrid because the EAL works as a whole part. The active power demand of three electrolytic aluminum production lines is listed in Table 2, demonstrating that the capacity of a single EAL is much larger than that of a single generator. Figure 2 demonstrates the frequency response when applying the load-shedding strategy. EAL1 is shed at 0.5 s after generator G5 being tripped off, leading the frequency rising to the peak of 50.72 Hz, which may activate the over-frequency protection of the generators and result in the further collapse of the system frequency. Obviously, it is difficult for the common methods, such as the primary frequency control and load-shedding strategy, to maintain the frequency stability in isolated industrial microgrids for the following reasons:

1. The reserve capacity from the thermal generators is insufficient for the large power imbalance caused by emergency faults. In addition, installing a high penetration of wind power (up to about 31%) and disconnecting from the utility power grid would further decrease the inertia of the microgrid.
2. Due to the special structure of the industrial microgrid, the installed capacity of a single generator or EAL is relatively huge compared with the total power demand capacity, which brings difficulties in achieving an active power balance with the common methods.

On the other hand, with the integration of a high penetration of wind power, the inertia of the isolated power system would decrease and the frequency stability characteristics would worsen. The active power imbalances caused by wind power fluctuation may result in rapid frequency change. The intensive wind fluctuation can reach up to ±15% of the installed capacity of wind farms in seconds. For cost reasons, wind turbine generators are generally encouraged to operate in maximum power point tracking (MPPT) mode, which provides no extra frequency regulation reserve. Except for the insufficient reserve capacity, the regulating ratio of the thermal generators may have difficulty in covering the intensive wind power fluctuation. Thus, integrating a high penetration of wind power in isolated industrial microgrids also brings severe frequency stability issues.

As a type of high-energy-consuming industrial load, the EAL has been proven to provide regulation potential, as mentioned in Section 1. Previous studies have also illustrated the control methods and regulation ability of EALs [16–20]. EALs are shown to provide regulation characteristics of a considerable capacity and a fast response, which will be discussed in detail in Section 3. Therefore, instead of traditional load-shedding strategies, the active power demand of EALs can be controlled accurately to provide reserve capacity to compensate for power imbalances. In view of the outcomes of the common control methods analyzed above, we propose a coordinated emergency frequency control scheme in this paper to combine the generation and demand sides in the presence of emergencies and intensive wind fluctuation scenarios. To obtain the theoretically feasible controller and desirable control effect, the output regulation theory will be applied to design the coordinate frequency control scheme in Section 4, aiming to achieve stable frequency recovery in the presence of large and intensive power disturbances.

3. Isolated Industrial Microgrid Modeling

3.1. Model and Control of EAL

In the case of the isolated industrial microgrid, the EAL is selected as the controlled plant for the demand-side control. Modern aluminum smelting production is usually generated in the electrolytic cell by using the Hall–Héroult Process to produce pure aluminum [28]. The high direct current (up to several hundred kiloamperes) is passed through to the electrolyzer and the aluminum oxide is smelted at high temperatures (over 950 °C). Then, the pure aluminum is decomposed between the carbon and aluminum electrodes.

The electrical diagram and equivalent circuit of the EAL have been investigated in detail in [16], which is shown in Figure 3. The EAL production line includes the AC bus, on-load tap changer (OLTC) transformer, rectification system, DC bus, and electrolyzer. All of the electrolyzers in a single EAL production line are connected in a series and supplied by the DC bus voltage U_{DC}. The AC bus voltage U_{AC} is regulated and rectified by the OLTC transformers (k_T is the transformer ratio) and rectification system. U_{SR} is the voltage drop on saturable reactors, with a normal working value of around 40 V. In [17], the electrolyzer is represented by a back electromotive force (EMF) E and a constant resistance R.

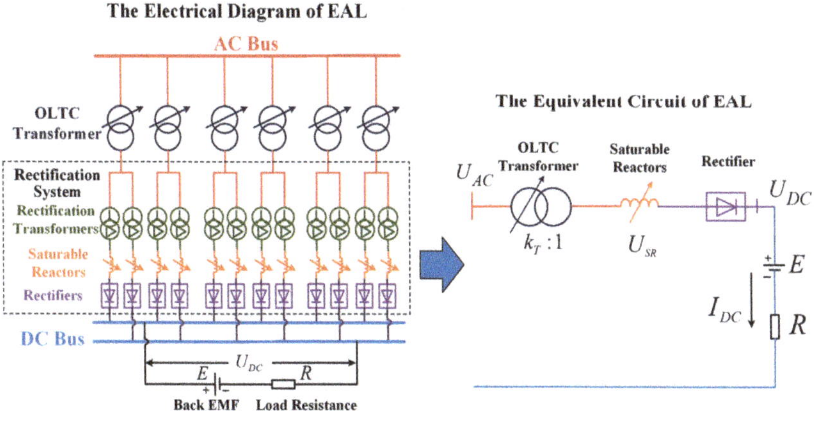

Figure 3. The electrical diagram and equivalent circuit of the EAL.

The DC current I_{DC} in the electrolyzer is calculated as [17]

$$I_{DC} = \frac{U_{DC} - E}{R} = \frac{1.35(U_{AC}/k_T - U_{SR}) - E}{R} \tag{1}$$

Thus, the active power consumption of the EAL P_{EAL} is calculated as

$$P_{EAL} = U_{DC}I_{DC} = (I_{DC}R + E)I_{DC} \tag{2}$$

In actual production, P_{EAL} can be regulated by the following two methods: (1) regulating the OLTC transformer ratio k_T; (2) changing the voltage drop on the saturable reactors U_{SR}. In practice, changing the voltage drop on the saturable reactors is preferable to regulating the OLTC transformer ratio for the following reasons:

1. Regulating the OLTC transformer ratio k_T is realized by changing the transformer tap changer positions, which takes minutes for one mechanical operation. In addition, the transformer tap changers are not allowed to be adjusted frequently for prolonging the life-span. Conversely, changing the voltage drop on saturable reactors U_{SR} with the electronic control only takes less than one second, which is much faster than the mechanical control, demonstrating the great advantages of the rapid response (millisecond level) in the presence of emergencies and wind power fluctuations.
2. Generally, the transformer tap changers have only five positions (i.e., $k_T = 1 \pm 2 \times 2.5\%$), indicating the step regulation on the active power control, while regulating the voltage drop on saturable reactors presents a much subtler way to achieve continuous regulation.

In [29], it was proven that P_{EAL} can be regulated from over -0.9 p.u. to 1.05 p.u. of the load capacity by controlling U_{SR} within its working limits (10–70 V), which means a considerable regulation capacity in terms of the megawatt-level loads. Thus, it is more feasible to adjust U_{SR} to achieve a rapid regulation.

Combining Equations (1) and (2), the relations between the active power deviation of the EAL ΔP_{EAL} and DC current deviation ΔI_{DC} voltage drop on saturable reactors ΔU_{SR} can be deduced as follows:

$$\Delta P_{EAL} = R\Delta I_{DC}^2 + R(I_{DC0} + E)\Delta I_{DC} \tag{3}$$

$$\Delta P_{EAL} = \frac{1.83\Delta U_{SR}^2 + (1.35E - 2.7\ U_{DC0})\Delta U_{SR}}{R} \tag{4}$$

where I_{DC0} and U_{DC0} are the initial values of the direct current and DC bus voltage.

Taking an actual EAL of 330 MW in [29] as an example, I_{DC0} and U_{DC0} are equal to 326.1 kA and 1.012 kV, respectively. E and R are equal to 354.6 V and 2.016 mΩ based on the field experiments. Thus, Equations (3) and (4) can be calculated as

$$\Delta P_{EAL} = 0.002016\Delta I_{DC}^2 + 1315.6\Delta I_{DC},\ \Delta P_{EAL} = 907\Delta U_{SR}^2 - 1.118 \times 10^6 \Delta U_{SR}$$

Obviously, the quadratic terms (ΔI_{DC}^2 and ΔU_{DC}^2) contain a slight proportion of and present a negligible influence on the value of ΔP_{EAL}. Thus, Equations (3) and (4) can be further simplified as

$$\Delta P_{EAL} \approx K_{PI}\Delta I_{DC},\ K_{PI} = R(I_{DC0} + E) \tag{5}$$

$$\Delta P_{EAL} \approx K_{PV}\Delta U_{SR},\ K_{PV} = (1.35E - 2.7\ U_{DC0})/R \tag{6}$$

where K_{PI} and K_{PV} are the gain coefficients, which are determined by the initial values of the EALs.

Based on the discussion above, the demand-side control model of the EAL is demonstrated in Figure 4. The control model includes the transfer functions of the electrolyzer and the saturable reactor control circuits. The time constants T_{ES} and T_{SR} are the response time of the electrolyzer and the saturable reactor control circuits, which can be approximated based on the dynamic response of each section. K_p, K_i, and K_{SR} are the gain coefficients in the saturable reactor control circuits. $\Delta U_{SR\max}$ and $\Delta U_{SR\min}$ are the upper and lower limits for ΔU_{SR}. The demand-side control model indicates that the active power consumption changes in the EAL ΔP_{EAL} can be regulated by providing the active power reference ΔP_{EALref}. Furthermore, the dynamic response of the demand-side control can be calculated as Equation (7), as follows:

$$G_{EAL}(s) = \frac{K_{PV} K_{SR}(K_i + sK_p)}{sK_{PI}(1+sT_{ES})(1+sT_{SR})} \tag{7}$$

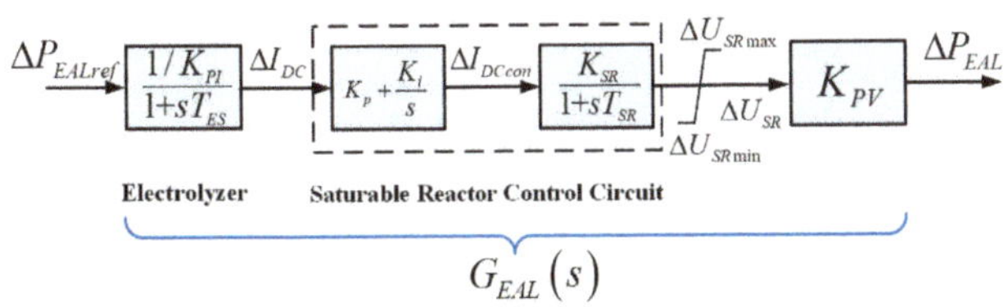

Figure 4. The demand-side control model on the EAL.

3.2. Coordinated Emergency Frequency Control Model of Isolated Industrial Microgrid

In this paper, a coordinated emergency frequency control model is proposed to combine the generation and demand side to balance the active power disturbance in the isolated industrial microgrid, and, if possible, to recover the frequency deviation Δf back to zero. Considering that the isolated industrial microgrid contains m thermal generators and n EALs, the block diagram of the coordinated emergency frequency control model is given as Figure 5. M and D are the equivalent constants of the inertia and load damping in the isolated industrial microgrid, respectively. ΔP_D is the active power disturbances, such as power imbalances or wind power fluctuations. For the i-th thermal generator and j-th EAL, the notations are presented in Table 4.

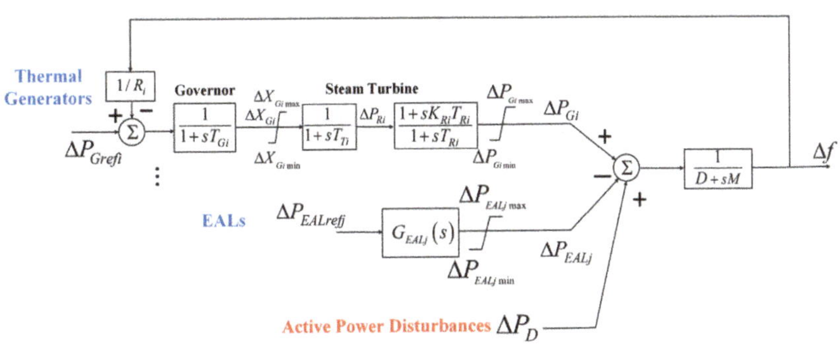

Figure 5. Coordinated emergency frequency control model of the isolated industrial microgrid.

Based on the block diagram in Figure 5, the state space model of the isolated industrial microgrid is described as

$$\begin{cases} \dot{\Delta f} = -\frac{D}{M}\Delta f + \frac{1}{M}\Delta P_{Gi} - \frac{1}{M}\Delta P_{EALj} + \frac{1}{M}\Delta P_D \\ \Delta \dot{P}_{Gi} = -\frac{1}{T_{Ri}}\Delta P_{Gi} + \left(\frac{1}{T_{Ri}} - \frac{K_{Ri}}{T_{Ti}}\right)\Delta P_{Ri} + \frac{K_{Ri}}{T_{Ti}}\Delta X_{Gi} \\ \Delta \dot{P}_{Ri} = -\frac{1}{T_{Ti}}\Delta P_{Ri} + \frac{1}{T_{Ti}}\Delta X_{Gi} \\ \Delta \dot{X}_{Gi} = -\frac{1}{T_{Gi}R_i}\Delta f - \frac{1}{T_{Gi}}\Delta X_{Gi} + \frac{1}{T_{Gi}}\Delta P_{Grefi} \\ \Delta \dot{I}_{DCj} = -\frac{1}{T_{ESj}}\Delta I_{DCj} + \frac{1}{T_{ESj}K_{PIj}}\Delta P_{EALrefj} \\ \Delta \dot{I}_{DCconj} = \left(K_{ij} - \frac{K_{pj}}{T_{ESj}}\right)\Delta I_{DCj} + \frac{K_{pj}}{T_{ESj}K_{PIj}}\Delta P_{EALrefj} \\ \Delta \dot{P}_{EALj} = \frac{K_{PVj}K_{SRj}}{T_{SRj}}\Delta I_{DCconj} - \frac{1}{T_{SRj}}\Delta P_{EALj} \end{cases} \quad (8)$$

Since the aim of the proposed coordinated emergency frequency control strategy is to eliminate the frequency deviation, the output of control system y is given as

$$y = \Delta f \quad (9)$$

Combining Equations (8) and (9), the coordinated emergency frequency control model can be summarized as

$$\begin{cases} \dot{x} = Ax + Bu + Pd \\ y = Cx \end{cases} \quad (10)$$

where x, u, y, and d are the variables of the system state, control input, output, and disturbance, respectively, expressed as in Equation (11). The detailed expressions of the matrices A, B, C, and P can be deduced according to Equations (8) and (9).

$$\begin{aligned} x &= \left[\Delta f, \Delta P_{Gi}, \Delta P_{Ri}, \Delta X_{Gi}, \Delta I_{DCj}, \Delta I_{DCconj}, \Delta P_{EALj}\right]^{\mathrm{T}} \in R^{(1+3i+3j)\times 1} \\ u &= \left[\Delta P_{Grefi}, \Delta P_{EALrefj}\right]^{\mathrm{T}} \in R^{(i+j)\times 1},\ d = \Delta P_D \end{aligned} \quad (11)$$

Therefore, the space state equations of the coordinated frequency control model of the isolated industrial microgrid are described as Equations (10) and (11). On the basis of that, the coordinated emergency frequency control strategy aims to increase the rapid frequency regulation ability and recover the frequency to the normal value when power disturbances exist.

Table 4. The notations in the coordinated emergency frequency control model.

Notation	Meaning
R_i	droop control coefficient
ΔX_{Gi}	valve position deviation
T_{Ri}	time constant of the reheater
ΔP_{Ri}	output power deviation of the reheater
K_{Ri}	turbine power fraction
ΔP_{Gi}	output power deviation of the steam turbine
$\Delta P_{Gimax}/\Delta P_{Gimin}$	maximal/minimal value of ΔP_{Gi}
T_{Gi}	time constant of the speed governor
T_{Ti}	time constant of the steam turbine
ΔP_{Grefi}	control input of the generator
ΔP_{EALj}	active power regulation of the EAL
$\Delta P_{EALjmax}/\Delta P_{EALjmin}$	maximal/minimal value of ΔP_{EALj}
$\Delta P_{EALrefj}$	control input of the EAL

4. Coordinated Emergency Frequency Control Through Output Regulation

4.1. Output Regulation via Error Feedback

In the presence of a power disturbance ΔP_D, the control objective of the coordinated emergency frequency control strategy is to regulate the active power output of the generators P_{Gi} and the power consumption of the EALs P_{EALj} to smooth ΔP_D. Thus, the coordinated frequency control is formulated as an output regulation issue, which can be described with the following linearization form:

$$\begin{cases} \dot{x} = Ax + Bu + Pd \\ \dot{d} = Sd \\ e = y - y_r = Cx + Qd \end{cases} \tag{12}$$

The first equation represents the relation between the system state $\boldsymbol{x}$, control input $\boldsymbol{u}$, and external disturbance $\boldsymbol{d}$. The second equation describes the characteristics of the exosystem with disturbances. The third equation defines an error variable $\boldsymbol{e}$ by introducing the output $y = Cx$ and the reference $y_r = Qd$. The detailed block diagram of the closed-loop control plant in Equation (12) is presented in Figure 6.

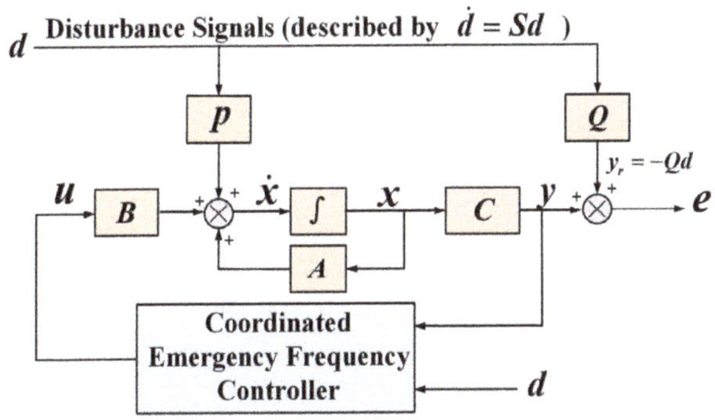

Figure 6. Block diagram of the coordinated emergency frequency control plant.

The coordinated emergency frequency controller aims to achieve error regulation and to maintain the system's internal stability. In other words, for the control system in Equation (12), find the gain matrices $\boldsymbol{K}$ and $\boldsymbol{L}$, if possible, to design the full-information feedback controller as

$$u = Kx + Ld \tag{13}$$

The feedback controller in Equation (13) is supposed to satisfy the following two requirements:

(R1) Internal stability. When the disturbance $\boldsymbol{d} = \boldsymbol{0}$, the closed-loop control system $\dot{x} = (\boldsymbol{A} + \boldsymbol{BK})x$ is asymptotically stable, which means that all of the eigenvalues of $(\boldsymbol{A} + \boldsymbol{BK})$ are on the open left-half plane;

(R2) Error regulation. When the disturbance $\boldsymbol{d} \neq \boldsymbol{0}$, for every initial condition $(\boldsymbol{x}_0, \boldsymbol{d}_0)$, the closed-loop control system achieves $\lim\limits_{t \to \infty} e(t) = 0$.

In order to solve the output regulation issues, the following two assumptions must be satisfied at first:

(A1) All of the eigenvalues of the matrix $\boldsymbol{S}$ are supposed to be located on the closed right-half plane;

(A2) The matrices pair $(\boldsymbol{A},\boldsymbol{B})$ is stabilizable.

Based on the regulator theory in [30], when the assumptions (A1) and (A2) are satisfied, the necessary and sufficient condition has been proven to solve the output regulation issues.

Proposition 1. *Under (A1) and (A2), the output regulation issues can be solvable by full-information feedback if and only if there exist matrices $\boldsymbol{\Pi}$ and $\boldsymbol{\Gamma}$, which solve the following regulator equations:*

$$\begin{cases} \Pi S = A\Pi + B\Gamma + P \\ 0 = C\Pi + Q \end{cases} \tag{14}$$

The gain matrix $\boldsymbol{L}$ can be computed as follows by the matrices $\boldsymbol{\Pi}$ and $\boldsymbol{\Gamma}$:

$$\boldsymbol{L} = \boldsymbol{\Gamma} - \boldsymbol{K\Pi} \tag{15}$$

Consequently, the feedback controller in Equation (13) can be designed as

$$u = \boldsymbol{K}x + (\boldsymbol{\Gamma} - \boldsymbol{K\Pi})d \tag{16}$$

Although the feedback controller can be constructed as Equation (16), it may be not feasible in practice because not all of the state variables x can be obtained as required. Thus, the feedback controller can be further refined by using only the measurable error signal $\boldsymbol{e}$. An observer is constructed to track $(\boldsymbol{x}, \boldsymbol{d})$ asymptotically based on the following assumption:

(A3) The matrices pair $\left(\begin{bmatrix} A & P \\ 0 & S \end{bmatrix}, \begin{bmatrix} C & Q \end{bmatrix}\right)$ is detectable.

Under (A3), the observer in the following equation is constructed to estimate $\boldsymbol{x}$ and $\boldsymbol{d}$ by using error signal $\boldsymbol{e}$:

$$\begin{bmatrix} \dot{\hat{x}} \\ \dot{\hat{d}} \end{bmatrix} = \left(\begin{bmatrix} A & P \\ 0 & S \end{bmatrix} - \begin{bmatrix} G_1 \\ G_2 \end{bmatrix}\begin{bmatrix} C & Q \end{bmatrix}\right)\begin{bmatrix} \hat{x} \\ \hat{d} \end{bmatrix} + \begin{bmatrix} G_1 \\ G_2 \end{bmatrix} e + \begin{bmatrix} B \\ 0 \end{bmatrix} \tag{17}$$

The matrices pair (G_1, G_2) is designed so that all of the eigenvalues of $\left(\begin{bmatrix} A & P \\ 0 & S \end{bmatrix} - \begin{bmatrix} G_1 \\ G_2 \end{bmatrix}\begin{bmatrix} C & Q \end{bmatrix}\right)$ are on the open left-half plane.

With the help of the observer in Equation (17), the error feedback controller can be designed with $\boldsymbol{\xi} = \mathrm{col}(\hat{\boldsymbol{x}}, \hat{\boldsymbol{d}})$ and the gain matrices $(\boldsymbol{F}, \boldsymbol{G}, \boldsymbol{H})$, given as

$$\begin{cases} \dot{\boldsymbol{\xi}} = \boldsymbol{F\xi} + \boldsymbol{G}e \\ u = \boldsymbol{H\xi} \end{cases} \tag{18}$$

The gain matrices $(\boldsymbol{F}, \boldsymbol{G}, \boldsymbol{H})$ can be deduced by replacing $(\boldsymbol{x}, \boldsymbol{d})$ in Equation (16) with $(\hat{\boldsymbol{x}}, \hat{\boldsymbol{d}})$ after solving Equation (14), with the following expressions:

$$\begin{gathered} \boldsymbol{F} = \begin{bmatrix} A - G_1C + BK & P - G_1Q + B(\Gamma - K\Pi) \\ -G_2C & S - G_2Q \end{bmatrix} \\ \boldsymbol{G} = \begin{bmatrix} G_1 \\ G_2 \end{bmatrix} \quad \boldsymbol{H} = \begin{bmatrix} K & (\Gamma - K\Pi) \end{bmatrix} \end{gathered} \tag{19}$$

By using the error signal, the feedback controller in Equation (18) is able to achieve the control requirements (R1) and (R2). On the basis of that, the control system output $\boldsymbol{y} = \boldsymbol{Cx}$ can be the controlled to follow the reference $\boldsymbol{y_r} = \boldsymbol{Qd}$ via error feedback.

4.2. Implementation of Coordinated Emergency Frequency Control Scheme

The design procedure of the coordinated emergency frequency control scheme based on the output regulation is presented as follows:

1. *Model the control plant.* At time t_k, the state variables $\boldsymbol{x}$, disturbance $\boldsymbol{d}$, and frequency deviation Δf in the isolated industrial microgrid are collected via the wide area measurement system (WAMS) [31]. The coordinated frequency control model that considers the generation and demand-side participation is constructed as in Equations (8)–(10).
2. *Check the assumptions* (A1)–(A3). Since the collected active power disturbance ΔP_D changes with the step signal at each sampling time, the value of ΔP_D is assumed as constant. Then, $\boldsymbol{d} = \Delta P_D$ satisfies $\dot{\boldsymbol{d}} = \boldsymbol{Sd} = \boldsymbol{0}$, which means that the exosystem is Poisson stable and the assumption (A1) holds in this work. (A2) and (A3) can be

checked by using the Popov–Belevitch–Hautus (PBH) test, with the rank conditions presented as

$$\begin{aligned} &Rank\left(\begin{bmatrix}\lambda \boldsymbol{I_A} - \boldsymbol{A} & \boldsymbol{B}\end{bmatrix}\right) = n_A \\ &Rank\left(\begin{bmatrix}\mu \boldsymbol{I_{APS}} - \begin{bmatrix}\boldsymbol{A} & \boldsymbol{P} \\ \boldsymbol{0} & \boldsymbol{S}\end{bmatrix} \\ \begin{bmatrix}\boldsymbol{C} & \boldsymbol{Q}\end{bmatrix}\end{bmatrix}\right) = n_{APS} \end{aligned} \tag{20}$$

where λ and μ are the eigenvalues of the matrices $\boldsymbol{A}$ and $\begin{bmatrix}\boldsymbol{A} & \boldsymbol{P} \\ \boldsymbol{0} & \boldsymbol{S}\end{bmatrix}$ sitting on the closed right-half plane. $\boldsymbol{I_A}$ and $\boldsymbol{I_{APS}}$ are the identity matrices, and n_A and n_{APS} are the dimensions, respectively.

3. *Select gain matrices.* Once the assumptions (A1)–(A3) hold, find a gain matrix $\boldsymbol{K}$, such that all of the eigenvalues of $\boldsymbol{A} + \boldsymbol{BK}$ are located on the left-half plane. A linear quadratic (LQ) method is applied in this work to obtain a reasonable gain matrix $\boldsymbol{K}$, achieving the control objectives with a satisfactory performance by the controllers. The cost function for the LQ method is given as follows:

$$J_{lq} = \int_0^{\infty} \left(\boldsymbol{y}^{\mathrm{T}} \boldsymbol{Q_{lq}} \boldsymbol{y} + \boldsymbol{u}^{\mathrm{T}} \boldsymbol{R_{lq}} \boldsymbol{u}\right) \mathrm{d}t \tag{21}$$

The selection of the weighting matrices $\boldsymbol{Q_{lq}}$ and $\boldsymbol{R_{lq}}$ means a trade-off between the eliminating error (the frequency deviation Δf in this case) and optimizing the control performance on the thermal generators/EALs. Thus, the power system inertia and flexibility in designing the coordinated control scheme is judged and weighted by solving following equations and computing the gain matrix $\boldsymbol{K}$:

$$\boldsymbol{A}^{\mathrm{T}} \boldsymbol{P_{lq}} + \boldsymbol{P_{lq}^{\mathrm{T}}} \boldsymbol{A} - \boldsymbol{P_{lq}} \boldsymbol{B} \boldsymbol{R_{lq}^{-1}} \boldsymbol{B}^{\mathrm{T}} \boldsymbol{P_{lq}} + \boldsymbol{C}^{\mathrm{T}} \boldsymbol{Q_{lq}} \boldsymbol{C} = 0 \tag{22}$$

$$\boldsymbol{K} = -\boldsymbol{R_{lq}^{-1}} \boldsymbol{B}^{\mathrm{T}} \boldsymbol{P_{lq}} \tag{23}$$

4. Similarly, it is also supposed that the gain matrix $\boldsymbol{G}$ satisfies that all of the eigenvalues of $\left(\begin{bmatrix}\boldsymbol{A} & \boldsymbol{P} \\ \boldsymbol{0} & \boldsymbol{S}\end{bmatrix} - \begin{bmatrix}\boldsymbol{G_1} \\ \boldsymbol{G_2}\end{bmatrix}\begin{bmatrix}\boldsymbol{C} & \boldsymbol{Q}\end{bmatrix}\right)$ are on the open left-half plane. In this paper, the pole assignment based on the interval linear matrix inequality method is applied to select the gain matrices $\boldsymbol{K}$ and $\boldsymbol{G}$.
5. *Construct the coordinated frequency controller.* The regulator issues in Equation (14) are solved and the solutions of matrices $\boldsymbol{\Pi}$ and $\boldsymbol{\Gamma}$ are found. Then, the observer is constructed as in Equation (17), and the error feedback controller in Equation (18) can be designed by the matrices $(\boldsymbol{F}, \boldsymbol{G}, \boldsymbol{H})$.
6. *Give the control orders.* Once the error feedback controller is constructed, the control input $\boldsymbol{u} = \left[\boldsymbol{\Delta P_{Grefi}}, \boldsymbol{\Delta P_{EALrefj}}\right]^{\mathrm{T}}$ is optimized at time t_k. Then, the optimal values of ΔP_{Grefi} and $\Delta P_{EALrefj}$ are transmitted to the control units of the thermal generators and EALs. Thus, the active power of the generation and demand sides are regulated to eliminate the frequency deviation in the isolated industrial microgrid.
7. *Repeat the above steps or finish.* Once the control objective is achieved, i.e., the frequency recovers to the normal value with the help of the proposed controller, the coordinated frequency control scheme is finished. Otherwise, repeat the above steps at the next time t_{k+1}.

In summary, once the proposed frequency control scheme is activated, the state variables $\boldsymbol{x}$, disturbance $\boldsymbol{d}$, and frequency deviation Δf in the isolated industrial microgrid are measured by the WAMS. Then, the coordinated frequency control model in Equation (10) is constructed as the foundation. After checking assumptions, selecting gain matrices, and solving regulator equations, the error feedback controller is designed by using the output regulation theory. Then, the optimized control inputs $\boldsymbol{u} = \left[\boldsymbol{\Delta P_{Grefi}}, \boldsymbol{\Delta P_{EALrefj}}\right]^{\mathrm{T}}$ are obtained and given as the regulation commands to thermal generators and EALs, so that the output $y = \Delta f$ can be eliminated by compensating for the power disturbance $\boldsymbol{d}$ caused by emergencies or wind power fluctuations. Based on that, the emergency frequency control problem is solved by applying the output regulation approach.

The effectiveness of the proposed frequency control scheme will be validated via time domain simulation in the next section.

5. Simulation Results

To validate the performance of the proposed coordinated frequency controller based on the output regulation approach, the isolated industrial microgrid presented in Figure 1 is taken as the study case, which is conducted on the real-time digital simulation (RTDS) platform and in MATLAB. The RTDS was developed for the electromagnetic transient simulation of power systems based on the real-time hardware computation platform. The synchronous generator model in the RTDS platform was applied for the simulation of thermal generators in the industrial microgrid. An aggregated model of a 2-MW doubly fed induction generator (DFIG) was used to represent wind farms based on the modeling method presented in [32]. The EALs were conducted, as in Figure 4, with the controllable load function in the RTDS platform.

When the coordinated frequency control strategy is activated, the state variables $\boldsymbol{x}$, disturbance $\boldsymbol{d}$, and error $\boldsymbol{e}$ of the industrial microgrid are collected and then transmitted to MATLAB during each time step. Thus, the error feedback controller in Equation (18) is calculated in MATLAB based on the procedures provided in Section 4.2, and the control input $\boldsymbol{u}$ is fed into the RTDS platform. Consequently, the thermal generator and EAL models will carry out the frequency control scheme as the optimized control input $\boldsymbol{u}$, which means that the output power of the thermal generators and EALs will be regulated to achieve error tracking. The whole simulation process between the RTDS platform and MATLAB, shown in Figure 7, will repeat in the next time step.

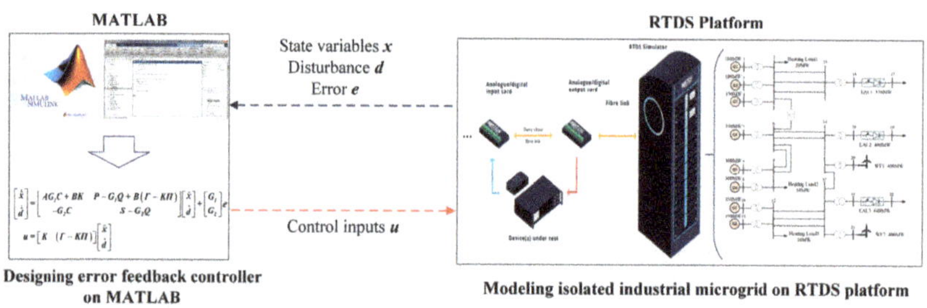

Figure 7. The simulation process between the RTDS platform and MATLAB.

The installed capacity of the generators and the power demand of the EALs are listed in Tables 1 and 2. In this section, we consider two cases, i.e., the "N-1" and wind power fluctuation scenarios, to verify the effectiveness and robustness of the coordinated frequency control scheme. To further prove the improved performance of the proposed control scheme, PI control methods are also conducted as a comparison.

5.1. "N-1" Scenario

In this case, the selected typical working situation of the industrial microgrid is listed in Table 3. The wind power fluctuates within a slight range of less than 1% of the installed wind farm capacity, which can be covered by the ESSs equipped in the wind farms. As mentioned in Section 2, the isolated industrial microgrid operation is supposed to meet the "N-1" principle. Therefore, we considered a more severe situation than in Section 2 (where G5 is out of service) so to verify the performance of the proposed coordinated frequency control scheme. In this case, the thermal generator G7 is tripped off due to a short-circuit fault at $t = 2$ s, leading to an active imbalance of 191.5 MW. The control strategies applied in this case are presented in Table 5.

Table 5. The control schemes in "N-1" scenarios.

Control Strategy	Method
Strategy A	Only primary frequency control on the rest of the thermal generators
Strategy B	Load-shedding strategy on EAL1
Strategy C	PI control methods on the generators and EALs
Strategy D	Proposed coordinated frequency control

In Strategy A, the power disturbance of 191.5 MW is only compensated by the primary frequency reserve of 72.5 MW from the rest of the thermal generators. With the traditional load-shedding strategy, the EAL1 of 330 MW is shed as a whole part in Strategy B, as mentioned in Section 2. The PI-based control strategy is designed by using the frequency deviation Δf for the thermal generators and EALs. The signal of Δf is collected and given as the input of the PI controllers. Then, the active power regulation of the thermal generators and EALs, ΔP_{Grefi} and $\Delta P_{EALrefj}$, can be calculated as follows:

$$\begin{aligned} \Delta P_{Grefi}(s) &= \left(K_{GPi} + \tfrac{K_{GIi}}{s}\right)\Delta f(s) \\ \Delta P_{EALrefj}(s) &= \left(K_{LPj} + \tfrac{K_{LIj}}{s}\right)\Delta f(s) \end{aligned} \tag{24}$$

where K_{GPi} and K_{GIi}, and K_{LPj} and K_{LIj}, are the gain coefficients of the PI controllers for the thermal generators and EALs, with the setting methods introduced in [33,34].

The simulation results are demonstrated in Figure 8. When the thermal generator G7 is out of service, the primary frequency reserve capacity from the rest of the generators is apparently insufficient for the active power imbalance, leading to the frequency collapse of the industrial microgrid within 30 s, as shown in Figure 8a. The total power deviations of the thermal generators and EALs are demonstrated in Figures 8b and 8c, respectively.

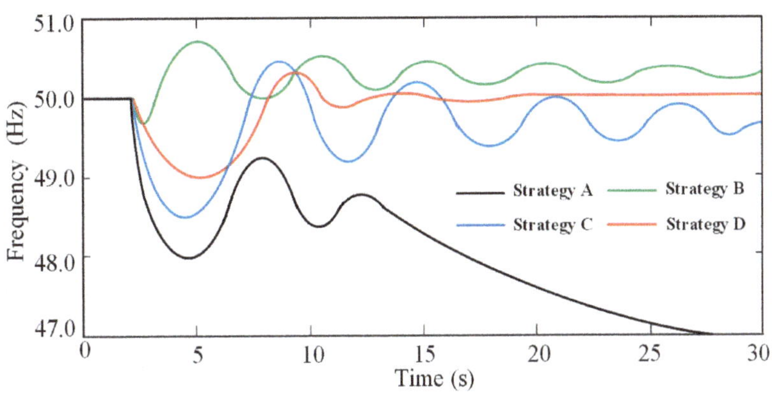

(a)

Figure 8. *Cont.*

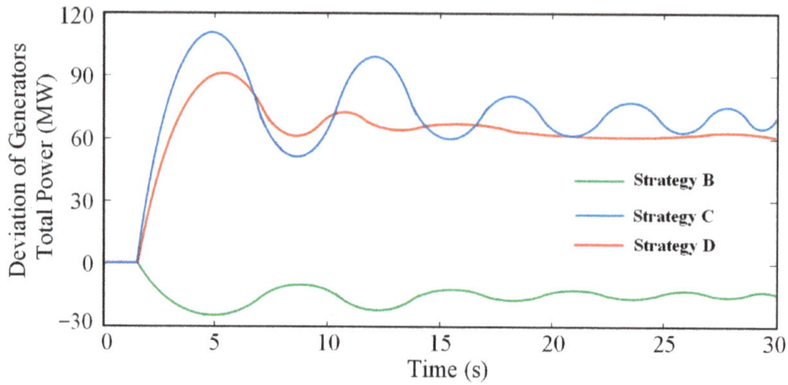

(b)

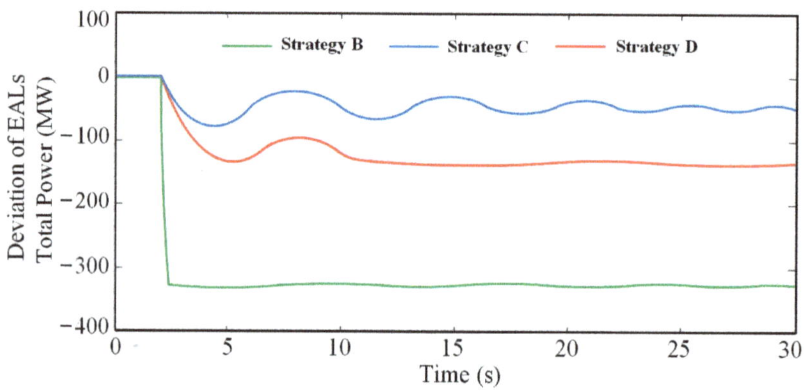

(c)

Figure 8. Simulation results of applying different control strategies in the "N-1" scenario, where G7 is tripped off. (**a**) The frequency response; (**b**) The total power deviation of the generators; (**c**) The total power deviation of the EALs.

Due to the large installed capacity of single aluminum production lines, the shedding of EAL1 of 330 MW is much higher than the active power disturbance. As a result, the frequency with Strategy B rises to over 50.5 Hz, which is not allowed for the industrial microgrid operation. When the PI control method is used on the generators and EALs, the industrial microgrid frequency decreases to 48.5 Hz at t = 4 s. Although the active power balance is able to be compensated by the generation and demand sides to a certain extent, the power reserve distribution among the generators and EALs is ambiguous because it lacks coordination with the PI controllers. Therefore, the PI control strategy presents an unsatisfactory robustness when reflecting on the frequency fluctuation within a certain range.

When the coordinated frequency control scheme is applied in Strategy D, the frequency response, such as the ROCOF and nadir, are obviously improved when compared to Strategies B and C. In addition, the frequency response presents a much stabler way to recover the normal working value. As discussed in Section 3, the EALs can be regulated from over −0.9 p.u. to 1.05 p.u. of the installed capacity. By virtue of the fast response, the EALs provide a reserve capacity of about 130 MW after G7 is out of service, preventing the further decrease in the frequency. The thermal generators provide a power reserve of about 90 MW in total. Then, under the coordination between the generators and EALs, the microgrid frequency recovers to 50 Hz within 20 s. Therefore, the proposed coordinated frequency control strategy is effective in restoring the frequency to a normal value when emergencies occur, which is significant for guaranteeing the frequency stability of the isolated industrial microgrid.

5.2. Wind Power Fluctuation Scenario

As presented in Table 3, the installed capacity of the wind power is about 30% of the total generation capacity in the industrial microgrid. Therefore, the influence of wind power fluctuation on the microgrid frequency control is studied. Normally, wind farms are equipped with ESSs; however, here, we consider a severe scenario with an extreme short-term wind power fluctuation without any assistance from the ESS.

Total wind power fluctuation in a period of 600 s is demonstrated in Figure 9, and the maximum deviation of the total wind power reaches about 180 MW. Due to the intermittency and variability of the wind power, the industrial microgrid frequency fluctuates as well, which is unsatisfactory for the frequency stability.

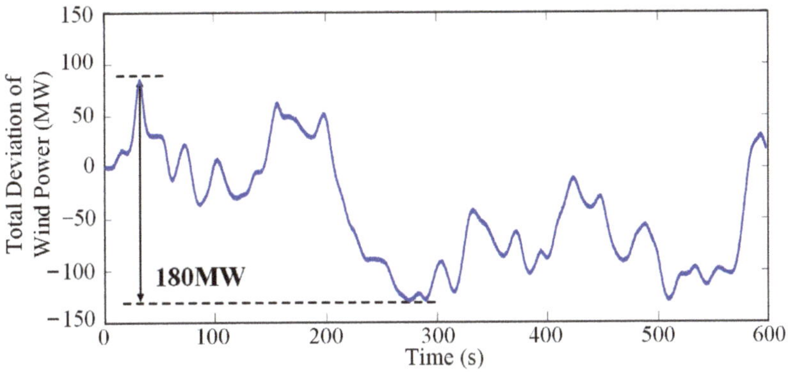

Figure 9. The total wind power fluctuation in a period of 600 s.

The control strategies applied in this case are presented in Table 5. Figure 10 illustrates the simulation results of the frequency response of different control strategies. With the primary frequency control of the generators, the frequency deviation caused by the wind power fluctuation changes from over −0.5 Hz to 0.3 Hz, which is not safe for the operation of the industrial microgrid. With the help of the PI control strategy, the wind power fluctuation can be partly compensated by the thermal generators and EALs. As a result, the frequency deviation is able to be restricted to ±0.3 Hz, which presents a positive effect of dealing with a power imbalance.

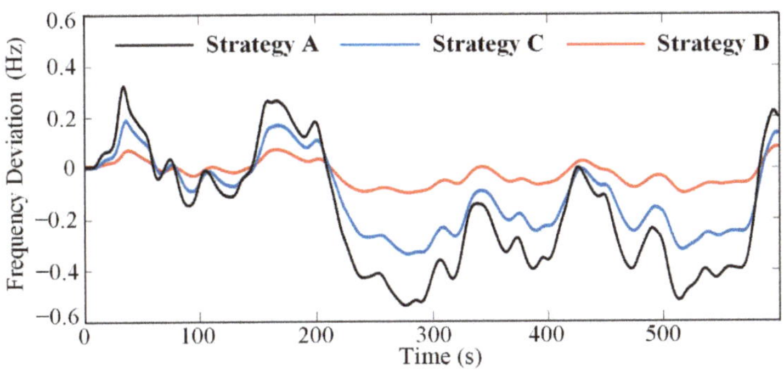

Figure 10. The frequency response with different control strategies.

Compared with the PI control method, the coordinated frequency control strategy shows a better performance in smoothing the intensive wind power fluctuation. Consequently, the frequency fluctuates within ±0.1 Hz by virtue of the rapid response of the EALs. The power deviation of the generators and EALs are demonstrated in Figure 11. The power deviation of the EALs reach to about 160 MW, while the thermal generators provide a power reserve of less than 30 MW for the wind power fluctuation, revealing that the wind power fluctuation is mainly compensated by the EALs instead of the generators. The proposed coordinated control scheme presents a satisfactory robustness against the wind power fluctuation, guaranteeing a stable frequency for the industrial microgrid.

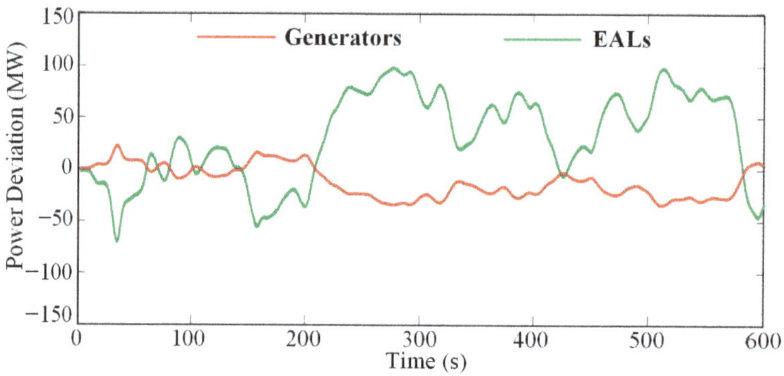

Figure 11. The power deviation of the generators and EALs with the coordinated frequency control strategy.

6. Discussion and Conclusions

This paper studied emergency frequency control issues in an actual isolated aluminum production industrial microgrid with a high proportion of wind power. As a type of high-energy-consuming industrial load, the EAL shows great potential for power regulation due to the fast response and huge capacity, which is able to participate in the frequency control of the industrial microgrid. The regulation method and control model of the EAL have been investigated in this work. On the basis of that, a coordinated frequency control model which combines the thermal generators and EALs was developed to deal with emergency scenarios in the microgrid. The coordinated frequency control issue was formulated as an output regulation problem. With the help of the output regulation theory, the error feedback controller for the isolated industrial microgrid was designed to regulate the output power of the generators and EALs, achieving coordinated frequency control when a large power disturbance occurs.

Compared with the traditional PI controller, the main advantages of the error feedback controller include the tracking of the power disturbance $\boldsymbol{d}$ and the achievement of steady error regulation ($\lim_{t\to\infty}\boldsymbol{e}(t) = 0$). As a type of model-based controller, the error feedback controller guarantees more desirable control effects in terms of frequency recovery. However, the limitations of the proposed approach in practical applications are also obvious, which mainly focus on the time-delay issues of the WAMS. The communication and operational delays might present a negative impact on the control effects. In practical applications, the evaluation model of the communication delay was applied so to decrease the negative effects. The operational delay also needs to be measured and analyzed in the process of designing the error feedback controller.

The simulation results conducted on the actual industrial microgrid have illustrated the following conclusions:

1. The proposed coordinated frequency control strategy is able to realize frequency recovery to normal values in a short amount of time in the presence of a large power imbalance.

2. Compared with the PI control method, the proposed frequency control method based on the output regulation approach presents better effectiveness in frequency recovery and robustness against the wind power fluctuations.
3. The EALs present a much faster response than the thermal generators in smoothing wind power fluctuations, which is applicable to improving the frequency stability in extreme emergency scenarios.

In summary, by developing the regulation potential of EALs, the proposed coordinated frequency control scheme based on output regulation theory presents satisfactory effectiveness and robustness in dealing with power disturbances caused by emergencies and wind power fluctuations, which significantly improves the frequency reliability and stability in industrial microgrids.

The working constraints on thermal generators and EALs, such as the regulating ratio and production limitations, may influence the practical control effect of the coordinated frequency control scheme based on the output regulation approach. Future studies will focus on the further optimization of the regulation between the generators and EALs so to achieve lower control costs. In addition, time-delay issues will be considered for the practical applications of the proposed frequency control scheme so to achieve the more accurate online computation of the power reserve distribution.

Author Contributions: Conceptualization, X.D. and S.Z.; methodology, X.D.; software, X.D.; validation, X.D. and S.Z.; formal analysis, X.D.; investigation, S.Z.; resources, X.D.; data curation, X.D.; writing—original draft preparation, S.Z.; writing—review and editing, X.D.; visualization, X.D.; supervision, X.D.; project administration, X.D.; funding acquisition, X.D. All authors have read and agreed to the published version of the manuscript.

Funding: This work was supported by the Technology Project of China Southern Power Grid (0500002022030301GH00163).

Data Availability Statement: Data are available on request from the corresponding author rather than via direct availability.

Conflicts of Interest: Author Sujie Zhang was employed by the State Grid Shanghai Municipal Electric Power Company, and remaining authors declare that no conflicts of interest.

References

1. Wei, Y.; Chen, K.; Kang, J.; Chen, W.; Wang, X.; Zhang, X. Policy and management of carbon peaking and carbon neutrality: A literature review. *Engineering* **2022**, *14*, 52–63. [CrossRef]
2. Zhuang, G.; Zhou, H.; Ni, W. *China's Road to Carbon Peaking and Carbon Neutrality*; Springer: Singapore, 2023; pp. 1–20.
3. Xu, B. Exploring the sustainable growth pathway of wind power in China: Using the semiparametric regression model. *Energy Policy* **2023**, *183*, 113845. [CrossRef]
4. National Energy Administration Released the Statistical Data of the National Power Industry in 2023. Available online: https://www.nea.gov.cn/2024-01/26/c_1310762246.htm (accessed on 26 January 2024).
5. Liao, S.; Xu, J.; Sun, Y.; Bao, Y. Local utilization of wind electricity in isolated power systems by employing coordinated control scheme of industrial energy-intensive load. *Appl. Energy* **2018**, *217*, 14–24. [CrossRef]
6. Li, H.; Eseye, A.T.; Zhang, J.; Zheng, D. Optimal energy management for industrial microgrids with high-penetration renewables. *Prot. Control Mod. Power Syst.* **2017**, *2*, 12. [CrossRef]
7. Wang, Z.; Yang, B.; Wei, W.; Zhu, S.; Guan, X.; Sun, D. Multi-Energy Microgrids: Designing, operation under new business models, and engineering practices in China. *IEEE Electrif. Mag.* **2021**, *9*, 75–82. [CrossRef]
8. Fernández-Guillamón, A.; Gómez-Lázaro, E.; Muljadi, E.; Molina-García, Á. Power systems with high renewable energy sources: A review of inertia and frequency control strategies over time. *Renew. Sustain. Energy Rev.* **2019**, *115*, 109369. [CrossRef]
9. Mohandes, B.; El Moursi, M.S.; Hatziargyriou, N.; El Khatib, S. A review of power system flexibility with high penetration of renewables. *IEEE Trans. Power Syst.* **2019**, *34*, 3140–3155. [CrossRef]
10. Davari, M.; Gao, W.; Jiang, Z.-P.; Lewis, F.L. An optimal primary frequency control based on adaptive dynamic programming for islanded modernized microgrids. *IEEE Trans. Autom. Sci. Eng.* **2020**, *18*, 1109–1121. [CrossRef]
11. Mottaghizadeh, M.; Aminifar, F.; Amraee, T.; Sanaye-Pasand, M. Distributed robust secondary control of islanded microgrids: Voltage, frequency, and power sharing. *IEEE Trans. Power Deliv.* **2021**, *36*, 2501–2509. [CrossRef]
12. Wang, C.; Mei, S.; Dong, Q.; Chen, R.; Zhu, B. Coordinated load shedding control scheme for recovering frequency in islanded microgrids. *IEEE Access* **2020**, *8*, 215388–215398. [CrossRef]

13. Li, J.; Chen, S.; Wang, X.; Pu, T. Load shedding control strategy in power grid emergency state based on deep reinforcement learning. *CSEE J. Power Energy* **2021**, *8*, 1175–1182.
14. Yu, L.; Wu, J.; Cheng, Y.; Meng, G.; Chen, S.; Lu, Y.; Xu, K. Control Strategy for Wind Farms-Energy Storage Participation in Primary Frequency Regulation Considering Wind Turbine Operation State. *Energies* **2024**, *17*, 3547. [CrossRef]
15. Arrigo, F.; Bompard, E.; Merlo, M.; Milano, F. Assessment of primary frequency control through battery energy storage systems. *Int. J. Electr. Power Energy Syst.* **2020**, *115*, 105428. [CrossRef]
16. Jiang, H.; Lin, J.; Song, Y.; Gao, W.; Xu, Y.; Shu, B.; Li, X.; Dong, J. Demand side frequency control scheme in an isolated wind power system for industrial aluminum smelting production. *IEEE Trans. Power Syst.* **2013**, *29*, 844–853. [CrossRef]
17. Xu, J.; Liao, S.; Sun, Y.; Ma, X.; Gao, W.; Li, X.; Gu, J.; Dong, J.; Zhou, M. An isolated industrial power system driven by wind-coal power for aluminum productions: A case study of frequency control. *IEEE Trans. Power Syst.* **2014**, *30*, 471–483. [CrossRef]
18. Ding, X.; Xu, J.; Sun, Y.; Liao, S.; Zheng, J. A demand side controller of electrolytic aluminum industrial microgrids considering wind power fluctuations. *Prot. Control Mod. Power Syst.* **2022**, *7*, 49. [CrossRef]
19. Liao, S.; Xu, J.; Sun, Y.; Bao, Y.; Tang, B. Control of energy-intensive load for power smoothing in wind power plants. *IEEE Trans. Power Syst.* **2018**, *33*, 6142–6154. [CrossRef]
20. Bao, Y.; Xu, J.; Feng, W.; Sun, Y.; Liao, S.; Yin, R.; Jiang, Y.; Jin, M.; Marnay, C. Provision of secondary frequency regulation by coordinated dispatch of industrial loads and thermal power plants. *Appl. Energy* **2019**, *241*, 302–312. [CrossRef]
21. Babahajiani, P.; Shafiee, Q.; Bevrani, H. Intelligent demand response contribution in frequency control of multi-area power systems. *IEEE Trans. Smart Grid* **2016**, *9*, 1282–1291. [CrossRef]
22. Zhao, J.; Lyu, X.; Fu, Y.; Hu, X.; Li, F. Coordinated microgrid frequency regulation based on DFIG variable coefficient using virtual inertia and primary frequency control. *IEEE Trans. Energy Convers.* **2016**, *31*, 833–845. [CrossRef]
23. Jiang, H.; Lin, J.; Song, Y.; Hill, D.J. MPC-based frequency control with demand-side participation: A case study in an isolated wind-aluminum power system. *IEEE Trans. Power Syst.* **2014**, *30*, 3327–3337. [CrossRef]
24. Xu, J.; Fu, H.; Liao, S.; Xie, B.; Ke, D.; Sun, Y.; Li, X.; Peng, X. Demand-side management based on model predictive control in distribution network for smoothing distributed photovoltaic power fluctuations. *J. Mod. Power Syst. Clean Energy* **2022**, *10*, 1326–1336. [CrossRef]
25. Cui, T.; Lin, W.; Sun, Y.; Xu, J.; Zhang, H. Excitation voltage control for emergency frequency regulation of island power systems with voltage-dependent loads. *IEEE Trans. Power Syst.* **2015**, *31*, 1204–1217. [CrossRef]
26. Zhao, H.; Hong, M.; Lin, W.; Loparo, K.A. Voltage and frequency regulation of microgrid with battery energy storage systems. *IEEE Trans. Smart Grid* **2017**, *10*, 414–424. [CrossRef]
27. Sun, Y.; Lin, J.; Song, Y.; Xu, J.; Li, X.; Dong, J. An industrial system powered by wind and coal for aluminum production: A case study of technical demonstration and economic feasibility. *Energies* **2012**, *5*, 4844–4869. [CrossRef]
28. Principles of the Hall-Héroult Process. Available online: http://www.peter-entner.com/e/theory/PrincHH/PrincHH.aspx (accessed on 26 January 2005).
29. Bao, Y.; Xu, J.; Liao, S.; Sun, Y.; Li, X.; Jiang, Y.; Ke, D.; Yang, J.; Peng, X. Field verification of frequency control by energy-intensive loads for isolated power systems with high penetration of wind power. *IEEE Trans. Power Syst.* **2018**, *33*, 6098–6108. [CrossRef]
30. Isidori, A.; Byrnes, C.I. Output regulation of nonlinear systems. *IEEE Trans. Autom. Control* **1990**, *35*, 131–140. [CrossRef]
31. Phadke, A.G.; Bi, T. Phasor measurement units, WAMS, and their applications in protection and control of power systems. *J. Mod. Power Syst. Clean Energy* **2018**, *6*, 619–629. [CrossRef]
32. Fernández, L.M.; Jurado, F.; Saenz, J.R. Aggregated dynamic model for wind farms with doubly fed induction generator wind turbines. *Renew. Energy* **2008**, *33*, 129–140. [CrossRef]
33. Sönmez, S.; Ayasun, S. Stability region in the parameter space of PI controller for a single-area load frequency control system with time delay. *IEEE Trans. Power Syst.* **2015**, *31*, 829–830. [CrossRef]
34. Jiang, L.; Yao, W.; Wu, Q.; Wen, J.; Cheng, S. Delay-dependent stability for load frequency control with constant and time-varying delays. *IEEE Trans. Power Syst.* **2011**, *27*, 932–941. [CrossRef]

energies

MDPI

Article

Power Oscillation Source Location Based on the Combination of Energy Function and Normal Distribution in a Fully Data-Driven Approach

Shujia Guo [1,*], Xu Liu [2], Chao Jiang [3] and Jing Cong [3]

1 School of Automation and Electrical Engineering, Zhejiang University of Science and Technology, Hangzhou 310023, China
2 State Grid Pinghu Power Supply Company, Jiaxing 314299, China; liuxgjdw@126.com
3 School of Electrical Engineering, Northeast Electric Power University, Jilin 132012, China; 20202974@neepu.edu.cn (C.J.); congjing@xinliangxiang.com (J.C.)
* Correspondence: 123065@zust.edu.cn; Tel.: +86-13904348753

Abstract: With the deepening of national efforts toward green energy transformation, the power system is evolving into one characterized by "double high"—a high proportion of new energy integration and a high level of power electronic systems. This results in a more complex system topology, necessitating improvements in various prevention and control measures. Traditional model-based methods for locating power oscillation disturbance sources in power systems are no longer sufficient to meet the operational demands of modern power systems. With the rapid development of wide-area measurement systems (WAMS), there is growing interest in disturbance source localization using system response data. System dynamics provide a wealth of easily extractable data that can accurately reflect the power system's behavior under normal conditions. This paper proposes a numerical method for locating disturbance sources, combining energy functions with normal distribution identification, based on power oscillation mechanisms and system response data. The method identifies potential disturbance sources, including small random load fluctuations and large forced power oscillations. The innovation lies in the introduction of a 3 Sigma value criterion to pinpoint the disturbance source location, addressing the limitations of traditional energy function methods that require manual intervention. By quantifying the localization of power oscillation disturbance sources, this method significantly improves both efficiency and accuracy.

Keywords: disturbance source location; system response data; energy function; normal distribution; 3 Sigma

1. Introduction

In recent years, the identification of disturbance sources in modern power systems has become an essential task due to the integration of renewable energy, distributed energy resources, and increasingly complex grid structures. Disturbances, such as oscillations and voltage fluctuations, can lead to serious reliability issues if not detected and localized accurately [1,2]. Identifying the source of external periodic forces that disrupt the normal operation of power systems is a crucial first step in designing and implementing effective countermeasures to mitigate the negative impacts of these oscillations on system performance [3]. The growing deployment of real-time measurement devices, such as wide-area measurement systems (WAMS) and Phasor Measurement Units (PMUs) [4,5], has proven to be an invaluable tool in addressing this challenge. In recent years, phasor measurements have provided researchers with data that enable the development of strategies allowing system operators to locate the sources of forced oscillations (FOs) in real-time [6]. FOs typically refer to a system's response to external periodic disturbances [7], which can stem from various causes, such as control valve malfunctions [8], and result in harmful effects like power quality degradation [9,10]. However, locating the exact source of these oscillations remains a complex problem, as the underlying causes may be any external periodic input disrupting the system, with their nature often unknown.

Currently, methods for locating the source of low-frequency oscillations in power systems fall into several categories, including hybrid dynamic simulation, traveling wave detection, frequency domain analysis, and energy function-based methods. These can be broadly classified into three groups: methods based on physical models of disturbance sources, those relying on measurement response data for source localization, and the hybrid method, which is the combination of physical models and measurement response data. Traditional methods of disturbance detection, which rely on signal processing and time–domain analysis, have been widely employed but face limitations in modern systems characterized by renewable energy variability and system nonlinearity [11].

Physical model-based approaches include hybrid dynamic simulation and frequency domain analysis. Meanwhile, methods based on response signals primarily consist of traveling wave detection and energy function techniques. Hybrid dynamic simulation methods [12,13], which combine system models with measurement data, use principles such as superposition and dynamic equivalence to locate disturbance sources. By simulating multiple small power grids isolated from a larger grid, these methods identify sources of disturbance by comparing anomalies in simulated and measured data. Systems without a disturbance source undergo iterations using equivalent power nodes, while those with a source are refined until the disturbance is located. However, hybrid dynamic simulation is complex, requires detailed system models, gradually narrows the search scope, and is time-consuming, making it unsuitable for real-time applications.

The frequency domain method uses response signals and transfer functions to locate disturbance sources. It calculates eigenvalues and eigenvectors based on the power system model and measurement data (e.g., generator power angle and speed) and uses these to estimate the disturbance source amplitude [14]. By calculating the normalized disturbance magnitude for each generator and analyzing errors in frequency domain assumptions, the location of the disturbance source can be determined [15]. Literature [16] uses the amplitude response of high-frequency transfer functions and the phase frequency response of low-frequency transfer functions to pinpoint disturbance sources. Another recent innovation is the application of Dynamic Mode Decomposition (DMD), which is capable of analyzing complex oscillatory patterns in power systems [17]. DMD has shown promise in identifying disturbances by breaking down system dynamics into distinct modes of behavior, allowing for clear isolation of disturbance sources [18].

The energy function method locates disturbance sources by analyzing potential energy changes at key nodes and their connected branches. This method is particularly effective for identifying inter-area oscillation sources. Literature [19] uses positive and negative slopes of aperiodic components to determine the propagation direction of disturbances, while literature [20] filters specific frequency bands in PMU data using a modified energy function, improving disturbance localization robustness through dissipated energy flow analysis. Literature [21,22] introduces a Transient Energy Function (TEF)-based approach for analyzing power system oscillations, with literature [23–25] extending the TEF concept from generators to network branches, introducing the concept of branch potential energy (BPE). Literature [26] first applies TEF and BPE to identify forced oscillation sources by monitoring energy flow and detecting the generator injecting energy into the network. Further studies confirm that energy dissipation correlates with damping torque under certain assumptions, enabling the identification of the disturbance source as the component producing energy [27–29]. The Port-Controlled Hamiltonian theory introduces an energy function for generators, incorporating excitation system and governor models [30].

Traveling wave methods utilize the principle of electromechanical wave propagation to identify oscillation sources [31]. If accurate arrival times of oscillations at different locations and wave speed maps are available, source location can be precisely determined [32]. The earliest detected peak of oscillation typically occurs at the generator, causing forced oscillations [33]. Literature [34] confirms this phenomenon by showing that buses far from the disturbance experience lagging oscillation phases compared to those closer to the disturbance. A similarity function can be used to detect time differences between oscillation arrivals at different generators connected to a common bus, identifying the leading generator as the source [35]. For large-scale systems, additional devices like Voltage Measuring Units (VMUs) can address data quality issues in PMUs [36,37]. Assuming uniform wave speeds across the network, sources are identified based on abnormal wave speeds. However, this method requires additional equipment and may result in errors if wave speeds vary or the source is near a VMU. A similar method uses PMU data to estimate time differences of oscillation arrivals across locations via least mean square methods [38].

Traveling wave-based methods offer the advantage of rapid response by relying on the initial periods of oscillations to determine arrival times. However, practical implementation requires addressing challenges such as PMU data quality, detection accuracy of arrival times, variable wave speeds, and adaptability to different network topologies. In addition, several other methods have been proposed for low-frequency oscillation localization in power systems, such as the Luenberger observer method [39], generator impedance matrix method [40,41], mode estimation method [42], damping torque method [43], equivalent circuit method [44], and generator learning method [45]. Although existing disturbance source location methods can solve problems in certain cases, there are few universal methods. The characteristics, applicability, advantages, and disadvantages of the current methods are summarized in Table 1.

Table 1. Summary of the existing disturbance source location methods.

Locating Methods	Occasion for Application	Online Performance	Pros	Cons	Accuracy
Hybrid simulation method	Forced oscillation	Low	Conceptual clarity	More complex, need to gradually narrow the scope, need system detailed model	Higher
Energy function method	Forced and negatively damping oscillations	High	Clear concept for practical application	Contributing "negative damping" to the generator and the resistance in the load may also produce transient state energy	High
Frequency domain relation method	Forced oscillation	High	Easy and easy	Requires the entire grid model or phase relationships to be researched	Higher
Observer method	Forced oscillation	Higher	Clear theory	Need the whole grid mode, can not locate excitation side disturbance source	High
Traveling wave method	Forced oscillation	High	Fast	Lag in response between nodes, the relationship remains to be studied	Higher
Generator impedance matrix method	Forced oscillation	High	Clear concept	Cannot locate the disturbance source set to prepare	High
Modal estimation method	No resonance occurs; forced oscillation	Higher	Suitable for multi-mode oscillations	Unexplained misjudgments	General
Damping torque method	Mainly negative damping oscillations	High	Appropriate physical interpretation	For the forced oscillation effect, it is not suitable	General

In this paper, the normal distribution identification method is applied to locate the source of power oscillation disturbances. The variables monitored by WAMS, such as the angular velocity, angular frequency, and generator's active power, are calculated using energy functions. Key energy function information, including generator outlet energy and load energy, is used as an input to quantitatively identify the disturbance source. This method has shown effective results in the WSCC 3-Generator 9-Bus System, the New England 10-Generator 39-Bus System, and China's 197-Bus system. The remainder of the paper is organized as follows: Section 2 outlines the mechanisms of power oscillation, the energy function, and the normal distribution identification method. Section 3 presents three simulation analyses, and finally, Section 4 provides conclusions and discusses areas for future research.

2. Mathematical and Mechanism Analysis

2.1. Analysis of Power Oscillation Mechanism

The energy function coupled with the normal distribution of the 3 Sigma model is used to locate power oscillation sources. Firstly, the mechanism and classification of power system power oscillation are introduced. There are two main types of power oscillation: negative damping oscillation and forced power oscillation. When there is power oscillation in the system, the output active power, bus voltage, generator power angle, and power speed will produce some fluctuation. The active power P_{ij} of the line from i to j on the system tie line is derived from Equation (1).

$$P_{ij} = \sum_{i=1}^{n} k_i \cdot x_{i0} e^{\lambda_i t} \tag{1}$$

where i is the number of modes. k_i is the participating factor of tie-line oscillation mode i. t is the time domain. x_{i0} denotes the ith oscillation mode of the system in its general form:

$$\lambda_{1,2} = \alpha \pm j\omega \tag{2}$$

where α denotes the attenuation characteristic of the power oscillation, ω is the frequency characteristic of the response oscillation, and the damping ratio of the oscillation mode is defined as follows [46]:

$$\xi = \frac{-\alpha}{\sqrt{\alpha^2 + \omega^2}} \tag{3}$$

When a system is set to a value of $\xi < 0$, any small change can trigger the oscillation pattern and change the state of the whole system. The power system forced power oscillation (PSO) is caused by the system resonance and large amplitude oscillation when the power system has a constant frequency of small periodic disturbances ($\xi > 0$) and the frequency of the natural oscillation is the same as the system natural oscillation frequency. It has the characteristics of rapid oscillation, constant amplitude oscillation, and rapid attenuation after the oscillation source is lost.

Generally speaking, the negative damping power oscillation presents many kinds of oscillation modes, and its oscillation waveform is very complex. However, the oscillating waveform, which only has certain oscillating modes, can be obtained by using the mode decomposition technique. Especially in the negative damped oscillation mode, the energy is much higher than the positive damped oscillation, so it is easier to decompose. When the external interference of the multi-generator system and its internal specific suppression mechanism cancel each other, many different incentive mechanisms can be produced, which makes the incentive mechanism of the multi-machine system reach balance, realizing the balance of the multi-computer system. Under certain conditions, the degree of oscillation may be small despite the influence of external forces.

By decomposition, two or more different oscillation curves can be obtained, which have independent oscillation modes. Studying their characteristics separately can better understand the connection between them. In some cases, when there is a significant change in the frequency of two different negative damping oscillation modes, they can be decomposed into two independent oscillation modes. When the frequencies of two oscillating signals are very close to each other, they can be easily separated by modal decomposition. Moreover, since they are so similar in character that, in some respects, they are almost identical, there is no clear boundary between them. Therefore, the influence of power oscillation signals in extreme power systems will be ignored in this study.

2.1.1. Negative Damped Oscillation

In general, the expression of the negative damping oscillation curve in a single oscillation mode is as follows:

$$\xi = \frac{-\alpha}{\sqrt{\alpha^2 + \omega^2}} \tag{4}$$

The expression for the envelope is $f_L(t) = A_0 e^{at}$, and the second derivative of the envelope is obtained:

$$f_L''(t) = a^2 A_0 \cdot e^{at} \tag{5}$$

From Equation (5), the second derivative of the envelope function on the oscillatory curve with negative damping is always positive regardless of the positive or negative damping of the system.

2.1.2. Forced Power Oscillation

The state space model of a power system under periodic disturbances can be expressed as follows:

$$\begin{cases} \dot{x}(t) = A \cdot x(t) + B \cdot u(t) \\ y(t) = C \cdot x(t) \end{cases} \tag{6}$$

where $x(t)$ and $y(t)$ represent the vector inputs and outputs of the system state variables, respectively. The input vector $u(t)$ includes periodic disturbances of mechanical components, excitation systems, and loads. The solution of response $y(t)$ can be obtained by eigenvalue decomposition. In addition, A represents the state matrix, B represents the gain matrix, and C represents the output matrix. This study focuses on the frequency domain solution in the frequency domain system model as follows:

$$y(s) = C(s \cdot I - A)^{-1} * Bu(s) = H(s) * u(s) \tag{7}$$

There is a zero-state solution for the system with sinusoidal input. Let $u(t)$ and $y(t)$ be connected by the stationary system in Equation (6). Suppose A has no eigenvalue on the line $Re(s) = 0$. If $u(t)$ follows from Equation (8):

$$u(t) = u_0 \cdot \sin \omega t \tag{8}$$

There is a unique periodic response $y(t)$, derived from Equation (9):

$$y(t) = \mathrm{Re}\{H(j\omega)u_0\} \sin \omega t + \mathrm{Im}\{H(j\omega)u_0\} \cos \omega t \tag{9}$$

In Equation (9), u_0 is the perturbed frequency vector (rad/s) and amplitude vector. $Re(\cdot)$ and $Im(\cdot)$ denote the real and imaginary parts of a complex number, respectively. Equations (8) and (9) are general equations for evaluating the time–domain response of power systems to sinusoidal disturbances. Suppose there is $u(t)$, as shown in Equation (10):

$$u(t) = \sum_{i=1}^{n} (u_i \cdot \sin \omega_i t + v_i \cdot \cos \omega_i t) \tag{10}$$

Derive $y(t)$ by undetermined coefficient method:

$$y(t) = \sum_{i=1}^{n} \mathrm{Re}\{H(j\omega)u_i + jH(j\omega)v_i\} \sin \omega_i t + \sum_{i=1}^{n} \mathrm{Im}\{H(j\omega)u_i + jH(j\omega)v_i\} \cos \omega_i t \tag{11}$$

The input vector includes the periodic signal forcing power oscillation.

When a disturbance source exists in a power system and causes forced power oscillation, the source continuously inputs oscillating energy to the system, which makes the active power oscillate continuously. Therefore, when the mechanical power output of the prime mover and its speed regulating system fluctuates, and the forced power oscillation occurs, the transient energy in the generator is exported from the prime mover and its speed regulating system to the power grid. The damping characteristics of the system depend on the transient energy consumption of each generator in the grid, which can be obtained by quantitative analysis. Therefore, it is possible to determine whether the generator is the source of power oscillation by analyzing the process of energy change in the generator, including the consumption of transient energy in the prime mover, its governing system, and the excitation system, and further identify whether it is a forced disturbance source. According to the above analysis, for the multi-machine system whose damping ratio comes from the energy consumption of each generator, its damping characteristics can be evaluated quantitatively by the energy consumption of each generator, for example, the excitation system and the speed control system of the generator will reduce the overall damping of the system.

2.2. Energy Function Mechanism Analysis

In the derivation of the energy function, considering the single-machine infinite-bus system, the typical model of the generator is as follows:

$$\frac{d\Delta\delta}{dt} = \omega \cdot \Delta\omega \tag{12}$$

$$\frac{d\Delta\omega}{dt} = \frac{1}{T_J}(\Delta P_T - \Delta P_E - D \cdot \Delta\omega) \tag{13}$$

Integrate Equation (13):

$$\frac{d\Delta\omega}{dt}\Delta\omega \cdot \omega \cdot d\tau + \int_0^t \Delta P_E \Delta\omega \cdot \omega \cdot d\tau = \int_0^t \Delta P_T \Delta\omega \cdot \omega \cdot d\tau + \int_0^t D\Delta\omega \cdot \Delta\omega \cdot \omega \cdot d\tau \tag{14}$$

In the energy function calculation:

$\Delta V_{KE} = \int_0^t T_J \frac{d\Delta\omega}{dt} \Delta\omega\omega d\tau$ is the kinetic energy function of the system;

$\Delta P_E = \int_0^t \Delta P_E \Delta\omega\omega d\tau$ is the potential energy function;

$\Delta V_T = \int_0^t \Delta P_T \Delta\omega\omega d\tau$ is the energy function of the disturbance source;

$\Delta V_D = \int_0^t D\Delta\omega\Delta\omega\omega d\tau$ is the function of energy dissipation for damping.

In a power system, the energy flowing into the system from the disturbance source is equal to the dissipative energy of the system. When a disturbance source acts on a generator, part of the injected energy is converted into the potential energy of the generator, and part is dissipated by damping. Since the energy injected by the source cannot be completely dissipated by damping, the potential energy of the electric generator increases. Since the potential energy cannot be fully converted, the increased potential energy flows into the network through the transmission line and is passed on to another electric generator. The potential energy of a line must increase to compensate for its energy consumption. In order to balance the incoming potential energy, the damping energy dissipation of the electric generator with no disturbance source increases continuously, which reduces the potential energy of the electric generator with no disturbance source. Therefore, during the whole system oscillation process, the potential energy of the disturbance source unit increases, while the potential energy of the unit without the disturbance source unit decreases.

$$\Delta V_{PEi}(t) = \int_0^t \Delta P_{ij} \Delta\phi_i d\tau \tag{15}$$

$$\Delta V_{PEj}(t) = \int_0^t \Delta P_{ji} \Delta\phi_j d\tau \tag{16}$$

where ΔP_{ij} and ΔP_{ji} are the changes of $i \rightarrow j$ and $j \rightarrow i$ power flow active power, $\Delta\phi_i$ and $\Delta\phi_j$ are the changes of phase angle of node i and node j; $\Delta V_{PEi}(t)$ and $\Delta V_{PEj}(t)$ are potential energy functions of the i-end and j-end of line L_{ij}, respectively.

By analyzing the potential energy of "Flow in" and "Flow out" nodes, we can infer that the potential energy of each node is conserved; that is, the potential energy of each node is generated by the branch it connects. Therefore, the potential energy of the generator can be expressed by the potential energy of the "Flow in" and "Flow out" nodes. The energy conversion process in the system includes two aspects: one is the periodic conversion between kinetic energy and potential energy of each generator, and the other is that the injected energy propagates in the network and is consumed by each energy-consuming element. With the second aspect, the direction of energy flow can be better understood, and the source of disturbance can be located more accurately.

In power oscillation, the dissipative energy of the system is generated by the components with negative damping, and the resulting oscillatory energy flows to the network and is consumed in the components with positive damping. When the dissipation energy transmitted from Bus i to Bus j on Line L_{ij} is positive, it shows that the disturbance source is closer to the side of Bus. The large-scale power system is divided into subsystems from the mth-zone to the nth-zone, and the dissipative energy of the tie-line in each region is calculated. The mth-zone subsystem is subdivided again to further reduce the suspect range of the disturbance source. When the cut set contains only the generator if the dissipative energy transmitted from the generator terminal is positive, the generator is a disturbance source; if the dissipative energy is negative, the generator is a non-disturbance source.

The above conclusions extend to multi-machine systems and still hold true. When forced power oscillation occurs in a multi-machine system, the unit where the disturbance source is located will continuously inject energy into the system and propagate in the form of potential energy. By observing the change trend of the potential energy function of each unit in the system (the unit where the disturbance source is located is obviously different from other units), the unit where the disturbance source is located can be located. When the system is in steady-state oscillation, there are two kinds of energy conversion processes: one is the mutual conversion of kinetic energy and potential energy of periodic oscillation, and the other is the energy injected by disturbance source propagating in the system. The latter can reflect the flow direction of potential energy in the whole system so that the source of disturbance can be located by judging the flow direction of energy.

2.3. Normal Distribution Identification Algorithm

A normal distribution is a distribution of a continuous random variable with two parameters μ and σ^2: The first parameter μ is the mean of a normally distributed random variable. The second parameter, σ^2, is the variance of this random variable, so the normal distribution is recorded as N (μ, σ^2), where μ is the position parameter of the normal distribution, describing the central tendency position of the normal distribution. The law of probability is that the probability of taking a value adjacent to μ is high, and the probability of taking a value farther away from μ is smaller. The normal distribution takes $X = \mu$ as the axis of symmetry, and it is completely symmetrical to the left and right. The expectation, mean, median, and mode of normal distribution are the same, all equal to μ. σ describes the degree of dispersion of the data distribution of the normal distribution data; the larger the σ, the more dispersed the data distribution, and the smaller the σ, the more concentrated the data distribution. Also known as the shape parameter of the normal distribution, the larger the σ, the flatter the curve; on the contrary, the smaller the σ, the thinner and taller the curve. For any random variable X subject to N (μ, σ^2) distribution, it can be transformed into a standard normal distribution with $\mu = 0$, $\sigma = 1$ after the following transformation. The conversion equation is as follows:

$$z = \frac{X - \mu}{\sigma} \tag{17}$$

The 3 Sigma model is the main model for normal distribution identification. 3 Sigma is used to detect abnormality. Generally, if it exceeds a threshold, it can be roughly regarded as abnormal data. The threshold value definition range is generally divided into two types. The first is defined by empirical values, such as setting a threshold > 1000. The second is to assume that each event is independent and without contextual correlation, and the 3 Sigma model can be used to detect it.

The 3 Sigma data need to obey a normal distribution. In principle, a 3 Sigma value that exceeds three standard deviations can be considered a 3 Sigma value. The probability of positive and negative is 99.7%, and the probability of occurrence of values other than the average value is shown in Equation (17), which is a very small probability event. If the data do not obey a normal distribution, it can also be described by how many times the standard deviation is away from the mean.

Detect the statistical data for this period of time, and if it conforms to a normal distribution, calculate the mean and variance. If the subsequent statistic is not within 3 Sigma of this range, the value can be considered a 3 Sigma value.

$$P(|x - u|3\sigma) \geq 0.003 \tag{18}$$

The 3 Sigma principle is also known as the Laida criterion. Specifically, the criterion is to first assume that a set of test data contains only random errors, calculate and process the original data to obtain the standard deviation, and then determine an interval according to a certain probability and consider the error. Anything that exceeds this interval is a 3 Sigma value. The normal distribution numerical distribution diagram is shown in Figure 1. As mentioned earlier, the 3 Sigma principle can be simply described as follows: if the data follow a normal distribution, the 3 Sigma value is defined as the value in a set of result values that deviates from the mean by more than three standard deviations. That is, under the assumption of normal distribution, the probability of occurrence of values other than three times the mean value (standard deviation) is very small, so it can be considered as a 3 Sigma value, which is shown in Equation (19).

$$P(|x - u|3\partial) \leq 0.003 \tag{19}$$

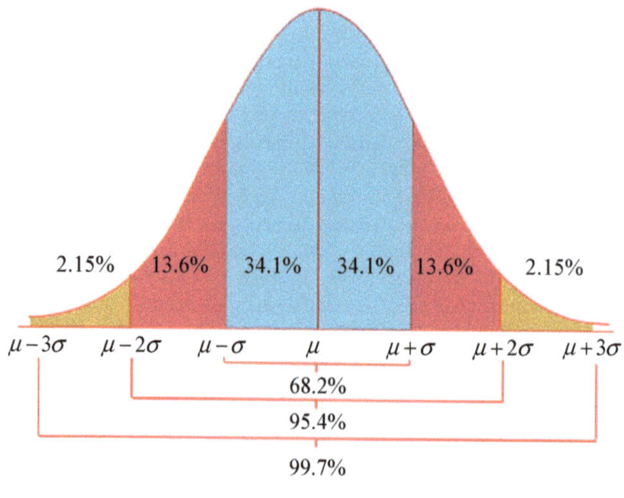

Figure 1. Normal distribution numerical distribution plot.

If the data do not obey the normal distribution, it can also be described by how many times the standard deviation is away from the mean (this makes this principle applicable to different business scenarios, but the k value in k Sigma needs to be determined empirically, this k value can be considered as the threshold).

2.4. Core Idea of Combination Method of Energy Function and Normal Distribution

First, through data cleaning and filtering, the input response data are determined and preprocessed to ensure its accuracy. On the generator side, the active power of the generator and the nodal voltage frequency can be selected as inputs based on the actual situation; on the load side, the active power of the load and the nodal voltage frequency can also be chosen as inputs depending on the specific conditions.

The branch potential energy formula is used to calculate the generator's output energy and the load potential energy, selecting all n energy calculation results as variables:

$$\left\{x_1^i, x_2^i, \ldots, x_n^i\right\} \tag{20}$$

The mean and variance of m samples are calculated as follows:

$$\mu_j = \frac{1}{m}\sum_{i=1}^{m} x_j^{(i)} \tag{21}$$

$$\sigma^2 = \frac{1}{m}\sum_{i=1}^{m}\left(x_j^{(i)} - \mu_j\right)^2 \tag{22}$$

A monitoring point x is set, and the disturbance source value $p(x)$ is calculated as follows:

$$p(x) = \prod_{j=1}^{n} p\left(x_j, \mu_j, \sigma_j^2\right) \tag{23}$$

Based on the above formulas, the Gaussian distribution (probability density function) for each dimension can be calculated, and the number of samples $p(x_i)$ with disturbance and non-disturbance source values is selected by analyzing the 3 Sigma distribution within the range of $(\mu - 3\sigma, \mu + 3\sigma)$.

The probability of values falling within $(\mu - \sigma, \mu + \sigma)$ is 0.6827.

The probability of values falling within $(\mu - 2\sigma, \mu + 2\sigma)$ is 0.9545.

The probability of values falling within $(\mu - 3\sigma, \mu + 3\sigma)$ is 0.9973.

Where μ is the mean value, and σ is the standard deviation.

In general, the values of data Y are concentrated within $(\mu - 3\sigma, \mu + 3\sigma)$. The likelihood of exceeding normal values within the given range is very small, with a probability of 0.3%. Therefore, this paper first calculates the active power of each generator unit, the loads, and the frequency of each node using a traditional energy function. The generator output energy and composite potential energy data are analyzed using the 3 Sigma model, which follows a normal distribution.

By classifying and validating different operating conditions, the results of energy function calculations for two types of oscillations, small disturbance load random fluctuation and forced power oscillation, are validated using the 3 Sigma model. From the energy function result curves, one or several curves that exhibit significant inconsistency can be identified. The location of the disturbance source is determined by assessing the number of values that conform to the normal distribution. The overall procedure of this model is shown in Figure 2.

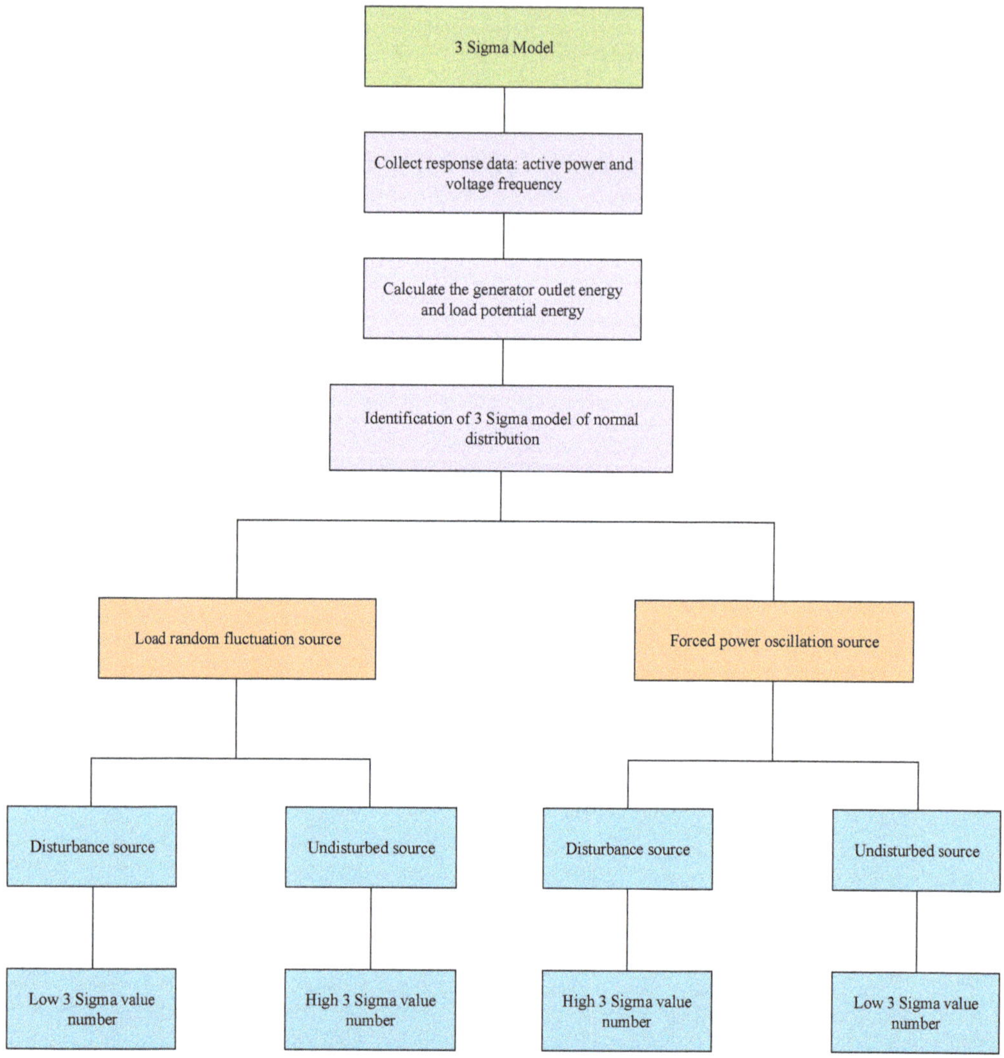

Figure 2. Flowchart of locating disturbance sources in the 3 Sigma model.

3. Power Oscillation Simulation Analysis

3.1. WSCC 3-Generator 9-Bus System Simulation Analysis

In the first simulation, the WSCC 3-Generator 9-Bus system is used to validate the method of disturbance source localization, which combines the energy function method with the normal distribution identification. The numerical example of the system is built through the DIgSILENT PowerFactory version 2023 SP4. The specific parameters are in Appendix A, and the single-line diagram of the WSCC 3-Generator 9-Bus system is shown in Figure 3. In order to ensure the normal operation of the generator excitation system and speed control system, short-term periodic disturbances must be found and eliminated in time. However, due to the random fluctuation of the load, these disturbances are difficult to suppress completely. Therefore, their amplitude and main frequency must be studied in order to better prevent the occurrence of forced power oscillation.

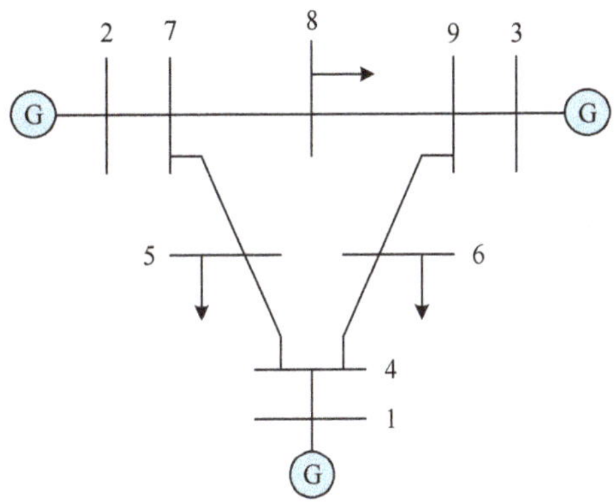

Figure 3. Single-line diagram of WSCC 3-Generator 9-Bus System. G represents the Generator. Number is the Bus number.

In the simulation of a WSCC 3-Generator 9-Bus system, first of all, the method proposed in this paper can identify the small fluctuation amplitude of the load random disturbance source. Because the fluctuation amplitude is small, it is necessary to evaluate the single load by the normal distribution. The potential energy data of the load with a 3 Sigma value is a non-disturbance source; the load potential energy data without a 3 Sigma value is the disturbance source, so the disturbance source on the load side can be identified. The location of the disturbance source can be determined by selecting the disturbance source with normal distribution for a single load.

The active power on the load side is selected, and the 100 s data of the load potential energy on the load side of the system are calculated by using the energy function formula; the load potential energy of each load is analyzed separately. The results of the three typical Scenarios are shown in Tables 2–4.

Table 2. 3 Sigma value classification in Scenario 1.

Load	Load A	Load B	Load C
3 Sigma value number (Count)	132	318	337
Whether it is a disturbance source	No	No	No

Table 3. 3 Sigma value classification in Scenario 2.

Load	Load A	Load B	Load C
3 Sigma value number (Count)	339	Low	336
Whether it is a disturbance source	No	Yes	No

Table 4. 3 Sigma value classification in Scenario 3.

Load	Load A	Load B	Load C
3 Sigma value number (Count)	Low	355	Low
Whether it is a disturbance source	Yes	No	Yes

Scenario 1. *No load random fluctuation.*

Scenario 2. *Load B fluctuates randomly.*

Scenario 3. *Load A and Load C fluctuate randomly at the same time.*

As can be seen from Tables 2–4, a number of 3 Sigma values are present on Load A, Load B, and Load C in Scenario 1, and therefore no disturbance source is present. In Scenario 2, Load B is located by a normal distribution disturbance source, and the result of calculation by the 3 Sigma model showed that the number of 3 Sigma values of Load B is low, and that of Load A is 339, the number of 3 Sigma values of Load C is 336, thus load B is identified as the disturbance source by the disturbance source localization method which combines the load potential energy with the normal distribution identification. In Scenario 3, the number of 3 Sigma values of Load A and Load C is low, and the number of 3 Sigma values of Load B is 355. It can be concluded that Load A and Load C are identified as disturbance sources by the method of load potential energy and normal distribution identification. From the above analysis, it can be concluded that the disturbance source location method, which combines the load potential energy and the normal distribution identification, can accurately identify the disturbance source of the load random fluctuation.

3.2. *New England 10-Generator 39-Bus System Analysis*

This paper verifies the applicability of the disturbance source localization method, which combines the energy function and the normal distribution identification in the forced power oscillation on the generator side. The simulation uses a New England 10-Generator 39-Bus System. The numerical example of the system is built through the DIgSILENT PowerFactory version 2023 SP4. The specific parameters are in Appendix B, and the single-line diagram is shown in Figure 4. The active power of each electric generator, the active power of each load, and the response data of the electric frequency of each node are collected, and the energy function formula is used to calculate the outlet energy and load potential energy of each generator, the disturbance source is located by the normal distribution identification method.

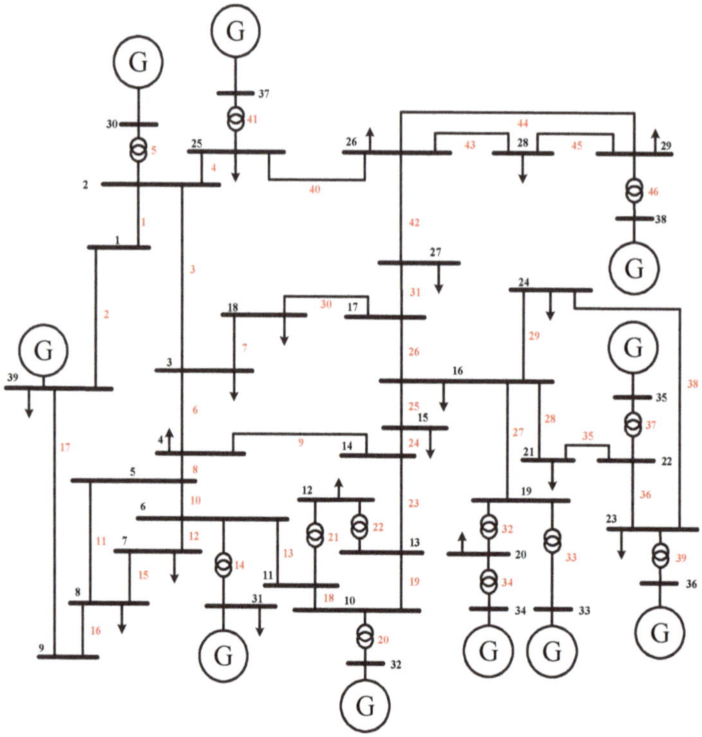

Figure 4. New England 10-Generator 39-Bus system structure diagram. G represents the Generator. Black number is the Bus number and red number is Transmission Line number.

In New England 10-Generator 39-Bus System, forced power oscillation with 10% amplitude is added to the side of the generator G2, considering the influence of the random fluctuation of the load side on the forced power oscillation of the generator side, the following 6 Scenarios are analyzed. Because of the large amplitude of the forced power oscillation of the generator, all the electric generator data are calculated in a unified way for better results, and random load fluctuations are added to three of the Scenarios. The Scenario information is as follows:

Scenario 4. *A forced power oscillation with an amplitude of 10% is applied to generator G2, with no random fluctuations on the load side.*

Scenario 5. *A forced power oscillation with an amplitude of 10% is applied to generator G7, with no random fluctuations on the load side.*

Scenario 6. *Forced power oscillations with an amplitude of 10% are applied simultaneously to generators G2 and G6, with no random fluctuations on the load side.*

Scenario 7. *A forced power oscillation with an amplitude of 10% is applied to generator G2, with random load fluctuations added to load 15 on the load side.*

Scenario 8. *A forced power oscillation with an amplitude of 10% is applied to generator G7, with random load fluctuations added to load 15 on the load side.*

Scenario 9. *Forced power oscillations with an amplitude of 10% are applied simultaneously to generators G2 and G6, with random load fluctuations added to load 15 on the load side.*

As shown in Table 5, the method of disturbance source location combined with energy function and normal distribution identification can accurately identify the forced power oscillation of a single electric generator, such as Scenarios 4 and 5. The method can accurately identify the forced power oscillation at generators G2 and G7. For Scenario 6, when generators G2 and G6 produce forced power oscillation at the same time, this method can identify the position of the disturbance source as generators G2 and G6, in which the position of the disturbance source of generator G2 is more obvious (more 3 Sigma values). The influence of the generator's random fluctuation on the forced power oscillation is evaluated. The results of Scenarios 7–9 show that the method has no effect on the location of the disturbance source of the forced power oscillation on the generator side due to the random fluctuation of the load side; the source of forced power oscillation can be identified accurately. No load side random fluctuation is added in Scenarios 4–6, and load random fluctuation is added in Load 15 in Scenarios 7–9; the identification results of the load side random fluctuation disturbance source by the method of energy function and normal distribution are shown in Table 6.

Table 5. Results of 3 Sigma values on the generator side of the forced power oscillation of the generator.

Gen	Sc4 (Count)	Sc5 (Count)	Sc6 (Count)	Sc7 (Count)	Sc8 (Count)	Sc9 (Count)
G1	Low	Low	Low	Low	Low	Low
G2	4197	Low	3928	4187	Low	3928
G3	Low	Low	Low	Low	Low	Low
G4	Low	Low	Low	Low	Low	Low
G5	Low	Low	Low	Low	Low	Low
G6	Low	Low	439	Low	Low	439
G7	Low	2303	Low	Low	2303	Low
G8	Low	Low	Low	Low	Low	Low
G9	Low	Low	Low	Low	Low	Low
G10	Low	Low	Low	Low	Low	Low

Where: Sc is the Scenario. Gen is the Generator Number.

Table 6. Result table of 3 Sigma value on the load side of generator forced power oscillation.

Load	Sc4 (Count)	Sc5 (Count)	Sc6 (Count)	Sc7 (Count)	Sc8 (Count)	Sc9 (Count)
Load 3	Low	Low	Low	Low	Low	Low
Load 4	Low	Low	Low	Low	Low	Low
Load 7	Low	Low	Low	Low	Low	Low
Load 8	Low	Low	Low	Low	Low	Low
Load 12	Low	Low	Low	Low	Low	Low
Load 20	Low	Low	Low	Low	Low	Low
Load 21	Low	Low	Low	Low	Low	Low
Load 23	Low	Low	Low	Low	Low	Low
Load 24	Low	Low	Low	Low	Low	Low
Load 25	Low	Low	Low	Low	Low	Low
Load 14	Low	Low	Low	Low	Low	Low
Load 15	Low	Low	Low	4140	4140	4140
Load 16	Low	Low	Low	Low	Low	Low
Load 17	Low	Low	Low	Low	Low	Low
Load 18	Low	Low	Low	Low	Low	Low
Load 19	Low	Low	Low	Low	Low	Low

Table 6 shows that, despite the influence of forced power oscillation on the generator side, the source location method combined with energy function and normal distribution identification can accurately identify the source of random fluctuation on the load side. In Scenarios 4–6, the method can accurately identify the random disturbance sources at Load 15.

In this section, two scenarios are selected to compare the method combining the energy function with the normal distribution against two other methods. The comparison methods chosen are the Chi-square distribution and Student's *t*-distribution, which are applied to Scenarios 4 and 7 for analysis and validation. The results are shown in Table 7. From the analysis of the results in the Table, it is evident that only the simulation results using the normal distribution show significant distinguishability. In contrast, the data from the other two methods do not exhibit such prominence, making it impossible to identify the location of the disturbance source.

In order to verify the adaptability of the 3 Sigma model for normal distribution identification in the system with new energy sources, 505 MW wind farms are connected to the outside node of generator G4. When forced power oscillations occur on the generator side, 3 Sigma values on the generator side and 3 Sigma values on the load side are shown in Tables 7 and 8. The Scenario information is as follows.

Scenario 10. *A forced power oscillation with an amplitude of 10% is applied to generator G2, with no random fluctuations on the load side, and a 250 MW wind turbine set is connected.*

Scenario 11. *A forced power oscillation with an amplitude of 10% is applied to generator G7, with no random fluctuations on the load side, and a 250 MW wind turbine set is connected.*

Scenario 12. *Forced power oscillations with an amplitude of 10% are applied simultaneously to generators G2 and G6, with no random fluctuations on the load side, and a 250 MW wind turbine set is connected.*

Scenario 13. *A forced power oscillation with an amplitude of 10% is applied to generator G2, with random load fluctuations added to Load 15 on the load side, and a 250 MW wind turbine set is connected.*

Scenario 14. *A forced power oscillation with an amplitude of 10% is applied to generator G7, with random load fluctuations added to Load 15 on the load side, and a 250 MW wind turbine set is connected.*

Scenario 15. *Forced power oscillations with an amplitude of 10% are applied simultaneously to generators G2 and G6, with random load fluctuations added to Load 15 on the load side, and a 250 MW wind turbine set is connected.*

From Tables 8 and 9, the results of 3 Sigma values on the generator side and the load side of the system change slightly when the generator side is forced to oscillate after the wind farm is connected, but the numerical results are consistent with those before the wind farm is connected. Therefore, the numerical analysis method proposed in this paper is not affected by the new energy sources, and the method based on the energy function and the normal distribution identification is also suitable for the new energy sources.

Table 7. Comparison of normal distribution (3 Sigma), Chi-square, and Student's *t*-distribution for Scenarios 4 and 7.

Gen	Sc4 (Count)			Sc7 (Count)		
	Chi-Square Distribution	Student's *t*-Distribution	Normal Distribution with 3 Sigma	Chi-Square Distribution	Student's *t*-Distribution	Normal Distribution with 3 Sigma
G1	Low	Low	Low	Low	Low	Low
G2	Low	Low	4197	Low	Low	4187
G3	Low	Low	Low	Low	Low	Low
G4	Low	Low	Low	Low	Low	Low
G5	Low	Low	Low	Low	Low	Low
G6	Low	Low	Low	Low	Low	Low
G7	Low	Low	Low	Low	Low	Low
G8	Low	Low	Low	Low	Low	Low
G9	Low	Low	Low	Low	Low	Low
G10	Low	Low	Low	Low	Low	Low

Table 8. Results of 3 Sigma values on the generator side of forced power oscillation after wind turbine connection.

Gen	Sc10 (Count)	Sc11 (Count)	Sc12 (Count)	Sc13 (Count)	Sc14 (Count)	Sc15 (Count)
G1	Low	Low	Low	Low	Low	Low
G2	3997	Low	3540	3997	Low	3540
G3	Low	Low	Low	Low	Low	Low
G4	Low	Low	Low	Low	Low	Low
G5	Low	Low	Low	Low	Low	Low
G6	Low	Low	271	Low	Low	271
G7	Low	2481	Low	Low	2481	Low
G8	Low	Low	Low	Low	Low	Low
G9	Low	Low	Low	Low	Low	Low
G10	Low	Low	Low	Low	Low	Low

Table 9. Results of 3 Sigma values on the load side of generator forced power oscillation after wind turbine connection.

Load	Sc10 (Count)	Sc11 (Count)	Sc12 (Count)	Sc13 (Count)	Sc14 (Count)	Sc15 (Count)
Load 3	Low	Low	Low	Low	Low	Low
Load 4	Low	Low	Low	Low	Low	Low
Load 7	Low	Low	Low	Low	Low	Low
Load 8	Low	Low	Low	Low	Low	Low
Load 12	Low	Low	Low	Low	Low	Low
Load 20	Low	Low	Low	Low	Low	Low
Load 21	Low	Low	Low	Low	Low	Low
Load 23	Low	Low	Low	Low	Low	Low
Load 24	Low	Low	Low	Low	Low	Low
Load 25	Low	Low	Low	Low	Low	Low
Load 14	Low	Low	Low	Low	Low	Low
Load 15	Low	Low	Low	3872	3872	3872
Load 16	Low	Low	Low	Low	Low	Low
Load 17	Low	Low	Low	Low	Low	Low
Load 18	Low	Low	Low	Low	Low	Low
Load 19	Low	Low	Low	Low	Low	Low

In a power system, during power oscillations, forced power oscillations may not only occur on the generator side but can also originate from the load side. Therefore, it is necessary to verify the applicability of the energy function and normal distribution disturbance source localization method when forced power oscillations occur on the load side. The New England 10-Generator 39-Bus system is used for simulation verification, with a 10% amplitude disturbance added to load 20. Considering the active power of the load and the electrical frequency of the nodes connected to the load, disturbance source localization is carried out for Scenarios 16–18. The results of the disturbance source localization analysis for the three scenarios are shown in Table 10.

Table 10. Results of 3 Sigma values on the load side for load forced power oscillation for Scenario 16, 17 and 18.

Load	Sc16 (Count)	Sc17 (Count)	Sc18 (Count)
Load 3	Low	Low	Low
Load 4	Low	Low	Low
Load 7	Low	Low	Low
Load 8	Low	Low	Low
Load 12	Low	Low	Low
Load 15	Low	Low	Low
Load 16	Low	Low	Low
Load 18	Low	Low	Low
Load 20	56	56	56
Load 21	Low	Low	Low
Load 23	Low	Low	Low
Load 24	Low	Low	Low

Table 10. *Cont.*

Load	Sc16 (Count)	Sc17 (Count)	Sc18 (Count)
Load 25	Low	Low	Low
Load 14	Low	Low	Low
Load 15	Low	Low	Low
Load 16	Low	Low	Low
Load 17	Low	Low	Low
Load 18	Low	Low	Low
Load 19	Low	Low	Low

Scenario 16. *A forced power oscillation is generated by adding a 10% amplitude sinusoidal disturbance to Load 20.*

Scenario 17. *A forced power oscillation is generated by adding a 10% amplitude sinusoidal disturbance to Load 20, with random fluctuations also added to Load 20.*

Scenario 18. *A forced power oscillation is generated by adding a 10% amplitude sinusoidal disturbance to Load 20, with random fluctuations added to Load 15.*

As shown in Table 10, in Scenarios 16–18, the energy function and normal distribution disturbance source localization method accurately identified the disturbance source as Load 20, with a 3 Sigma value of 56, while the 3 Sigma values for other loads are low. Additionally, in all three scenarios, the change in the disturbance source of the random load fluctuations had no impact on the combined energy function and normal distribution method, which still accurately identified Load 20 as the source of the forced power oscillation. Moreover, the added random fluctuations on different load sides did not affect the localization of the forced power oscillation source. Since both the forced power oscillation and random load fluctuations occurred on the load side, the minor random load fluctuations are masked by the forced power oscillation; hence, the random fluctuation disturbance source is not identified.

A 197-Bus Actual System Analysis

In the third simulation analysis, a 197-Bus system is built with the equivalent parameters of a practical example in China. The simulation example is constructed by PASAP; the typical description of the approach is as follows:

There are 197 buses (85 nodes in a 500 KV main grid) in the example, which are divided into five regions. The generators in each region are as follows. The single-line diagram of the system is shown in Figure 5, and the system is divided into five regions, which are:

Zones DZ: Gendz-1, Gendz-2, Gendz-3, Gendz-4.

Area 1A: Send End.

Region FF: Gen1A-1, Gen1A-2, Gen1A-3, Gen1A-4, Gen1A-5.

Area 1B: receiver.

Regional FH: Gen1B-1, Gen1B-2, Gen1B-32, Gen1B-41, Gen1B-5, Gen1B-61, Gen1B-62, new energy units.

Firstly, the active power and angular frequency data of each synchronous generator are collected, and the simulation analysis and calculation are carried out by PSASP version 5.0. The active power time where the domain diagram of the selected generator is shown in Figure 6. Generator Gendz-1, generator Gen1A-1, and generator Gen1B-1 are set up as the balancing units. All the generators take the balancing units as the reference units to collect the power angle data. The fault mode of this example is the n-1 fault of the Bus1A-17~Bus1A-13 line in the high-proportion hydropower delivery system, and the system has the problem of ultra-low frequency oscillation. This example is analyzed and verified by the normal distribution disturbance source localization method proposed in this paper.

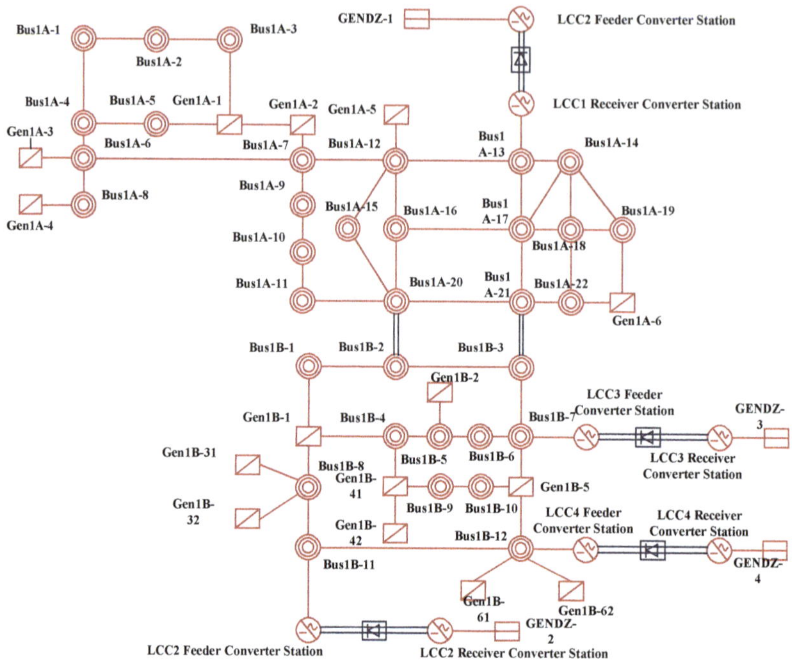

Figure 5. Single-line diagram of the 197-Bus actual power system.

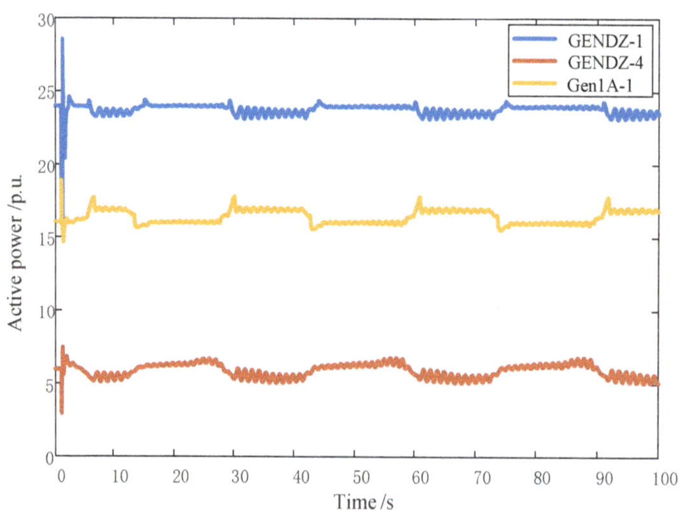

Figure 6. Time–domain diagram of generator active power.

Using the Power System Analysis Software Package (PSASP) for simulation, the generator units GENDZ-1, Gen1A-1, and Gen1B-1 are set as reference units, while all other generator units are referenced to these balanced units. The active power on the load side and the electrical frequency at each node are collected, with forced power oscillation set on the load side. Measurement data for two Scenarios (Scenarios 19 and 20) are analyzed.

Scenario 19. *Forced power oscillation occurred on Load 1A-14.*

Scenario 20. *Forced power oscillation occurred simultaneously on Loads 1A-10 and 1A-14.*

As shown in Table 11, in Scenario 19, the number of 3 Sigma values for Load 1A-14 is 1671, while the number of 3 Sigma values for all other loads is low, indicating that the disturbance source is Loads 1A-14. In Scenario 20, the number of 3 Sigma values for Load 1A-10 is 359, and for Load 1A-14, it is 1377, with the number of 3 Sigma values for all other loads being low. This indicates that the disturbance sources are Load 1A-10 and Load 1A-14. Thus, the disturbance source localization method combines the energy function and normal distribution identification to accurately locate disturbance sources in measured systems.

Table 11. Results of the 3 Sigma values on the load side for load forced power oscillation for Scenario 19 and 20.

Load	Sc19 (Count)	Sc20 (Count)
Load 1A-1	Low	Low
Load 1A-10	Low	359
Load 1A-11	Low	Low
Load 1A-12	Low	Low
Load 1A-13	Low	Low
Load 1A-14	1671	1377
Load 1A-15	Low	Low
Load 1A-16	Low	Low
Load 1A-17	Low	Low
Load 1A-18	Low	Low
Load 1A-19	Low	Low
Load 1A-2	Low	Low
Load 1A-20	Low	Low
Load 1A-21	Low	Low
Load 1A-22	Low	Low
Load 1A-3	Low	Low
Load 1A-4	Low	Low
Load 1A-5	Low	Low
Load 1A-7	Low	Low
Load 1A-8	Low	Low
Load 1A-9	Low	Low
Load 1B-1	Low	Low
Load 1B-10	Low	Low
Load 1B-2	Low	Low
Load 1B-3	Low	Low
Load 1B-5	Low	Low
Load 1B-6	Low	Low
Load 1B-8	Low	Low
Load 1B-9	Low	Low

4. Conclusions

Traditional energy function methods can locate disturbances by calculating the oscillating energy between active power and voltage frequency, with the energy function trend identified through manual observation. However, this approach struggles to adapt to the complex topology of modern power systems and is challenging to apply online. Therefore, this paper proposes an enhanced method that combines the traditional energy function with normal distribution identification to analyze and locate disturbance sources. Based on simulations of the WSCC 3-Generator 9-Bus system, the New England 10-Generator 39-Bus system, and the analysis of a 197-Bus test system in China, the following conclusions can be drawn:

(1) By recording the active power and voltage frequency of each node on the generator and load sides and calculating the outlet energy and load potential energy of each generator, accurate disturbance source location is achievable. This is enabled by driven data from WAMS and energy function calculations, which are crucial prerequisites for evaluating the accuracy of disturbance source identification;

(2) Under various scenarios, the proposed method can accurately locate the source of random fluctuations on the load side as well as forced power oscillations on the generator side. This demonstrates that combining the energy function with normal distribution provides strong robustness across different conditions. Even in new energy scenarios, such as when wind farms are integrated into the system, the method continues to accurately locate disturbance sources, unaffected by the presence of new energy inputs;

(3) The numerical method combining energy functions with normal distribution identification, as presented in this paper, is reliable and can accurately identify units or loads with the largest oscillation amplitudes. By using the 3 Sigma analysis, the severity of each disturbance source can be quantitatively evaluated, allowing for precise identification of the power oscillation disturbance source. This approach, built entirely on response-driven data, eliminates the risk of misjudgment associated with manual decision-making, leading to accurate disturbance identification.

While numerical analysis based on the energy function requires the accumulation of data over a period of time before calculations can be performed, further improvements in calculation efficiency are necessary. Additionally, the normal distribution-based method for disturbance source localization requires tailored strategies to handle different types of data.

Author Contributions: Data curation, S.G. and J.C.; Formal analysis, X.L.; Methodology, J.C.; Software, C.J.; Validation, J.C.; Writing—original draft, S.G.; Writing—review and editing, S.G. All authors have read and agreed to the published version of the manuscript.

Funding: This research received no external funding.

Data Availability Statement: The original contributions presented in the study are included in the article, further inquiries can be directed to the corresponding author.

Conflicts of Interest: Author Xu Liu was employed by the company State Grid Pinghu Power Supply Company. The remaining authors declare that the research was conducted in the absence of any commercial or financial relationships that could be construed as a potential conflict of interest.

Appendix A. WSCC 3-Generator 9-Bus System Parameters

Table A1. Parameters of detailed generator model for WSCC 3-Generator 9-Bus System.

Gen	H	X_d	X_q	X'_d	X'_q	T'_{d0}	T'_{q0}	T''_{d0}	T''_{q0}	$X''_d = X''_q$
1	23.64	0.438	0.2907	0.1824	0.2907	8.96	0.0	0.04	0.06	0.18
2	6.4	1.7916	1.729	0.2378	0.3938	6.00	0.54	0.033	0.078	0.178
3	3.01	1.313	1.258	0.1813	0.25	5.89	0.6	0.033	0.07	0.107

Table A2. Parameters of transmission lines and transformers for WSCC 3-Generator 9-Bus System.

First Node	End Node	R (pu/m)	X (pu/m)	B (pu/m)	k
1	4	0.0	0.0576	0.0	1
2	7	0.0	0.0625	0.0	1
3	9	0.0	0.0586	0.0	1
4	5	0.0100	0.0680	0.1760	0
4	6	0.0170	0.0920	0.1580	0
5	7	0.0320	0.1610	0.3060	0
6	9	0.0390	0.1738	0.3580	0
7	8	0.0085	0.0576	0.1490	0
8	9	0.0119	0.1008	0.2090	0

The excitation of each unit in the simulation system uses IEEE Type I (Figure A1), and the parameters are the same, as shown in Table A3.

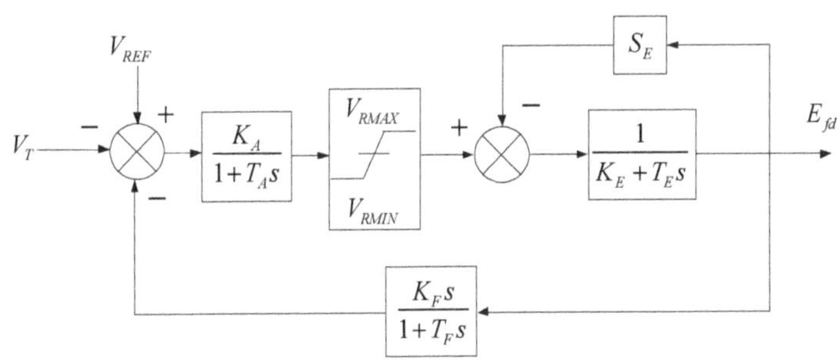

Figure A1. Model of IEEE I for WSCC 3-Generator 9-Bus System.

Table A3. Excitation system model for WSCC 3-Generator 9-Bus System.

K_A	T_A	V_{RMAX}	V_{RMIN}	K_E	T_E	K_F	T_F	S_{E1}	S_{E2}
20	0.05	5	−5	0.85	0.6	0.025	0.9	0.06	0.85

Appendix B. New England 10-Generator 39-Bus System Parameters

Table A4. Parameters of transmission lines and transformers for New England 10-Generator 39-Bus System.

First Node	End Node	R (pu/m)	X (pu/m)	B (pu/m)	k
1	2	0.0035	0.0411	0.6987	0
1	39	0.0010	0.025	0.7500	0
2	3.	0.0013	0.0151	0.2572	0
2	25	0.0070	0.0086	0.1460	0
3	4	0.0013	0.0213	0.2214	0
3	18	0.0011	0.0133	0.2138	0
4	5	0.0008	0.0128	0.1342	0
4	14	0.0008	0.0129	0.1382	0
5	6	0.0002	0.0026	0.0434	0

Table A4. *Cont.*

First Node	End Node	R (pu/m)	X (pu/m)	B (pu/m)	k
5	8	0.0008	0.0112	0.1476	0
6	7	0.0006	0.0092	0.1130	0
6	11	0.0007	0.0082	0.1389	0
7	8	0.0004	0.0046	0.0780	0
8	9	0.0023	0.0363	0.3804	0
9	39	0.0010	0.0250	1.2000	0
10	11	0.0004	0.0043	0.0729	0
10	13	0.0004	0.0043	0.0729	0
13	14	0.0009	0.0101	0.1723	0
14	15	0.0018	0.0217	0.3660	0
15	16	0.0009	0.0094	0.1710	0
16	17	0.0007	0.0089	0.1342	0
16	19	0.0016	0.0195	0.3040	0
16	21	0.0008	0.0135	0.2548	0
16	24	0.0003	0.0059	0.0680	0
17	18	0.0007	0.0082	0.1319	0
17	27	0.0013	0.0173	0.3216	0
21	22	0.0008	0.0140	0.2565	0
22	23	0.0006	0.0096	0.1846	0
23	24	0.0022	0.0350	0.3610	0
25	26	0.0032	0.0323	0.5130	0
26	27	0.0014	0.0147	0.2396	0
26	28	0.0043	0.0474	0.7802	0
26	29	0.0057	0.0625	1.0290	0
28	29	0.0014	0.0151	0.0249	0
12	11	0.0016	0.0435	0.000	1.006
12	13	0.0016	0.0435	0.000	1.006
6	31	0.025	0.0000	1.070	1.070
10	32	0.000	0.02	0.000	1.070
19	33	0.0007	0.0142	0.000	1.070
20	34	0.0009	0.018	0.000	1.009
22	35	0.0000	0.0143	0.000	1.025
23	36	0.0005	0.0272	0.000	1.000
25	37	0.0006	0.0232	0.000	1.025
2	30	0.0000	0.0181	0.000	1.025
29	38	0.0008	0.0156	0.000	1.025
19	20	0.0007	0.0138	0.000	1.060

Table A5. Parameters of the detailed generator model for New England 10-Generator 39-Bus System.

Gen	H	X_d	X_q	X'_d	X'_q	T'_{d0}	T'_{q0}	$X''_d = X''_q$
1	500.0	0.02	0.019	0.006	0.008	7.0	0.003	0.003
2	30.3	0.295	0.282	0.0697	0.17	6.56	0.035	0.035
3	35.8	0.2495	0.237	0.0531	0.0876	5.7	0.0304	0.0304
4	28.6	0.262	0.258	0.0436	0.166	5.69	0.0295	0.0295
5	26.0	0.67	0.62	0.132	0.166	5.4	0.054	0.054
6	34.8	0.254	0.241	0.05	0.0814	7.3	0.0224	0.0224
7	26.4	0.295	0.292	0.049	0.186	5.66	0.0322	0.0322
8	24.3	0.29	0.28	0.057	0.0911	6.7	0.028	0.028
9	34.5	0.2106	0.205	0.057	0.0587	4.79	0.0298	0.0298
10	42.0	0.1	0.069	0.031	0.008	10.2	0.0125	0.0125

The excitation of each generator in the simulation system adopts IEEE Type I (Figure A2), and the parameters are shown in Table A6.

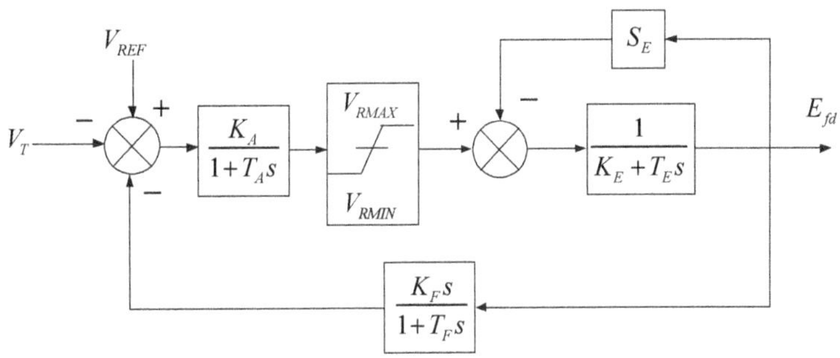

Figure A2. Model of IEEE I for New England 10-Generator 39-Bus System.

Table A6. Parameters of the excitation system model for New England 10-Generator 39-Bus System.

Gen	K_A	T_A	V_{RMAX}	V_{RMIN}	K_E	T_E	K_F	T_F	S_{E1}	S_{E2}
1	5.0	0.06	−1	1	0	0.25	0.04	0.1	0.265	0.08
2	6.2	0.05	−5	5	0	0.405	0.057	0.5	0.88	0.666
3	5.0	0.06	−1	1	0	0.5	0.08	1	0.34	0.13
4	5.0	0.06	−1	1	0	0.5	0.08	1	0.314	0.08
5	40	0.02	−10	10	1	0.785	0.03	1	0.91	0.07
6	5.0	0.02	−1	1	0	0.471	0.075	1.25	0.251	0.064
7	40.0	0.02	−6.5	6.5	1	0.73	0.03	1	0.74	0.53
8	5.0	0.02	−1	1	0	0.528	0.085	1.26	0.282	0.072
9	40.0	0.02	−10.5	10.5	1	1.4	0.03	1	0.85	0.62
10	5.0	0.06	−1	1	−0.0485	0.25	0.04	1	0.26	0.08

References

1. Zhang, Y.; Li, X.; Chen, Z. A Novel Approach for Disturbance Source Location in Power Grids with High Penetration of Renewable Energy. *IEEE Trans. Smart Grid* **2022**, *13*, 987–995.
2. Wang, J.; Xu, Q.; Huang, Y. Advanced Techniques for Disturbance Detection and Localization in Modern Power Systems with Renewable Integration. *Electr. Power Syst. Res.* **2021**, *199*, 107394.

3. Maslennikov, S.; Wang, B.; Zhang, Q.; Ma, F.; Luo, X.; Sun, K.; Litvinov, E. A Test Cases Library for Methods Locating the Sources of Sustained Oscillations. In Proceedings of the IEEE PES General Meeting, Boston, MA, USA, 17–21 July 2016; pp. 17–21.
4. Zhao, B.; Zhang, L.; Chen, Q. An Overview of Wide-Area Measurement Systems in Modern Power Systems: Challenges and Opportunities. *IEEE Trans. Power Deliv.* **2022**, *37*, 759–769.
5. Khan, A.A.; Ali, U. Enhancing the Performance of Phasor Measurement Units through Advanced Signal Processing Techniques. *J. Mod. Power Syst. Clean Energy* **2023**, *11*, 723–733.
6. Wang, B.; Sun, K. Location methods of oscillation sources in power systems: A survey. *J. Mod. Power Syst. Clean Energy* **2016**, *5*, 151–159. [CrossRef]
7. Ghorbaniparvar, M. Survey on forced oscillations in power system. *J. Mod. Power Syst. Clean Energy* **2016**, *5*, 671–682. [CrossRef]
8. Hubbard, L. *Reliability Guideline: Forced Oscillation Monitoring and Mitigation*; Tech. Rep., NERC; North American Electric Rail Corporation: Atlanta, GA, USA, 2017.
9. Sarmadi, S.A.N.; Venkatasubramanian, V.; Salazar, A. Analysis of november 29, 2005 western american oscillation event. *IEEE Trans. Power Syst.* **2016**, *31*, 5210–5211. [CrossRef]
10. Vanfretti, L.; Bengtsson, S.; Peri, V.S.; Gjerde, J.O. Effects of forced oscillations in power system damping estimation. In Proceedings of the IEEE International Workshop on Applied Measurements for Power Systems (AMPS) Proceedings, Aachen, Germany, 26–28 September 2012; pp. 1–6.
11. Liu, Y.; Zhang, H.; Wang, Z. A Review of Disturbance Detection and Localization in Modern Power Systems. *IEEE Trans. Power Syst.* **2021**, *36*, 4060–4072.
12. Ma, J.; Zhang, P.; Fu, H. Application of phasor measurement unit on locating disturbance source for low-frequency oscillation. *IEEE Trans. Smart Grid* **2010**, *1*, 340–346. [CrossRef]
13. Wu, H.; Duan, Q.; Ma, J. Disturbance source self-diagnosis of the smart grid. In Proceedings of the Spring Congress on Engineering and Technology, Xi'an, China, 27–30 May 2012; pp. 27–30.
14. Cabrera, I.R.; Wang, B.; Sun, K. A method to locate the source of forced oscillations; based on linearized model and system measurements. In Proceedings of the IEEE Power Energy Society General Meeting, Chicago, IL, USA, 16–20 July 2017; pp. 1–5.
15. Agrawal, U.; Pierre, J.W.; Follum, J.; Duan, D.; Trudnowski, D.; Donnelly, M. Locating the source of forced oscillations using PMU measurements and system model information. In Proceedings of the IEEE Power Energy Society General Meeting, Chicago, IL, USA, 16–20 July 2017; pp. 1–5.
16. Zhou, N.; Ghorbaniparvar, M.; Akhlaghi, S. Locating sources of forced oscillations using transfer functions. In Proceedings of the IEEE Power Energy Society General Meeting, Champaign, IL, USA, 23–24 February 2017; pp. 1–8.
17. Cai, G.; Guo, S.; Liu, C. Parameter Identification of electro mechanical oscillation mode in power system driven by data: A quasi-real-time method based on Randomized-DMD-Multilayer Artificial Neural Network. *Front. Energy Res.* **2022**, *10*, 908937. [CrossRef]
18. Shi, T.; Wang, F.; Jiang, L. Dynamic Mode Decomposition-Based Disturbance Localization in Power Systems with Renewable Energy Sources. *IET Renew. Power Gener.* **2022**, *16*, 658–667.
19. Tang, F.; Wang, B.; Liao, Q.; Pisani, C.; Dong, C.; Jia, J.; Guo, K. Research on forced oscillations disturbance source locating through an energy approach. *Int. Trans. Electr. Energy Syst.* **2016**, *26*, 192–207. [CrossRef]
20. Maslennikov, S.; Wang, B.; Litvinov, E. Dissipating energy flow method for locating the source of sustained oscillations. *Int. J. Electr. Power Energy Syst.* **2017**, *88*, 55–62. [CrossRef]
21. Jinag, C.; McCalley, J.; Kommareddy, M. An energy approach to analysis of interarea oscillations in power systems. *IEEE Trans. Power Syst.* **1996**, *11*, 734–740. [CrossRef]
22. Messina, A.; Ochoa, M.; Barocio, E. Use of energy and power concepts in the analysis of the inter-area mode phenomenon. *Electr. Power Syst. Res.* **2001**, *59*, 111–119. [CrossRef]
23. Moon, Y.; Cho, B.; Lee, Y.; Kook, H. Derivation of energy conservation law by complex line integral for the direct energy method of power system stability. In Proceedings of the 38th IEEE Conference on Decision and Control, Phoenix, AZ, USA, 7–10 December 1999; pp. 7–10.
24. Cai, G.; Mu, G.; Liu, Z.; Lin, Z. Network transient energy function defined on output trajectory. *Autom. Electr. Power Syst.* **1999**, *23*, 28–32.
25. Cai, G.; Mu, G.; Chan, K.; Lu, F. Branch potential energy method for power system transient stability assessment based on network dynamic variables. *Proc. CSEE* **2004**, *24*, 1–6.
26. Yu, Y.; Min, Y.; Chen, L.; Zhang, Y. Disturbance source location of forced power oscillation using energy functions. *Autom. Electr. Power Syst.* **2010**, *34*, 1–6.
27. Chen, L.; Min, Y.; Hu, W. Low frequency oscillation analysis and oscillation source location based on oscillation energy part one mathematical foundation and energy flow computation. *Autom. Electr. Power Syst.* **2012**, *36*, 22–27.
28. Chen, L.; Chen, Y.; Min, Y.; Hu, W.; Zhang, K. Low frequency oscillation analysis and oscillation source location based on oscillation energy part two method for oscillation source location and case studies. *Autom. Electr. Power Syst.* **2012**, *36*, 1–5.
29. Chen, L.; Min, Y.; Hu, W. An energy-based method for location of power system oscillation source. *IEEE Trans. Power Syst.* **2013**, *28*, 828–836. [CrossRef]
30. Li, Y.; Shen, C.; Liu, F. An energy-based methodology for locating the source of forced oscillations in power systems. In Proceedings of the IEEE International Conference on Power System Technology (POWERCON), Auchland, NewZealand, 30 October–2 November 2012; p. 6.

31. Thorp, J.S.; Seyler, C.E.; Phadke, A.G. Electromechanical wave propagation in large electric power systems. *IEEE Trans. Circuits Syst. I Fundam. Theory Appl.* **1998**, *45*, 614–622. [CrossRef]
32. Markham, P.N.; Liu, Y. Electromechanical speed map development using FNET/GridEye frequency measurements. In Proceedings of the IEEE PES General Meeting | Conference & Exposition, National Harbor, MD, USA, 27–31 July 2014; Volume 5, pp. 27–31.
33. Hu, N.; Li, X.; Yang, Y.; Guo, X.J.; Li, K.; Xiao, J. Research for space-time variational features of frequency for power system forced oscillations and disturbance source location. *J. Sichuan Univ. (Eng. Sci. Edn.)* **2013**, *45*, 135–142.
34. Geng, T.; Zhang, Z.; Xiang, L.; Bao, Y.; Bai, S.; Fang, Y. A locating and splitting scheme for disturbance source of forced power oscillation based on the propagation characteristic. *Power Syst. Prot. Control* **2015**, *43*, 98–103.
35. Dong, Q.; Zhang, L.; Yan, X. Automatic location of generation units behaving as disturbance source of low-frequency oscillation. *Power Syst. Technol.* **2012**, *36*, 265–269.
36. Dong, Q.; Liang, J.; Yan, X.; Yang, R. Locating method of disturbance source of low frequency oscillation in large scale power grid. *Proc. CSEE* **2012**, *32*, 78–83.
37. Dong, Q.; Zhang, L.; Yan, X.; Liu, X. Automatic locating source method of compelled resonance low frequency oscillation in power grid. *Proc. CSEE* **2012**, *32*, 68–75.
38. Zhu, Z.; Liu, D.C.; Liao, Q.F.; Tu, L.; Han, X.; Zhao, Q. Disturbance source location of forced power oscillation based on LMS time delay estimation method. *Autom. Electr. Power Syst.* **2015**, *39*, 58–62.
39. Li, S.Y.; Luan, M.D.; Gan, D.Q.; Wu, D. A model-based decoupling observer to locate forced oscillation sources in mechanical power. *Int. J. Electr. Power Energy Syst.* **2018**, *103*, 127–135. [CrossRef]
40. Chevalier, S.; Vorobev, P.; Turitsyn, K. Using effective generator impedance for forced oscillation source location. *IEEE Trans. Power Syst.* **2018**, *33*, 6264–6277. [CrossRef]
41. Chevalier, S.; Vorobev, P.; Turitsyn, K. A Bayesian Approach to Forced Oscillation Source Location Given Uncertain Generator Parameters. *IEEE Trans. Power Syst.* **2019**, *34*, 1641–1649. [CrossRef]
42. Sarmadi, S.A.N.; Venkatasubramanian, V. Inter-area resonance in power systems from forced oscillations. *IEEE Trans. Power Syst.* **2016**, *31*, 378–386. [CrossRef]
43. Gao, Y.; Liu, D.C.; Huang, G.B.; Shi, Q.Y. Locating method of disturbance source of forced power oscillation based on prony anyasis. In Proceedings of the 2012 China International Conference on Electricity Distribution, Shanghai, China, 10–14 September 2012; pp. 10–14.
44. Yang, Y.; Liu, T.; Li, X.; Hu, N.; Shi, H. An equivalent circuit approach to locate souce of power system forced power oscillation. *Proc. CSEE* **2012**, *36*, 101–108.
45. Jiang, C.; Liu, J.; Liu, Y.; Gou, J.; Chen, C.; Chen, R.; Bazargan, M. Online forced oscillation detection and identification based on wide area measurement system and CELL theory. *Electr. Power Autom. Equip.* **2015**, *35*, 125–132.
46. Guo, S.; Cai, G.; Yang, D.; Jiang, C.; Liu, C. Compressed- DMD for Electromechanical Mode Estimation from High-Dimensional Ambient Data. *Int. Trans. Electr. Energy Syst.* **2022**, *2022*, 1832313. [CrossRef]

energies

MDPI

Article

A Rotating Tidal Current Controller and Energy Router Siting and Capacitation Method Considering Spatio-Temporal Distribution

Junqing Jia [1,2], Jia Zhou [3], Yuan Gao [4], Chen Shao [5,*], Junda Lu [5] and Jiaoxin Jia [5]

1 Key Laboratory of Intelligent Power Grid Simulation Enterprise of Inner Mongolia Autonomous Region, Hohhot 010020, China; nmdky1234@163.com
2 Inner Mongolia Electric Power (Group) Co., Ltd., Inner Mongolia Electric Power Research Institute, Hohhot 010020, China
3 Navigation Inspection Branch, Inner Mongolia Power (Group) Co., Ltd., Hohhot 010020, China; zhoujia1313139@163.com
4 Ordos Power Supply Branch, Inner Mongolia Power (Group) Co., Ltd., Ordos 017004, China; gaoyuan1312@163.com
5 Hebei Key Laboratory of Distributed Energy Storage and Microgrid, North China Electric Power University, Baoding 071003, China; lujd777@163.com (J.L.); jiajx33@163.com (J.J.)
* Correspondence: shaochen_336@163.com

Abstract: As the proportion of new energy access increases year by year, the resulting energy imbalance and voltage/trend distribution complexity of the distribution network system in the spatio-temporal dimension become more and more prominent. The joint introduction of electromagnetic rotary power flow controller (RPFC) and energy router (ER) can improve the high proportion of new active distribution network (ADN) consumption and power supply reliability from both spatial and temporal dimensions. To this end, the paper proposes an ADN expansion planning method considering RPFC and ER access. A two-layer planning model for RPFC and ER based on spatio-temporal characteristics is established, with the upper model being the siting and capacity-setting layer, which takes the investment and construction cost of RPFC and ER as the optimization objective, and the lower model being the optimal operation layer, which takes the lowest operating cost of the distribution network as the objective. The planning model is solved by a hybrid optimization algorithm with improved particle swarm and second-order cone planning. The proposed planning model and solving algorithm are validated with the IEEE33 node example, and the results show that the joint access of RPFC and ER can effectively improve the spatial-temporal distribution of voltage in the distribution network and has the lowest equivalent annual value investment and operation cost.

Keywords: active distribution network expansion planning; rotating current controller; energy router; improved particle swarm algorithm; second order cone planning

1. Introduction

With the urgent need to accommodate a high proportion of distributed renewable energy integration, it is essential to enhance the capacity and reliability of active distribution networks (ADN) from both spatial and temporal dimensions [1]. Current distribution systems face challenges such as weak grid structures, uneven local load distribution, and inadequate power supply capabilities. As the share of renewable energy increases each year, issues of energy imbalance, complex voltage and power flow distribution across spatial and temporal dimensions within the distribution network become increasingly prominent [2]. Flexible Interconnection Devices (FID) based on power electronics can enable flexible interconnections in medium- and low-voltage distribution networks, such as through smart soft switches (SOP), electromagnetic rotary power flow controller (RPFC), and energy routers (ER). These devices facilitate loop operations between AC and DC distribution feeders, enabling functions such as power transfer, load balancing, loss reduction, fault isolation, and power restoration. However, power-electronic-based FIDs also have drawbacks, including limited shock tolerance, challenging maintenance, and high costs, which significantly restrict their wider application in medium- and low-voltage distribution networks. The electromagnetic rotary power flow controller (RPFC) [3], using a wound motor structure, offers advantages of high shock resistance and reliability, providing the same functions as power-electronic-based FIDs within distribution networks. It also has distinct benefits: a response time on the order of seconds, simple control, strong continuity, no harmonic output, and lower costs. Therefore, establishing a flexible AC/DC active distribution network (ADN) using RPFC and ER [4] presents clear advantages over ADN solutions based solely on power-electronic devices [5]. To fully leverage the benefits that widespread integration of RPFC and ER can bring to ADNs, it is crucial to study and develop expansion planning models that consider RPFC and ER integration [6].

The RPFC is typically installed between two looped lines in a distribution system, enabling adjustable power transfer of any magnitude and direction between these looped lines to optimize the operation of the distribution network [7]. In terms of flexible interconnections within the distribution network, research on the siting and sizing of RPFC remains limited. Existing planning and configuration schemes based on soft open points (SOP) in interconnected distribution networks provide a useful reference. Reference [8] proposes an optimized configuration method for SOPs in active distribution networks, utilizing an improved sensitivity analysis approach. This method employs a second-order cone programming algorithm to solve the siting and sizing optimization model for SOPs under enhanced sensitivity calculations. The case study results demonstrate that optimized SOP configuration effectively reduces the annual operational costs of the distribution network and improves system voltage stability. Building on this, Reference [9] considers the loss characteristics of transformers and SOPs within a two-layer model to determine SOP locations. The upper layer minimizes annual total costs, while the lower layer minimizes system losses. This approach was validated on interconnected systems with IEEE 22-bus and IEEE 15-bus nodes, showing that the proposed control strategy significantly reduces system losses while maintaining minimal annual investment costs, thereby enhancing both economic efficiency and reliability. Reference [10] further develops an expansion planning study for ADNs that incorporates SOP access. This research addresses the coordinated siting and sizing of new or expanded substations, new lines, SOPs, distributed generation (DG), energy storage systems (ESS), and reactive power compensation devices.

The above literature indicates that reasonable planning for the location and capacity of the RPFC is crucial for improving ADN operational efficiency and promoting flexible development in distribution networks. However, the RPFC can only optimize voltage and power flow distribution across the active distribution network spatially [11]. The ER, on the other hand, provides flexibility support to external systems on a node basis. By enabling optimized scheduling of flexible resources at its ports, the ER addresses energy imbalances in the time dimension, power quality issues, and the reliability of power supply for critical loads. Coordinated operation of RPFC and ER enhances the ADN's ability to accommodate high proportions of renewable energy and ensures reliable power supply from both spatial and temporal dimensions [12,13]. References propose an optimal planning method for ER clusters in distribution networks, which includes the unified planning of ER groups and other distributed resources within the network [14,15]. Simulation validations confirm the method's effectiveness, providing a planning reference for distribution networks integrating various new types of generation and load.

Currently, extensive research has been conducted on the siting and planning of flexible interconnection devices and ERs [16,17]. However, in terms of coordinated spatiotemporal planning, Reference [18] proposes an ESS-SOP network collaborative planning method that considers the integration of flexible loads. This approach takes into account the temporal characteristics of distributed generation and flexible loads, constructing a collaborative planning model for ESS-SOP and network architecture, which was validated on an IEEE 33-bus system. Reference [19] further integrates the operational characteristics of ESS and SOP, introducing a three-layer model for ESS and SOP. The upper and middle layers aim to minimize the combined investment and operational costs for siting and capacity planning, while the lower layer targets daily average flexibility in optimizing operational scheduling. A hybrid optimization algorithm was employed to solve this model, with case studies validating the model's rationality and effectiveness for SOP and ESS coordinated planning. To date, however, no research team has extended the collaborative function of RPFC and ER into ADN planning scheme design.

Based on existing research, this paper proposes an ADN expansion planning method that considers the spatiotemporal regulation characteristics of RPFC and ER. First, an ADN two-stage expansion planning model is developed with the objective function of minimizing the annualized comprehensive cost, taking into account the integration of RPFC and ER. Given that the proposed model represents a nonlinear programming problem, a hybrid algorithm combining improved particle swarm optimization (IPSO) and second-order cone programming (SOCP) is introduced to solve it. The feasibility and effectiveness of the proposed planning model and solution algorithm are then analyzed and validated through a case study on a 33-bus ADN system.

The main innovations of this paper are as follows:

(1) Model Construction: This work presents the first construction of a grid planning model that includes the RPFC output model and associated constraints, providing a reference framework for subsequent RPFC planning methodologies.
(2) Application Scenario Innovation: The paper introduces a novel siting and planning model that incorporates the hybrid integration of RPFC and ER into distribution networks, offering a new reference for specific distribution network planning approaches.
(3) Solution Design for the Dual-Layer Model: A hybrid method combining improved particle swarm optimization (IPSO) and second-order cone programming (SOCP) is applied to solve the RPFC and ER dual-layer model. This approach is validated through simulation on an IEEE 33-bus system, demonstrating the effectiveness of the proposed solution.

2. Two-Layer Planning Modeling of RPFC and ER Based on Spatio-Temporal Properties

The ADN expansion planning scheme with RPFC and ER integration is shown in Figure 1. During the planning process, branch flexibility primarily considers the optimal location and capacity of the RPFC, while node flexibility focuses on the optimal location and capacity of the ER (with a four-port ER being the primary focus of this analysis). This approach aims to maximize the system's flexibility demand and flexibility supply functions.

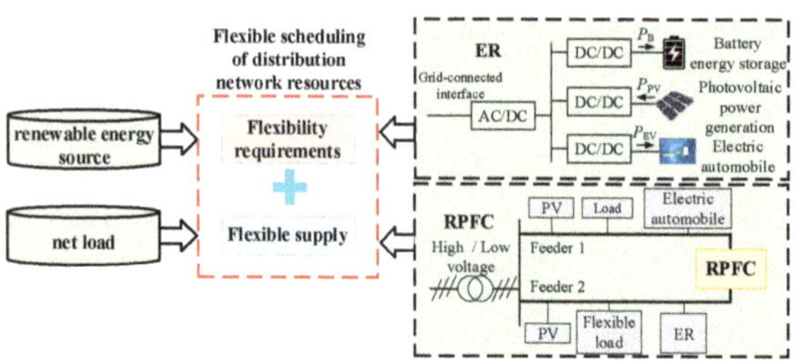

Figure 1. ADN expansion planning scheme considering RPFC and ER accesses.

Based on the access requirements shown in Figure 1, a dual-layer planning scheme is designed in Figure 2. First, the siting and sizing of the RPFC and ER are treated as the upper-layer model, which optimizes the equivalent annual investment cost for various capacities and locations. When device capacity is low, it does not sufficiently enhance system flexibility; however, larger capacities significantly improve operational flexibility but result in higher costs. Therefore, the upper-layer model aims to minimize the combined investment cost for both RPFC and ER, providing initial siting and sizing results to the lower layer.

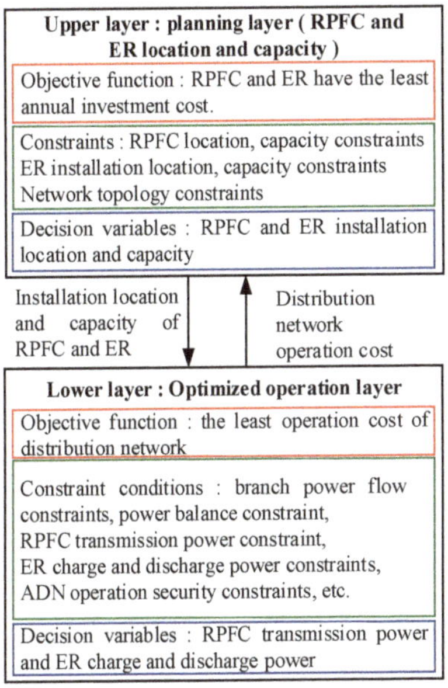

Figure 2. Framework of the RPFC and ER two-tier coordinated planning models.

The lower-layer model focuses on optimizing operational flexibility in the distribution network. Based on the RPFC and ER configuration plan from the upper layer, it minimizes operational costs across multiple typical daily scenarios by solving for the transmission power in RPFC-connected lines and the charge/discharge power at ER-connected nodes within the interconnected distribution network. These operational costs are then communicated back to the upper layer. This iterative process between the two layers ultimately yields an optimal solution that satisfies both operational and configuration requirements.

2.1. *Upper Layer Model*

2.1.1. Upper Model Objective Function

The objective function of the upper-layer model is to minimize the annualized investment costs over the planning period for the Active Distribution Network (ADN). This includes the investment cost for the RPFC F_{RPFC}, the investment cost of ER F_{ER}(including the construction cost of the ER F_{ERD} and the supporting DG construction cost F_{DG}, ES construction cost F_{ES}, V2G construction cost F_{V2G}), as well as the civil engineering cost, F_{CE}. The expression is as follows:

$$\min F_1 = \theta^{\mathrm{RPFC}} F_{\mathrm{RPFC}} + \theta^{\mathrm{ER}} F_{\mathrm{ER}} + \theta^{\mathrm{CE}} F_{\mathrm{CE}} \tag{1}$$

$$F_{\mathrm{ER}} = F_{\mathrm{ERD}} + F_{\mathrm{DG}} + F_{\mathrm{ESS}} + F_{\mathrm{V2G}} \tag{2}$$

$$\theta = \frac{d(1+d)^{L_{\mathrm{T}}x}}{(1+d)^{L_{\mathrm{T}}x} - 1}, x \in \{\mathrm{RPFC}, \mathrm{ER}, \mathrm{CE}\} \tag{3}$$

where: θ is the equal-year-value operator, which is used to convert total costs to equal-year-value costs; where d and L_{T} are the discount rate and the life cycle of the asset, respectively.

2.1.2. Upper Model Constraints

(1) RPFC installation location, capacity and power constraints:

$$\left\{\begin{array}{l} P_{i,t}^{\mathrm{RPFC}}+P_{j,t}^{\mathrm{RPFC}}+P_{loss,t}^{\mathrm{RPFC}}=0 \\ Q_{i,t}^{\mathrm{RPFC}}+Q_{j,t}^{\mathrm{RPFC}}=0 \\ \sqrt{(P_{ij,t}^{\mathrm{RPFC}})^2+(Q_{ij,t}^{\mathrm{RPFC}})^2}\leq S_{ij}^{\mathrm{RPFC}} \\ P_{\min}^{\mathrm{RPFC}}\leq P_{ij,t}^{\mathrm{RPFC}}\leq P_{\max}^{\mathrm{RPFC}} \\ Q_{\min}^{\mathrm{RPFC}}\leq Q_{ij,t}^{\mathrm{RPFC}}<Q_{\max}^{\mathrm{RPFC}} \\ z^{\mathrm{RPFC}}\in z_{ij} \end{array}\right. \tag{4}$$

where: $P_{i,t}^{\mathrm{RPFC}}$, $P_{j,t}^{\mathrm{RPFC}}$, $Q_{i,t}^{\mathrm{RPFC}}$, $Q_{j,t}^{\mathrm{RPFC}}$ and $P_{loss,t}^{\mathrm{RPFC}}$ are the active power, reactive power and transmission losses flowing through nodes i and j at time t, respectively; $P_{ij,t}^{\mathrm{RPFC}}$, $Q_{ij,t}^{\mathrm{RPFC}}$, S_{ij}^{RPFC} are the active power, reactive power and installed capacity of branch ij transmission, respectively; $P_{\min}^{\mathrm{RPFC}}$, $P_{\max}^{\mathrm{RPFC}}$, $Q_{\min}^{\mathrm{RPFC}}$, $Q_{\max}^{\mathrm{RPFC}}$ are the maximum active power regulation and maximum reactive power of the RPFC, respectively. ER installation location, rate regulation range; z^{RPFC} is the RPFC access node and z_{ij} is the RPFC allowed access line.

(2) Capacity and power constraints:

$$\left\{\begin{array}{l} P_{\mathrm{ESS},t}^{\mathrm{ER}}+P_{\mathrm{DG},t}^{\mathrm{ER}}+P_{\mathrm{V2G},t}^{\mathrm{ER}}+P_{\mathrm{loss},t}^{\mathrm{ER}}=P_{\mathrm{AC},i,t}^{\mathrm{ER}} \\ Q_{\mathrm{ESS},t}^{\mathrm{ER}}+Q_{\mathrm{DG},t}^{\mathrm{ER}}+Q_{\mathrm{V2G},t}^{\mathrm{ER}}=Q_{\mathrm{AC},i,t}^{\mathrm{ER}} \\ \sqrt{\left(P_{\mathrm{AC},i,t}^{\mathrm{ER}}\right)^2+\left(Q_{\mathrm{AC},i,t}^{\mathrm{ER}}\right)^2}\leq S_i^{\mathrm{ER}} \\ \{S_{\mathrm{ESS}}^{\mathrm{ER}},S_{\mathrm{V2G}}^{\mathrm{ER}},S_{\mathrm{DG}}^{\mathrm{ER}}\}\leq S_i^{\mathrm{ER}} \\ P_{\mathrm{AC,min}}^{\mathrm{ER}}\leq P_{\mathrm{AC},i,t}^{\mathrm{ER}}\leq P_{\mathrm{AC,max}}^{\mathrm{ER}} \\ Q_{\mathrm{AC,min}}^{\mathrm{ER}}\leq Q_{\mathrm{AC},i,t}^{\mathrm{ER}}\leq Q_{\mathrm{AC,max}}^{\mathrm{ER}} \\ \mathrm{z}^{\mathrm{ER}}\in \mathrm{z_i} \end{array}\right. \tag{5}$$

where: $P_{\mathrm{AC},i,t}^{\mathrm{ER}}$ and $Q_{\mathrm{AC},i,t}^{\mathrm{ER}}$ are the AC active and reactive power of energy router access node i in time period t; $P_{\mathrm{ESS},t}^{\mathrm{ER}}$, $P_{\mathrm{DG},t}^{\mathrm{ER}}$, $P_{\mathrm{V2G},t}^{\mathrm{ER}}$, $Q_{\mathrm{ESS},t}^{\mathrm{ER}}$, $Q_{\mathrm{DG},t}^{\mathrm{ER}}$, $Q_{\mathrm{DG},t}^{\mathrm{ER}}$ and $P_{\mathrm{loss},t}^{\mathrm{ER}}$ are the active and reactive exchange power of energy router internal storage, PV, V2G, and network losses, respectively; $S_{\mathrm{ESS}}^{\mathrm{ER}}$, $S_{\mathrm{V2G}}^{\mathrm{ER}}$, $S_{\mathrm{DG}}^{\mathrm{ER}}$, and S_i^{ER} are the ESS, V2G, DG, and ER capacities, respectively; $P_{\mathrm{AC,min}}^{\mathrm{ER}}$, $Q_{\mathrm{AC,min}}^{\mathrm{ER}}$, $P_{\mathrm{AC,max}}^{\mathrm{ER}}$, and $Q_{\mathrm{AC,max}}^{\mathrm{ER}}$ are the ER maximum active and maximum reactive power regulation ranges, respectively; and z^{ER} is the ER access node. z_i is the permitted access node of the distribution network.

2.2. *Lower Level Model*

2.2.1. Lower Model Objective Function

The lower layer model aims to meet the lowest cost under the safe power supply operation of the distribution network, including the main grid power purchase cost F_{buy}, O&M cost of RPFC $F_{\mathrm{O\text{-}RPFC}}$, O&M cost of ER $F_{\mathrm{O\text{-}ER}}$, penalty cost of light abandonment F_{cur}, and loss cost F_{loss}:

$$\min F_2=F_{\mathrm{buy}}+F_{\mathrm{O-RPFC}}+F_{\mathrm{O-ER}}+F_{\mathrm{cur}}+F_{\mathrm{loss}} \tag{6}$$

$$F_{\mathrm{buy}}=365\sum_{u\in\Omega_u}D_{\mathrm{u}}\sum_{i\in N}\sum_{t\in 24}S_{\mathrm{buy},i}\cdot C_{\mathrm{EC},t} \tag{7}$$

$$F_{\mathrm{cur}}=365\sum_{u\in\Omega_u}D_{\mathrm{u}}\sum_{i\in N}\sum_{t\in 24}c^{\mathrm{PVG}}\cdot P_{u,i,t}^{\mathrm{PVG}} \tag{8}$$

$$F_{\mathrm{O-RPFC}}=365\sum_{u\in\Omega_u}D_{\mathrm{u}}\sum_{i\in N}\sum_{t\in 24}k_1S_{ij}^{RPFC}\cdot C_{\mathrm{EC},t} \tag{9}$$

$$F_{\mathrm{O-ER}}=365\sum_{u\in\Omega_u}D_{\mathrm{u}}\cdot\sum_{i\in N}\sum_{t\in 24}(k_2S_i^{\mathrm{ER}}+k_3S_{\mathrm{ESS}}^{\mathrm{ER}}+k_4S_{\mathrm{V2G}}^{\mathrm{ER}}+k_5S_{\mathrm{DG}}^{\mathrm{ER}})\cdot C_{\mathrm{EC},t} \tag{10}$$

$$F_{\text{loss}} = 365 \sum_{u \in \Omega_u} D_{\text{u}} \sum_{i \in N} \sum_{t \in 24} P^{\text{total}}_{\text{loss},t} \cdot C_{\text{EC,t}} \tag{11}$$

$$P^{\text{total}}_{\text{loss},t} = P^{z}_{loss,t} + P^{\text{RPFC}}_{loss,t} + P^{\text{ER}}_{loss,t} \tag{12}$$

$$P^{z}_{loss,t} = \sum_{(i,j) \in N} \left(I^{t}_{ij}\right)^2 r_{ij} \tag{13}$$

where: W_u is the set of typical day scenarios; u denotes the current typical day scenario; D_{u} is the probability of the uth typical day scene.; N is the set of nodes in the distribution network; t is the corresponding time period of the typical day; $S_{\text{buy},i}$ is the load size of node i; $C_{\text{EC},t}$ is the peak and valley tariff of the corresponding time period of t; c^{PVG} is the penalty cost of DG per unit of discarded light, and $P^{\text{PVG}}_{u,i,t}$ is the discarded power; k_1, k_2, k_3, k_4, k_5 are the O&M cost coefficients of RPFC, ER, ESS, V2G, and DG, respectively. cost coefficients; $P^{\text{total}}_{\text{loss},t}$ is the total network loss at time t; $P^{z}_{loss,t}$, $P^{\text{RPFC}}_{loss,t}$, $P^{\text{ER}}_{loss,t}$ are the line, RPFC and ER operating losses at time t, respectively; I^{t}_{ij} is the line ij current at time t; and r_{ij} is the resistance of line ij.

2.2.2. Lower Model Constraints

The lower layer model, while satisfying the constraints of Equations (4) and (5), should also satisfy the constraints on the safe operation of the active distribution network, which mainly include:

(1) ADN Safe Operation Constraints

$$\begin{cases} \sum\limits_{ij \in \Omega_N} \left[P_{ij,t} - r_{ij}\left(I^{t}_{ij}\right)^2\right] + P_{j,t} = \sum\limits_{jk \in \Omega_N} P_{jk,t} \\ \sum\limits_{ij \in \Omega_N} \left[Q_{ij,t} - x_{ij}\left(I^{t}_{ij}\right)^2\right] + Q_{j,t} = \sum\limits_{jk \in \Omega_N} Q_{jk,t} \end{cases} \tag{14}$$

$$(U_{i,t})^2 - (U_{j,t})^2 + (r^2_{ij} + x^2_{ij})\left(I^{t}_{ij}\right)^2 - 2(r_{ij}P_{ij,t} + x_{ij}Q_{ij,t}) = 0 \tag{15}$$

$$U^2_{i,t}\left(I^{t}_{ij}\right)^2 = (P_{ij,t})^2 + (Q_{ij,t})^2 \tag{16}$$

$$U_{i.\min} \le U_{i,t} \le U_{i.\max} \tag{17}$$

$$0 \le \left|I^{t}_{ij}\right| \le I_{ij.\max} \tag{18}$$

where: $P_{ij,t}$ and $Q_{ij,t}$ are the power flowing through node ij at time t of line ij, respectively; x_{ij} is the reactance of line ij; W_{N} is the set of branch circuits; $U_{i,t}$ and $U_{j,t}$ are the voltages of nodes i and j at time t, respectively; $U_{i.\min}$ and $U_{i.\max}$ are the lower and upper voltage limits of node i, respectively; $I_{ij.\max}$ is the maximum value of line current.

(2) ER internal port ESS operation constraints

ER satisfies Equation (5) port energy balance constraints, its internal port ESS satisfies Equation (19) constraint [20]:

$$E^{\text{ER}}_{\text{ESS},t+1} = \eta_{\text{ESSc}} P^{\text{ER}}_{\text{ESSc},t}\Delta t - \eta_{\text{ESSd}} P^{\text{ER}}_{\text{ESSd},t}\Delta t + E^{\text{ER}}_{\text{ESS},t} - \varepsilon_{\text{ESS}}\Delta t \tag{19a}$$

$$E^{\text{ER}}_{\text{ESS,min}} \le E^{\text{ER}}_{\text{ESS},t} \le E^{\text{ER}}_{\text{ESS,max}} \tag{19b}$$

$$P^{\text{ER}}_{\text{ESSc,min}} \le P^{\text{ER}}_{\text{ESS,c},t} \le P^{\text{ER}}_{\text{ESSc,max}} \tag{19c}$$

$$P^{\text{ER}}_{\text{ESSd,min}} \le P^{\text{ER}}_{\text{ESS,d},t} \le P^{\text{ER}}_{\text{ESSd,max}} \tag{19d}$$

$$\delta_{\text{ESSc}} + \delta_{\text{ESSd}} \le 1 \tag{19e}$$

where: $E^{\text{ER}}_{\text{ESS},t}$ and $E^{\text{ER}}_{\text{ESS},t+1}$ are the ESS capacity at time t and t + 1, respectively; h_{ESSc} is the ESS charging efficiency; h_{ESSd} is the ESS discharging efficiency; $P^{\text{ER}}_{\text{ESSc},t}$ is the charging power; $P^{\text{ER}}_{\text{ESSd},t}$ is the ESS discharging power; ε_{ESS} is the ESS self-discharging rate; $E^{\text{ER}}_{\text{ESS,min}}$

and $E^{ER}_{ESS,min}$ are the ESS minimum capacity and maximum capacity, respectively; $P^{ER}_{ESSc,min}$ and $P^{ER}_{ESSc,max}$ are the ESS minimum charging power and maximum charging power, respectively; $P^{ER}_{ESSd,min}$ and $P^{ER}_{ESSd,max}$ are the ESS minimum and maximum discharging power, respectively; and δ_{ESSc} and δ_{ESSd} denote the ESS discharging and charging states (state variables 0/1).

(3) ER internal port V2G operation constraints

Electric vehicles (EVs) can be regarded as a combination of energy storage systems (ESS) and flexible loads, with variable times for connecting to and disconnecting from the grid. Therefore, analyzing the state of EVs from three perspectives—initial charge, travel time, and parking probability—provides essential references for optimizing scheduling decisions [21]:

$$f_{D(s)} = \frac{1}{s\sigma_D\sqrt{2\pi}}\exp[-\frac{(\ln s - u_D)^2}{2\sigma_D^2}] \tag{20a}$$

$$f_{B(t)} = \lambda_1 e^{-(\frac{t-\alpha_1}{\beta_1})^2} + \lambda_2 e^{-(\frac{t-\alpha_2}{\beta_2})^2} \tag{20b}$$

$$f_{(T)} = \frac{1}{\sigma\sqrt{2\pi}}e^{-\frac{(T-u)^2}{2\sigma^2}} \tag{20c}$$

Equation (20a) is the probability density function of electric vehicle daily traveling mileage, which affects its initial power after arriving at the destination, where: s is the electric vehicle traveling mileage per day in km; u_D and s_D are the expectation and variance of the function, respectively. Equation (20b) is the probability density function of electric vehicle traveling moments, which affects the time period of connecting to the grid, where: $l_1 = 0.389$, $a_1 = 7.046$, $b_1 = 1.086$, $l_2 = 0.066$, $a_2 = 15.610$, and $b_2 = 9.667$ [21]. Equation (20c) is the probability density function of the automobile parking distribution, where: t is the time of parking and charging of the electric vehicle in units of h; u is the mean value; s is the standard deviation. At this time, the constraints to be satisfied for V2G to participate in grid regulation include:

$$E^{ER}_{V2G,t+1} = \eta_{V2Gc}P^{ER}_{V2Gc,t}\Delta t - \eta_{V2Gd}P^{ER}_{V2Gd,t}\Delta t + E^{ER}_{V2G,t} - \varepsilon_{V2G}\Delta t \tag{21a}$$

$$E^{ER}_{V2G,min} \le E^{ER}_{V2G,t} \le E^{ER}_{V2G,max} \tag{21b}$$

$$P^{ER}_{V2Gc,min} \le P^{ER}_{V2G,c,t} \le P^{ER}_{V2Gc,max} \tag{21c}$$

$$P^{ER}_{V2Gd,min} \le P^{ER}_{V2G,d,t} \le P^{ER}_{V2G,max} \tag{21d}$$

$$\delta_{V2Gc} + \delta_{V2Gd} \le 1 \tag{21e}$$

where: $E^{ER}_{V2G,t}$ and $E^{ER}_{V2G,t+1}$ are the V2G capacity at time t and t + 1, respectively; h_{V2Gc} is the V2G charging efficiency; h_{V2Gd} is the V2G discharging efficiency; $P^{ER}_{V2Gc,t}$ is the charging power; $P^{ER}_{V2Gd,t}$ is the V2G discharging power; ε_{V2G} is the V2G self-discharging rate; $E^{ER}_{V2G,min}$ and $E^{ER}_{V2G,min}$ are the V2G minimum and maximum capacity, respectively; $P^{ER}_{V2Gc,min}$ and $P^{ER}_{V2Gc,max}$ are the V2G minimum and maximum charging power, respectively; $P^{ER}_{V2Gd,min}$ and $P^{ER}_{V2Gd,max}$ are the V2G minimum and maximum discharging power, respectively; and δ_{V2Gc} and δ_{V2Gd} denote the V2G discharge and charging states (state variables 0/1). Discharging and charging states (state variables: 0/1).

(4) PV output constraints

$$\delta_{V2Gc} + \delta_{V2Gd} \le 1 \tag{22}$$

where: $P_{DG}(t)$ is the real-time PV output, and $P_{DG,max}$ is the PV access capacity.

(5) RPFC operational constraints

$$\begin{cases} S_{ij}^{\mathrm{RPFC}} > 0 \\ \sqrt{(P_{ij,t}^{\mathrm{RPFC}})^2 + (Q_{ij,t}^{\mathrm{RPFC}})^2} \leq S_{ij}^{\mathrm{RPFC}} \end{cases} \quad (23)$$

where: S_{ij}^{RPFC} is the rated adjusting capacity of RPFC.

3. Model Solving Based on Hybrid Optimization Algorithm

To address the configuration issues in AC/DC hybrid active distribution networks (ADN) with RPFC and ER integration, a hybrid optimization algorithm combining an improved particle swarm optimization (IPSO) and second-order cone programming (SOCP) is proposed. This approach decomposes the investment costs of the new RPFC and ER and the distribution network's operational optimization costs into a two-layer optimization problem. The upper-layer model serves as the planning model, where IPSO is used to determine the locations and capacities of the RPFC and ER in the network. The lower-layer model focuses on operational optimization, where the RPFC and ER models are processed using SOCP to obtain optimized operational targets. Based on the topology provided by the upper layer, the algorithm calculates the optimal operational results for RPFC and ER, allowing for an evaluation of the network's annualized comprehensive cost at this stage.The solving process of the bi-level programming model is shown in Figure 3.

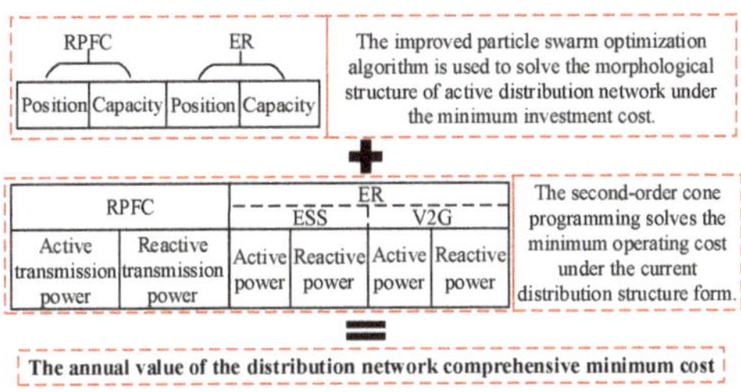

Figure 3. Solving two-layer planning model based on IPSO and SOCP.

3.1. *Improved Particle Swarm Algorithm with Time-Varying Coefficients*

In the classical particle swarm algorithm, individual particles are mainly composed of position and velocity parameters:

$$\begin{cases} v_{i+1} = \omega v_i + c_1 r_1 (pbest_i - x_i) + c_2 r_2 (gbest_i - x_i) \\ x_{i+1} = x_i + v_{i+1} \end{cases} \quad (24)$$

where: v_{i+1}, w, c_i, r_i, v_i, $pbest_i$, x_i, $gbest_i$ are the ith particle velocity vector, inertia weight, acceleration factor, interval [0, 1] random number, the ith particle velocity vector, the local optimal position of the ith particle individual, the ith particle position vector, and the global optimal position of the ith particle individual, respectively. The traditional particle swarm algorithm position update process is shown in Figure 4:

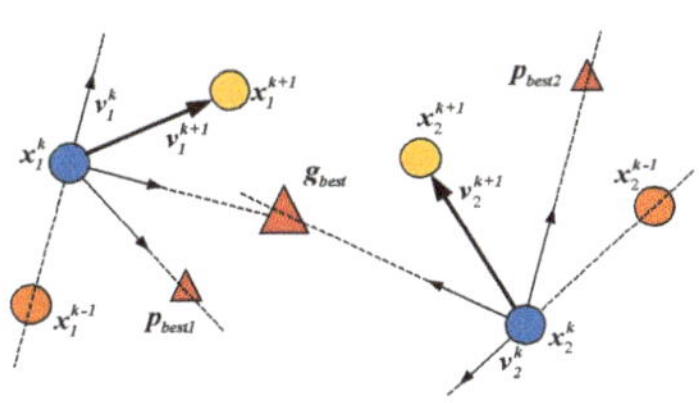

Figure 4. Classical particle swarm algorithm movement.

In this paper, for the defects of particle swarm algorithm which is easy to fall into local optimization, a time-varying coefficients Improved Particle Swarm Algorithm (IPSO) is used to solve the upper layer model, the key factors of the particle swarm algorithm, w and c_i are added into the time-varying features, and a new iterative optimization index is added into the calculation of the velocity vector of v_i in order to improve the performance of the computation as shown in Equation (25):

$$v_{i+1} = \omega v_i + c_1 r_1 (pbest_i - x_i) + c_2 r_2 (gbest_i - x_i) + c_3 r_3 (Ibest_i - x_i) \tag{25a}$$

$$\omega = \omega_{\max} - \frac{\omega_{\max} - \omega_{\min}}{i_{\max}} \times i \tag{25b}$$

$$c_1 = c_{\max} - \frac{c_{\max} - c_{\min}}{i_{\max}} \times i \tag{25c}$$

$$c_2 = c_{\min} + \frac{c_{\max} - c_{\min}}{i_{\max}} \times i \tag{25d}$$

$$c_3 = c_1 \times \left(1 - e^{(-c_2 \times i)}\right) \tag{25e}$$

where $w_{\min}$, $w_{\max}$, c_1, c_2, c_3, $c_{\min}$, $c_{\max}$ are the minimum and maximum values of the inertia weights, and the current, minimum, and maximum values of the learning factors, respectively; i is the current number of iterations; $i_{\max}$ is the maximum number of iterations; and $Ibest_i$ denotes the historical optimal solution of particle i.

3.2. Second-Order Cone Planning

Considering that the distribution network optimization and operation model is a nonlinear nonconvex model, the heuristic algorithm solution requires a large number of cyclic calculations and the solution is time-consuming, so the second-order cone relaxation is performed for Equations (4) and (5) in the RPFC and ER constraints and Equation (14) in the distribution network operation constraints:

$$\left\| \left[P_{ij,t}^{\mathrm{RPFC}} \quad Q_{ij,t}^{\mathrm{RPFC}} \right]^{\mathrm{T}} \right\|_2 \leq S_{ij}^{\mathrm{RPFC}} \tag{26}$$

$$\left\| \left[P_{\mathrm{AC},i,t}^{\mathrm{ER}} \quad Q_{\mathrm{AC},i,t}^{\mathrm{ER}} \right]^{\mathrm{T}} \right\|_2 \leq S_{i}^{\mathrm{ER}} \tag{27}$$

$$\left\| \left[2P_{ij,t} \quad 2Q_{ij,t} \quad I_{ij}^{t} - U_{i,t} \right]^{\mathrm{T}} \right\|_2 \leq I_{ij}^{t} + U_{i,t} \tag{28}$$

3.3. Optimize the Specific Flow of the Regulation Process

As shown in Figure 5, the solution process of the two-layer model begins by gathering essential parameters: network parameters, operational data for photovoltaics and loads, equipment parameters for RPFC and ER, storage parameters, and V2G forecast data. These initial conditions are essential for establishing and solving the model. The IPSO algorithm is then employed to determine the optimal locations and capacities of RPFC and ER, initializing the particle swarm algorithm parameters and generating an initial population for RPFC and ER. The configuration from the upper layer is passed to the lower layer, where the objective function of the upper-layer planning model is calculated.

Using the distribution network structure generated from the upper-layer model, the lower layer solves for the optimal power flow by employing the SOCP model, considering the outputs of RPFC and ER. The solution to the lower-layer model yields the objective function for the operational model, resulting in a local optimal solution at this stage. Based on the theoretical analysis from Section 2.1, particle velocities and positions are updated iteratively until the maximum number of iterations is reached, ultimately yielding the model's global optimal solution.

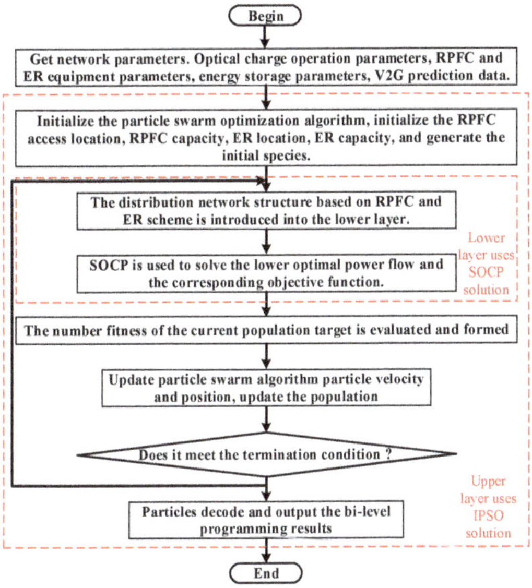

Figure 5. Flowchart for solving the bilayer model.

4. Simulation Analysis

To validate the effectiveness of the proposed RPFC and ER siting and sizing methods, the IEEE 33-bus distribution system is used as an example. The system structure is illustrated in Figure 6. The rated voltage is set at 12.66 kV. The RPFC operates by connecting to four normally open branches, while the ER is connected to nodes ranging from Node 2 to Node 33. The parameters of the network topology and equipment models are detailed in Table 1, and the line loads are calculated based on proportional scaling of real measured data.

Table 1. Device model parameters.

Parameter Type	Numerical Value	Parameter Type	Numerical Value
F_{RPFC}	0–200 kVA: \$110/kVA; 200–500 kVA: \$85/kVA; 500–700 kVA: \$55/kVA; 700–1000 kVA: \$30/kVA; 1 MW and above: \$20/KVA.	F_{ERD}	0–50 kVA: \$280/kVA; 50–100 kVA: \$140/kVA; 100–200 kVA: \$105/kVA; 200 kVA and above: \$85/kVA.
$F_{\text{O-RPFC}}$	$k_1 = 0.002$	$F_{O\text{-}ER}$	$k_2 = 0.003$; $k_3 = 0.001$; $k_4 = 0.002$; $k_5 = 0.001$
$L_{\text{T-ERD,DG,ES,V2G}}$	20 years	$L_{\text{T-RPFC}}$	30 years
d	0.08	$P_{loss,t}^{\text{RPFC}}$	$0.03\ P_{i,t}^{\text{RPFC}}$
$F_{\text{buy-peak period}}$	\$0.08	$F_{\text{buy-valley period}}$	\$0.04
c^{PVG}	\$0.04/kW·h	F_{CE}	\$28,000

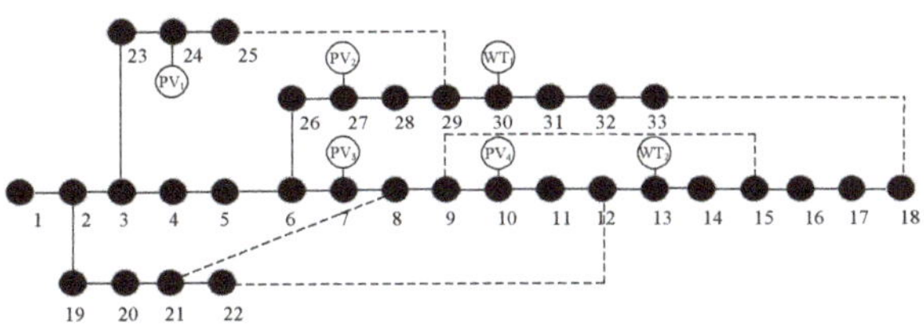

Figure 6. IEEE33 node active distribution network with DGs.

4.1. RPFC and ER Joint Planning Results

To validate the effectiveness of the proposed method, the following comparison scenarios were set up:

Scenario 1: Neither RPFC nor ER is installed. In this case, across several typical daily scenarios, high levels of distributed photovoltaic (PV) integration result in voltage limit violations within the distribution network, as shown in Figure 7.

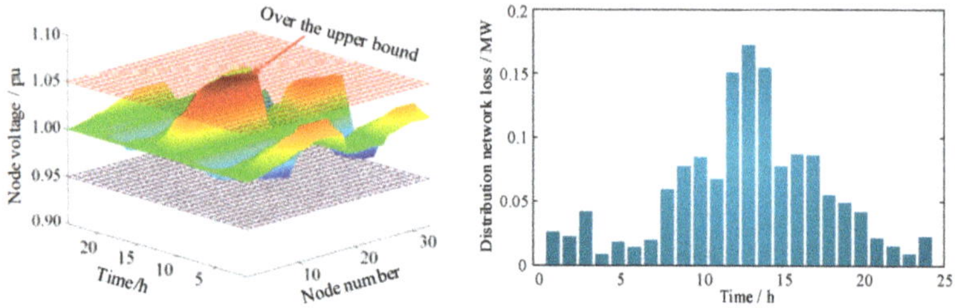

(a)

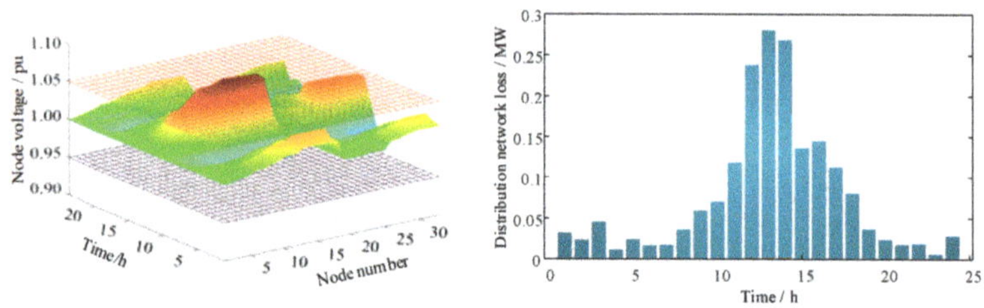

(b)

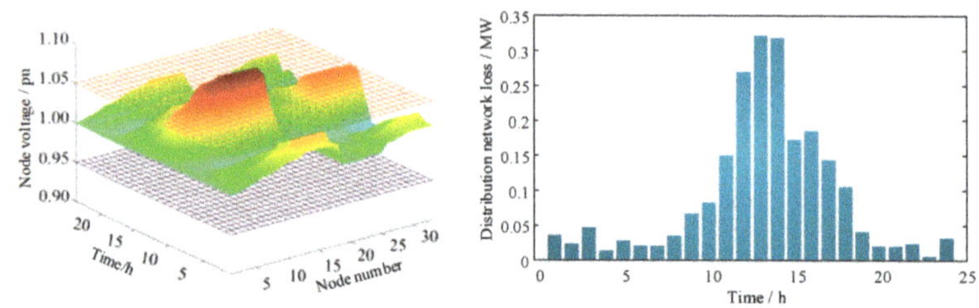

(c)

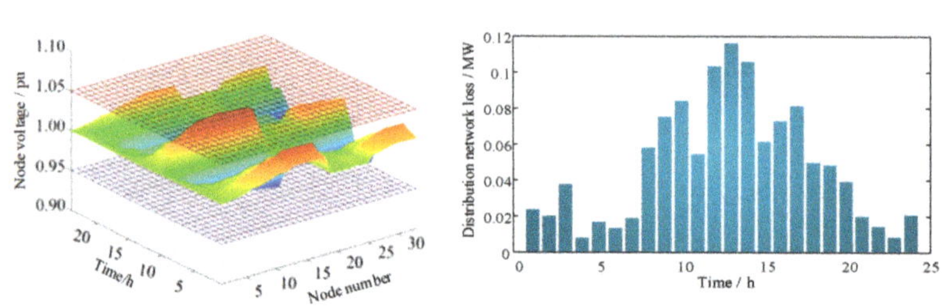

(d)

Figure 7. Distribution network voltage and line loss without RPFC and ER participation. (**a**) Typical day 1; (**b**) Typical day 2; (**c**) Typical day 3; (**d**) Typical day 4.

Scenario 2: Only RPFC is installed. Compared to Scenario 1, the inclusion of RPFC helps mitigate voltage limit violations to some extent and effectively reduces line losses, as illustrated in Figure 8.

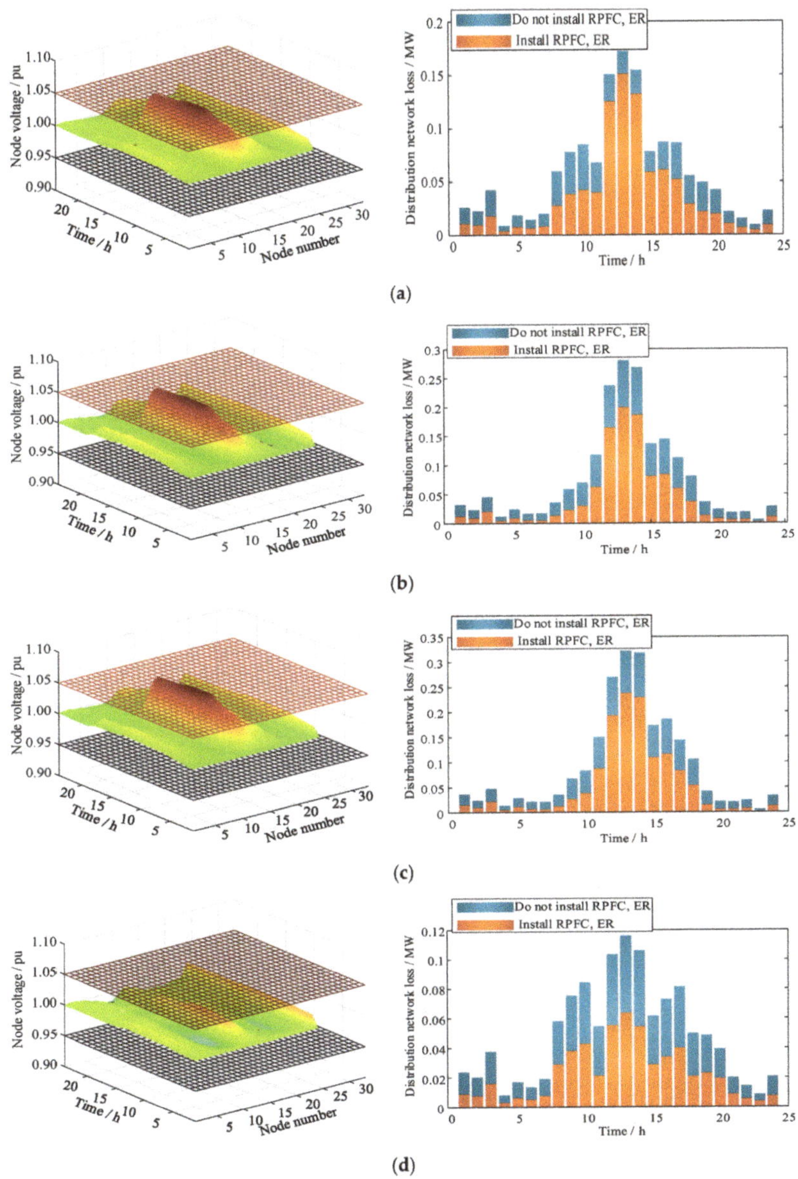

Figure 8. Voltage and line loss of the distribution network with only RPFC installed. (**a**) Typical day 1; (**b**) Typical day 2; (**c**) Typical day 3; (**d**) Typical day4.

Scenario 3: Both RPFC and ER are installed. In contrast to Scenario 2, the combined installation of RPFC and ER addresses voltage and line loss issues from both spatial and temporal dimensions, further reducing voltage deviations and line losses, as demonstrated in Figure 9.

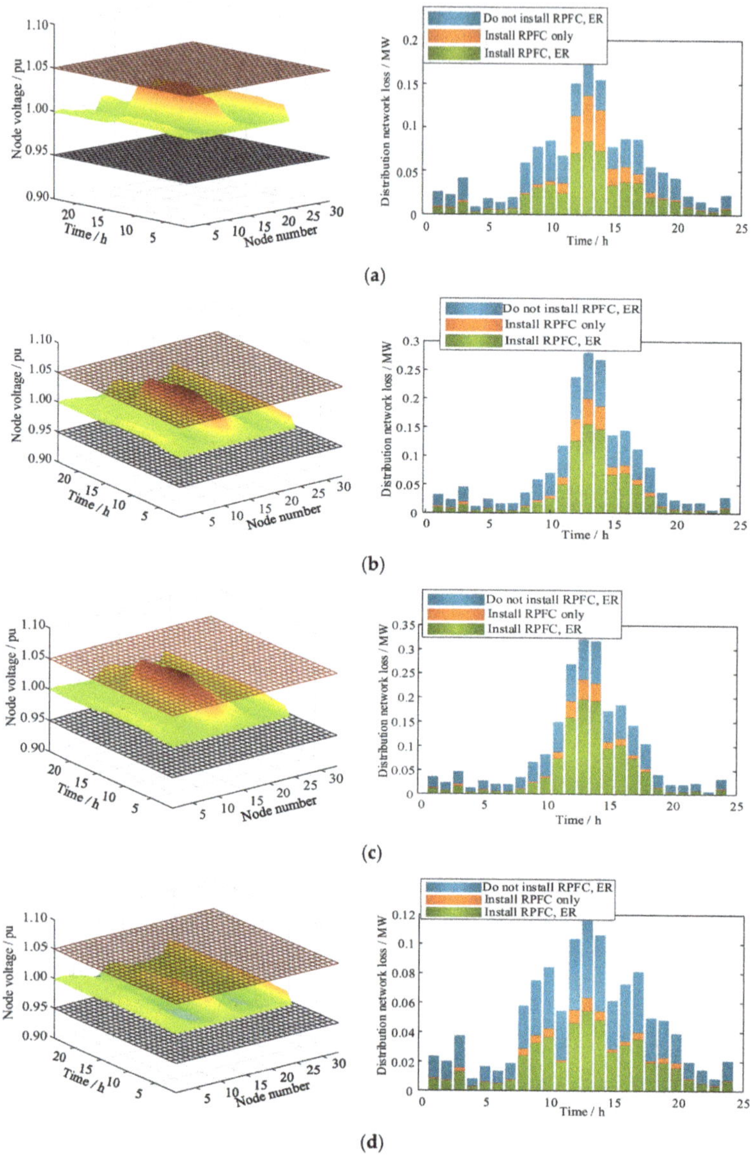

Figure 9. Voltage and line loss of the distribution network with RPFC and ER installed together. (**a**) Typical day 1; (**b**) Typical day 2; (**c**) Typical day 3; (**d**) Typical day 4.

The annual operating costs of the different scenarios are shown in Table 2. Since more wind and light rejection exists under the operation of Scenario 1, Scenarios 2 and 3 not only reduce the voltage deviation and the size of the line loss by introducing RPFC and ER, but also reduce the operating cost of its equivalent annual value compared to Scenario 1.

Table 2. RPFC and ER planning results.

		Option 1	Option 2	Option 3
Decision variables	RPFC	-	(9, 15), (0.8 WM)	(25, 29), (0.5 WM)
	ER	-	-	(15), (0.3 WM)
$F_1/\$$	$q^{\mathrm{RPFC}}F_{\mathrm{RPFC}}$	-	4290	3431
	$q^{\mathrm{ER}}F_{\mathrm{ER}}$	-	-	2574
$F_2/\$$	$F_{\text{O-RPFC}}$	-	7071	3247
	$F_{\text{O-ER}}$	-	-	2480
	F_{cur}	43,458	-	-
	F_{loss}	41,248	23,426	19,203
Total cost/$		84,706	34,788	30,937

4.2. Analysis of RPFC and ER Response Results

The optimization process is analyzed using Typical Day 1 as an example. Figure 10 shows the time-series graph of active and reactive power at the RPFC ports in Scheme 2. As seen in Figures 5–10, during periods of high photovoltaic generation, power support between Node 9 and Node 15 reduces the power transmission path within the IEEE 33-bus system, effectively lowering system operating losses and voltage deviation.

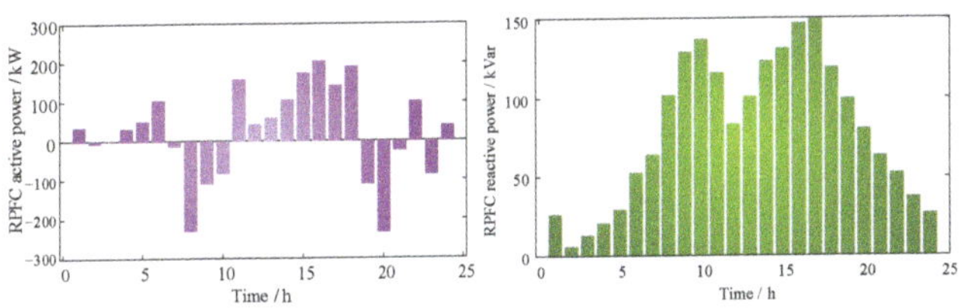

(a) (b)

Figure 10. RPFC operation regulation under optimized regulation strategy. (**a**) Interconnection node active power; (**b**) Interconnection node reactive power.

As shown in Figure 11 for the optimization and regulation results of Scheme 3, it can be seen that at the moment of large PV output, in addition to the RPFC participating in the power mutualization at nodes 25 and 29, the ER participates in the energy management at node 15 on the time scale, which further reduces the system network loss.

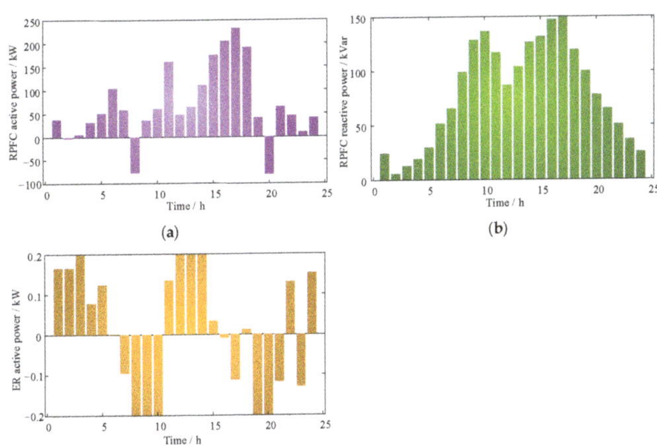

Figure 11. RPFC and ER operation under optimized regulation strategy. (**a**) Interconnection node active power; (**b**) Interconnection node reactive power; (**c**) Power exchange at ER access nodes.

4.3. Algorithm Performance Comparison

The classic Particle Swarm Optimization (PSO) algorithm was compared with the proposed improved PSO algorithm for location planning in Scenario 3. As shown in Figure 12, the convergence curves of the two algorithms reveal a distinct difference: the classic PSO algorithm converges at a value of 46,000, indicating a limited capacity for further exploration in a larger solution space and resulting in a restriction to local optima. In contrast, the convergence curve of the improved PSO algorithm demonstrates a stronger global optimization capability, converging to a significantly lower value. This result indicates that the time-varying parameter enhancements in the improved PSO algorithm enable it to more effectively escape local optima and accelerate convergence.

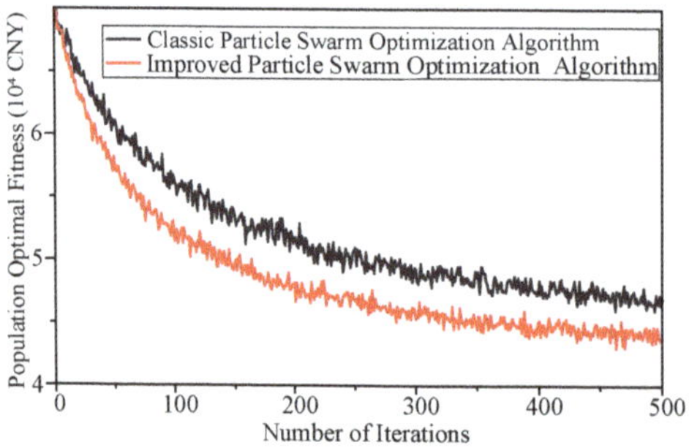

Figure 12. Comparison of Algorithm Convergence Curves.

Figure 13 shows a box plot comparison of the optimization results for the two algorithms. It can be observed that, compared to PSO, IPSO demonstrates a wider exploration range and performs local optimization around the optimal solution, enhancing both global and local optimization capabilities. In contrast, PSO expends significant computation on the local optimum without successfully escaping it.

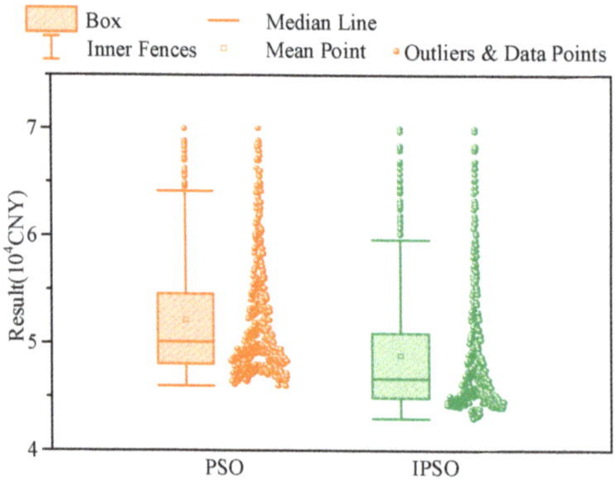

Figure 13. Comparison of Algorithm Box Plots.

5. Conclusions

This paper presents an expansion planning method for active distribution networks (ADN) based on the spatiotemporal regulation characteristics of RPFC and ER, with validation through case studies on the IEEE 33-bus system. The main conclusions are as follows:

(1) The study establishes a two-stage expansion planning model for ADN, incorporating RPFC and ER, and develops a hybrid algorithm based on IPSO and SOCP. This approach optimizes the flexible interconnection of RPFC and ER within a typical distribution network and reduces system line loss costs.

(2) RPFC and ER effectively redistribute power across spatial and temporal dimensions. By appropriately configuring capacities and connection points, they significantly mitigate voltage limit violations at distribution network nodes and reduce line losses.
(3) Current heuristic-based improvement strategies are limited by computational speed, allowing only for siting and capacity planning of devices like RPFC and ER. These strategies fall short of meeting real-time, minute-level optimization and dispatch requirements. Future research should focus on artificial intelligence techniques to enable rapid scheduling and control of flexible devices.

Author Contributions: Methodology, J.Z.; formal analysis, J.J. (Junqing Jia); data curation, Y.G.; writing—original draft preparation, C.S.; writing—review and editing, J.L.; funding acquisition, J.J. (Jiaoxin Jia); All authors have read and agreed to the published version of the manuscript.

Funding: Science and technology project funding of Inner Mongolia Power Group: 2024-4-47; Supported by the National Natural Science Foundation of China: 51777162.

Data Availability Statement: Data are unavailable due to privacy restrictions.

Conflicts of Interest: Author Junqing Jia was employed by the company Key Laboratory of Intelligent Power Grid Simulation Enterprise of Inner Mongolia Autonomous Region and Inner Mongolia Electric Power (Group) Co., Ltd., Inner Mongolia Electric Power Research Institute Branch. Author Jia Zhou was employed by the company Navigation Inspection Branch of Inner Mongolia Power (Group) Co., Ltd., Inner Mongolia Autonomous Region. Author Yuan Gao was employed by the companyOrdos Power Supply Branch, Inner Mongolia Power (Group) Co., Ltd. The remaining authors declare that the research was conducted in the absence of any commercial or financial relationships that could be construed as a potential conflict of interest.

Nomenclature

RPFC	rotary power flow controller	ER	energy router
ADN	active distribution network	SOP	soft open point
ESS	Energy storage system	ISOP	improved particle swarm optimization
SOCP	second-order cone programming	F^{grid}	system's electricity purchasing cost
F_{RPFC}	the investment cost of RPFC	F_{ER}	the investment cost of ER
F_{CE}	construction cost	$P^{RPFC}_{i,t}$	the active power flowing through node i in time period t
$P^{RPFC}_{j,t}$	the active power flowing through node j in time period t	$Q^{RPFC}_{i,t}$	the transmission loss of t time period flowing through node i
$Q^{RPFC}_{j,t}$	The transmission loss of t time period flowing through node j	$P^{RPFC}_{loss,t}$	transmission loss
$P^{RPFC}_{ij,t}$	branch ij transmits active power	$Q^{RPFC}_{ij,t}$	branch ij transmits reactive power
S^{RPFC}_{ij}	installation capacity of branch ij	P^{RPFC}_{min}	RPFC minium active power control
P^{RPFC}_{max}	RPFC maximum active power control	Q^{RPFC}_{min}	RPFC minium active power control
Q^{RPFC}_{max}	RPFC maximum reactive power control	z^{RPFC}	RPFC access node
z_{ij}	RPFC allows access to the line	$P^{ER}_{AC,i,t}$, $Q^{ER}_{AC,i,t}$	AC active power and reactive power of energy router access node i in t period
$P^{ER}_{ESS,t}$, $Q^{ER}_{ESS,t}$	the active and reactive power exchange power of energy storage inside the energy router	$P^{ER}_{DG,t}$, $Q^{ER}_{DG,t}$	the active and reactive power exchange power of photovoltaic inside the energy router
$P^{ER}_{V2G,t}$, $Q^{ER}_{DG,t}$	the active and reactive power exchange of V2G in the energy router	$P^{ER}_{loss,t}$	internal network loss of energy router
S^{ER}_{ESS}	the capacity of ESS	S^{ER}_{V2G}	the capacity of V2G
S^{ER}_{DG}	the capacity of DG	S^{ER}_{i}	the capacity of ER
$P^{ER}_{AC,min}$, $Q^{ER}_{AC,min}$	ER minimum active power regulation and minimum reactive power regulation range	$P^{ER}_{AC,max}$, $Q^{ER}_{AC,max}$	ER maximum active power regulation and maximum reactive power regulation range
z^{ER}	ER access node	z_i	distribution network allows access nodes

F_{buy}	main network electricity purchase cost	$F_{\text{O-RPFC}}$	operation and maintenance cost of RPFC
$F_{\text{O-ER}}$	the operation and maintenance cost of ER	F_{cur}	abandonment penalty cost
F_{loss}	cost of depreciation	W_u	typical day scene set
D_{u}	probability of the uth typical day scenario	N	distribution network node set
$S_{\mathrm{buy},i}$	node i load size	$C_{\mathrm{EC},t}$	Peak-valley electricity price corresponding to t period
c^{PVG}	DG unit light abandonment penalty cost	$P_{u,i,t}^{\mathrm{PVG}}$	abandoned light power
k_1	the operation and maintenance cost coefficients of RPFC.	k_2	the operation and maintenance cost coefficients of ER.
k_3	the operation and maintenance cost coefficients of ESS.	k_4	the operation and maintenance cost coefficients of V2G.
k_5	the operation and maintenance cost coefficients of DG.	$P_{\mathrm{loss},t}^{\mathrm{total}}$	total network loss in period t
$P_{loss,t}^{z}$	T period line operation loss	$P_{loss,t}^{\mathrm{RPFC}}$	RPFC operating loss
$P_{loss,t}^{\mathrm{ER}}$	ER operating loss	I_{ij}^{t}	line ij current in period t
r_{ij}	the resistance of line ij	$P_{ij,t}$, $Q_{ij,t}$	the power of line ij flowing through node ij in time period t
x_{ij}	reactance of line ij	W_{N}	branch set
$U_{i,t}$, $U_{j,t}$	nodes i and j voltage at time t	$U_{i.\min}$, $U_{i.\max}$	the lower and upper voltage limits of node i
$I_{ij.\max}$	maximum line current	$E_{\mathrm{ESS},t}^{\mathrm{ER}}$, $E_{\mathrm{ESS},t+1}^{\mathrm{ER}}$	ESS capacity in t and $t+1$ periods.
h_{ESSc}	ESS charging efficiency	h_{ESSd}	ESS discharge efficiency
$P_{\mathrm{ESSc},t}^{\mathrm{ER}}$	charging power	$P_{\mathrm{ESSd},t}^{\mathrm{ER}}$	discharge power
$\varepsilon_{\mathrm{ESS}}$	ESS self-discharge rate	$E_{\mathrm{ESS,min}}^{\mathrm{ER}}$, $E_{\mathrm{ESS,max}}^{\mathrm{ER}}$	ESS minimum capacity and maximum capacity
$P_{\mathrm{ESSc,min}}^{\mathrm{ER}}$, $P_{\mathrm{ESSc,max}}^{\mathrm{ER}}$	ESS minimum charging power and maximum charging power	$P_{\mathrm{ESSd,min}}^{\mathrm{ER}}$, $P_{\mathrm{ESSd,max}}^{\mathrm{ER}}$	ESS minimum discharge power and maximum discharge power
δ_{ESSc}, δ_{ESSd}	ESS discharge and charge state	$E_{\mathrm{V2G},t}^{\mathrm{ER}}$, $E_{\mathrm{V2G},t+1}^{\mathrm{ER}}$	V2G capacity in t and $t+1$ periods.
h_{V2Gc}	V2G charging efficiency	h_{V2Gd}	V2G charging efficiency
$P_{\mathrm{V2Gc},t}^{\mathrm{ER}}$	charging power	$P_{\mathrm{V2Gd},t}^{\mathrm{ER}}$	discharge power
$\varepsilon_{\mathrm{V2G}}$	V2G self-discharge rate	$E_{\mathrm{V2G,min}}^{\mathrm{ER}}$, $E_{\mathrm{V2G,max}}^{\mathrm{ER}}$	V2G minimum capacity and maximum capacity
$P_{\mathrm{V2Gc,min}}^{\mathrm{ER}}$, $P_{\mathrm{V2Gc,max}}^{\mathrm{ER}}$	V2G minimum charging power and maximum charging power	$P_{\mathrm{V2Gd,min}}^{\mathrm{ER}}$, $P_{\mathrm{V2Gd,max}}^{\mathrm{ER}}$	V2G minimum discharge power and maximum discharge power
δ_{V2Gc}, δ_{V2Gd}	V2G charging and discharge	v_{i+1}	the (i + 1) th particle velocity vector
ω	the i-th particle inertia weight	c_i	the i-th particle acceleration factor
r_i	the ith particle interval [0, 1] random number	v_i	the i-th particle velocity vector
$pbest_i$	the local optimal position of the ith particle individual	x_i	the i th particle position vector
$gbest_i$	the global optimal position of the ith particle individual	$w_{\min}$, $w_{\max}$	the minimum and maximum values of inertia weight
c_1, c_2, c_3	the current value of the learning factor	$c_{\min}$, $c_{\max}$	The minimum and maximum values of the learning factor
$i_{\max}$	maximum number of iterations	$Ibest_i$	the historical optimal solution of particle i
PV	photovoltaic		

References

1. Omri, M.; Jooshaki, M.; Abbaspour, A.; Fotuhi-Firuzabad, M. Modeling Microgrids for Analytical Distribution System Reliability Evaluation. *IEEE Trans. Power Syst.* **2024**, *39*, 6319–6331. [CrossRef]
2. Zhu, W.; Pan, B.; Liu, Q.; Wang, Y.; Shi, D. An accounting method of REDD reduction of renewable energy based on power flow distribution matrix. *Int. J. Environ. Technol. Manag.* **2024**, *27*, 232–241. [CrossRef]
3. Yan, X.; Peng, W.; Shao, C.; Jia, J.; Li, T. User-Side Voltage Regulation Method Based on Rotary Power Flow Controller. *Diangong Jishu Xuebao/Trans. China Electrotech. Soc.* **2023**, *38*, 70–79. [CrossRef]
4. Nie, Y.; Nasr Esfahani, M.; Hu, Y.; Li, X.; Alkahtani, M. Multi-Port Energy Router in Mobile Energy Storage for Emergency Power Outage in Urban Cities. *Energies* **2024**, *17*, 2927. [CrossRef]

5. Li, Z.; He, J.; Wang, X.; Yip, T.; Luo, G. Active control of power flow in distribution network using flexible tie switches. In Proceedings of the 2014 International Conference on Power System Technology, Chengdu, China, 20–22 October 2014; pp. 1224–1229. [CrossRef]
6. Solat, S.; Aminifar, F.; Safdarian, A.; Shayanfar, H. An expansion planning model for strategic visioning of active distribution network in the presence of local electricity market. *IET Gener. Transm. Distrib.* **2023**, *17*, 5410–5429. [CrossRef]
7. Shao, C.; Yan, X.; Yang, Y.; Aslam, W.; Jia, J.; Li, J. Multiple-Zone Synchronous Voltage Regulation and Loss Reduction Optimization of Distribution Networks Based on a Dual Rotary Phase-Shifting Transformer. *Sustainability* **2024**, *16*, 1029. [CrossRef]
8. Yang, L.; Li, Y.; Zhang, Y.; Xie, Z.; Chen, J.; Qu, Y. Optimal allocation strategy of SOP in flexible interconnected distribution network oriented high proportion renewable energy distribution generation. *Energy Rep.* **2024**, *11*, 6048–6056. [CrossRef]
9. Chen, Y.; Yang, G.; Song, Z.; Sun, M.; Zhou, S. Optimal configuration method of soft open point considering flexibility of distribution system. In Proceedings of the 2022 IEEE 5th International Conference on Automation, Electronics and Electrical Engineering (AUTEEE), Shenyang, China, 18–20 November 2022; pp. 524–529. [CrossRef]
10. Bai, H. An optimization method for operational efficiency of ADN with four-port SSOP. *Energy Rep.* **2022**, *8*, 472–479. [CrossRef]
11. Yan, X.; Wang, Q.; Bu, J. High Penetration PV Active Distribution Network Power Flow Optimization and Loss Reduction Based on Flexible Interconnection Technology. *SSRN* 2023, *preprint*. [CrossRef]
12. Xie, H.; Wang, W.; Wang, W.; Tian, L. Optimal Dispatching Strategy of Active Distribution Network for Promoting Local Consumption of Renewable Energy. *Front. Energy Res.* **2022**, *10*, 826141. [CrossRef]
13. Xia, S.; Wang, Z.; Gao, X.; Li, W. Optimal planning of mobile energy storage in active distribution network. *IET Smart Grid* **2024**, *7*, 1–12. [CrossRef]
14. Liu, X. Energy station and distribution network collaborative planning of integrated energy system based on operation optimization and demand response. *Int. J. Energy Res.* **2020**, *44*, 4888–4909. [CrossRef]
15. Kumar, N.; Kumar, T.; Nema, S.; Thakur, T. A multiobjective planning framework for EV charging stations assisted by solar photovoltaic and battery energy storage system in coupled power and transportation network. *Int. J. Energy Res.* **2022**, *46*, 4462–4493. [CrossRef]
16. Wang, X.; Guo, Q.; Tu, C.; Li, J.; Xiao, F.; Wan, D. A two-stage optimal strategy for flexible interconnection distribution network considering the loss characteristic of key equipment. *Int. J. Electr. Power Energy Syst.* **2023**, *152*, 109232. [CrossRef]
17. Rediske, G.; Siluk JC, M.; Gastaldo, N.G.; Rigo, P.D.; Rosa, C.B. Determinant factors in site selection for photovoltaic projects: A systematic review. *Int. J. Energy Res.* **2019**, *43*, 1689–1701. [CrossRef]
18. Jia, Y.; Li, Q.; Liao, X.; Liu, L.; Wu, J. Research on the Access Planning of SOP and ESS in Distribution Network Based on SOCP-SSGA. *Processes* **2023**, *11*, 1844. [CrossRef]
19. Wang, J.; Wang, W.; Wang, H.; Wang, S.; Du, J. Two-stage coordinated planning of DG, SOP and ESS in an active distribution network considering violation risk. *Dianli Xitong Baohu Yu Kongzhi/Power Syst. Prot. Control* **2022**, *50*, 71–82. [CrossRef]
20. Wang, Y.; Wang, X.; Li, S.; Ma, X.; Chen, Y.; Liu, S. Optimization model for harmonic mitigation based on PV-ESS collaboration in small distribution systems. *Appl. Energy* **2024**, *356*, 122410. [CrossRef]
21. Lee, J.-G.; Jung, S.; Choi, J.-H.; Kim, Y.-H.; Yoon, Y.-B. A study on Energy Storage System (ESS) application for dynamic stability improvement and generation constraint reduction. *Trans. Korean Inst. Electr. Eng.* **2017**, *66*, 1554–1560. [CrossRef]

energies

MDPI

Article

Dynamic Modeling of Distribution Power Systems with Renewable Generation for Stability Analysis

Darko Madjovski, Ivan Dumancic and Carolina Tranchita *

Computer Science & Engineering, Frankfurt University of Applied Sciences, 60318 Frankfurt am Main, Germany; darko.madzovski@hotmail.com (D.M.); dumancic@stud.fra-uas.de (I.D.)
* Correspondence: tranchita@fb2.fra-uas.de

Abstract: This paper presents a comprehensive study on the dynamic modeling of distribution power systems with a focus on the integration of renewable energy sources (RESs) for stability analysis. Our research delves into the static and dynamic behavior of distribution systems, emphasizing the need for enhanced load modeling to mitigate planning and operational uncertainties. Using MATLAB/Simulink®, we simulate four distinct study cases characterized by varying load types and levels of distributed generation (DG), particularly solar PV, under both balanced and unbalanced conditions. Our findings highlight the critical role of DG in influencing voltage stability, revealing that deviations in voltage and current during grid imbalances remain within acceptable limits. The study underscores the importance of DG-based inverters in maintaining grid stability through reactive power support and sets the stage for future research on microgrid simulations and battery storage integration to further enhance system stability and performance.

Keywords: load modeling; renewable energy; distribution power system; static load model; dynamic load model; composite load model; stability analysis; distributed generation

1. Introduction

The European Green Deal and the "FIT for 55" package have established a legal framework that requires EU member states to meet stringent climate goals, primarily focusing on achieving carbon neutrality [1]. Two critical strategies for achieving these goals involve the electrification of heating and transport, as well as increasing the production of electricity from renewable energy sources (RESs). Over the past two decades, Europe has seen significant integration of RES into the power grid, but even more profound changes are anticipated in the coming years [2]. The growing adoption of Electric Vehicles (EVs) and new, highly efficient heating and air conditioning technologies will further reshape energy demand [3,4].

These transformations place significant pressure on distribution power systems, which must evolve to handle the dynamic conditions associated with variable generation and shifting loads [5]. Ensuring grid stability and security in this context will require increased flexibility in power system operations. To prevent congestion and ensure reliable power delivery, distribution systems must accommodate flexible solutions like prosumers (consumers who also produce energy), energy storage systems (ESSs), demand response (DR), and electric vehicles (EVs), which can contribute to balancing supply and demand.

The integration of these flexible energy resources, alongside a growing share of RESs, underscores the importance of developing a deep understanding of the static and dynamic behavior of distribution networks. Addressing the technical, economic, and environmental challenges posed by these new scenarios will be key to ensuring the long-term success of the EU's climate initiatives.

Load modeling is essential for both the planning and operation of power systems because it allows for a detailed understanding of how different types of loads behave under various system conditions. Accurate load models improve the prediction of power system performance in both steady-state and transient conditions [6–8]. This helps system operators and planners to anticipate how the power grid will respond to changes in load demand, which is increasingly important with the integration of renewable energy sources (RESs) and new, flexible power technologies like EVs and ESSs [9]. Accurate load modeling reduces planning and operational uncertainties, contributing to better system reliability and cost-efficiency. In distribution grids, static load models typically capture how loads respond to voltage changes, whereas at the transmission level, dynamic and composite load models are used to represent more complex behaviors, such as time-dependent variations in load response. These models are critical for understanding system performance under both steady-state and disturbed conditions [10]. Historically, transmission system operators (TSOs) modeled distribution grids as passive loads without dynamic responses for stability analysis [11,12]. The integration of RESs of flexible power directly into the distribution grid means that the usual approach of load modeling that was performed in the past by the TSOs has to be adapted in a certain way [10].

The main objective of this paper is to implement a proper dynamic modeling of the distribution power system with different percentages of generation based on renewable energy in order to perform effective stability analysis. Additionally, we conduct electromagnetic transient (EMT) simulations to capture the detailed dynamic behavior of the system under various operational conditions. The modeling includes a comprehensive dynamic load representation to ensure more accurate stability assessments. The paper is organized as follows: Section 2 illustrates the state of the art regarding distribution systems and their modeling, Section 3 describes the proposed modeling of the distribution grid for stability analysis, Section 4 presents four study cases that were simulated, and Section 5 gives the overall conclusion of this paper.

2. State of the Art of Distribution Power Systems

2.1. Latest Issues and Trends

The integration of RESs is driving the transition towards low-carbon energy systems. Current trends indicate that high levels of RES integration into distribution grids introduce numerous operational challenges. The reliability of distribution grids with high levels of RESs becomes critical to the overall operation and stability of the power grid.

Traditionally, synchronous generators (SGs), connected to transmission systems, provide significant system inertia, which determines the rate of change in system frequency after a disturbance and ensures stability in a significant portion of the power system. It is assumed that each SG connected to the grid contributes to maintaining the stability and security of the large power system [13]. However, RESs are integrated into the AC grid through power electronics. Inverter-based RESs are progressively replacing conventional SGs, decreasing the overall system inertia and the fault tolerance capability [14]. As more conventional SGs are decommissioned and DERs are integrated on a large scale, power systems may become increasingly vulnerable to sudden disturbances [15].

The integration of inverter-based RESs primarily at the distribution grid level presents several challenges for distribution system operators (DSOs). One major issue is voltage regulation, as the intermittent and variable nature of solar and wind energy causes fluctuations, making it difficult to maintain voltage levels within operational limits. In addition, ensuring the effectiveness of protection systems becomes more complex. Unlike traditional SGs, inverter-based RESs generate much lower fault currents, which can prevent protection devices from accurately detecting faults, leading to potential malfunctions. Furthermore, directional protection schemes that rely on negative-sequence components often fail due to altered phase angle characteristics in systems dominated by inverter-based RESs [16]. DSOs must ensure operational flexibility and adopt new planning strategies to accommodate RES intermittency, which involves investments in smart grid technologies and real-time monitoring systems.

These challenges underline the need for further analysis at the distribution grid level. The increased variability and dynamic behavior introduced by high levels of RES integration require more sophisticated modeling techniques to fully grasp the complexities involved. By deepening analysis at this level, DSOs will be better equipped to plan and operate their systems effectively, ensuring grid reliability.

Research in this field emphasizes the importance of load modeling to capture the complex behavior of loads and their impact on power system operations, ensuring reliable simulation results. It further highlights the need to incorporate modeling techniques that simulate the dynamic behavior of distribution grids, especially with a high penetration of RESs [10,11,17]. A thorough analysis of distribution grids requires applying appropriate generation and load modeling techniques [18,19]. In past times, when the system stability analysis was being assessed only by TSOs, the distribution networks were modeled as static loads without any dynamic response. However, with the increasing integration of decentralized RESs and new loads, there is a pressing need for new modeling approaches that treat distribution grids as active systems [11].

It is widely acknowledged that load modeling presents inherent complexities. The demand for electricity arises from various devices connected to the grid and is primarily influenced by human behavior. Due to this uncertainty, achieving the perfect modeling of this behavior is challenging. New loads such as EVs, the widespread use of power electronics, and the electrification of economies further complicate the process. Load modeling is never entirely accurate, as certain uncertainties always exist. It should be noted that the advanced modeling and forecasting of energy production from variable renewables has to some extent allowed, e.g., in Germany, a massive integration of RESs. In this way, a more accurate modeling of the load will help to deal with the new challenges on the path to climate neutrality. As the future operation of power systems will be based on conventional SGs, inverted-based resources, as well as on ESs, DR, and EVs, techniques to guarantee the stability and security of the system must be re-examined or even developed.

Numerous research efforts are underway to represent the behavior of renewable energy resources as well as inverter-based sources for stability studies [18,20].

Grid codes are evolving, requiring inverter-based generators to support power systems by equipping them with features such as fault ride through (FRT) and voltage and frequency support during disturbances.

2.2. Simulation of Distribution Power Systems

The increasing complexity of modern distribution power systems, driven by the widespread integration of DGs and renewable energy sources, presents significant challenges for accurate modeling and simulation. The interaction between grid infrastructure, varying generation sources, and loads necessitates the use of more advanced simulation methods that go beyond traditional tools.

To address these challenges, simulation trends are moving towards multi-processor systems optimized for real-time analysis, such as real-time simulators (RTSs). These simulators allow for the real-time progression of modeled systems, providing accurate time responses that closely mimic the behavior of actual distribution networks. By running simulations in real time, system responses to various scenarios can be analyzed in a practical context, making RTS a powerful tool for system validation. This capability is particularly useful when investigating scenarios that would be difficult or risky to replicate physically, as real-time simulations offer a secure and cost-effective alternative [21,22].

In this context, electromagnetic transient (EMT) simulations have become important for capturing the fast dynamics and transient behaviors of distribution systems, particularly those with a high penetration of inverter-based renewable energy technologies. EMT simulations provide finer temporal resolution compared to traditional root mean square (RMS) models, making them ideal for modeling switching events, harmonics, and rapid system responses. However, the computational intensity of EMT simulations, especially in large-scale distribution systems, can limit their feasibility for full-system studies.

To mitigate the limitations of EMT and RMS simulations, hybrid co-simulation models that combine both are being developed. These models integrate EMT simulations with root mean square (RMS) models, balancing the need for high-detail modeling in specific areas of the network, such as those with inverter-based resources, with the computational efficiency of RMS simulations for larger, slower dynamics. This co-simulation approach allows for the detailed modeling of the critical parts of the grid while maintaining simulation feasibility across the broader system.

Additionally, co-simulation techniques extend beyond hybrid EMT/RMS. They integrate multiple simulation environments, enabling the coupling of information and communication technology (ICT) with traditional power systems software, offering a more comprehensive view of system behavior. Multi-agent systems complement co-simulation by enabling decentralized modeling approaches well-suited to distributed energy resources and smart grid technologies, simulating multiple independent actors interacting within a shared grid environment.

2.3. Measurements in Distribution Power Systems

Another significant challenge in modern distribution power systems is the lack of synchronized measurements to perform state estimation and conduct further analysis on stability and security. Traditional distribution systems have few measurement points, and to estimate the true state of the system, it is necessary to calculate a high volume of pseudo-measurements, which have larger errors than real-time telemetered measurements [23]. To address instability concerns in distribution grids and avoid undesirable system states, the inclusion of synchrophasor measurements and new control strategies is necessary. Synchrophasor measurement units, also known as phasor measurement units (PMUs), provide a resilient tool for observing the dynamic behavior and performance of the power system. The technology has evolved significantly in the last decade and is now used on a global scale [14]. PMUs offer synchronized measurements of positive-sequence voltage and current to within a microsecond, as well as negative- and zero-sequence quantities. Additionally, PMUs measure local frequency, its rate of change, and individual phase voltages and currents. PMUs have a wide range of applications in electric power systems, and their use will continue to expand in the years ahead [24].

3. Stability Analysis

Power system stability is defined as the ability of a power system to remain in a state of equilibrium under normal operating conditions and to regain a state of equilibrium after being subjected to a disturbance [25,26]. Historically, stability analysis has predominantly focused on transmission networks, where rotor angle stability—the synchronism of large-scale synchronous generators—was the primary concern. This form of stability is influenced by the dynamics of rotor angles and power–angle relationships in transmission systems. However, the increasing integration of DG, particularly from RESs such as photovoltaic (PV) systems in distribution systems, has shifted the stability landscape. The power dynamics in such systems, which were traditionally considered passive, now play a critical role in overall system stability. As RES-based generators replace conventional synchronous generators, the need to study stability in distribution networks has grown.

In modern distribution networks with a high penetration of inverter-based RESs, stability challenges can differ significantly [27,28]. Voltage stability is a primary concern due to the fluctuating nature of renewable energy generation and the increased deployment of inverters. Voltage stability is defined as the ability of a power system to maintain voltages into the acceptable range at all the buses in the power system under normal operating conditions and after the occurrence of a disturbance [25,29,30]. Voltage instability arises when the system fails to sustain stable voltage levels, typically due to reactive power shortages or rapid fluctuations in generation and demand. Unlike traditional rotor angle instability, this type of instability can manifest while the system remains synchronized, highlighting the unique challenges faced by distribution networks with significant renewable energy integration. Therefore, addressing voltage stability in distribution grids is important to ensuring reliable operation, particularly as renewable energy sources continue to proliferate.

Inverter-based DERs such as PV systems do not inherently provide frequency support because they lack the rotating mass that traditional synchronous generators possess. This rotating mass in conventional generators offers inertial support, which helps maintain grid stability by naturally resisting frequency changes following disturbances. In contrast, inverters used in DERs typically operate with power electronics that disconnect them from the physical inertia of the grid. However, through advanced control strategies for inertia emulation, inverters can be programmed to provide frequency support by adjusting their power output in response to changes in grid frequency [31].

The focus of this paper is on voltage and frequency stability in distribution networks, driven by the increasing penetration of RESs. The inverters modeled in this study are grid-following inverters in distribution systems, lacking advanced control mechanisms dedicated to dynamic voltage or frequency support, reflecting the typical inverters currently being integrated into such low-voltage systems.

Reactive Power Demand

Reactive power in the power grid causes additional undesirable losses and can also lead to further system stability problems. Numerous studies suggest that local reactive power compensation has proven to be highly effective. Reactive power compensation has traditionally been performed with the help of synchronous generators, capacitor banks, and, more recently, flexible alternating current transmission system (FACTS) devices. Improving reactive power management is crucial for the reliable and efficient operation of the power grid. Since controlling reactive power via inverters is possible, it has become a favorable method, as no additional investment in new devices is required. As a result, the need for PV inverters to provide reactive power to the grid is becoming increasingly important. Recently, this requirement has been mandated for new photovoltaic inverters, often executed with proportional control of the power factor or droop control for local voltage regulation [32]. Due to constraints within the inverter model employed in the MATLAB/Simulink® library for renewables (specialized power systems), the provision and control of reactive power are currently beyond the scope of this paper. However, this topic is of particular interest for possible future studies.

4. Proposed Modeling for Dynamic Studies

4.1. Distribution Power System

The distribution power system is the last stage of the electrical grid. It distributes the electric power to the end-users. Small industrial, residential, and commercial customers are supplied by feeders in the distribution power system. It consists of lines, transformers, poles, and protection circuits [6]. Managing the distribution network ensures an acceptable power supply quality for the end-users. The electric power delivered from the distribution network must be within acceptable ranges for voltage, frequency, and current levels to ensure quality service to end-users. Effective planning and operation of the distribution network is essential to maintain these power supply standards. Additionally, to ensure safe operation, the distribution network must be resilient to specific disturbances, minimizing adverse impacts on end-users [20]. Recently, the high-level integration of DGs into the distribution power system has improved local voltage regulation and reduced technical losses in the system. However, it can also introduce negative side effects, such as decreasing grid stability and affecting existing protection schemes. In this context, the dynamic modeling of distribution power systems is needed to evaluate and predict both the positive and negative effects of RES integration into the distribution network.

4.2. Modeling of the Power Grid

The design and development of the equivalent distribution grid in this work were carried out using MATLAB/Simulink® tools. The representation of the outer grid was visualized by the use of the three-phase source block from the Simscape Electrical Specialized Power Systems library of MATLAB/Simulink®. It implements a balanced three-phase voltage source, with a base phase-to-phase voltage Vrms of 400 V and a frequency of 50 Hz. The integration of distributed RESs into a weak power system is a challenge with regard to the voltage control and reactive power compensation. For the present study, in order to properly simulate the weakness of the distribution network for the different scenarios, the internal impedance of the grid is adjusted by changing the value of the source internal resistance R. The three voltage sources are wye-connected with a neutral ground.

The whole distribution grid model in the MATLAB/Simulink® software, version R2023b (MATLAB R2023b), is shown in Figure 1.

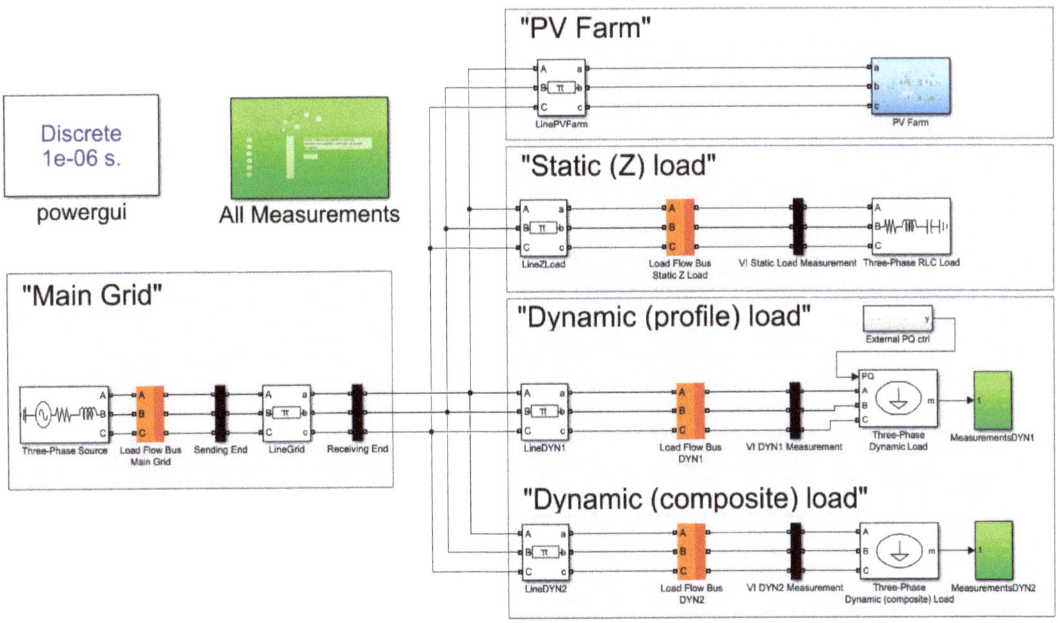

Figure 1. Distribution power grid modeled in MATLAB/Simulink®.

In Figure 1, it can be seen that the three-phase source block is on the left, providing power to the loads in the distribution grid, and that the grid topology is radial. In this paper, one of the objectives is to analyze the behavior of a weak distribution system. In order to represent the unbalanced characteristic of the power distribution system, the value of the parameters of the static load, represented in the three-phase RLC load block, have been accordingly adjusted for this purpose. In this block, the active and reactive power for each phase has been specified, loading phase A with more inductive reactive power than the other two phases, thus making the grid unbalanced.

For the connections between the three-phase source block and the loads, the three-phase PI section line block is used. The expected voltage drop across the grid is therefore modeled by the PI section line parameters. This block implements a balanced three-phase line model with the length of the line, which is 200 m each. The model consists of one set of RL elements connected between the input and output terminals and two sets of shunt capacitances lumped at both ends of the line. This is illustrated in Figure 2.

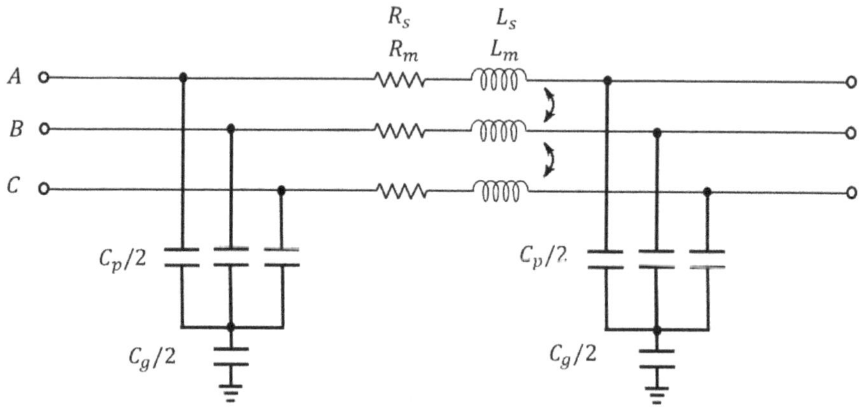

Figure 2. Three-phase PI section line modeled in MATLAB/Simulink®.

In Figure 2, Rs and Rm are the self and mutual resistances, whereas Ls and Lm are the self and mutual inductances of the coupled inductors. Cp and, correspondingly, Cg are the phase capacitances and the ground capacitances, which are deducted from the positive- and zero-sequence RLC parameters.

4.3. Load Modeling

The influence of load modeling on stability analysis is notably significant. A particular disturbance at a specific operating point may render the system stable under one load model, while causing instability under a different load. Multiple challenges make load modeling a difficult task. In real-world scenarios, the load always varies in terms of quantity and variety. Furthermore, under changing voltages and frequencies, the loads behave in a different way. In addition to this, the number of individual devices is huge and the computational requirements for simulating a whole distribution grid presents a hindrance; thus, most of the time, the load is modeled at the level of the transmission system [6]. In this work, composite load models are used to represent the load in a distribution power system. The objective of this modeling is to denote an aggregation of different types of loads that can be found in the residential, commercial, industrial, and agricultural sector at the distribution level and to perform a proper analysis of the effect that these composite loads have on the grid itself.

4.3.1. Static Load Model

Since the dynamics of the load is not of particular interest when modeling a static load, there is no need for the usage of differential equations to the describe the dynamics. Simple algebraic equations are sufficient to represent static load models. There are two main methods for describing how static loads respond to voltage variations: the polynomial static load model and the exponential static load model. The polynomial static load model, known as the ZIP model, represents the load as a mixture of constant impedance (Z), constant current (I), and constant power (P) [8]. It is illustrated with the help of Equation (1):

$$\begin{aligned} P &= P_0\left[p_1\left(\tfrac{V}{V_0}\right)^2 + p_2\left(\tfrac{V}{V_0}\right)^1 + p_3\left(\tfrac{V}{V_0}\right)^0\right] = P_0\left[p_1\left(\tfrac{V}{V_0}\right)^2 + p_2\tfrac{V}{V_0} + p_3\right] \\ Q &= Q_0\left[q_1\left(\tfrac{V}{V_0}\right)^2 + q_2\left(\tfrac{V}{V_0}\right)^1 + q_3\left(\tfrac{V}{V_0}\right)^0\right] = Q_0\left[q_1\left(\tfrac{V}{V_0}\right)^2 + q_2\tfrac{V}{V_0} + q_3\right] \end{aligned} \quad (1)$$

where V_0 is the per-unit (pu) bus voltage for the initial condition from the power flow solution, P_0 and Q_0 are the real and reactive pu loads, p_1, p_2, p_3 are the percentages of P_0

assigned as constant impedance, constant current, and constant power, in that order, and q_1, q_2, q_3 are the percentages for Q_0. The conditions in Equation (2) have to be respected:

$$p_1 + p_2 + p_3 = 1.0 \qquad q_1 + q_2 + q_3 = 1.0 \tag{2}$$

The exponential static model expresses the real and reactive components of the load, as shown in Equation (3), where the parameters are the coefficients p and q and the exponents α and β.

$$P = P_0\left[p\left(\frac{V}{V_0}\right)^{\alpha}\right] Q = Q_0\left[q\left(\frac{V}{V_0}\right)^{\beta}\right] \tag{3}$$

In this paper, the ZIP model is used as the static load model. Specifically, the three-phase series RLC load block from the Simscape Electrical Specialized Power Systems library in MATLAB/Simulink® is employed, connected to the VI static load measurement block, as shown in Figure 1, under the area labeled "Static (Z) Load Block". The load block is wye-connected with a neutral ground. It represents a Z-type load, which means that at the given frequency the load has a constant impedance, and the active and reactive power consumed by the load is proportional to the square of the applied voltage. This means that the values of p_2, p_3 and, respectively, q_2, q_3 from Equation (2) are zero. Therefore, Equation (1) can be re-written in the following form:

$$P = P_0\left[p_1\left(\frac{V}{V_0}\right)^{2}\right] Q = Q_0\left[q_1\left(\frac{V}{V_0}\right)^{2}\right] \tag{4}$$

4.3.2. Dynamic Load Model

To represent the aggregation of various load types and their dynamic behavior in the distribution grid, a dynamic load model from the Simscape library of MATLAB/Simulink® was adopted. In Figure 3, the modeling of the dynamic load is illustrated.

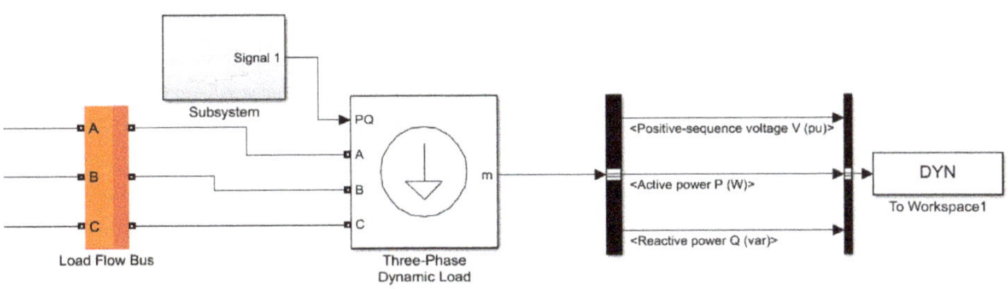

Figure 3. Modeling of the dynamic load in MATLAB/Simulink®.

In the regions labeled "Dynamic Load" in Figure 1, the three-phase dynamic load block from the Simscape Electrical Specialized Systems library is used. It implements a three-phase, three-wire dynamic load, connected to the load flow bus, as shown in Figure 3. The voltage on the dynamic load block is positive-sequence, which means that only the positive sequence is considered for the output calculation of the block. However, the output calculation of the block has an impact on the negative- and zero-sequence quantities of the whole simulated system, although they are not available for analysis. This document primarily focuses on voltage analysis, with specific reference to the insights provided by source [6]. According to the authors, the study of negative- and zero-sequence components of voltages and currents is considered less relevant in the context of stability investigations. Consequently, any potential limitations related to the negative- and zero-sequence components within the MATLAB/Simulink® model are of lesser concern for the purpose of this research. In Figure 4, the block subsystem that is connected to the PQ signal input of the dynamic load block is shown.

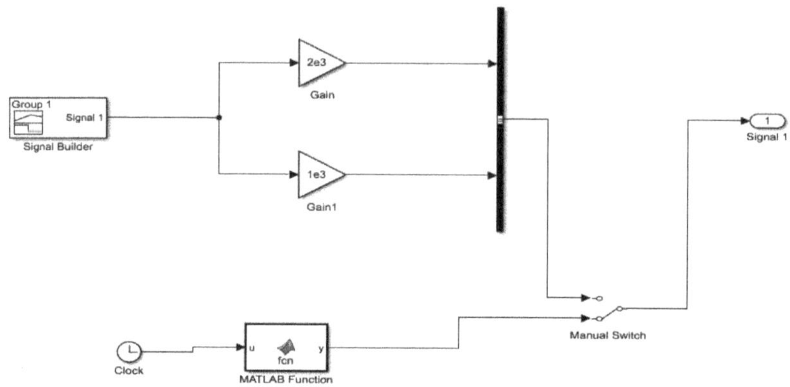

Figure 4. Subsystem for controlling the active power P and reactive power Q in MATLAB/Simulink®.

The active power P and the reactive power Q are controlled externally with the desired load profile with the help of this system, shown in Figure 4. There are two possible options to create signal 1 that serves this purpose. As shown in Figure 4, it can be seen that this signal can either be generated through a MATLAB function that contains the code for the desired load profile or using the signal builder, where the signal is generated and then amplified using the gain blocks. The desired option can be selected by changing the position of the manual switch.

Another dynamic block is also used in the modeled network. In this block, the active power P and the reactive power Q are not controlled externally. Instead, when the terminal voltage exceeds the minimum value V_{min}, the active power P and reactive power Q of the load vary according to Equation (5):

$$P(s) = P_0\left(\frac{V}{V_0}\right)^{n_p}\frac{1+T_{p1}s}{1+T_{p2}s} \tag{5}$$

where V_0 is the initial positive-sequence voltage, P_0 and Q_0 are the initial active and reactive powers at the initial voltage V_0, and V is the positive-sequence voltage. T_{p1} and T_{p2} are time constants controlling the dynamics of the active power P, and T_{q1} and T_{q2} are time constants controlling the dynamics of the reactive power Q. The active and reactive components n_p and n_q are exponents controlling the nature of the load, meaning that adjusting these values enables the modeling of different load behaviors. This is illustrated in section composite load models, and the values of the exponents n_p and n_q can be found in Table 1 below.

Table 1. The values of n_p and n_q for different load models [33].

Constant Impedance	Constant Current	Constant Power	Industrial	Commercial	Residential	Load Type
2	1	0	0.18	1.51	0.92	n_p
2	1	0	6	3.4	4.04	n_q

Using this setup, however, means that it is not possible to set a desired load profile. Although the name of the block is the dynamic load block, it may not constantly behave as a dynamic load. If the terminal voltage V of the load is lower than a specified value V_{min}, then the load impedance is kept constant; therefore, in the time interval when this specific condition is active, the load behaves like a static Z load.

4.3.3. Composite Load Model

In practical application and as commonly observed in the literature, the modeling of distribution grids often entails the representation of loads as constant current, constant power, or constant impedance, primarily due to data limitations. Because of the varying behavior of loads nowadays, it is evident that outcomes derived from these conventional models may not faithfully mirror the actual picture. More accurate results can be delivered when modeling with the help of the composite load model [33,34]. It combines the various kinds of loads that can be found in the power grid. It can be a combination of static loads and different types of dynamic loads such as three-phase induction motors, single-phase induction motors, power electronic loads, and even distributed generation [25]. In this work, by varying the exponents n_P and n_q from Equation (1) that are implemented in dynamic load block 2 used in MATLAB/Simulink®, composite load models are created. In Table 1, various kinds of loads that can be found in a distribution grid are shown: the industrial, commercial, residential, constant power, constant current and constant impedance load types in correlation with the values of n_P and n_q.

4.4. Modeling of the Distributed Generation

The high-level integration of DERs in the existing distribution grid may lead to voltage fluctuations and overloads of different elements within the grid. The power factor may also be seriously affected, and additional unbalanced voltage conditions may occur as well [35]. The results in this work also point to the importance of the power factor problem. Consequently, to comprehensively assess the influence of DGs on the distribution grid, it is imperative to undertake a proper and accurate modeling of DG systems. In this work, photovoltaics (PVs) are integrated into the modeled distribution network. This is achieved by using the subsystem grid-connected PV array. In Figure 5, an illustrated view of the subsystem is given.

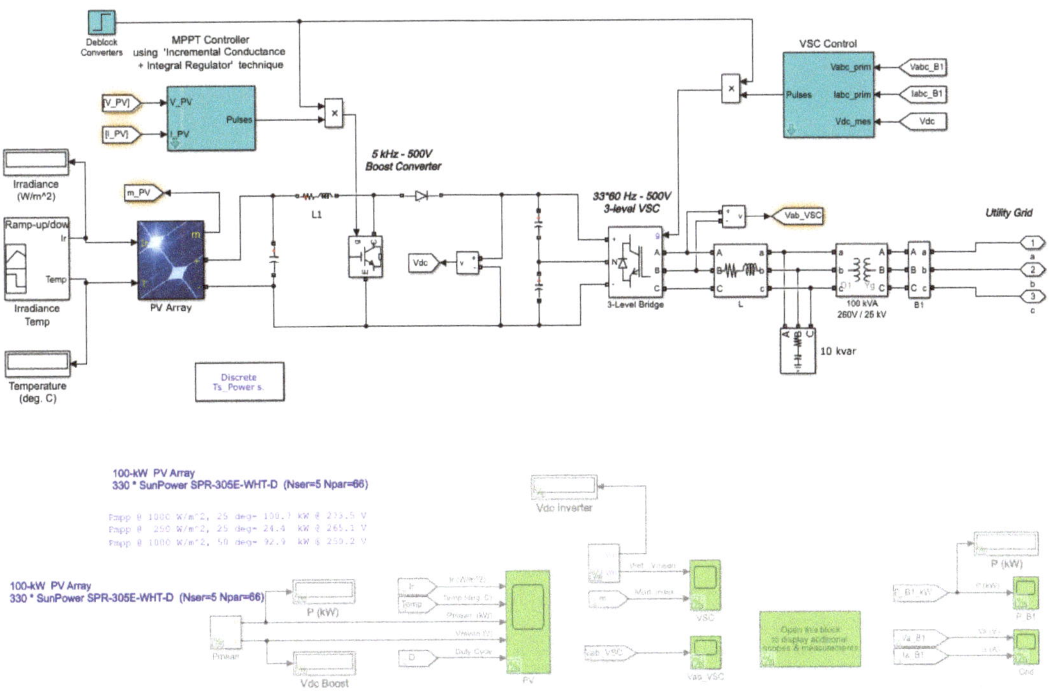

Figure 5. Grid-connected PV array in MATLAB/Simulink®.

In our implementation, we use the inverter model employed for the low-voltage PV farm simulation in the MATLAB/Simulink® library for renewables (specialized power systems). This inverter operates as a grid-following device, where the control strategy is based on measuring the AC-side voltages and currents, which are transformed into the dq0 reference frame. This allows for the regulation of both the DC bus voltage and the output currents. A phase-locked loop (PLL) is employed to synchronize the inverter to the grid frequency, while a maximum power point tracking (MPPT) algorithm, specifically the perturb and observe (P&O) method, is utilized to optimize power extraction from the PV panels [36,37]. The MPPT adjusts the output voltage periodically and compares the power obtained during each cycle, optimizing the system's performance under varying climatic conditions. However, although the inverter synchronizes with the grid and follows its frequency, it lacks control functionalities that support grid stability, such as transient voltage regulation through reactive power control (e.g., Q–V droop) or frequency stabilization through active power control (e.g., frequency–Watt droop). As a result, it is not expected to contribute to dynamic voltage or frequency regulation in the grid. Despite this, inverters of this type are commonly used at the low-voltage level in many countries. This presents a challenge, as future grid codes—particularly with the growing penetration of renewable energy sources—may require inverters to not only maximize power generation but also actively participate in grid support functionalities.

For study cases 3 and 4 in this paper, the number of PV modules is changed, so different scenarios—with low, average, and high integration of DGs—can be tested, adjusting the load ratio with respect to the nominal load to 50%, 75%, and 100%. This can be adjusted in the block PV array in the subsystem, as shown in Figure 6.

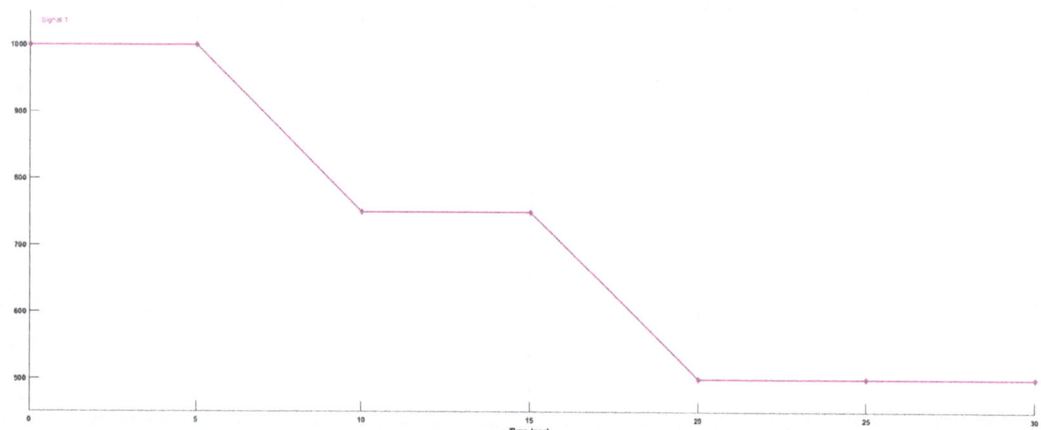

Figure 6. Signal given to the PV array in MATLAB/Simulink®.

The x-axis represents the time duration of 30 s. The y-axis corresponds to the sun irradiance that is given in W/m^2. In real-world applications, the created signal would represent a sudden change in the sun irradiance from 1000 W/m^2 to 750 W/m^2, from the 5th to the 10th s, further decreasing from 750 W/m^2 to 500 W/m^2, from the 15th to the 20th s. These three values of sun irradiance correspond to 100 kW, 75 kW, and 50 kW of generation from the PV farm. The transformer in the subsystem was also adjusted accordingly with respect to the nominal conditions of the modeled grid. The PV farm subsystem offers a multitude of possibilities and encompasses several variables that can be changed, including the choice of the module manufacturer and its specifications, sun irradiance levels, ambient temperature conditions, and converter specifications. This flexibility renders the PV subsystem a highly valuable resource, presenting ample opportunities for in-depth investigations within this domain.

4.5. Study Cases

In this work, four distinct study cases have been formulated, each further subdivided into two main subcases: the balanced and the unbalanced subcase. The introduction of the imbalance in the grid is achieved by increasing the capacitive reactive power of the static RLC load in phase A by 50% with respect to its nominal power in the balanced case.

4.5.1. Study Case 1

In the first study case, the system is modeled using only a static Z load without dynamic load modeling or renewable energy integration. The simulated grid, as shown in Figure 1, includes only the "Main Grid" and the "Static Z load", with all other elements deactivated. The power factor of the grid is 0.9, with a capacitive (leading) nature. The results, illustrated in Figure 7, compare two subcases: balanced and unbalanced scenarios.

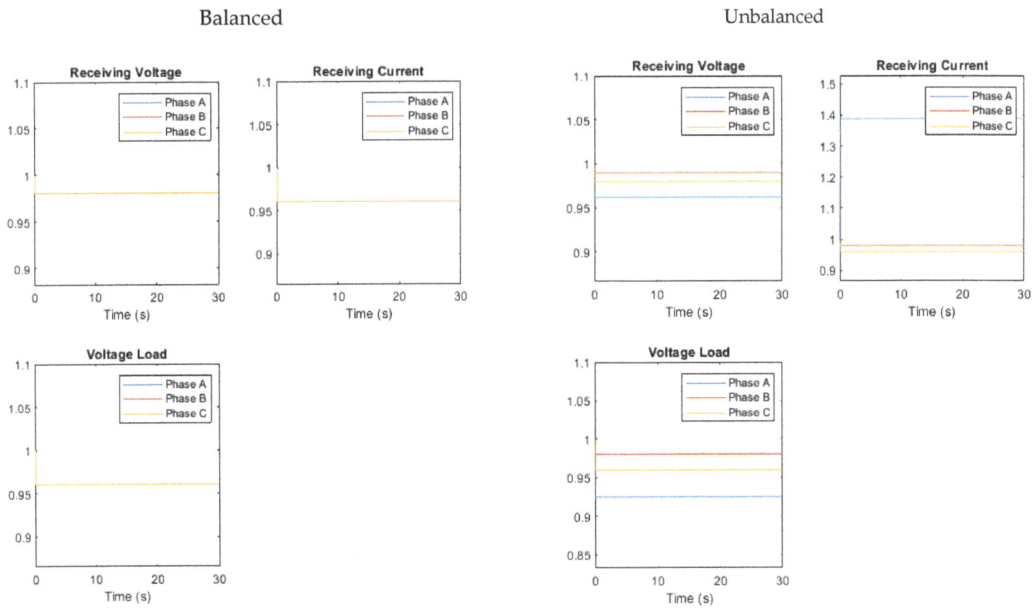

Figure 7. Study case 1.

In the unbalanced subcase, the voltage of the static load bus for phase A decreases to 0.925 pu, while the voltage for phase B slightly increases to 0.98 pu, and phase C remains constant at 0.96 pu. In contrast, in the balanced case, the voltage for all three phases is 0.96 pu.

In the unbalanced subcase, phase A of the static load bus experiences a voltage drop to 0.925 pu, while phase B shows a slight increase to 0.98 pu, and phase C remains constant at 0.96 pu. For the balanced scenario, all phases maintain a voltage of 0.96 pu. A similar behavior is observed at the voltage receiving bus. In the balanced subcase, the voltage across all phases is 0.98 pu. In contrast, in the unbalanced subcase, phase A decreases to 0.97 pu, phase B rises to 0.99 pu, and phase C remains unchanged at 0.98 pu.

Further analysis of the current measured at the receiving bus shows an increase across all three phases, with phase A exhibiting the largest rise to 1.4 pu. This increase is primarily due to the 50% rise in capacitive reactive power in phase A with respect to the nominal power. All the voltages in the three phases A, B, and C differ from each other, which is proof of the unbalanced behavior of the system.

In Figure 8, the active and reactive power measured from the receiving current bus, which flows from/to the grid in both balanced and unbalanced cases, is shown.

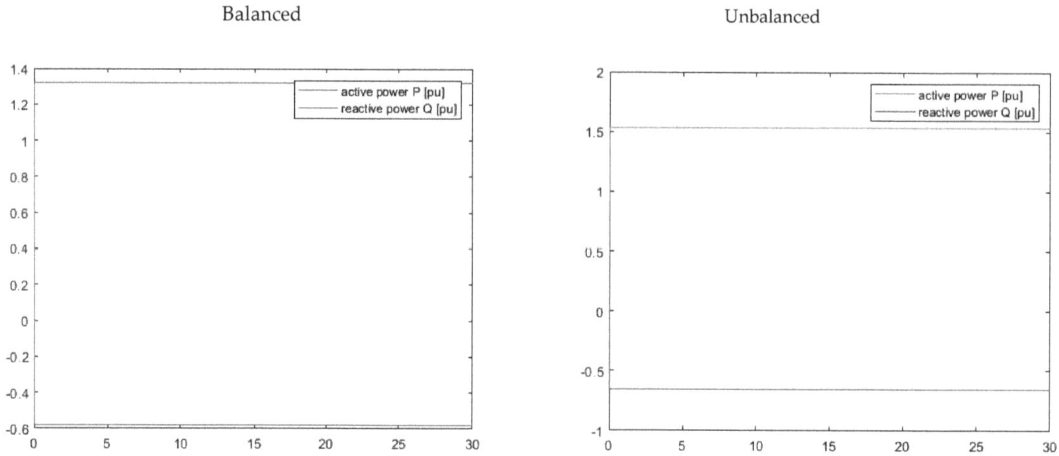

Figure 8. Active and reactive power flow.

Figure 8 presents the active and reactive power flow measured at the receiving bus. In the balanced case, the active power flowing from the grid to the load is around 1.32 pu. In contrast, in the unbalanced case, the active power demand increases slightly to 1.53 pu. The negative sign in the reactive power indicates that the capacitive load is feeding reactive power back to the grid. In the balanced scenario, this reactive power feedback is around 0.58 pu, while in the unbalanced case, due to the increase of capacitive reactive power in phase A, the reactive power flowing back to the grid increases to 0.65 pu.

This first case study highlights the impact of unbalanced conditions on system voltages and currents, as well as the increased demand on active and reactive power in the unbalanced scenario, providing critical insights into the system's response to unbalanced load conditions.

4.5.2. Study Case 2

In study case 2, the grid model incorporates both static Z load and dynamic load modeling, but without integrating renewable energy sources. The components of the grid used in this case, as shown in Figure 1, include the "Main Grid", "Static Z load", and both "Dynamic (profile) load" and "Dynamic (composite) load". All other components are deactivated. This case also explores both balanced and unbalanced conditions in the network, with three further subdivisions to represent different types of composite loads: industrial, residential, and commercial. The total power supplied by the grid is evenly divided, with 50% consumed by the static load and 50% allocated to the dynamic composite load model.

Study Case 2.1

In this particular study case, along with the simulated static Z load, there are also simulated industrial loads that can be commonly encountered in real-world scenarios, by setting the values of n_P to 0.18 and n_q to 6 into the "dynamic (composite) load" part. The results are shown in Figure 9.

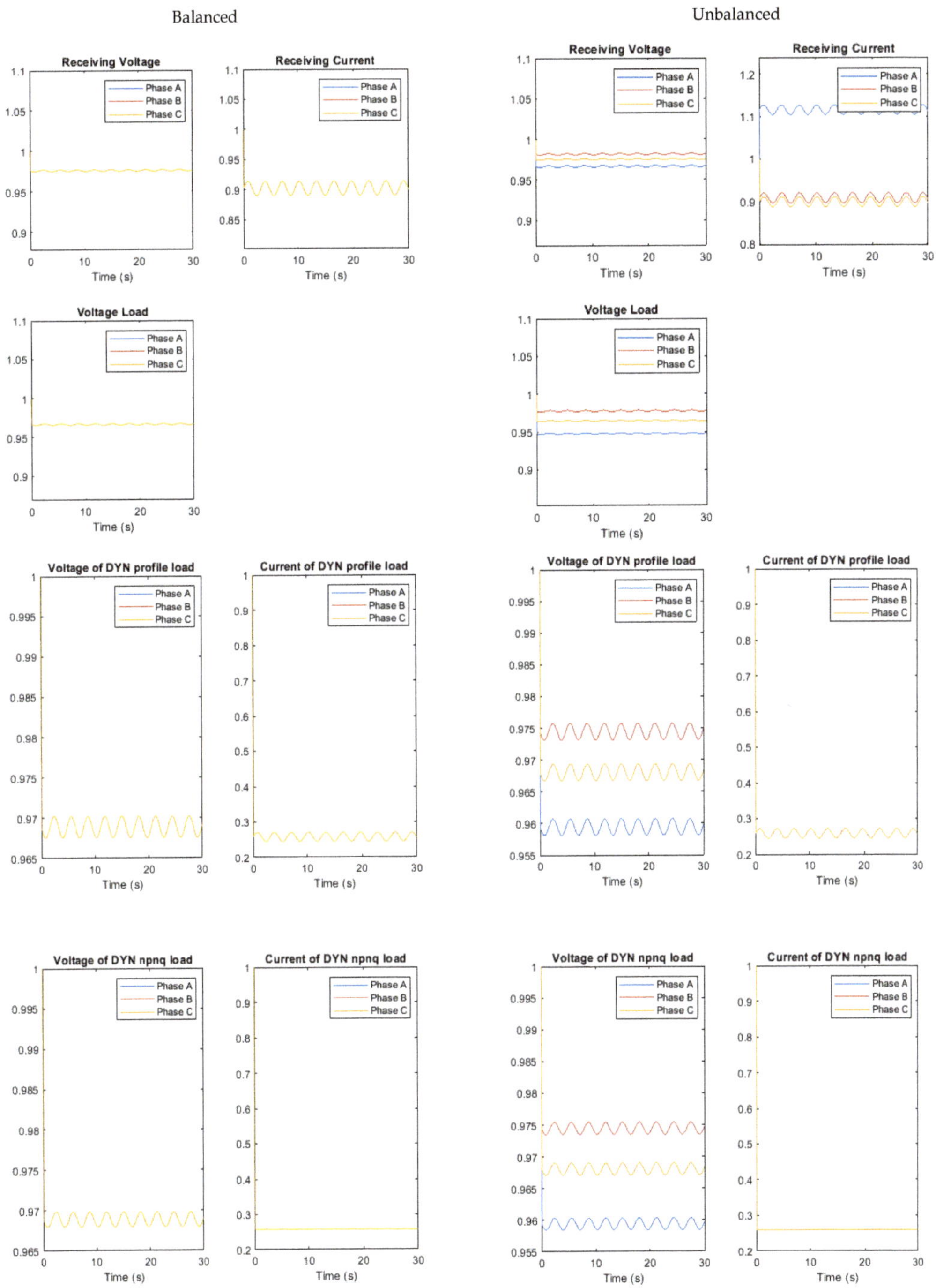

Figure 9. Study case 2 (industrial load type).

As in the first study case, a clear distinction between the balanced and unbalanced scenarios is observed. In the unbalanced subcase, the capacitive reactive power in phase A is increased by 50% relative to the nominal value. This leads to an increase in the receiving current, most significantly in phase A, where the impact of the reactive power increase is most pronounced. Analyzing the voltages at the load side, the unbalanced case results in a voltage drop in phase A, while phase B experiences a slight voltage rise, and phase C remains constant.

With the addition of dynamic and composite load models, the voltages and currents at these buses show behavior similar to the static load buses but with more noticeable variations due to the dynamic nature of the loads. Figure 9 illustrates that the voltages at the buses connected to the dynamic and composite loads exhibit a more pronounced variation in waveforms compared to the static load scenario. However, these voltage fluctuations remain within acceptable limits, indicating that the system remains stable despite the unbalanced conditions.

Interestingly, the currents at the dynamic and composite loads remain symmetrical even though the system is unbalanced. In practical situations, some deviation from this ideal behavior are expected. This situation is due to the limitations of modeling. The inherent characteristics of the "Three-Phase Dynamic Load" block from the Simscape library, as detailed in the Dynamic Load Model chapter, explain this phenomenon. The block does not account for negative- and zero-sequence currents, ensuring that the load currents remain balanced even when the voltage is unbalanced.

This study case underscores the importance of dynamic load modeling in analyzing real-world industrial scenarios, where the load behavior can significantly affect the overall system's voltage and current characteristics under unbalanced conditions.

Study Case 2.2

In this particular study case, the simulated loads can be typically encountered in commercial areas. This is achieved by configuring the specific parameter value of n_p to 1.51 and n_q to 3.4 into the "dynamic (composite) load" part. The results are shown in Figure 10.

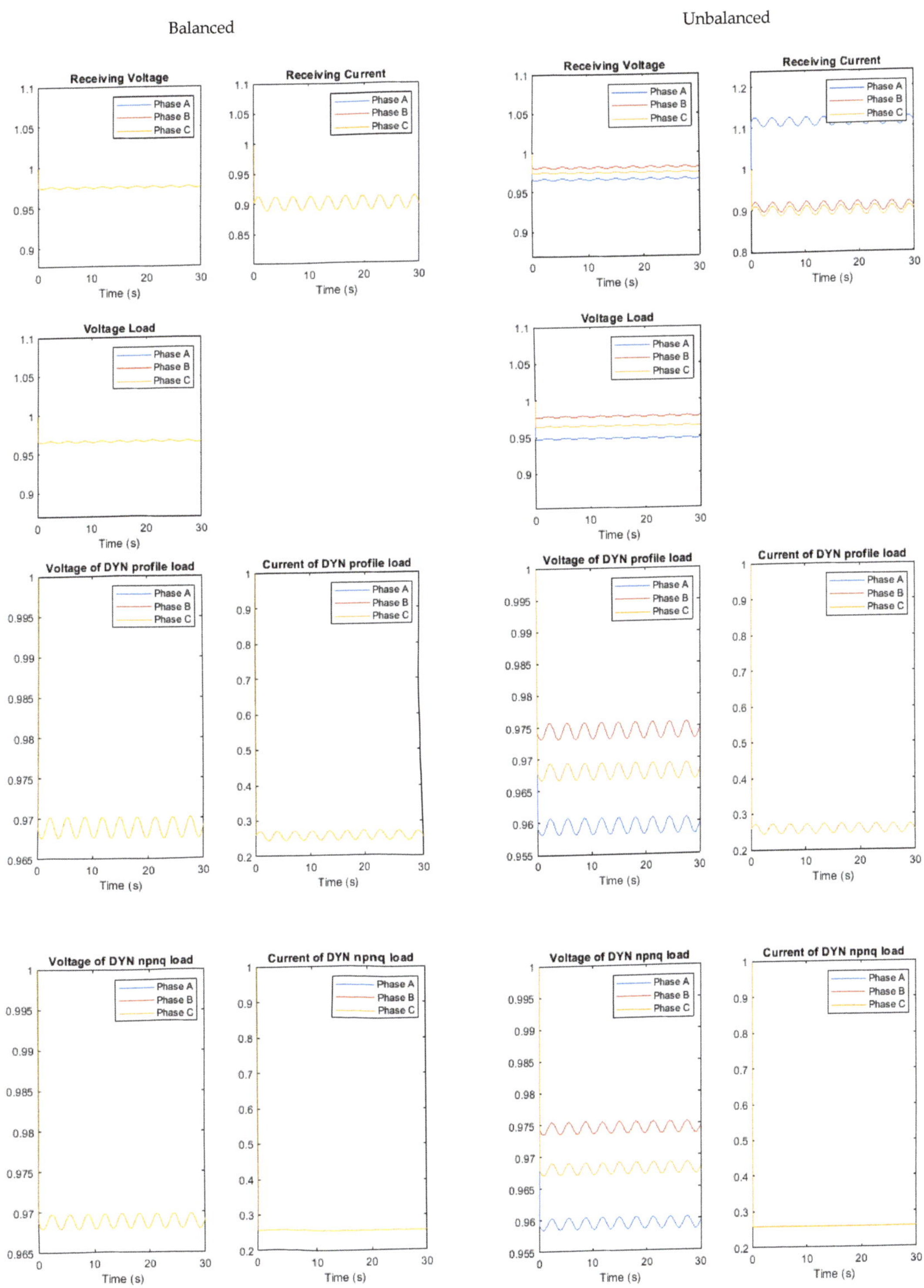

Figure 10. Study case 2 (commercial load type).

The key distinction between study case 2.1 and study case 2.2 lies in the parameter values of n_P and n_q in the three-phase dynamic block 2, which are changed in order to simulate a commercial load type. As the measurements from that block are depicted in Figure 10 under the labels "Voltage DYN npnq Load" and "Current DYN npnq Load", respectively, for the balanced and the unbalanced scenarios, a small difference is barely noticeable. In this case, the voltage and current levels have insignificantly decreased.

Study Case 2.3

In this study case, the loads in residential areas are simulated by setting the values of n_P to 0.92 and n_q to 4.04 into the "dynamic (composite) load" part to reflect the behavior of such loads. The results of this simulation are shown in Figure 11.

In Figure 11, the measured voltage and current, represented as "Voltage of DYN profile Load" and "Current of DYN profile Load", exhibit slight differences compared to the previous study cases. The variations in voltage and current between the balanced and unbalanced cases are relatively small, indicating that while residential loads introduce more fluctuations compared to industrial loads, these differences remain moderate.

Comparison of Study Cases "Static Z Load"

The key distinction among study case 2.1 (industrial load), study case 2.2 (commercial load), and study case 2.3 (residential load) lies in the adjustment of the parameters n_P and n_q in the three-phase dynamic block. In study case 2.1, $n_P = 0.18$ and $n_q = 6$, simulating industrial loads that typically exhibit stable and predictable power demand with minimal sensitivity to voltage variations. In contrast, study case 2.2 models commercial loads, with n_P =1.51 and n_q =3.4, which reflect more variable usage patterns due to lighting, HVAC systems, and office equipment. For study case 2.3, residential loads are simulated using n_P =0.92 and $n_q = 4.04$, capturing the fluctuating power consumption caused by varying household activities, such as appliance usage and heating/cooling systems.

The measurements from each case, named the voltage DYN npnq load and current DYN npnq load, respectively, for the balanced and the unbalanced cases, show small differences across all load types. Figure 12 provides a detailed comparison of the total power (both active and reactive) for each load type, highlighting how these small variations affect overall system performance.

The main distinction between study case 2.3 (residential load) and the previous cases—study case 2.1 (industrial load) and study case 2.2 (commercial load)—is the adjustment of n_P and n_q in the three-phase dynamic load block. These values simulate the characteristics of residential loads, which tend to fluctuate based on varying household activities such as appliance usage, lighting, and heating/cooling systems.

In study case 2.1, which represents the industrial load type, the value of n_P is 0.18, which is close to zero. This explains why it can be seen that the blue curve bares more of a resemblance to a straight line than a curve (from Table 1 it can be seen that for the value of 0 for n_P, it represents a constant power load type, a straight line). In the commercial load case, where $n_P = 1.51$, the active power exhibits noticeable oscillations over time, indicating a higher sensitivity to voltage variations compared to the industrial load case (with $n_P = 0.18$), where the active power remains nearly constant. This is consistent with the operational behavior of commercial loads, which typically experience greater fluctuations due to more variable usage patterns. The residential load case (where n_P =0.92) also shows oscillations in active power, but to a lesser extent than the commercial load. However, these oscillations remain within operational limits, ensuring grid stability is maintained despite the load variability.

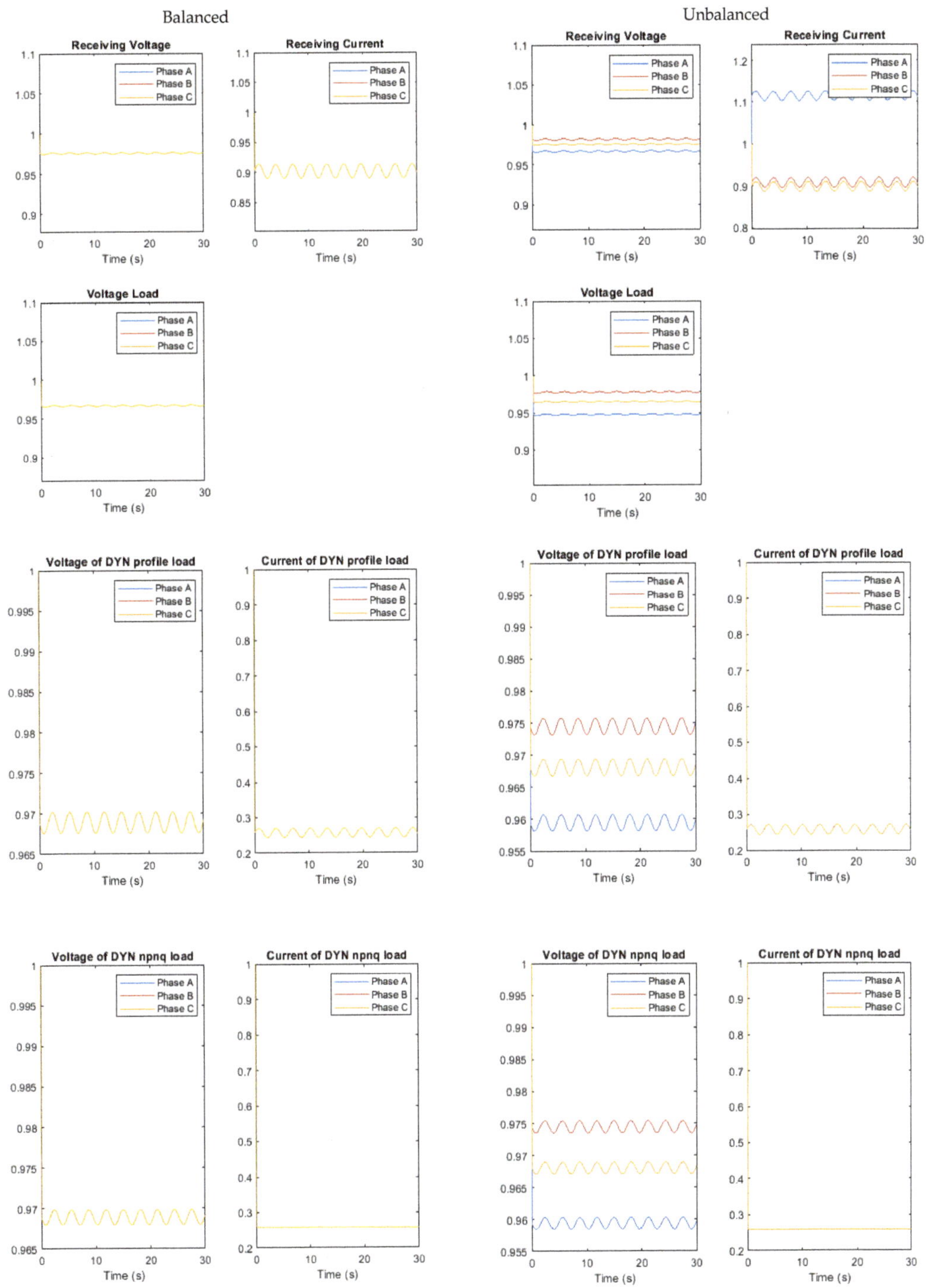

Figure 11. Study case 2 (residential load type).

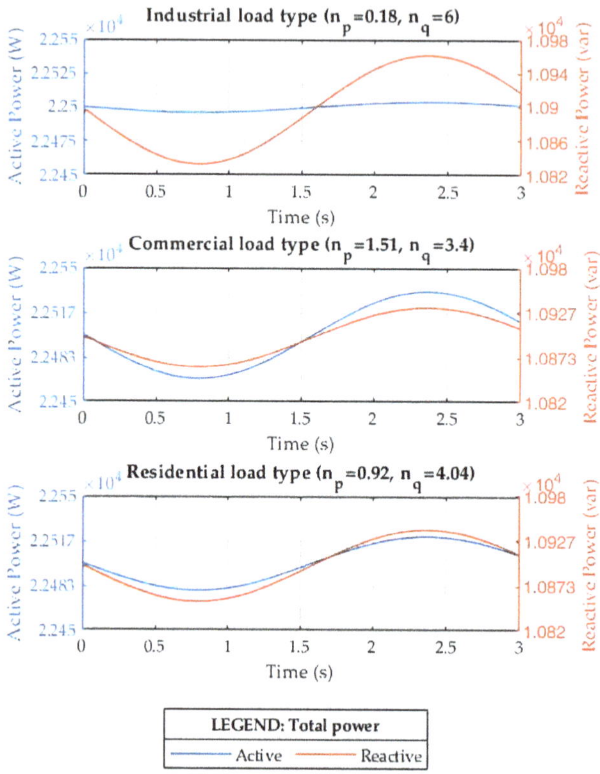

Figure 12. Study case 2 total power (for the industrial, commercial, and residential load type).

Similarly, the reactive power for the commercial load shows slight variation compared to the industrial load. The higher value of n_q reflects the reactive power demand typical of commercial environments with a lot of inductive loads. In the residential load case, slight fluctuations in reactive power are observed but remain within acceptable limits.

The comparison of these curves highlights how reactive power demands differ between the load types, with industrial loads being the most stable and predictable, and commercial and residential loads showing more variability. While the reactive power does fluctuate, the magnitude of these changes is small and does not significantly affect the system's voltage stability.

4.5.3. Study Case 3

In study case 3, the static Z load model is implemented without dynamic load modeling, but this case includes the integration of renewable energy into the distribution network. As shown in Figure 1, the modeled grid incorporates the "Main Grid", "Static Z Load", and a PV farm, while other elements are deactivated in the simulation. This case also divides the network into balanced and unbalanced scenarios. The electric power supplied by the PV farm follows a generation profile, varying over time and reflecting the integration of 50%, 75%, and 100% of the power consumed by the loads, as illustrated in Figure 13.

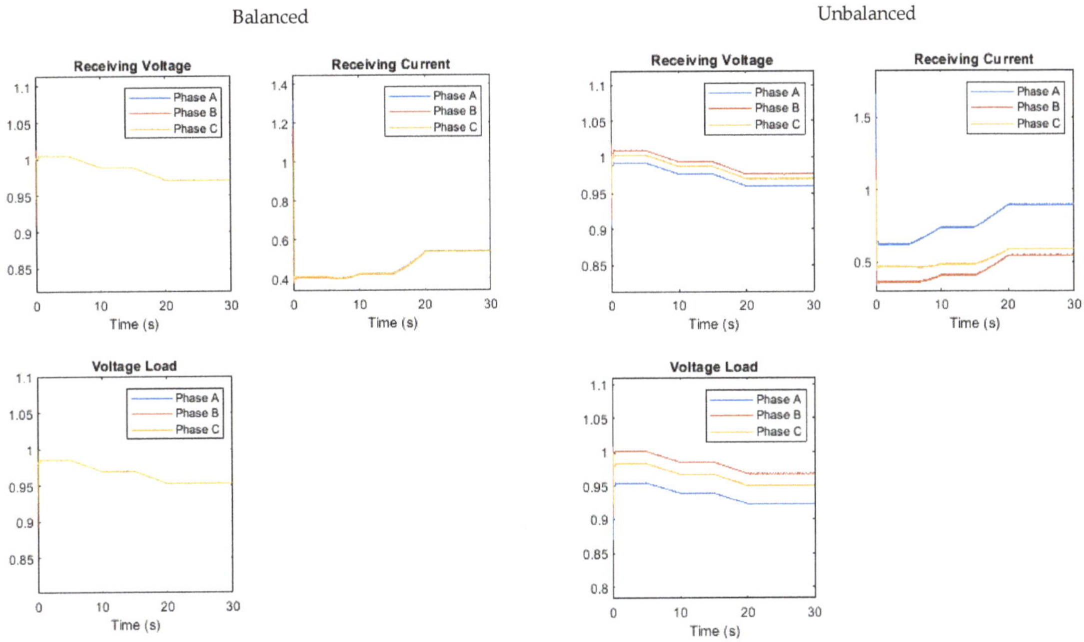

Figure 13. Study case 3.

In the first 5 s of the simulation time, the power consumed by the loads in the network comes 100% from the PV farm. Then, in the next 5 s, it gradually decreases to 75% and stays constant. It again starts to decrease in the 15th s of the simulation until it reaches 50% in the 20th s of the simulation, and it stays at this level for 10 s more. Once again, the imbalance of the grid is achieved by increasing the capacitive reactive power in phase A by 50% with respect to the nominal power, and just like in the previous study cases, the values of the voltage for phase A in the static load bus expectedly decrease, whereas the values of the voltage for phase B slightly increase, and the voltage of phase C maintains the same value. A similar situation can be observed in the receiving voltage bus.

Of particular interest is the behavior of the receiving current bus under unbalanced conditions. The integration of the PV farm affects the outer grid, as seen in the behavior of the current in phase A during the initial 5 s. The current value in phase A is about 0.7 pu, while phase C remains around 0.5 pu, and phase B hovers at approximately 0.2 pu. During this time, the PV farm generates 100% of the power consumed by the loads, which would typically suggest that the current flowing from the grid would be near 0 pu. However, the current remains notably higher. The direction of the current, which flows from the static load back to the grid, is noteworthy. This phenomenon occurs due to the increased capacitive reactive power in phase A, which is raised by 50% relative to the nominal power.

Figure 14 further illustrates the active and reactive power observed at the receiving bus. The most significant observation is the inability of the PV farm's inverter to provide reactive power to support the grid. The active power closely follows the sun's irradiance profile, while the reactive power remains constant at approximately 0.6 pu in the balanced case and 0.65 pu in the unbalanced case. This constancy indicates that the reactive power surplus or demand is independent of the PV farm, and the inverter used cannot provide reactive power support to the grid.

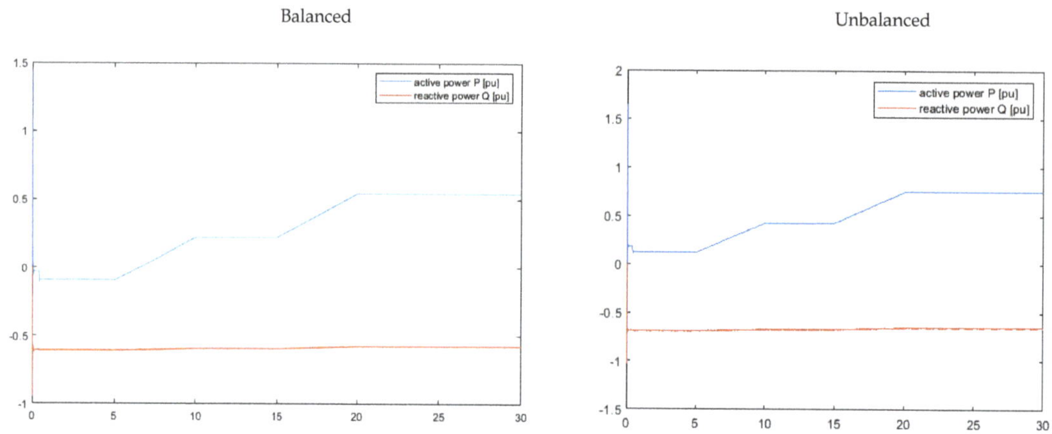

Figure 14. Study case 3 active and reactive power observed from the receiving bus.

4.5.4. Study Case 4

In study case 4, the static Z load model is implemented along with dynamic load modeling and the integration of renewable energy into the network. All components shown in Figure 1 are active in this simulation, including the main grid, static Z load, dynamic (composite and profile) load, and the PV farm. The network is divided between balanced and unbalanced conditions, as in previous study cases. The composite load model is configured to represent industrial, residential, and commercial loads, as detailed in Table 1. Half of the total power is consumed by the static load, while the other half is consumed by the dynamic load model. The PV farm provides power according to a generation profile that varies over time, supplying 50%, 75%, and 100% of the total load, as seen in study case 3.

Study Case 4.1

In this scenario, the composite load is modeled to represent industrial loads, combining dynamic and composite load characteristics. The key difference between study case 4 and study case 3 is the inclusion of dynamic load behavior.

Figure 15 illustrates the current and voltage behavior, with the dynamic load undergoing rapid fluctuations every 2 s, ranging between approximately 0.24 pu and 0.26 pu. These fluctuations impact the entire grid, influencing the voltage on the static load as well as the sending and receiving buses. Compared to study case 3, the graphs show more pronounced fluctuations, due to the dynamic loads' behavior. Nevertheless, the overall effect of this load on the grid in the voltage remains within acceptable limits. The "Receiving current" measurements are minimal during nominal power injection from the PV system, indicating that the system is predominantly supplying the loads. Only a small amount of power is being drawn from the external grid. Consequently, the curves can be interpreted as primarily driven by the power injected by the PV system. At the 5th s, under balanced conditions, the load is fully supplied (100%) by the PV system.

As in previous cases, the grid imbalance is introduced by increasing the capacitive reactive power in phase A by 50% relative to nominal power. This is evident in the current receiving block, where phase A shows a higher current magnitude compared to phases B and C.

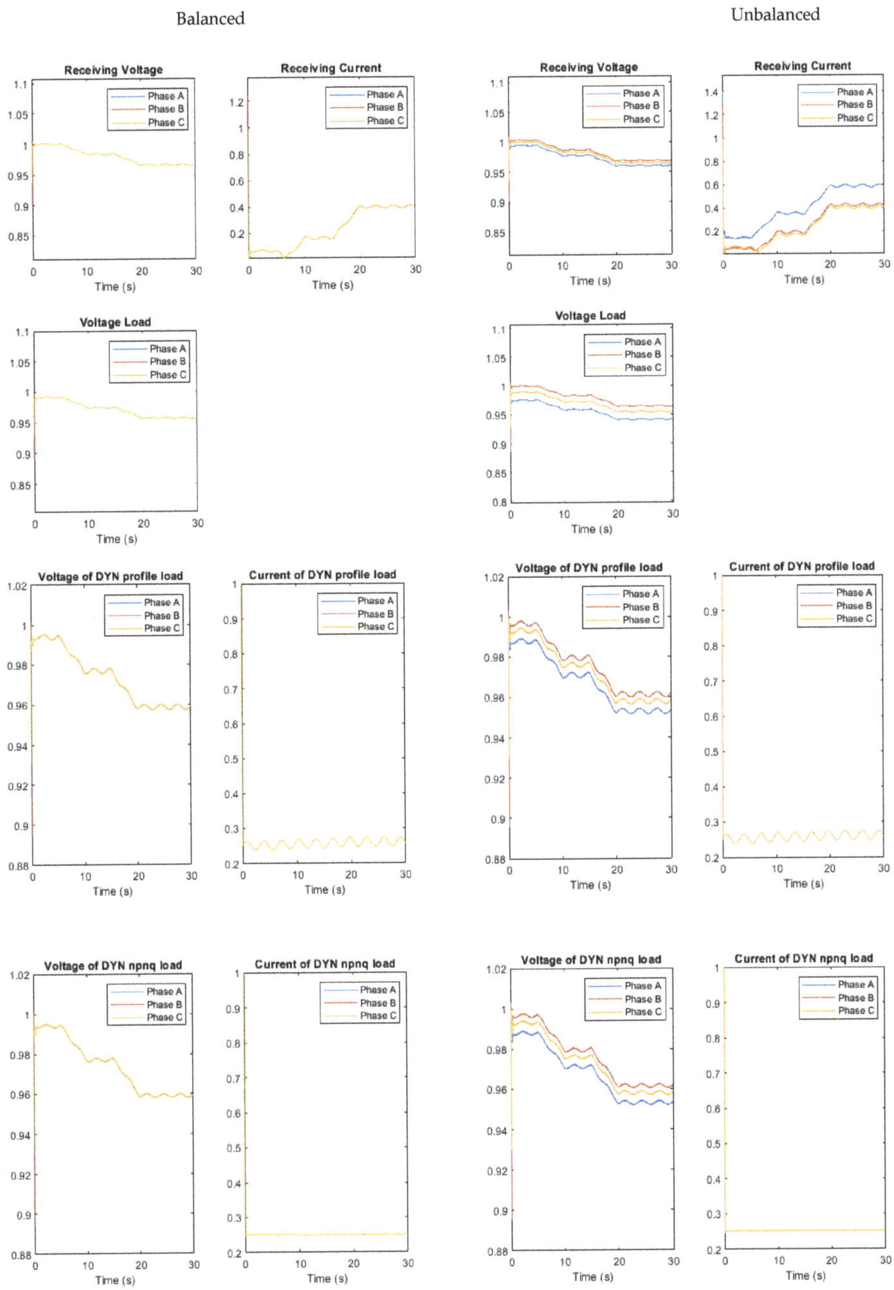

Figure 15. Study case 4 with industrial load type.

Study Case 4.2

In this study case, the composite load is modeled to represent loads in the commercial sector. Upon comparing the graphs in study case 4.1 and 4.2, it can be seen relatively minor differences. As the composite load model consists of only 25% of the total load, it can be concluded that it makes no significant difference whether the composite load model represents loads in the industrial or commercial sector. The results are represented in Figure 16.

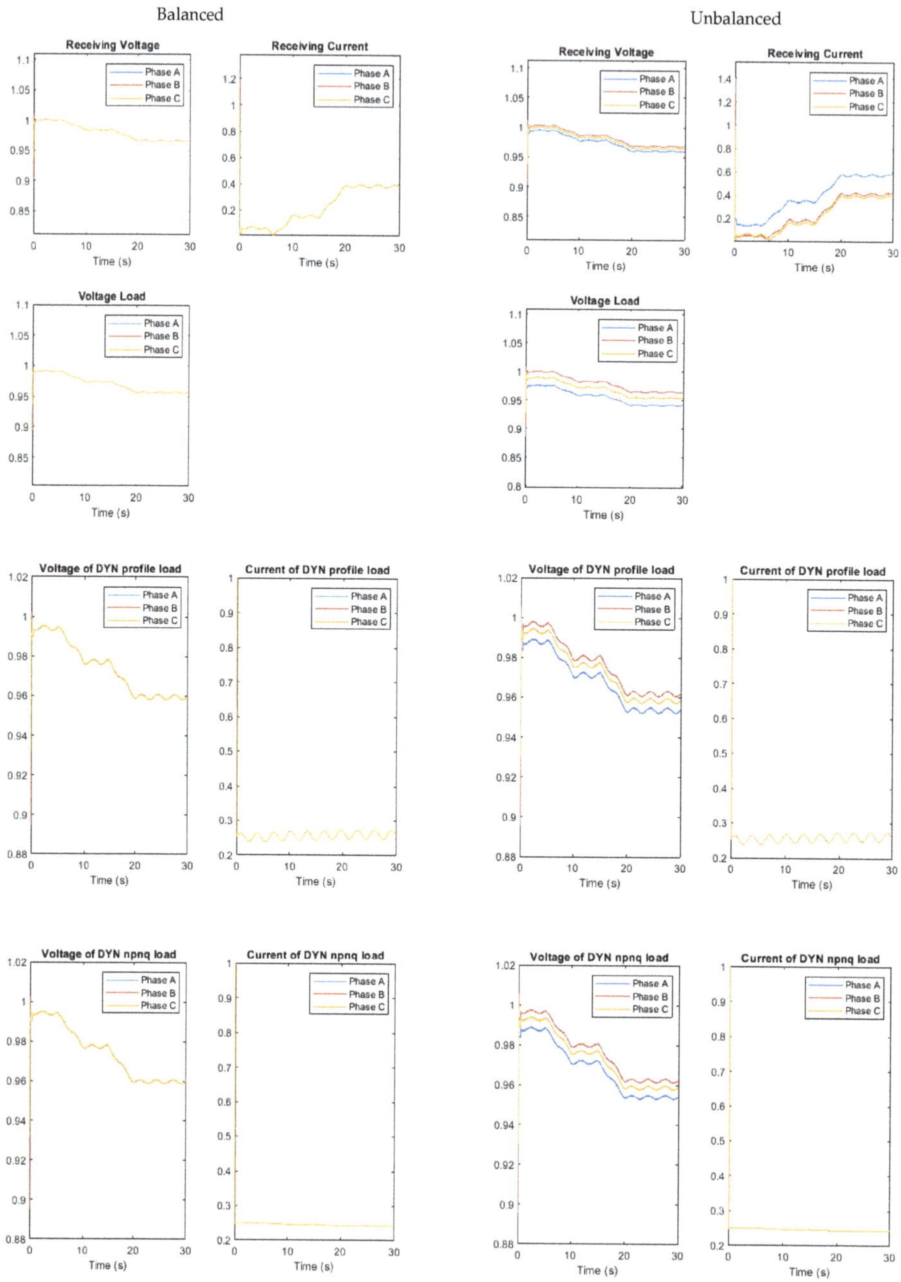

Figure 16. Study case 4 with commercial load type.

Study Case 4.3

In this study case, the composite load is modeled to represent loads in the residential sector. The results from this study case are shown in Figure 17.

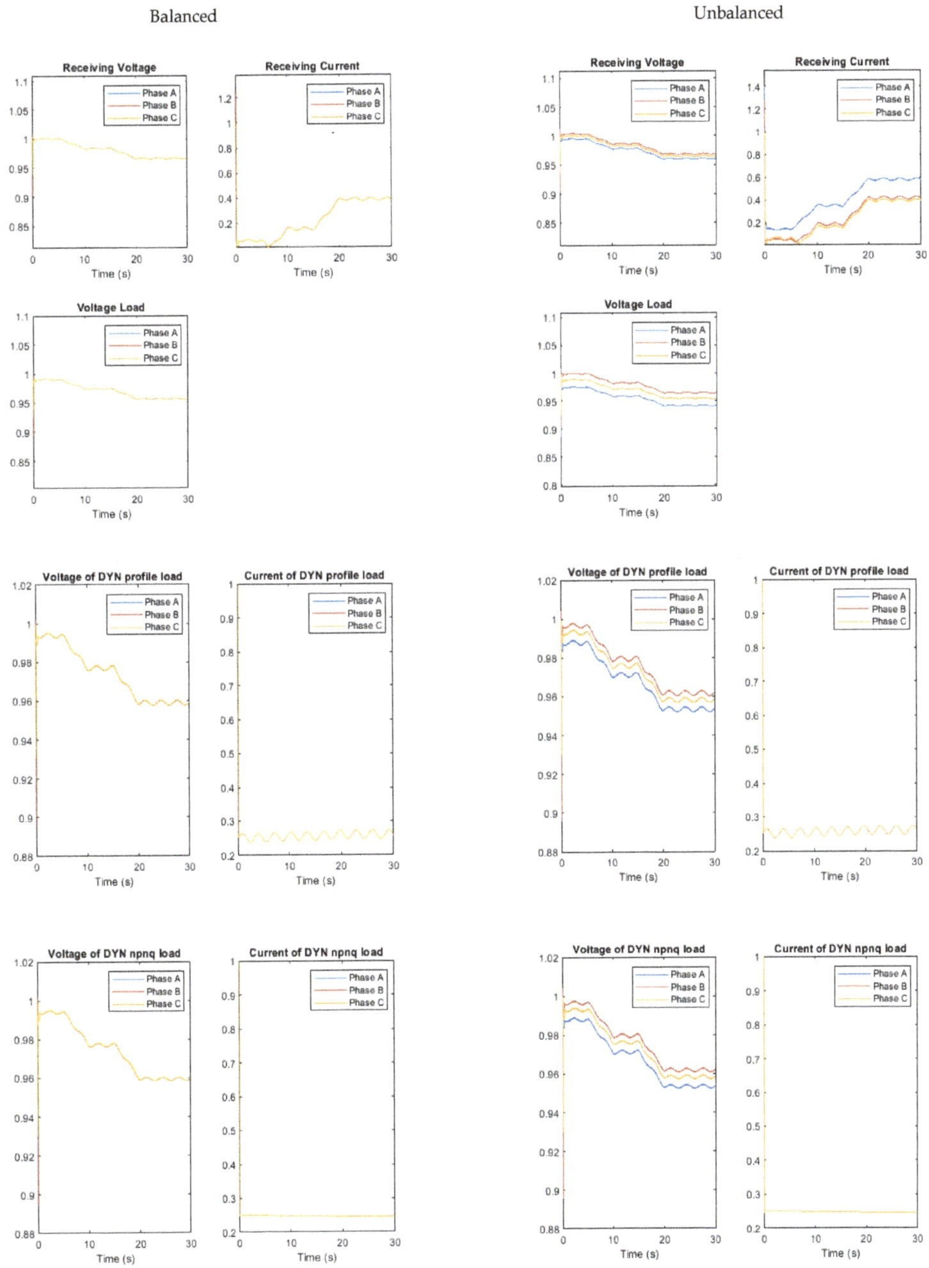

Figure 17. Study case 4 with residential load type.

Comparison of Study Cases "Static Z Load, Dynamic Load and RES"

When comparing the graphs from previous three study cases, relatively minor differences in behavior can be observed. However, to facilitate a clearer comparison, Figure 18 illustrates the total power consumed by the composite load in each case. The blue curve represents the active power consumption, while the orange curve represents the reactive power consumption across the industrial, commercial, and residential load scenarios.

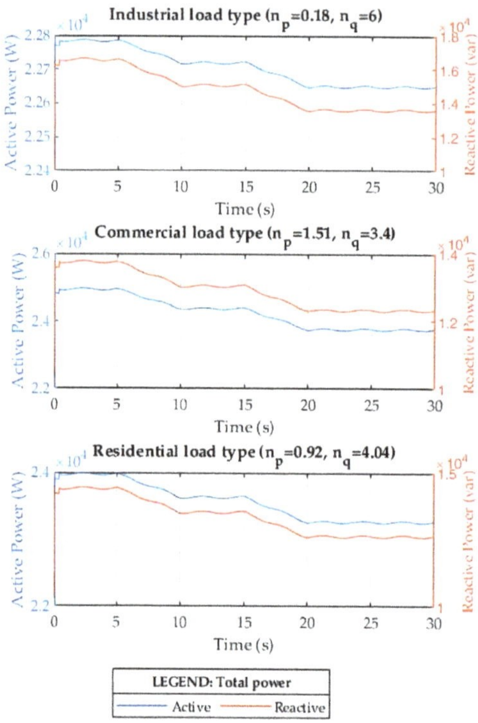

Figure 18. Study case 4—total power.

Figure 18 shows that in the industrial and residential load types, the active power consumption is higher than the reactive power. In contrast, for the commercial load type, reactive power consumption is more significant than active power. This behavior highlights the distinct power consumption characteristics of each load type. Additionally, the power consumption of the composite load is influenced by the generation profile of the PV farm. As the PV generation decreases, it leads to a voltage drop in the grid, which causes a corresponding reduction in power consumption from the composite load.

Due to the composite load being a mixture of static ZIP loads and various dynamic loads, the decrease in power consumption does not occur at the same rate as the voltage drop caused by the reduced PV generation. The power consumption of the composite load adjusts, but not in a direct proportion to the generation decrease. As the PV farm generates less power, the composite load compensates by drawing more power from the main grid. This can be seen in Figures 15–17, where the measurements from the receiving bus show an increase in the current drawn from the main grid as the PV generation decreases.

The total power consumed by the composite load is influenced by both load type and the generation profile of the PV farm. While industrial and residential loads consume more active power than reactive power, commercial loads show a higher proportion of reactive power consumption. Furthermore, the decrease in PV generation leads to increased reliance on the main grid to meet the power demand of the composite load.

4.6. Comparison of All Study Cases

The four study cases collectively highlight the impact of different load types, dynamic modeling, and renewable integration on grid stability under balanced and unbalanced conditions and are summarized in Table 2.

Table 2. Simulated Study Cases.

Study Case	Static Load Used	Dynamic Load Used	RES Used
1	Yes	No	No
2	Yes	Yes	No
3	Yes	No	Yes
4	Yes	Yes	Yes

Study case 1, which only includes static Z loads, serves as a baseline and shows that even modest imbalances in the grid can result in voltage and current asymmetries. Study case 2 demonstrates that adding dynamic loads introduces more variation in voltage behavior, especially under unbalanced conditions, but the system remains largely stable due to the inherent properties of the dynamic load models.

Study case 3 introduces renewable energy into the mix, showing that the grid's response becomes more complex when renewable generation fluctuates. A notable limitation is the inability of the PV farm's inverter to support the grid with reactive power, a critical factor in maintaining voltage stability under unbalanced conditions. Finally, study case 4 reveals that the combination of dynamic load modeling and renewable integration leads to more significant current and voltage fluctuations, although these remain within acceptable operational limits.

In conclusion, the study cases underscore the importance of considering both dynamic loads and renewable integration when analyzing grid stability, particularly in systems where unbalanced conditions may arise. The findings suggest that while renewable integration enhances grid resilience, it also requires advanced inverter technologies capable of providing both active and reactive power support to ensure stable operation.

5. Conclusions

This study offers a thorough analysis of the dynamic modeling of distribution power systems with integrated RESs, focusing on stability assessment. Using EMT modeling and detailed simulations conducted in MATLAB/Simulink®, we examined distribution system behavior across various levels of distributed generation (DG) penetration and diverse load models. Both balanced and unbalanced network conditions were analyzed to capture how distributed photovoltaic (PV) systems influence voltage stability and overall grid performance.

The results underscore the critical role of inverter control techniques in maintaining voltage stability, particularly through reactive power control. Significant limitations were observed in the reactive power compensation capabilities in the implemented inverters. This shortfall poses a substantial risk to stability in grids with high DG penetration. Without proper voltage and frequency control techniques implemented in inverters, voltage and frequency stability could be further compromised, especially as DG levels increase.

A contribution of this study lies in the modeling of the external grid's behavior. By adjusting the internal impedance of the equivalent upstream grid, we simulate its inherent weaknesses, particularly in upstream systems with high DG penetration and numerous dynamic loads (such as motors). The study demonstrates that grid stability could be at risk without proactive reinforcements. This finding emphasizes the urgency for strengthening distribution grids to handle increasing operational demands as DG integration continues to grow in both low and high voltage.

Dynamic load modeling proved crucial for understanding interactions between RESs and distribution networks. The study reveals that different load types—industrial, commercial, and residential—affect grid performance in distinct ways. The diverse behaviors of these loads, when combined with renewable integration, highlight the need for meticulous planning and operational management to ensure reliable and stable grid operation.

Future research will focus on modeling advanced voltage control techniques in inverters, integrating battery storage systems, and exploring islanding methods to further enhance grid stability. Addressing the limitations in reactive power control in current inverter technologies will be pivotal, especially as distribution grids evolve to accommodate higher levels of renewable energy.

This work establishes a robust foundation for ongoing advancements in dynamic modeling, offering critical insights into the challenges and opportunities presented by the transition to low-carbon energy systems. Proactive grid reinforcements, especially in the context of high DG integration, and the continued innovation of inverter technologies will be essential for ensuring future grid stability and reliability.

Author Contributions: Conceptualization, C.T.; methodology, C.T.; software, I.D. and D.M.; validation, D.M. and I.D.; formal analysis, D.M. and I.D.; investigation, D.M. and I.D.; resources, C.T. and I.D.; data curation, D.M. and I.D.; writing—original draft preparation, D.M., C.T. and I.D.; writing—review and editing, D.M., C.T. and I.D.; visualization, D.M. and I.D.; supervision, C.T.; project administration, C.T. All authors have read and agreed to the published version of the manuscript.

Funding: This research received no external funding.

Data Availability Statement: The data presented in this study are available on request from the corresponding author.

Conflicts of Interest: The authors declare no conflict of interest.

References

1. European Commission. Fit for 55: Delivering the EU's 2030 Climate Targets on the Way to Climate Neutrality. Available online: https://www.eesc.europa.eu/en/our-work/opinions-information-reports/opinions/fit-55-delivering-eus-2030-climate-target-way-climate-neutrality (accessed on 22 September 2024).
2. European Environment Agency. Renewable Energy in Europe: Key for Climate Objectives. Available online: https://www.eea.europa.eu/themes/energy/renewable-energy (accessed on 22 September 2024).
3. European Environment Agency. Electric Vehicles as a Key for Reducing Emissions. Available online: https://www.eea.europa.eu/themes/transport/electric-vehicles (accessed on 22 September 2024).
4. Asare-Bediako, B.; Woud, W.K.; Ribeiro, P.F. Future residential load profiles: Scenario-based analysis of high penetration of heavy loads and distributed generation. *Energy Build.* **2014**, *75*, 228–238. [CrossRef]
5. ElNozahy, M.S.; Salama, M.M.A. A Comprehensive Study of the Impacts of PHEVs on Residential Distribution Networks. *IEEE Trans. Sustain. Energy* **2014**, *5*, 332–342. [CrossRef]
6. Vittal, V.; McCalley, J.D.; Anderson, P.M.; Fouad, A.A. *Power System Control and Stability*, 3rd ed.; John Wiley & Sons: New York, NY, USA, 2019.
7. Electric Power Research Institute. *Measurement-Based Load Modeling*; Product ID: 1014402; Electric Power Research Institute: Palo Alto, CA, USA, 2006.
8. *IEEE Standard 2781*; IEEE Guide for Load Modeling and Simulations for Power Systems. IEEE Standard: Piscataway, NJ, USA, 2022.
9. Neagu, B.-C.; Gavrilas, M.; Ivanov, O.; Grigoras, G. Load Modeling Approaches in Smart Grids: An Overview. In *Springer Proceedings in Energy*; Springer: Berlin/Heidelberg, Germany, 2022. [CrossRef]
10. Rahman, M.T.; Hasan, K.N.; Sokolowski, P. Impact of Renewable Integration in Distribution Networks on Static Load Model Parameters. In Proceedings of the 2020 IEEE PES Innovative Smart Grid Technologies Europe (ISGT-Europe), The Hague, The Netherlands, 26–28 October 2020; pp. 725–729. [CrossRef]
11. Fulgêncio, N.; Moreira, C.; Carvalho, L.; Lopes, J.A. Equivalent Dynamic Model of Active Distribution Networks for Large Voltage Disturbances. In Proceedings of the 2019 IEEE Milan PowerTech, Milan, Italy, 23–27 June 2019; pp. 1–6. [CrossRef]
12. Zali, S.M.; Milanovic, J.V. Generic Model of Active Distribution Network for Large Power System Stability Studies. *IEEE Trans. Power Syst.* **2013**, *28*, 3126–3133. [CrossRef]
13. Tielens, P.; Hertem, D.V. The relevance of inertia in power systems. *Renew. Sustain. Energy Rev.* **2016**, *55*, 999–1009. [CrossRef]
14. Fan, L.; Miao, Z. *Modeling and Stability Analysis of Inverter-Based Resources*; CRC Press: Boca Raton, FL, USA, 2024.

15. IEEE/NERC Task Force on Short-Circuit and System Performance Impact of Inverter-Based Generation. Impact of Inverter-Based Generation on Bulk Power System Dynamics and Short-Circuit Performance. IEEE PES Technical Report PES-TR68. 2018. Available online: https://resourcecenter.ieee-pes.org (accessed on 23 September 2024).
16. Haddadi, A.; Farantatos, E.; Kocar, I.; Karaagac, U. Impact of Inverter Based Resources on System Protection. *Energies* **2021**, *14*, 1050. [CrossRef]
17. Boricic, A.; Rueda Torres, J.L.; Popov, M. Impact of Modelling Assumptions on the Voltage Stability Assessment of Active Distribution Grids. In Proceedings of the 2020 IEEE PES Innovative Smart Grid Technologies Europe (ISGT-Europe), The Hague, The Netherlands, 26–28 October 2020; pp. 1040–1044. [CrossRef]
18. Kyesswa, M.; Çakmak, H.; Kühnapfel, U.; Hagenmeyer, V. Dynamic Modelling and Control for Assessment of Large-Scale Wind and Solar Integration in Power Systems. *IET Renew. Power Gener.* **2021**, *14*, 4010–4018. [CrossRef]
19. Al Essa, M.J.M.; Cipcigan, L.M. Integration of Renewable Resources into Low Voltage Grids Stochastically. In Proceedings of the 2016 IEEE International Energy Conference (ENERGYCON), Leuven, Belgium, 4–8 April 2016; pp. 1–5. [CrossRef]
20. Mohapatra, G.R.; Kalam, A. Dynamic Stability Analysis of Renewable Energy Sources Interconnected to the Distribution Networks. In Proceedings of the 2008 Australasian Universities Power Engineering Conference, Sydney, Australia, 14–17 December 2008; pp. 1–4.
21. Musse, B.F.; Barbosa, M.W.; Carvalho, J.S.; da Silva, D.C.; de Oliveira, L.W.; de Oliveira, J.G. Real Time Simulation of PV System Integration to the Distribution Grid Using Dynamic Load Model. In Proceedings of the 2017 IEEE 8th International Symposium on Power Electronics for Distributed Generation Systems (PEDG), Florianopolis, Brazil, 17–20 April 2017; pp. 1–6. [CrossRef]
22. Li, F.; Wang, Y.; Wu, F.; Huang, Y.; Liu, Y.; Zhang, X.; Ma, M. Review of Real-Time Simulation of Power Electronics. *J. Mod. Power Syst. Clean Energy* **2020**, *8*, 796–808. [CrossRef]
23. Clements, K.A. The Impact of Pseudo-Measurements on State Estimator Accuracy. In Proceedings of the 2011 IEEE Power and Energy Society General Meeting, Detroit, MI, USA, 24–28 July 2011; IEEE: San Diego, CA, USA, 2011; pp. 1–4. [CrossRef]
24. Phadke, A.G.; Thorp, J.S. *Synchronized Phasor Measurements and Their Applications*; Springer Science & Business Media: Berlin/Heidelberg, Germany, 2008.
25. Kundur, P. *Power System Stability and Control*; McGraw-Hill Education: New York, NY, USA, 1994.
26. IEEE/CIGRE Joint Task Force on Stability Terms and Definitions. Definition and Classification of Power System Stability. *IEEE Trans. Power Syst.* **2004**, *19*, 1387–1401. [CrossRef]
27. Munkhchuluun, E.; Meegahapola, L.; Vahidnia, A. Long-term Voltage Stability with Large-Scale Solar-Photovoltaic (PV) Generation. *Int. J. Electr. Power Energy Syst.* **2020**, *117*, 105663. [CrossRef]
28. Gasperic, S.; Mihalic, R. Estimation of the Efficiency of FACTS Devices for Voltage-Stability Enhancement with PV Area Criteria. *Renew. Sustain. Energy Rev.* **2019**, *105*, 144–156. [CrossRef]
29. Cutsem, T.V.; Vournas, C. *Voltage Stability of Electric Power Systems*; Kluwer Academic Publishers: Boston, MA, USA, 1998.
30. Cutsem, T.V. Voltage Instability: Phenomena, Countermeasures, and Analysis Methods. *Proc. IEEE* **2000**, *88*, 208–227. [CrossRef]
31. Vartanian, C.; Bauer, R.; Casey, L.; Loutan, C.; Narang, D.; Patel, V. Ensuring System Reliability: Distributed Energy Resources and Bulk Power System Considerations. *IEEE Power Energy Mag.* **2018**, *16*, 52–63. [CrossRef]
32. Alenius, H.; Luhtala, R.; Messo, T.; Roinila, T. Autonomous Reactive Power Support for Smart Photovoltaic Inverter Based on Real-time Grid-impedance Measurements of a Week Grid. *Electr. Power Syst. Res.* **2020**, *182*, 106207. [CrossRef]
33. Eftekhari, S.; Oukati, M. The Effect of Load Modelling on Phase Balancing in Distribution Networks Using Search Harmony Algorithm. *Int. J. Electr. Comput. Eng. (IJECE)* **2019**, *9*, 1461. [CrossRef]
34. Bustillo, C.S.; Diaz, O.I.; Mendoza, J.; Rivera, D.A. Modeling and Simulation of Composite Load Using MATLAB/SIMULINK. In Proceedings of the 2019 IEEE 39th Central America and Panama Convention (CONCAPAN XXXIX), Guatemala City, Guatemala, 20–22 November 2019; pp. 1–6. [CrossRef]
35. Navarro-Espinosa, A.; Ochoa, L. *Deliverable 3.6 "What-If Scenario Impact Studies Based on Real LV Networks"*; eScholar: Niagara Falls, NY, USA, 2014. [CrossRef]
36. MathWorks. 250 kW Grid-Connected PV Array. Available online: https://de.mathworks.com/help/sps/ug/250-kw-grid-connected-pv-array.html (accessed on 21 September 2024).
37. Brito, M.A.G.d.; Sampaio, L.P.; Melo, G.J.d.A.e.; Canesin, C.A. Comparative analysis of MPPT techniques for PV applications. In Proceedings of the 2011 International Conference on Clean Electrical Power (ICCEP), Ischia, Italy, 14–16 June 2011; pp. 99–104.

energies

MDPI

Article

Current Compensation Method in a Distribution System Based on a Four-Leg Inverter under Unbalanced Load Conditions Using an Artificial Neural Network

Tae-Gyu Kim, Chang-Gyun An, Junsin Yi and Chung-Yuen Won *

Department of Electrical and Computer Engineering, Sungkyunkwan University, Suwon 16419, Republic of Korea; rlaxorb29@skku.edu (T.-G.K.); acg1643@skku.edu (C.-G.A.); junsin@skku.edu (J.Y.)

* Correspondence: woncy550@gmail.com; Tel.: +82-31-290-7169

Abstract: This study proposes an unbalanced current compensation method based on a four-leg inverter using an artificial neural network (ANN) under unbalanced load conditions. Distribution systems exhibit rapid load variations, and conventional filter-based control methods suffer from the drawback of requiring an extended time period to reach a steady state. To address this problem, an ANN is applied to calculate the unbalanced current reference and enhance dynamic performance. Additionally, because of the periodic incorrect output inherent in the ANN, applying it to a proportional–integral controller would result in an error being directly reflected in the current reference. In the aforementioned problem, an ANN is applied to the dq0 coordinate system current controller to compensate for the periodic incorrect output in the current reference calculation. The proposed ANN-based unbalanced current compensation method is validated through PSIM simulations and experiments.

Keywords: unbalanced current compensation; artificial neural network; four-leg inverter; grid-connected

1. Introduction

In recent years, power systems have become increasingly complex owing to continuous advancements. In particular, with the rising demand for power, the integration of renewable energy sources, and the introduction of electric vehicles and distributed power generation systems, power networks are now operating in dynamic and diverse environments. These conditions can deteriorate power quality, significantly affecting the stability and performance of power systems. Factors that contribute to the degradation of power system quality can be explained in terms of harmonics, reactive power, and power imbalances [1,2].

Harmonics arise owing to the discontinuous nature of rectifiers or AC/DC converters, causing malfunctions in precision control equipment and a decrease in power quality by reducing the power factor. Therefore, various approaches have been investigated to compensate for harmonics in power systems, focusing on mitigating their effects [3,4]. Reactive power arises from the use of inductive and capacitive load components, which can degrade the power quality by reducing the power factor of the system. Various methods for controlling reactive power in power systems have also been studied to improve the power quality [5,6]. Unbalanced power refers to situations in which the magnitudes and types of loads used in each phase are different, resulting in phase and magnitude imbalances in the current or voltage of the system. In particular, if voltage, current, and power imbalances occur due to an unbalanced load, they can cause power quality issues across power generation and distribution systems [7]. Awareness of the risks posed by unbalanced power is continually increasing, leading to a continuous increase in research aimed at minimizing the effects of unbalanced power [8]. Although LC filters can be used in power systems to compensate for unbalanced power, issues such as filter size and parameter selection arise [9]. Following the advancements in power electronics research, recently, the focus has shifted towards active power filters and load unbalanced compensation (LUC) to address harmonic distortions and non-linear loads, thereby improving the overall power quality of grid systems [10]. The fundamental approach in such compensation strategies involves integrating an inverter in parallel between the grid and the load to inject harmonic and unbalanced currents back into the grid [11]. As a result, elements requiring compensation, such as harmonics and unbalanced values, necessitate the computation of current references for the inverter, which acts as a compensating agent. Therefore, this study aims to investigate the process of determining current references using a specific methodology. Typically, the instantaneous reactive power theory method is employed [12,13]. However, this method has the disadvantage of being susceptible to noise. To address these limitations, stationary reference frame (STRF) approaches have been proposed, transforming the abc frame into a set of stationary coordinates of α-β-0 to mitigate unbalances [14,15]. Furthermore, various approaches based on the synchronous reference frame (SRF) have been proposed to alleviate unbalances by synchronizing with the voltage of the power system and transforming the abc frame into the dq0 synchronous frame [16,17]. Subsequently, research has been conducted to improve the performance of unbalanced power compensation by applying bandpass filters, high-pass filters (HPFs), and low-pass filters (LPFs) to STRF and SRF [18,19]. However, the filter-based method exhibits disadvantages in regards to time delay [20,21]. In power electronics-based approaches, research has also been conducted to apply artificial neural networks (ANNs) to replace controllers [22,23], and as well as to employ ANN-based nonlinear compensation in active power filters [24].

In this paper, we propose a current unbalance compensation method for distribution systems using ANN. Conventional filter-based methods require several cycles to reach a steady state. To address this, we apply ANN to calculate the current unbalance compensation component, reducing the time to reach a steady state to within one cycle. However, periodic errors occur during this process. By applying ANN controllers instead of traditional proportional–integral (PI) controllers, we mitigate the issue of oscillatory components. ANN is applied to two aspects, and to verify control performance, simulations and laboratory-scale experiments were conducted. The structure of this paper is as follows: Section 2 discusses representative conventional methods, namely the p-q theory and the synchronous frame theory. In Section 3, we describe the structure and training step of the ANN applied for the proposed method. Section 4 details the performed simulations of current compensation under unbalanced load conditions. Subsequently, Section 5 shows the results of the experiments conducted to validate the proposed method. Finally, Section 6 presents the conclusion.

2. Conventional Grid Current Compensation under Unbalanced Load Conditions

Figure 1 illustrates the scenario of an unbalanced load in the system connected to the inverter. When the inverter is not controlled, the load power is entirely supplied by the power grid. Such an unbalanced load causes distortions in the distribution system and deteriorates the overall power quality [25]. Figure 2 shows a vector diagram of the distorted current improved by the inverter. To address these issues, research has been conducted on the LUC method, which compensates for unbalanced loads. When the inverter compensates for the unbalanced loads, the current in the system is balanced. This LUC method comprises two parts: calculating the current reference to compensate for the unbalance current, and the current controller [24]. In this section, we describe the conventional method of calculating the current reference under an unbalanced load. Equation (1) mathematically describes the principle of current compensations.

$$\vec{i}_{g,abc}(balanced) = \vec{i}_{l,abc}(unbalanced) + \vec{i}_{c,abc}(compensation) \quad (1)$$

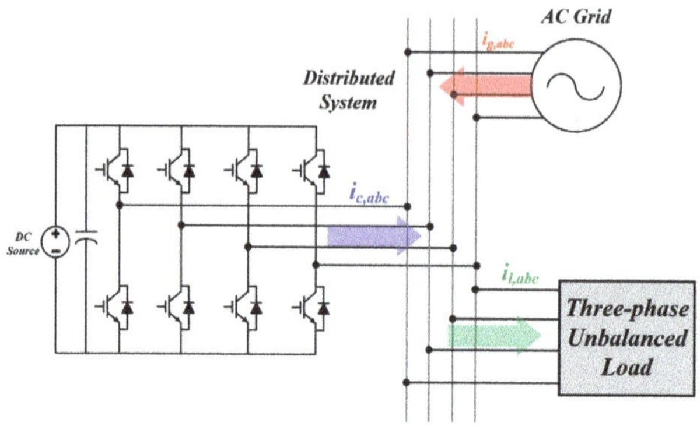

Figure 1. Unbalanced current compensation configuration.

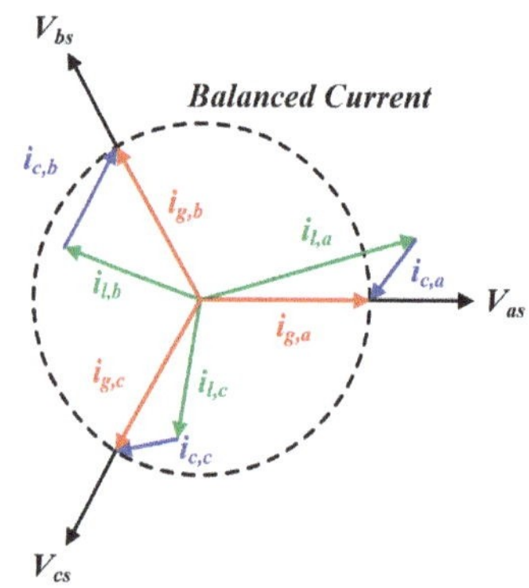

Figure 2. Unbalanced load current compensation vector diagram.

2.1. LUC Based on the P-Q Theory

Figure 3 depicts the block diagram for calculating the current reference based on the instantaneous power p-q theory. Calculations based on the p-q theory remain effective in both the steady and transient states, providing rapid response characteristics even under transient conditions [11]. This method utilizes a band pass filter to compute the current reference using the AC power component. Inputs for the calculation are obtained by transforming the abc coordinates to the stationary coordinates α-β-0, which is represented in Equations (2)–(4) [13].

$$\begin{bmatrix} V_0 \\ V_\alpha \\ V_\beta \end{bmatrix} = \sqrt{\frac{2}{3}} \begin{bmatrix} 1/\sqrt{2} & 1/\sqrt{2} & 1/\sqrt{2} \\ 1 & -1/2 & -1/2 \\ 0 & \sqrt{3}/\sqrt{2} & -\sqrt{3}/\sqrt{2} \end{bmatrix} \cdot \begin{bmatrix} V_a \\ V_b \\ V_c \end{bmatrix} \tag{2}$$

$$\begin{bmatrix} i_0 \\ i_\alpha \\ i_\beta \end{bmatrix} = \sqrt{\frac{2}{3}} \begin{bmatrix} 1/\sqrt{2} & 1/\sqrt{2} & 1/\sqrt{2} \\ 1 & -1/2 & -1/2 \\ 0 & \sqrt{3}/\sqrt{2} & -\sqrt{3}/\sqrt{2} \end{bmatrix} \cdot \begin{bmatrix} i_a \\ i_b \\ i_c \end{bmatrix} \tag{3}$$

$$\begin{bmatrix} p \\ q \end{bmatrix} = \begin{bmatrix} v_\alpha \, v_\beta \\ -v_\beta \, v_\alpha \end{bmatrix} \cdot \begin{bmatrix} i_\alpha \\ i_\beta \end{bmatrix} \tag{4}$$

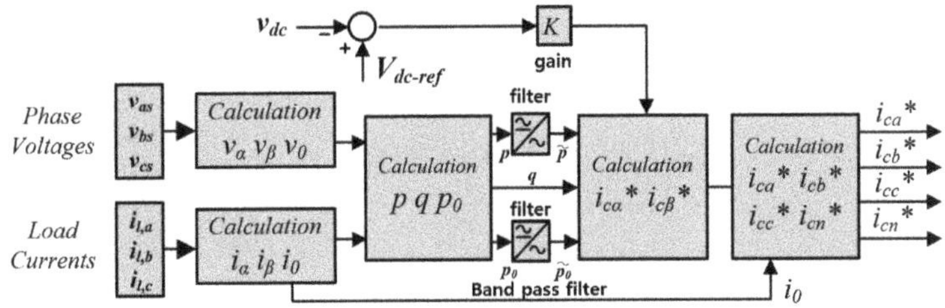

Figure 3. Current reference calculation based on the instantaneous p-q theory [13].

2.2. LUC Based on the Synchronous Frame Theory

Figure 4 illustrates a block diagram for calculating the current reference based on the synchronous reference frame theory. To compute the current reference, the current values must be transformed from the abc coordinates to the dq0 synchronous coordinates. Equation (5) represents the formula for transforming the abc coordinate system into the dq0 coordinate system. When the load is in an unbalanced state, the dq0 current can be expressed by Equation (6). Using this represented current, the unbalanced current reference of the inverter can be computed [26,27]. In Equation (6), the current is divided into DC and AC parts. To obtain a DC current component, an LPF with a very low cut-off frequency must be used. By extracting the constant component of the current through the LPF, the current reference in dq0 coordinates can be calculated using Equation (7). However, the performance of deriving the current reference from an LPF-based synchronous coordinate system is constrained by the time delay introduced by the filter [10,21]. Here, $\overline{i_d}$, $\overline{i_q}$, $\overline{i_0}$ and $\tilde{i}_d$, $\tilde{i}_q$, $\tilde{i}_o$ represent the constant and varying parts of the current, respectively.

$$\begin{bmatrix} i_d \\ i_q \\ i_0 \end{bmatrix} = \sqrt{\frac{2}{3}} \begin{bmatrix} \cos\theta & \cos\left(\theta - \frac{2}{3}\pi\right) & \cos\left(\theta + \frac{2}{3}\pi\right) \\ \sin\theta & \sin\left(\theta - \frac{2}{3}\pi\right) & \sin\left(\theta + \frac{2}{3}\pi\right) \\ \frac{1}{2} & \frac{1}{2} & \frac{1}{2} \end{bmatrix} \begin{bmatrix} i_a \\ i_b \\ i_c \end{bmatrix} \tag{5}$$

$$\begin{bmatrix} i_d \\ i_q \\ i_0 \end{bmatrix} = \begin{bmatrix} \overline{i_d} + \tilde{i}_d \\ \overline{i_q} + \tilde{i}_q \\ \overline{i_0} + \tilde{i}_o \end{bmatrix} \tag{6}$$

$$\begin{bmatrix} i_d^* \\ i_q^* \\ i_0^* \end{bmatrix} = \begin{bmatrix} i_d - \overline{i_d} \\ i_q - \overline{i_q} \\ i_0 - \overline{i_0} \end{bmatrix} \tag{7}$$

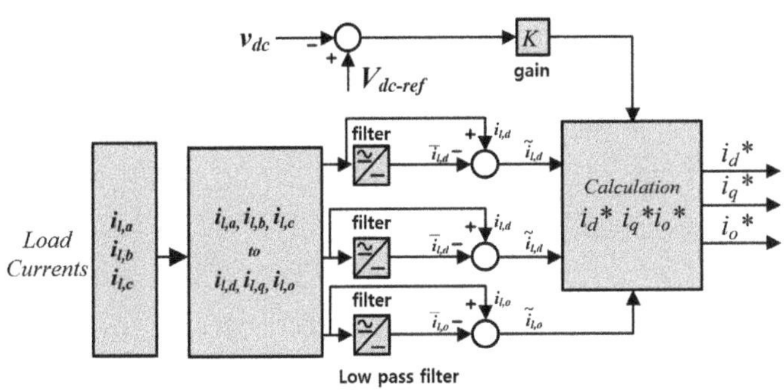

Figure 4. Current reference calculation based on the synchronous reference frame method [27].

3. Proposed Current Compensation Method Using ANN

Figure 5 shows the control block diagram of the inverter, outlining the proposed approach for compensating the grid current under unbalanced load conditions. Initially, an ANN was devised to calculate the reference current and compensate for the unbalanced current. The ANN was fed with the synchronous coordinate values of the AC current as the input. In addition, three supplementary ANNs were established to control the phase current. A conventional carrier-based pulse width modulation method was employed to govern the inverter, based on the output derived from the ANN training.

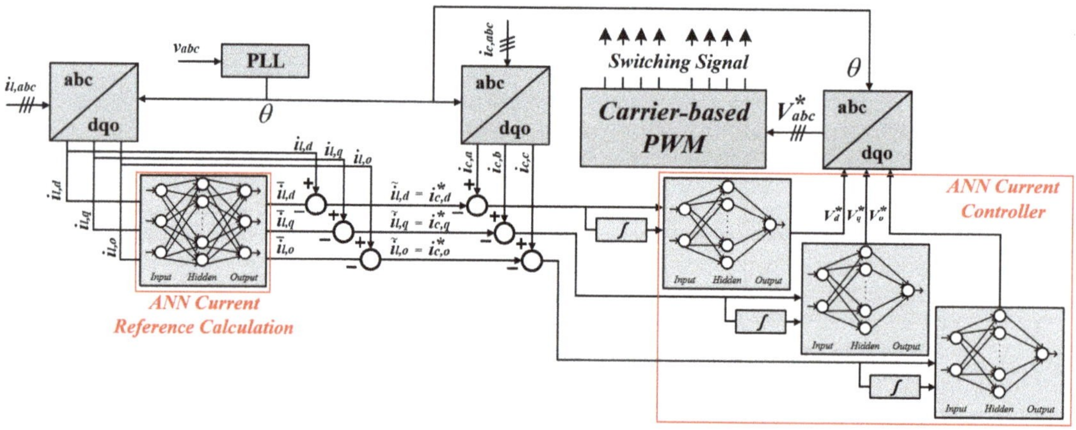

Figure 5. Proposed current compensating method control block diagram based on ANN.

3.1. Designing the ANN Structure using MATLAB

The structure of an ANN is similar to that of a human neural network. The architecture of the ANN applied in this paper is depicted in Figure 6, where x_k represents the input variables; w_k denotes the multiplication weights of the ANN; the activation function, the output function, and the bias represent the addition weights of the ANN; and y signifies the output value.

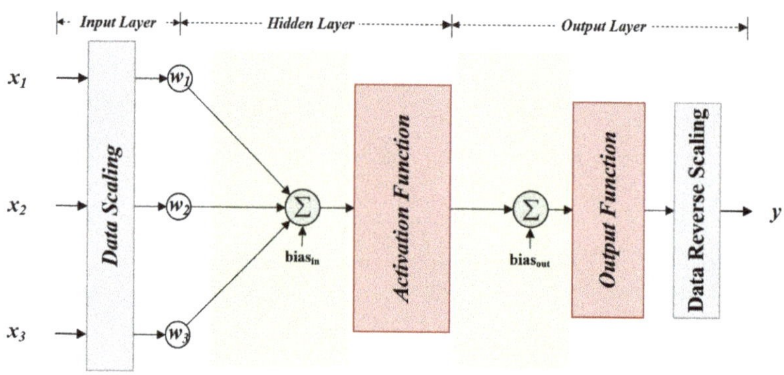

Figure 6. The structure of the ANN applied in this paper.

The ANN consists of three layers: the input, the hidden, and the output layers. The output of the activation function is formulated as the product and sum of the inputs, as expressed in the following equation, where $x_{k,min}$, $x_{k,max}$, y_{min}, and y_{max} are the minimum and maximum values of each sample of training input and output data, respectively.

$$y = \frac{y_{\max} - y_{\min}}{2}\left[f\left[\left\{2\left(\frac{x_1 - x_{1,\min}}{x_{1,\max} - x_{1,\min}}\right) - 1\right\}w_1 + \left\{2\left(\frac{x_2 - x_{2,\min}}{x_{2,\max} - x_{2,\min}}\right) - 1\right\}w_2 + \left\{2\left(\frac{x_3 - x_{3,\min}}{x_{3,\max} - x_{3,\min}}\right) - 1\right\}w_3 + bias_{in}\right] + bias_{out}\right] + y_{\min} \quad (8)$$

The activation function outputs large values for inputs above a certain threshold, without outputting small values. Owing to the characteristics of each activation function, they typically output values between 0 and 1 or between −1 and 1, necessitating the scaling of the input and output data. Although various activation functions exist, the three most commonly used functions include the sigmoid, the rectified linear unit (ReLU), and the hyperbolic tangent functions, which are represented in Figure 7.

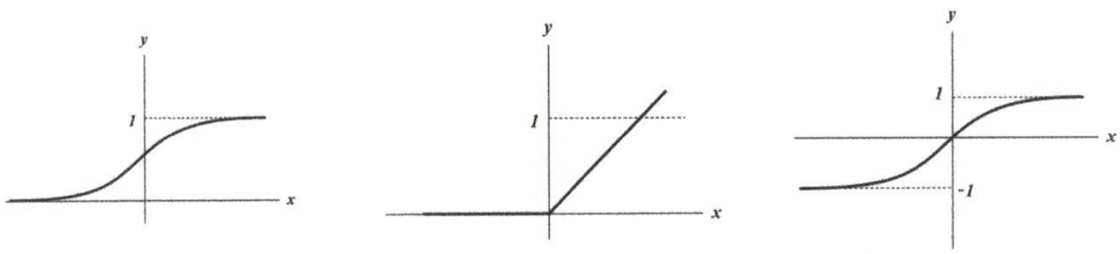

(**a**) Sigmoid (**b**) ReLU (**c**) Hyperbolic Tangent

Figure 7. Example of the activation function.

The reason for using nonlinear activation functions is to solve the problems of the model more effectively. When using linear functions, the presence of hidden layers may result in unnecessary outcomes because the relationship between the inputs and outputs is merely represented by addition and multiplication. Therefore, the activation functions must be non-linear. However, for the output function, it is necessary to reflect data passing through the hidden layers of the output. However, the application of a non-linear function to an output function may lead to data loss. In this study, we applied a sigmoid function as the activation function, which is reflected in the simulation results presented in the following section [28]. The MATLAB R2020b toolbox used to train the ANN applied in the paper was validated through three steps.

(1) Step 1: training data ratio validation

In this study, MATLAB's Neural Fitting (nftool), from MathWorks, was utilized for the ANN training, and the structure of the ANN was configured as shown in Figure 8. When using the training toolbox, the user can select the ratio of training data, validation data to monitor the training progress, and testing data to evaluate the performance of the trained ANN. In this study, 100 test datasets for the q-axis current controller were used to determine the data sample ratio. The results are presented in Table 1 and Figure 9, where it can be observed that the presence of appropriate ratios of validation and testing data is essential for achieving high accuracy.

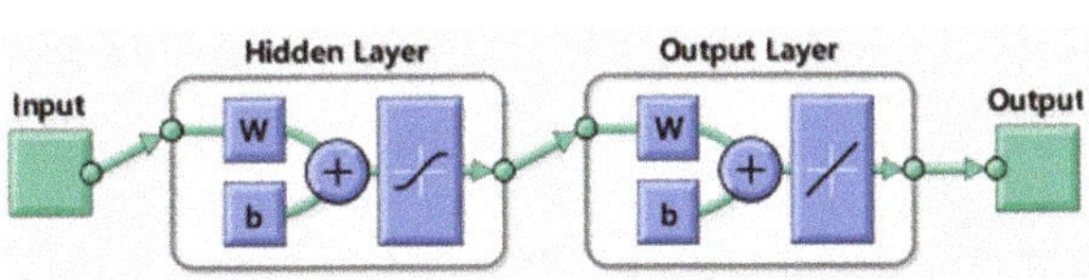

Figure 8. ANN structure in MATLAB.

Table 1. Result of training data ratio based on the q-axis current controller dataset.

		Samples (%)	MSE (Mean Square Error)	R
1 Set	Training	70	4.55392×10^{-8}	9.99999×10^{-1}
	Validation	15	2.22321×10^{-7}	9.99999×10^{-1}
	Testing	15	1.63285×10^{-7}	9.99999×10^{-1}
2 Set	Training	90	3.44847×10^{-7}	9.99999×10^{-1}
	Validation	5	1.24599×10^{-6}	9.99999×10^{-1}
	Testing	5	9.57606×10^{-7}	9.99999×10^{-1}

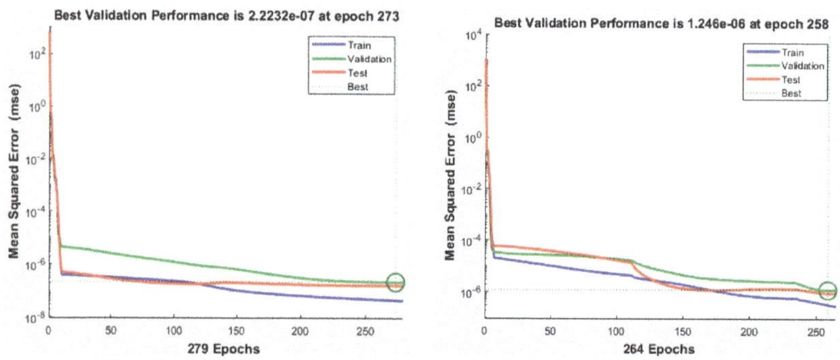

(**a**) Result of 1 set training. (**b**) Result of 2 set training.

Figure 9. MSE of ANN training according to dataset ratio.

(2) Step 2: training algorithm validation

Based on previous results, the ratio of the test data was configured as follows: training 70%, validation 15%, and testing 15%. In this study, ANN was trained using the backpropagation method, and MATLAB provided three algorithms: the Levenberg–Marquardt (LM) algorithm, Bayesian regularization (BR), and the scaled conjugate gradient (SCG) method, which can be summarized as follows:

- Levenberg–Marquardt algorithm: The LM method combines the Gauss–Newton and gradient descent methods to efficiently determine both global and local minima. Although this offers both stability and faster convergence, it requires significant computational effort to determine initial values when they are not provided.
- Bayesian regularization: This method trains a neural network to minimize errors only for the provided training data, which may lead to over-confidence. Although BR considers uncertainty and enhances robustness against noise or inputs beyond the training data, it tends to have a slower convergence. Additionally, it may produce different outputs for the same parameters owing to its consideration of uncertainty.
- Scaled conjugate gradient method: Introduced by Moller, the SCG method differs from other conjugate gradient algorithms in that it does not recalculate at each iteration and performs backpropagation with a second-order approximation of the error. This approach ensures the robustness and independence of the neural network from user-defined training data. However, the approach is complex and requires substantial computational resources and training times.

Figure 10 depicts the results of training the ANN current controller using three algorithms supported by MATLAB. All simulations involved 1000 iterations of identical training using the q-axis current controller dataset. When comparing the training performance and error histograms, as shown in Table 2, it is observed that the LM training algorithm exhibits the least amount of error.

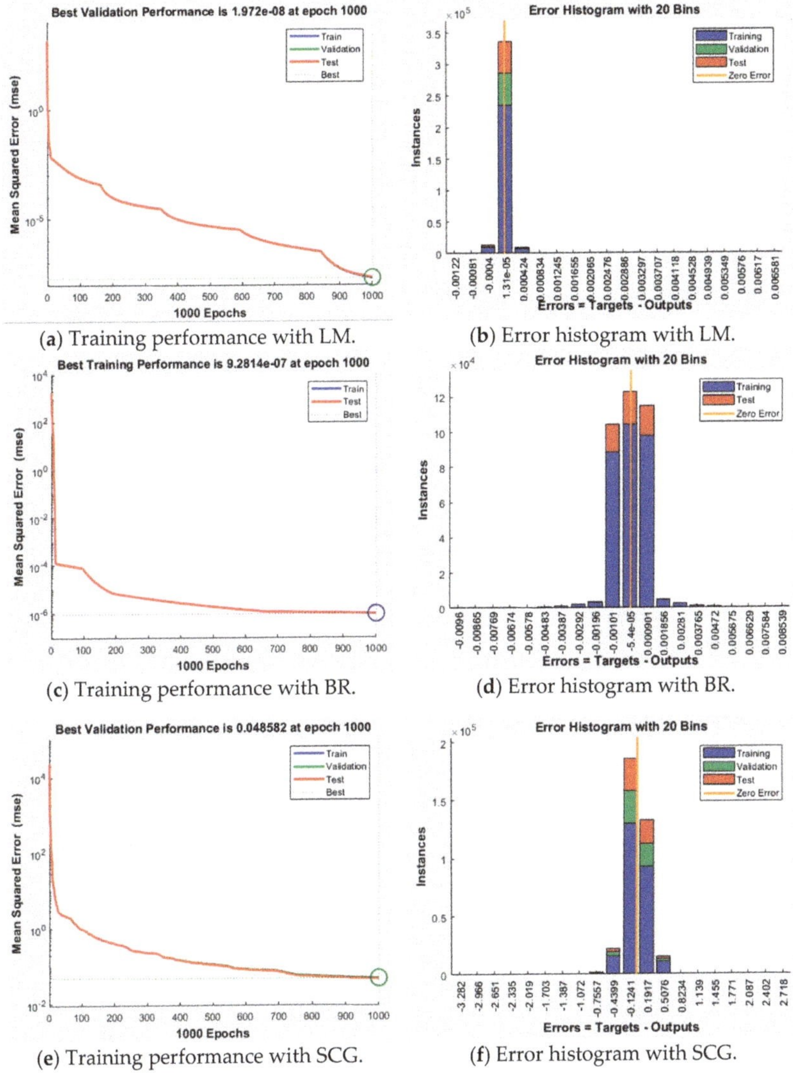

(**a**) Training performance with LM. (**b**) Error histogram with LM.

(**c**) Training performance with BR. (**d**) Error histogram with BR.

(**e**) Training performance with SCG. (**f**) Error histogram with SCG.

Figure 10. ANN training results according to the algorithm.

Table 2. Comparison of ANN training algorithms—Epoch 1000.

	Training Performance	Error Histogram
Levenberg–Marquardt	1.972×10^{-8}	1.31×10^{-5}
Bayesian Regularization	9.2814×10^{-7}	-5.4×10^{-5}
SCG Method	0.0048582	-0.1241

(3) Step 3: number of neural validations

Figure 11 illustrates the training performance and error histogram based on the number of hidden neurons in the ANN using the q-axis current controller dataset. The training performance of the ANN with 30 hidden neurons is superior; however, the error histogram of the ANN with 20 hidden neurons indicates that the majority of errors are dominantly distributed in one area. Such dominant errors make it easier to compensate for the results of the ANN through the feedforward method. Based on the simulation results, we applied an ANN with 20 hidden neurons trained using the LM method.

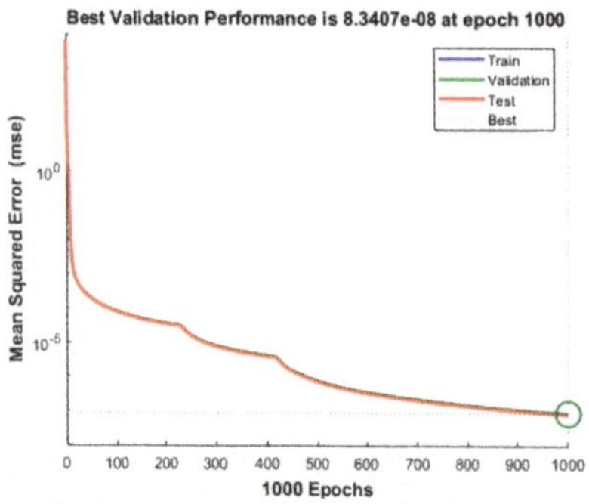

(**a**) Training performance with 10 neurons.

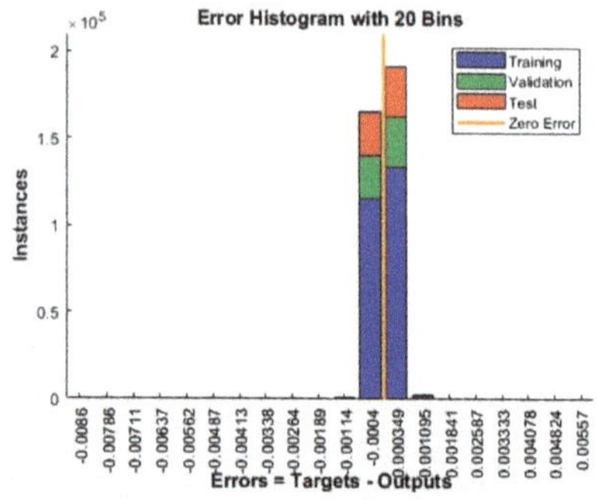

(**b**) Error histogram with 10 neurons.

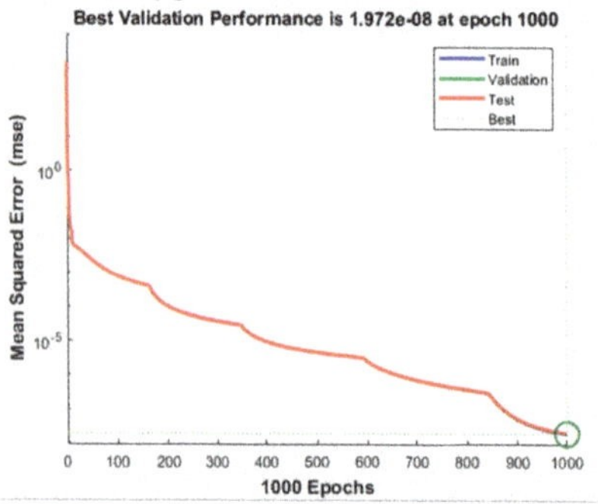

(**c**) Training performance with 20 neurons.

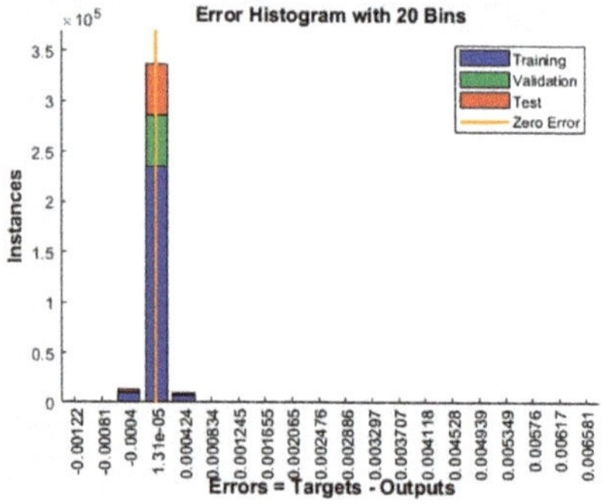

(**d**) Error histogram with 20 neurons.

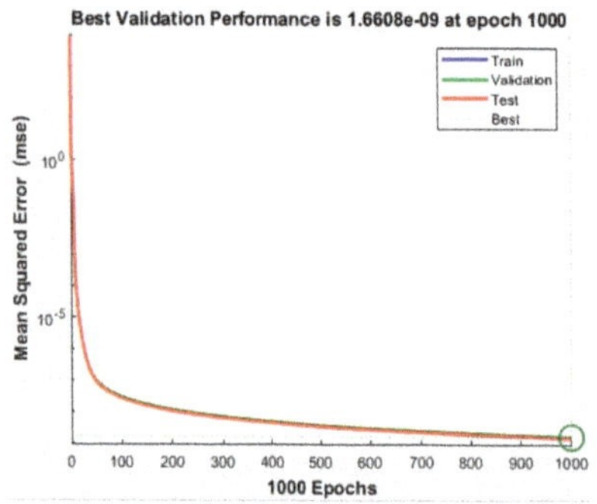

(**e**) Training performance with 30 neurons.

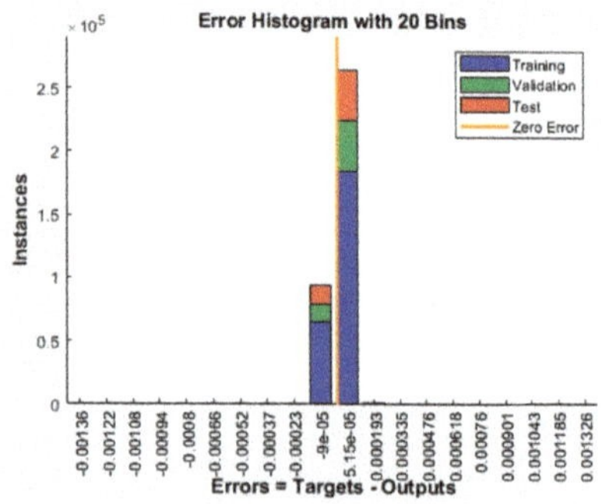

(**f**) Error histogram with 30 neurons.

Figure 11. ANN training results based on the number of hidden neurons.

3.2. *Designing an ANN for the Calculation of Current Reference under Unbalanced Load Conditions*

The structure of the ANN for calculating the current reference in the case of load unbalance is illustrated in Figure 12. The ANN input consisted of the dq0 axis synchronous reference frame currents of the load, while the output wass composed of the average values of the unbalanced magnitudes in the dq0 axis. The ANN was configured using three inputs, three outputs, and 20 hidden neurons. Figure 13 shows the training results of the ANN for unbalance calculation. Table 3 summarizes the approximately 40,000 samples of data used to train the ANN.

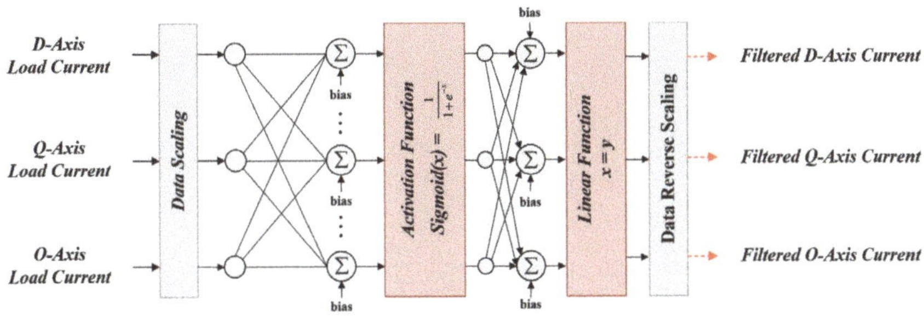

Figure 12. ANN configuration of current reference calculation.

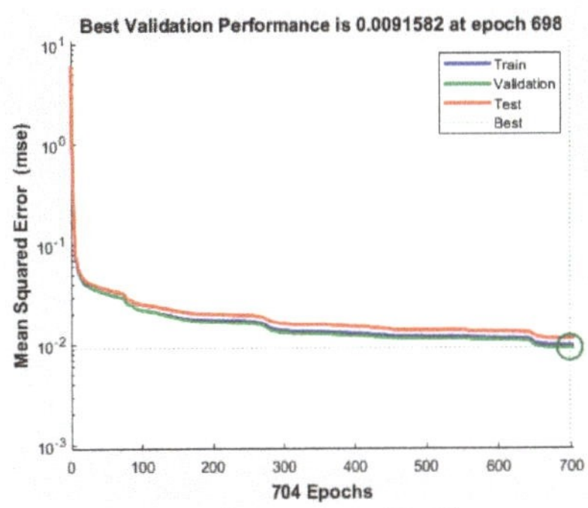

(**a**) Training performance with 20 neurons.

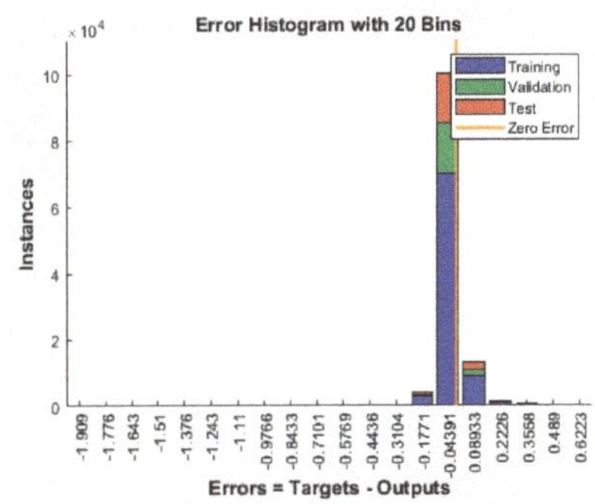

(**b**) Error histogram.

Figure 13. Training result of ANN for current reference calculation.

Table 3. ANN training data summary of current reference calculation.

Input Data			Output Data		
I_{de}	I_{qe}	I_{oe}	I_{de_ref}	I_{qe_ref}	I_{oe_ref}
−0.881433331	−9.342378775	−0.181107398	-6.01×10^{-1}	−10.781192	0.000155916
−0.661866636	−9.316670329	−0.070921802	-6.01×10^{-1}	−10.781192	0.000155916
−0.440898051	−9.324221858	0.039687159	-6.01×10^{-1}	−10.781192	0.000155916
		……			
−2.942782404	−9.855400085	−1.241158981	−0.77460625	−13.886127	0.000126828
−2.312665572	−9.575398404	−0.905744661	−0.77460625	−13.886127	0.000126828
−1.647599185	−9.393269653	−0.565187099	−0.77460625	−13.886127	0.000126828

3.3. *Designing an ANN for the Current Controller in the dq0 Axis*

The structure of the ANN for the dq0 axis current controller in the case of a load unbalance is illustrated in Figure 14. The input of the ANN consists of the proportional and integral error values of each axis, whereas the output is composed of the PI controller of each axis. The ANN was configured with two inputs, one output, and 20 hidden neurons. Figure 15 shows the training results of the ANN for the current controller of each axis. Table 4 summarizes approximately 85,000 samples of each axis of the data used to train the ANN. Table 5 summarizes the training results of the ANN conducted in this paper.

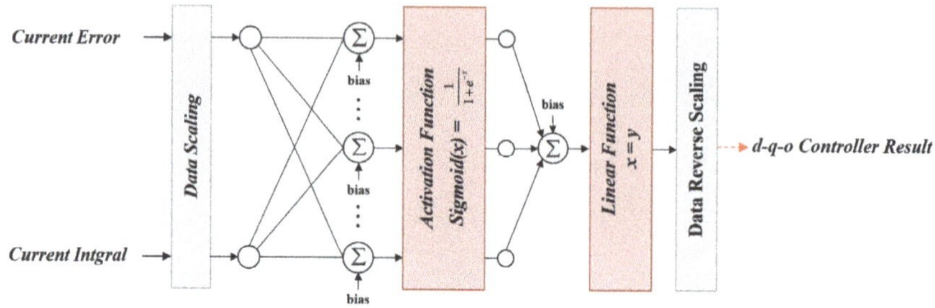

Figure 14. ANN configuration of dq0 axis.

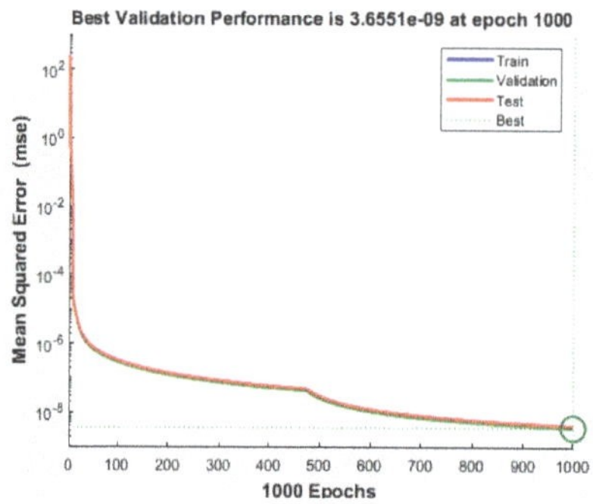

(**a**) D-axis training performance.

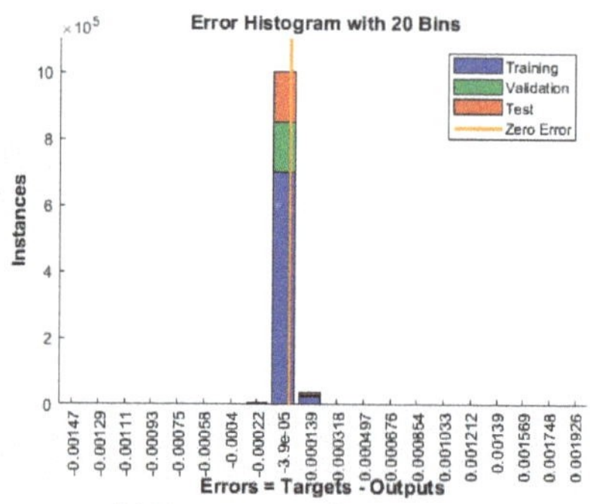

(**b**) D-axis error histogram.

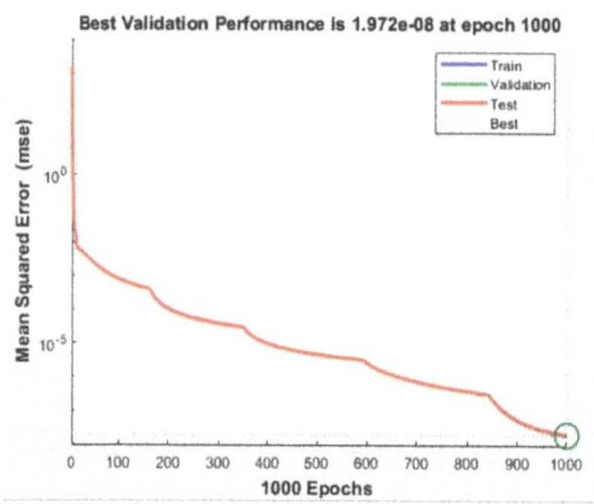

(**c**) Q-axis training performance.

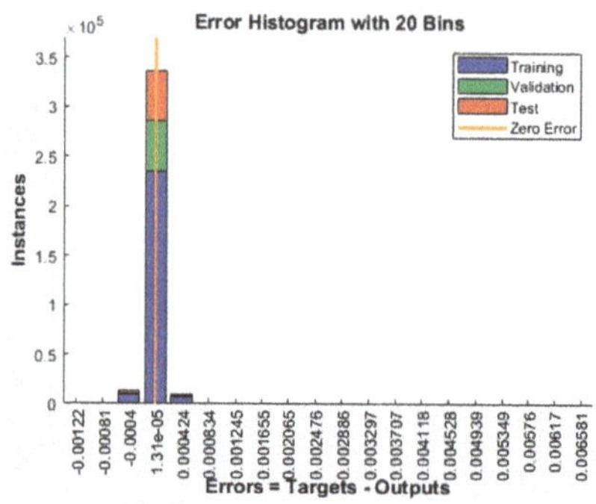

(**d**) Q-axis error histogram.

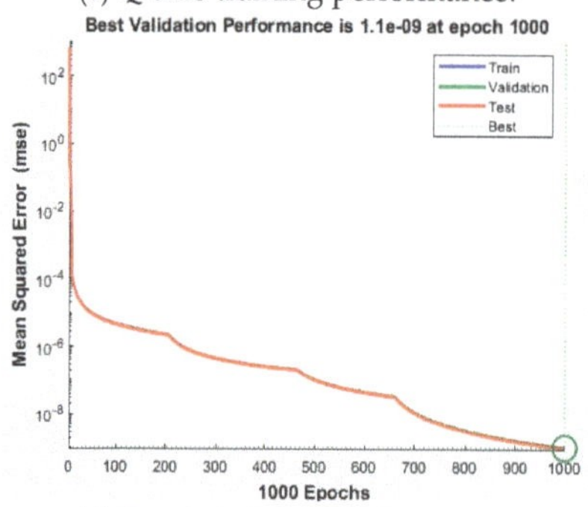

(**e**) O-axis training performance.

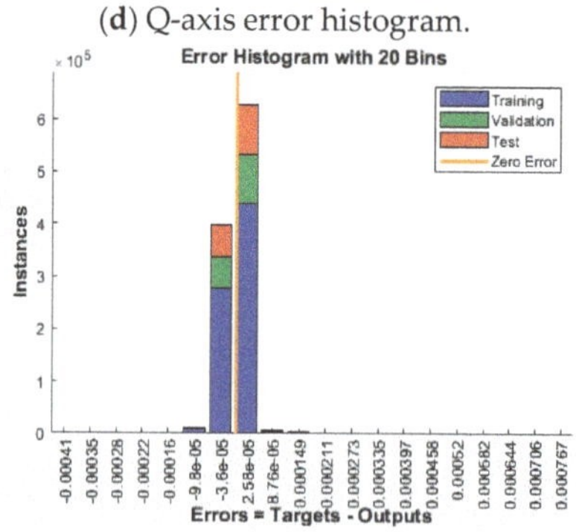

(**f**) O-axis error histogram.

Figure 15. Training results for an ANN for the current controller of dq0 axis.

Table 4. ANN training data summary of the current controller.

Input Data						Output Data		
Error D-Axis	Integral D-Axis	Error Q-Axis	Integral Q-Axis	Error O-Axis	Integral O-Axis	D-Axis	Q-Axis	O-Axis
0.065365	12.41371	−0.0188	−358.686	0.074611	0.390667	13.72101	−359.062	1.882889
0.07662	12.5265	−0.0413	−358.738	0.034672	0.45742	14.05891	−359.564	1.150865
0.041886	12.50608	−0.04018	−358.977	0.038226	0.616964	13.34381	−359.781	1.38149
				....				
0.059906	12.49728	−0.07878	−358.877	0.076294	0.651035	13.69541	−360.452	2.176914
0.06789	12.1754	0.006137	−358.662	−0.04231	−0.22027	13.53321	−358.54	−1.06654
−0.00376	11.72399	0.075598	−359.193	−0.06424	0.526357	11.64879	−357.681	−0.75839

Table 5. Summary of ANN learning results.

	Training Performance	Dominant Error Histogram
Unbalance calculation	0.0091582	−0.04391
d-axis controller	3.6551×10^{-9}	-3.9×10^{-5}
q-axis controller	1.972×10^{-8}	1.31×10^{-5}
o-axis controller	1.19×10^{-9}	2.58×10^{-5}

4. Simulation Results

Simulations were performed using the PSIM 9.1.4 program to validate the effectiveness of the proposed method. Three simulation cases are considered to conduct a comparative analysis using the conventional method. The initial case involved the application of the LPF-based dq0 reference generation and the PI current control. In the second case, an ANN-based dq0 reference generation, coupled with a PI controller, was incorporated. Finally, the proposed approach utilized an ANN-based dq0 reference generation and an ANN current controller for simulation. Figure 16 illustrates the configuration of the applied simulation and experimental setups. The parameters of the four-leg inverter and the load applied in the simulations and experiments are listed in Table 6. The unbalance loads for each phase are listed in Table 6, and the voltages and currents before the compensation are shown in Figure 17.

Table 6. Parameters of the simulation and experiment.

Parameters		Values	Unit
Grid Line to Line Voltage		220	V_{rms}
DC-Link Voltage		380	V_{dc}
AC Load	A Phase	1.014	kW
	B Phase	0.69	kW
	C Phase	0.69	kW
Inverter Rated Power		20	kW
LCL Filter	Grid Inductance	100	μH
	Filter Capacitance	22	μF
	Converter Inductance	1500	μH
DC-Link Capacitance		1200	μF

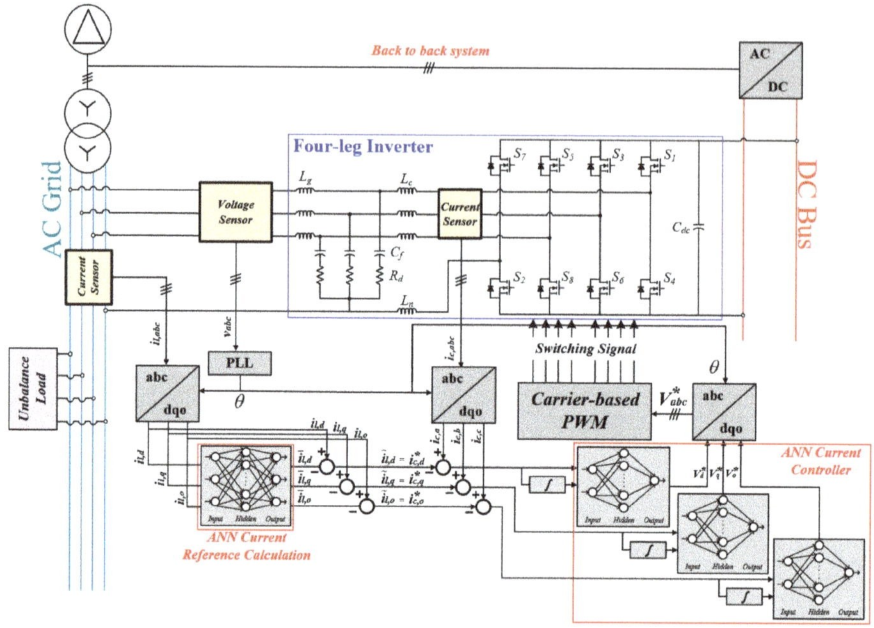

Figure 16. Simulation and experimental setup configuration.

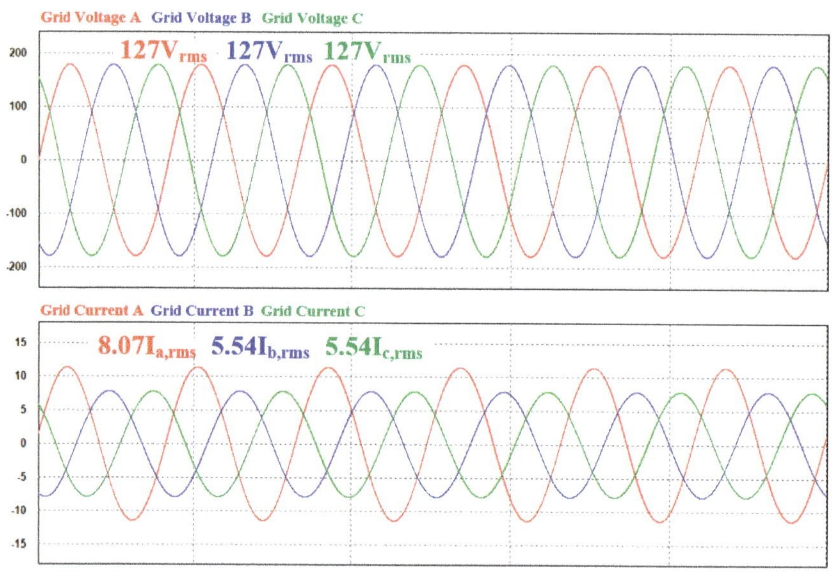

Figure 17. Nominal voltage and current under unbalanced load conditions.

4.1. Case 1 (LPF-Based dq0 Reference Generation and PI Current Controller)

The conventional method employs a synchronous reference frame LPF-based current reference calculation. This approach calculates the magnitude of the unbalance value using an LPF and compensates for the unbalanced current using an inverter with a PI current controller. Figure 18 shows the simulation waveforms obtained using the conventional method. At time t_1, a load is applied during the simulation, and at time t_2, the load is removed. The load was applied with 1.014 kW in phase A, 0.69 kW in phase B, and 0.69 kW in phase C, making an unbalanced current situation. The waveform depicts the current in the distribution system and the compensating phase current provided by the inverter. The next waveform represents the magnitude of the unbalance component calculated using the LPF. The final simulation waveform depicts the current reference for the compensation performed by the inverter. The simulation waveform on the left represents the situation in which the system transitions from a no-load to a moderate load condition, while the simulation waveform on the right represents the situation in which the system transitions from a loaded state to a no-load condition. The first waveform in Figure 18 represents the grid current under unbalanced load conditions. This current slowly follows the load variations owing to the delay in determining the unbalance magnitude using the LPF, as depicted in the third plot. When transitioning from a no-load to a loaded condition, it takes 0.125 s to reach an average q-axis current value of −10.788 in a steady state. Similarly, when transitioning from a loaded to a no-load condition, it takes 0.1 s to reach a steady state. Figure 19 shows enlarged waveforms of points A and B in Figure 18. The ripple magnitudes and total harmonic distortion (THD) of each phase current are listed in Table 7. However, for Point B, the ripple in the phase A current is negligible and is thus not indicated separately.

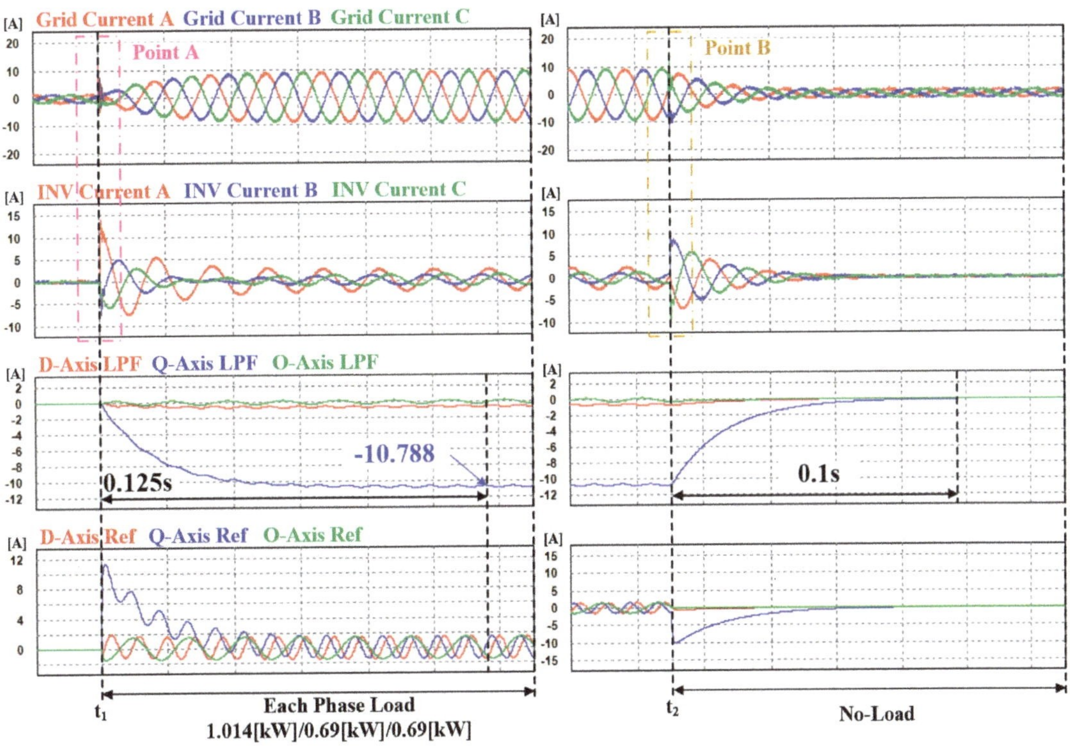

Figure 18. Case 1 simulation of waveforms with LPF-based dq0 reference generation and a PI current controller.

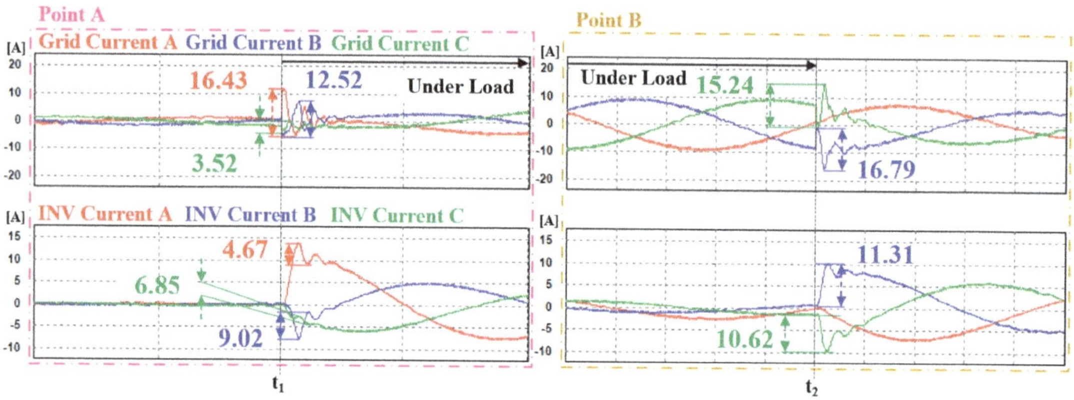

Figure 19. Enlarged transient state waveforms of each point with LPF-based dq0 reference generation and a PI current controller.

Table 7. Maximum ripple and THD of phase currents in the transient state owing to load fluctuations in Case 1.

		A Phase	B Phase	C Phase	Average
Point A [A]	Grid	16.43	12.52	3.52	10.82
	Inverter	4.67	9.02	6.85	6.85
Point B [A]	Grid	-	16.79	15.24	16.02
	Inverter	-	11.31	10.62	10.97
THD [%]	Grid	3.365	3.409	3.451	3.41
	Inverter	5.637	12.21	10.852	9.57

Such delays are unsuitable for distribution systems in which loads continuously fluctuate. To address this issue, an ANN was applied to the dq0 reference generation in the inverter, reducing the time required for compensation to reach a steady state.

4.2. *Case 2 (ANN dq0 Reference Generation and PI Current Controller)*

In Case 2, unlike in Case 1, the synchronous reference LPF-based current reference calculation was performed using an ANN. The current controller used the same PI current controller as that employed in Case 1. In the simulation, the load transitioned from no load to load at t_1, and from load to no load at t_2, as in Case 1. The load values were also the same as in Case 1, with 1.014 kW in phase A, 0.69 kW in phase B, and 0.69 kW in phase C. The simulation waveforms shown in Figure 20 are arranged in a manner similar to that in Case 1; however, the third waveform represents the magnitude of the unbalanced current calculated by the ANN. Figure 21 shows enlarged waveforms of points A and B from Figure 20. The maximum ripple magnitude and THD for each current are listed in Table 8. Similar to Case 1, for Point B, the ripple in the phase A current is negligible and is thus not indicated separately.

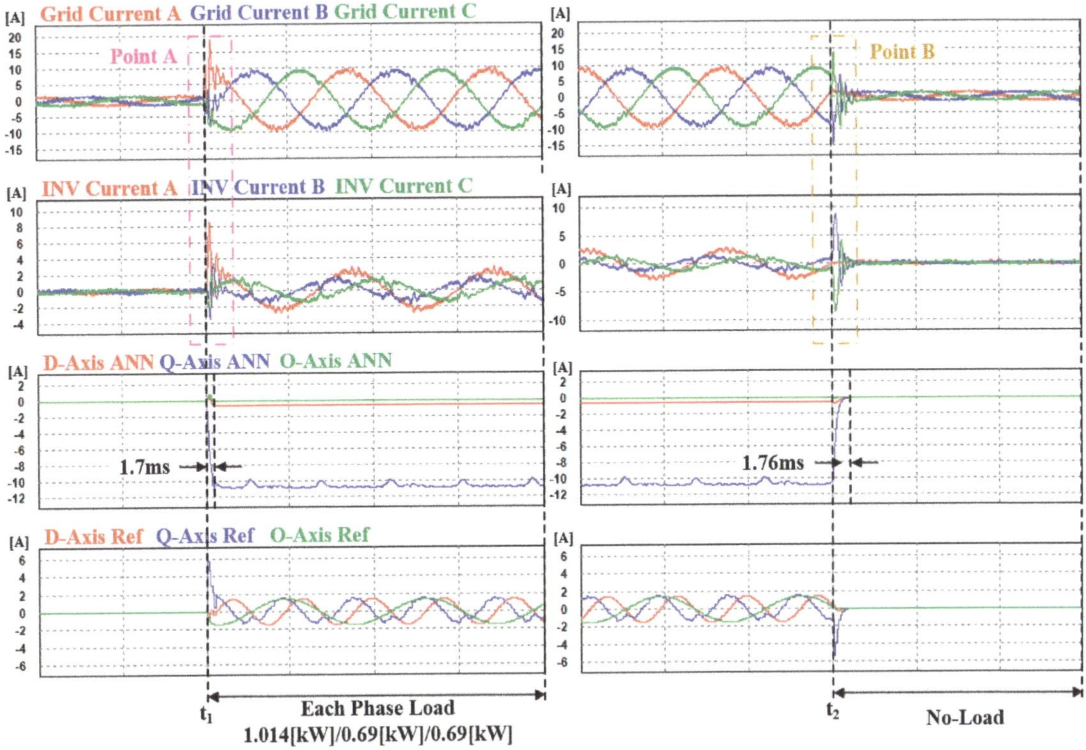

Figure 20. Case 2 simulation waveforms, with ANN dq0 reference generation and a PI current controller.

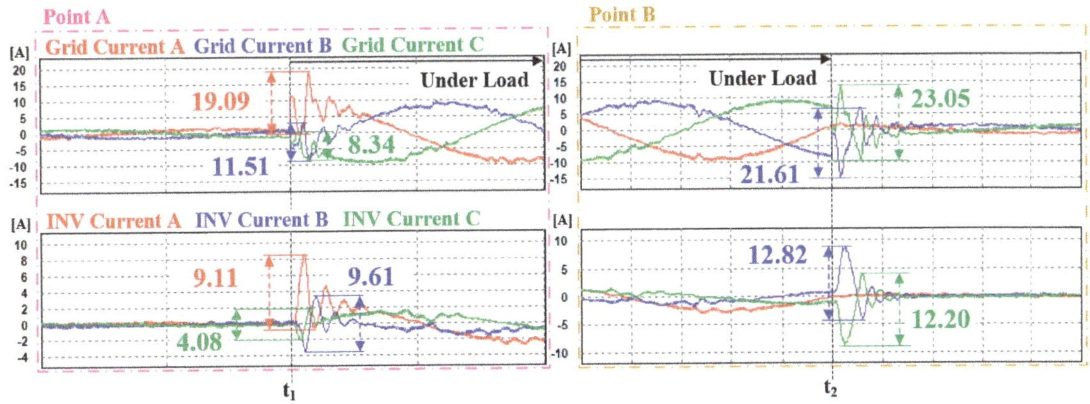

Figure 21. Enlarged transient state waveforms of each point, with ANN dq0 reference generation and a PI current controller.

Table 8. Maximum ripple and THD of phase currents in the transient state due to load fluctuations in Case 2.

		A Phase	B Phase	C Phase	Average
Point A [A]	Grid	19.09	11.51	8.34	12.98
	Inverter	9.11	9.61	4.08	7.6
Point B [A]	Grid	-	21.61	23.05	22.33
	Inverter	-	12.82	12.2	12.51
THD [%]	Grid	5.385	6.809	5.595	5.93
	Inverter	9.41	30.24	28.11	22.59

Instead of the synchronous reference LPF-based current reference calculation, applying an ANN significantly reduces the transition time from no-load to load conditions. Specifically, the time to reach the steady state decreases from 0.125 s to 1.7 ms, which is 73.5 times faster when transitioning from no-load to load conditions. Similarly, when transitioning from load to no-load conditions, the time reduces from 0.1 s to 1.76 s, which is 56.82 times faster for reaching the steady state. The simulation results revealed the following outcomes. The MSE for calculating the dq0 commands using an ANN was found to be exceedingly low. However, owing to the error and integral value changes in the PI current controller, unlearned input values distort the output of the ANN every 8.5 ms. Consequently, this decreases the THD of the currents in the inverter and grid. However, this problem arises because the periodic occurrence of the incorrect ANN output leads to an overall increase in the phase current ripple in both the grid and inverter. This ripple increase resulted in deteriorated THD levels. The outcomes of this study are presented in Table 8. Ultimately, at Point A, the average maximum ripple in the grid increased by 1.2 times, and for the inverter current, it increased by 1.12 times. At Point B, the average maximum ripple in the grid increased by 1.39 times, and for the inverter current, it increased by 1.14 times. The THD was calculated to be 1.74 times higher than the average current distortion at Point A and 2.36 times higher at Point B. In the no-load condition, the ANN unbalance calculation outputs values close to zero for the d, q, and o currents. In addition to these training results, considering that the three inputs (the dq0 axis currents) determine the three outputs (dq0 axis current references), all three values include AC components, resulting in a slight time delay in outputting the DC component values.

Combining the conventional PI current controller with the ANN reference calculation resulted in a reduction in the transient states. However, the average maximum ripple and current average THD decreased. To mitigate this periodic distortion, a separately trained ANN-based current controller, which learned the repeated incorrect output and transients of the ANN, was applied in Case 3.

4.3. *Case 3 (ANN dq0 Reference Generation and ANN Current Controller)*

In Case 3, both ANN-based reference calculations and current controllers were applied. The ANN-based current controller applied in Case 3 was trained to address the repetitive incorrect outputs of the ANN observed in Case 2, along with the transient and steady states, based on the PI current controller. Case 3 also applied the same load magnitude at the same time as in Cases 1 and 2 to compare the transient states at the same point in time. Figures 22 and 23 represent the simulation results for Case 3, with a simulation setup identical to that of Case 2. The maximum ripple magnitude and THD for each current are listed in Table 9. Similar to the previous cases, for Point B, the ripple in the phase A current is negligible, and is thus not indicated separately.

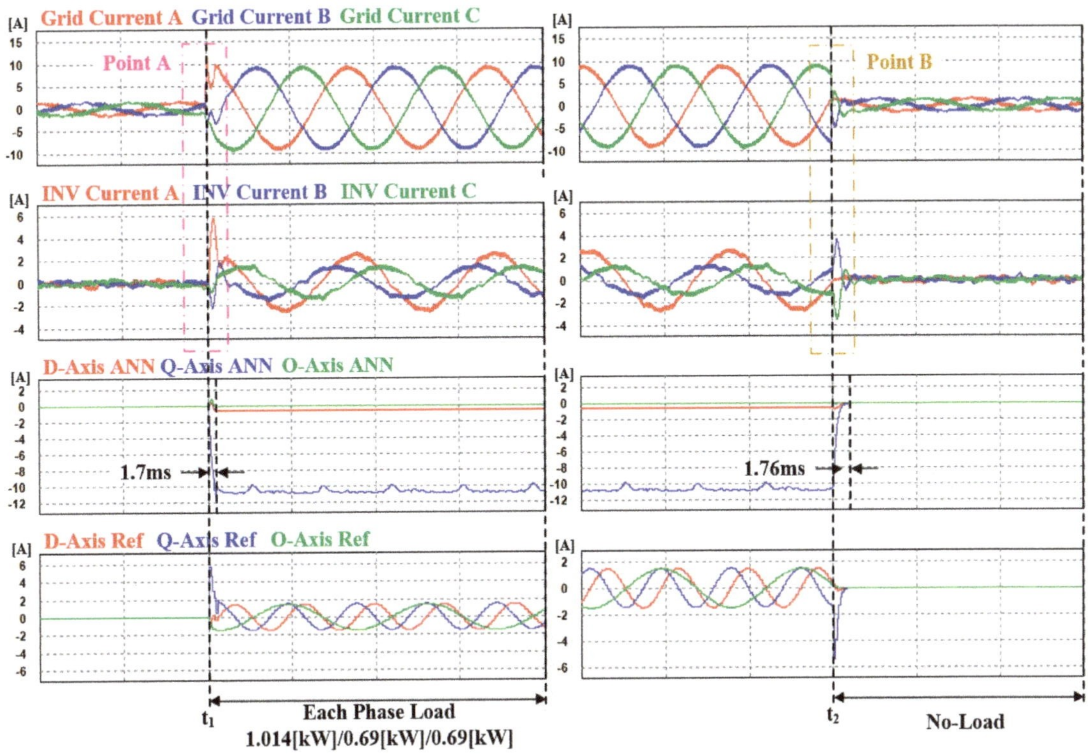

Figure 22. Case 3 simulation waveforms, with ANN dq0 reference generation and ANN current controller.

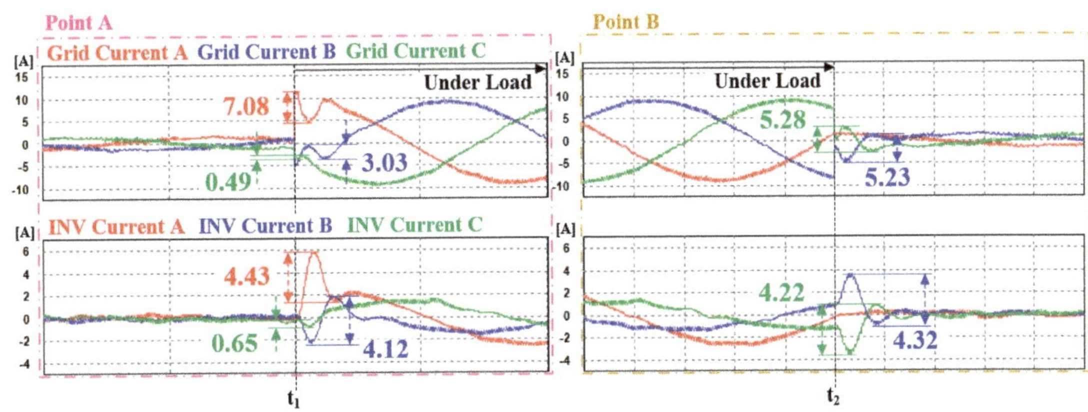

Figure 23. Enlarged transient state waveforms of each point, with ANN dq0 reference generation and ANN current controller.

Table 9. Maximum ripple and THD of phase currents in the transient state due to load fluctuations in Case 3.

		A Phase	B Phase	C Phase	Average
Point A [A]	Grid	7.08	3.03	0.49	3.53
	Inverter	4.43	4.12	0.65	3.07
Point B [A]	Grid	-	5.23	5.28	5.26
	Inverter	-	4.32	4.22	4.27
THD [%]	Grid	3.46	3.44	3.5	3.47
	Inverter	6.16	12.8	14.1	11.02

The time required for the current reference calculation to reach a steady state, based on the ANN, was confirmed through simulations to be the same as that in Case 2. As shown in Table 8, when the ANN and PI current controllers were applied, the maximum ripple and THD of the current deteriorated significantly. However, with the improvement achieved by the ANN current controller, compared to Case 1, the maximum ripples of the grid at points A and B decreased by 3.05 times and 2.57 times, respectively. For the inverter currents, they decreased by 3.06 times and 2.23 times, respectively, at each point. Simulation results confirm that compared to Case 2, the THD of both the grid and inverter currents improved by 1.71 times and 2.05 times, respectively, approaching a performance level similar to that of Case 1 in terms of THD.

Table 10 summarizes the average values of the three cases and compares their overall performance. The simulation results indicate that Case 3, which applies the proposed method, is advantageous for distribution system applications.

Table 10. Comparison of simulation average results for the three cases.

		Case 1	Case 2	Case 3
Point A maximum ripple [A]	Grid	10.82	12.98	3.53
	Inverter	6.85	7.6	3.07
Point B maximum ripple [A]	Grid	16.02	12.33	5.26
	Inverter	10.97	12.51	4.27
Average THD [%]	Grid	3.41	5.93	3.47
	Inverter	9.57	22.59	11.02
Average steady state reaching time [s]		0.1125	0.00173	0.00173

5. Experimental Results

Figure 24 shows the experimental configuration used to verify the proposed ANN-based compensation method under unbalanced load conditions. The parameters of the three-phase the four-leg inverter and the three-phase unbalanced resistance load are listed in Table 6. In the experiment, the current compensation method was applied to a four-leg inverter in a distributed system under unbalanced load conditions using an ANN.

The experiments were conducted in three cases, similar to the method used in the simulations. The AC voltage was maintained by the grid, while the DC voltage was controlled by the AC/DC converter connected in a back-to-back system linking to a delta-winding grid. The laboratory grid was in a delta-winding configuration, so three single-phase transformers were connected in a Y-Y configuration to perform the four-wire experiments. The experiments were conducted only on the no-load to loaded state sequence. Figure 25 depicts the voltage and current waveforms of the secondary side of the Y-Y transformer. Since the experiments were conducted in the laboratory, the grid was limited to the operation of the unbalanced experiments, and there is slight distortion in the voltage due to the presence of harmonics caused by the characteristics of the Y-Y transformer. The switching frequency of the four-leg inverter is 10 kHz, and the control period is 5 kHz, with experiments conducted accordingly.

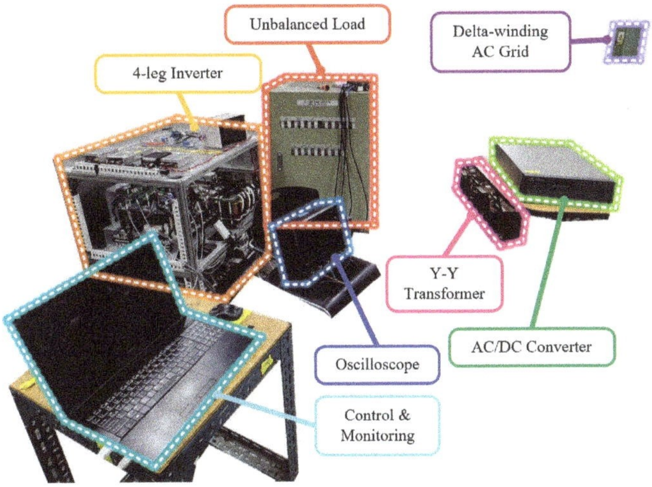

Figure 24. Experimental configuration of current compensation under the unbalanced load conditions.

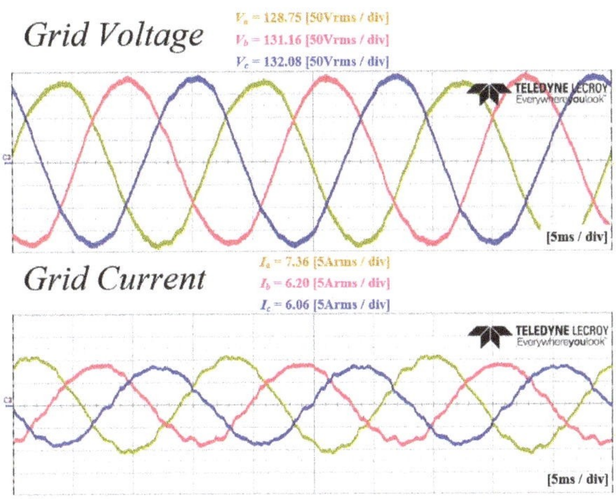

Figure 25. Nominal voltage and current under unbalanced load conditions in the experiment.

5.1. *Case 1 (LPF-Based dq0 Reference Generation and a PI Current Controller)*

In the Case 1 experiment, the conventional method was applied. The unbalanced current reference was calculated based on LPF, and a PI current controller was applied. Additionally, the transition to the load state occurred at time t_1. Figures 26 and 27, respectively, illustrate the grid current and converter current. Figure 28 represents the amount of unbalance in the dq0 axis using LPF. At time t_s, as shown in this figure, it reaches the average value, which takes approximately 81.1 ms. Figure 29 depicts the current reference in the dq0 axis generated by the calculation. Figures 30 and 31 show an enlarged waveform of Figures 26 and 27. The experimental results indicate that it requires more than several cycles to reach the overall steady state.

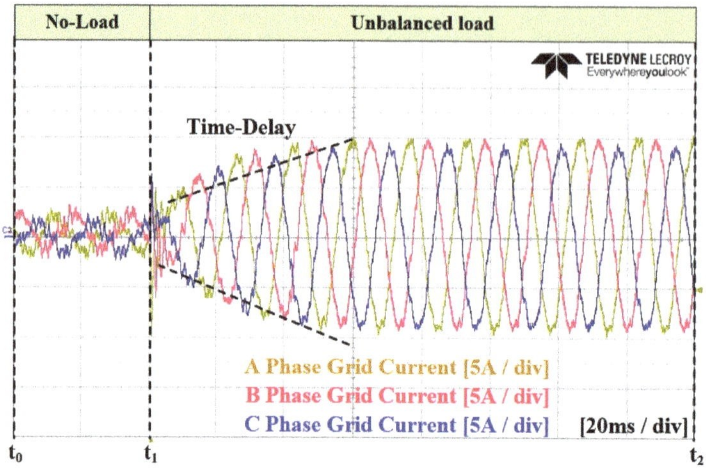

Figure 26. Experimental waveforms of the three-phase grid current in Case 1.

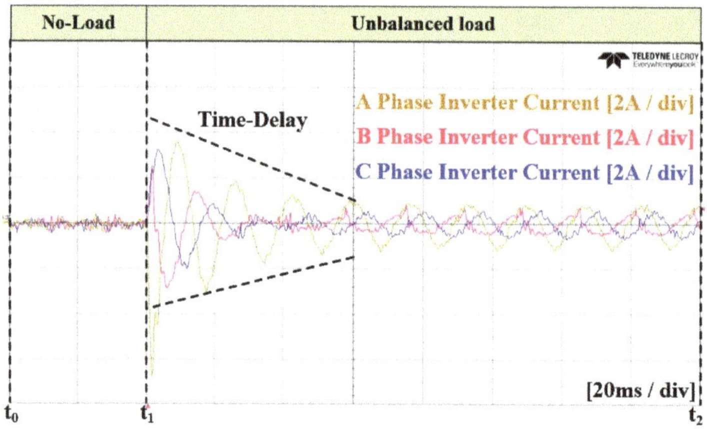

Figure 27. Experimental waveforms of the three-phase inverter current in Case 1.

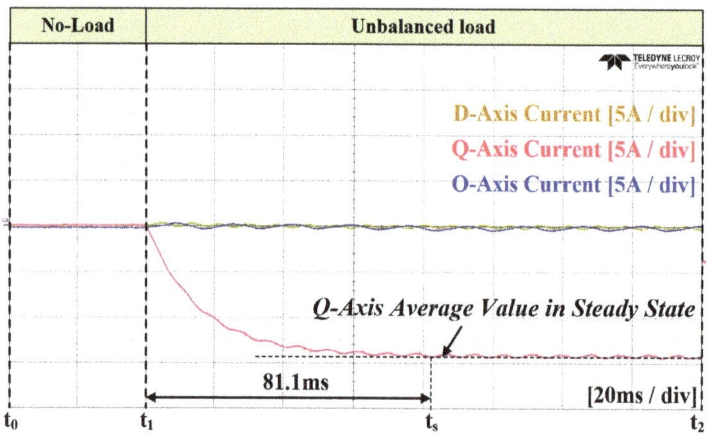

Figure 28. Experimental waveforms of the filtered dq0 value in Case 1.

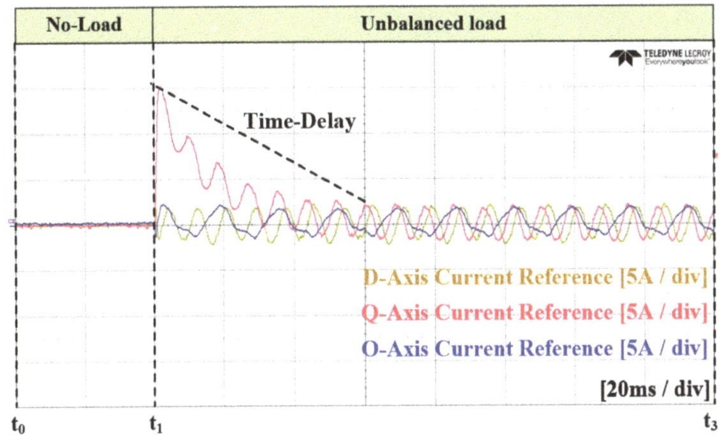

Figure 29. Experimental waveforms of the dq0 current reference in Case 1.

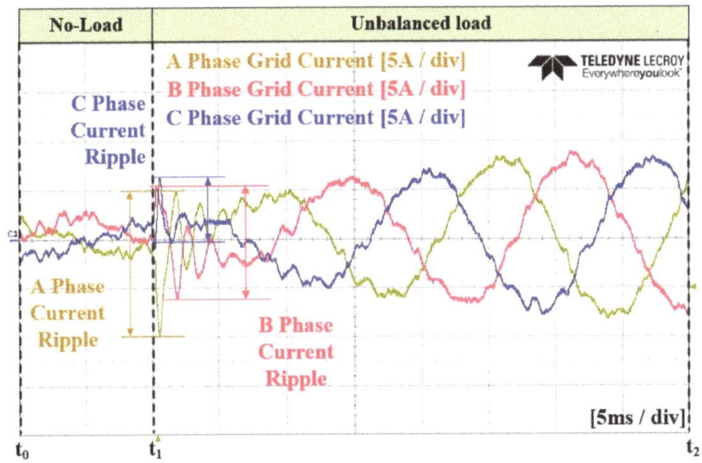

Figure 30. Enlarged waveform of the three-phase grid current in Case 1.

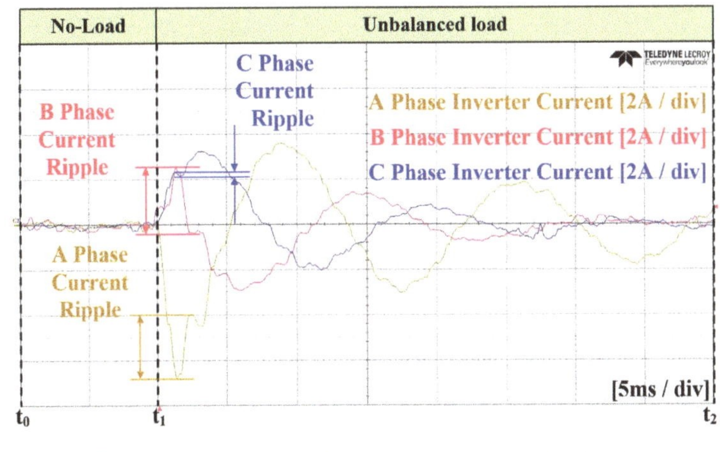

Figure 31. Enlarged waveform of the three-phase inverter current in Case 1.

5.2. Case 2 (ANN dq0 Reference Generation and PI Current Controller)

In Case 2 of the experiment, an ANN was applied instead of an LPF for calculating the magnitude of unbalance. The unbalanced current reference was calculated based on the ANN, and a PI current controller was applied. Similar to Case 1, the transition to the load condition occurred at t_1. Figures 32 and 33 represent the grid current and the converter current, respectively. Figure 34 illustrates the unbalance in the dq0 axis using an ANN. At t_s, it reaches the average value, taking approximately 6.7 ms. Figure 35 depicts the current reference on the dq0 axis calculated by the ANN. Figures 36 and 37 show the zoomed-in waveforms of Figures 32 and 33, respectively. Overall, the experimental results show the achievement of a steady state within one cycle. Additionally, it can be observed that the peak portion of the current exhibits more distortion compared to that noted when the conventional method was applied. However, unlike in the simulations, it appears challenging to quantitatively compare the size of the current ripples due to the inability to transition to the load condition at the same point.

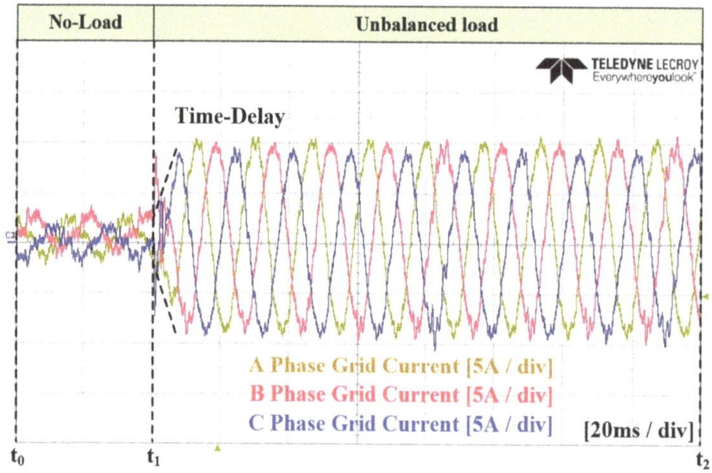

Figure 32. Experimental waveforms of the three-phase grid current in Case 2.

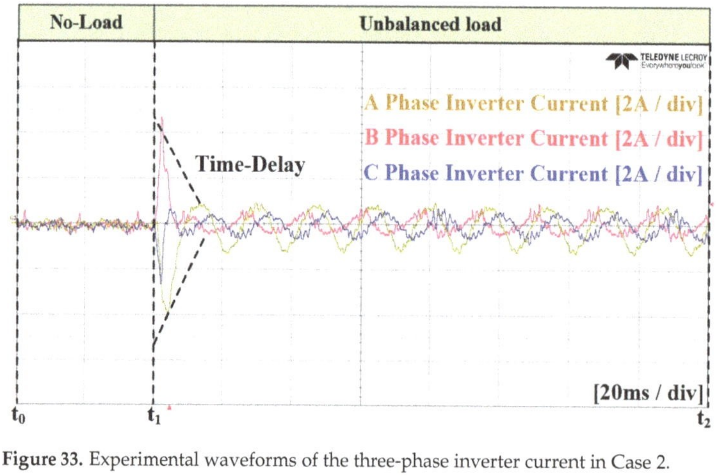

Figure 33. Experimental waveforms of the three-phase inverter current in Case 2.

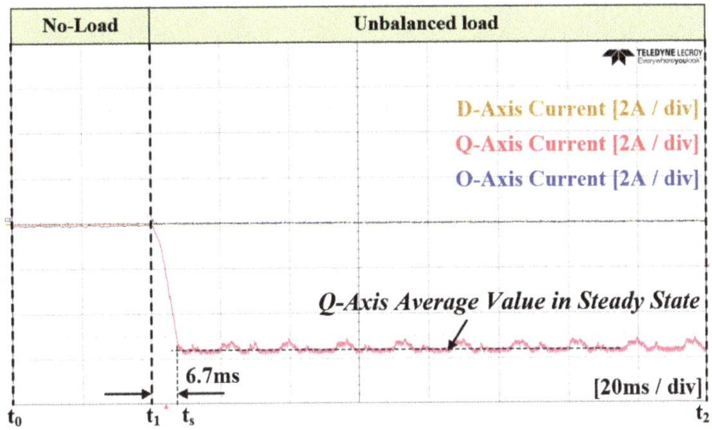

Figure 34. Experimental waveforms of the ANN output value in Case 2.

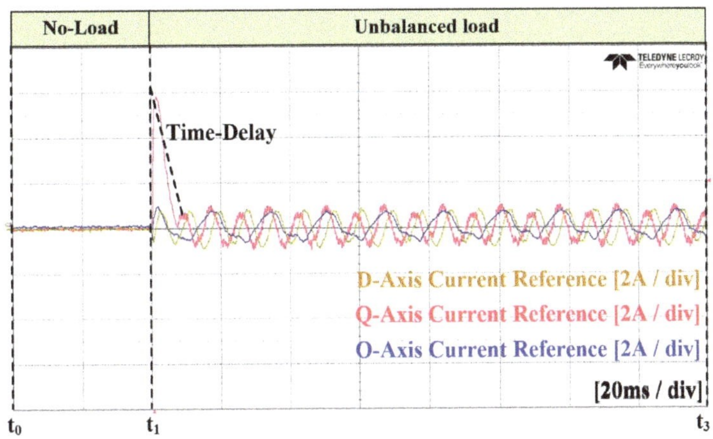

Figure 35. Experimental waveforms of the dq0 current reference in Case 2.

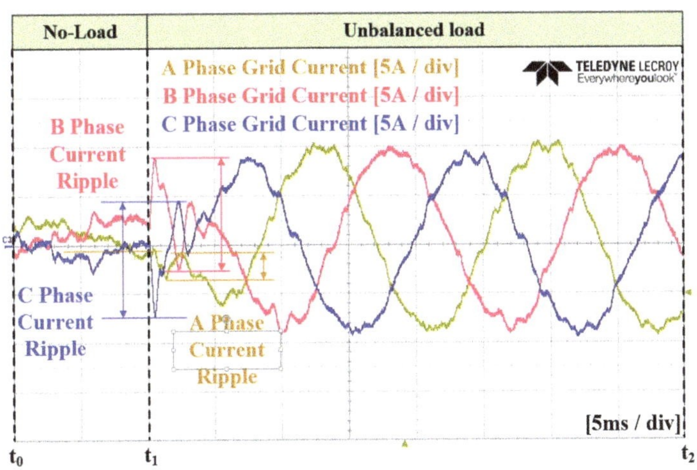

Figure 36. Enlarged waveform of the three-phase grid current in Case 2.

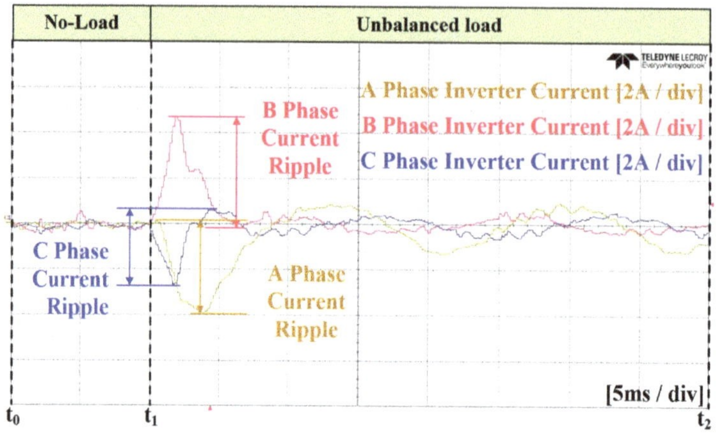

Figure 37. Enlarged waveform of the three-phase inverter current in Case 2.

5.3. Case 3 (ANN dq0 Reference Generation and ANN Current Controller)

In the Case 3 experiment, an ANN was applied for unbalanced magnitude calculation and for the current controller. Figures 38 and 39 represent the grid current and converter current, respectively. Figure 40 illustrates the amount of unbalance on the dq0 axis calculated using the ANN. In this figure, it reaches the average value at time t_s, which shows a similar time to that noted in Case 2. Figure 41 shows the current reference on the dq0 axis generated by the ANN. Figures 42 and 43 display the enlarged waveforms of Figures 38 and 39, respectively. The experimental results confirm a similar attainment of steady state as that noted in Case 2, with an improvement in distortion at the peak portion of the current.

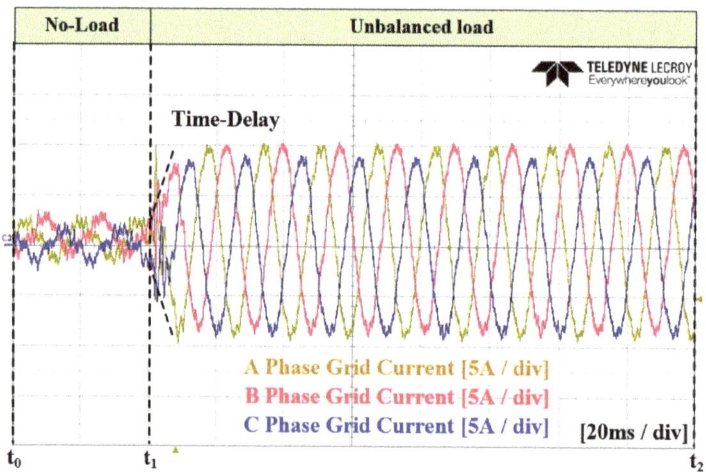

Figure 38. Experimental waveforms of the three-phase grid current in Case 3.

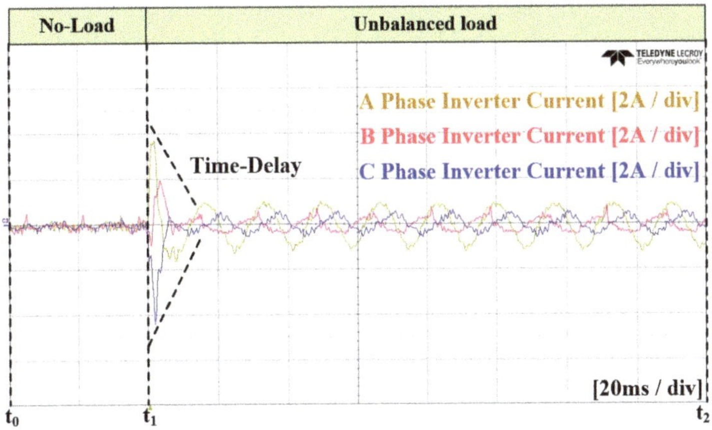

Figure 39. Experimental waveforms of the three-phase inverter current in Case 3.

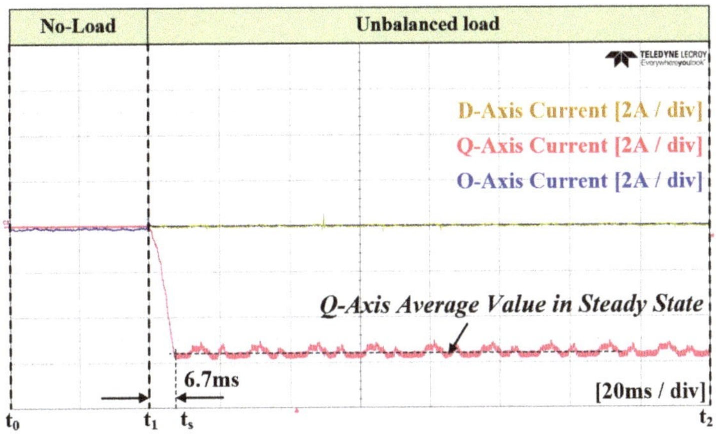

Figure 40. Experimental waveforms of the ANN output value in Case 3.

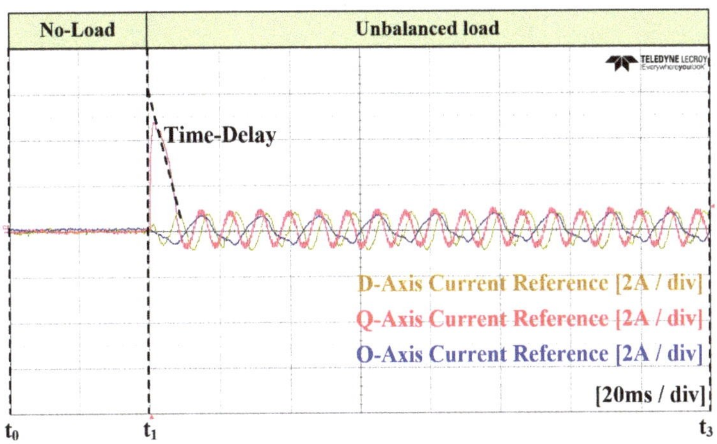

Figure 41. Experimental waveforms of the dq0 current reference in Case 3.

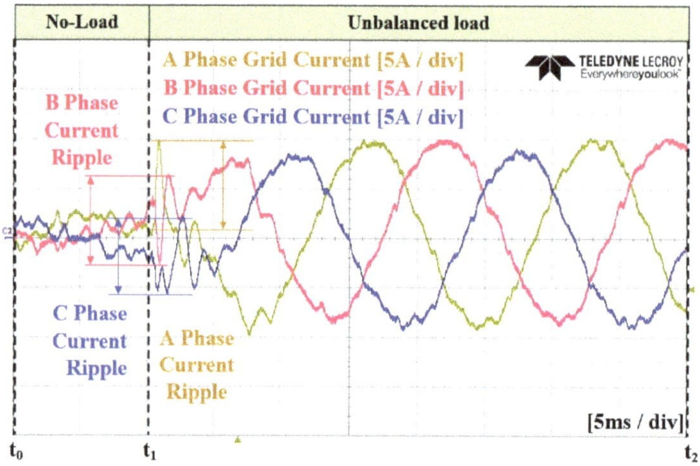

Figure 42. Enlarged waveform of the three-phase grid current in Case 3.

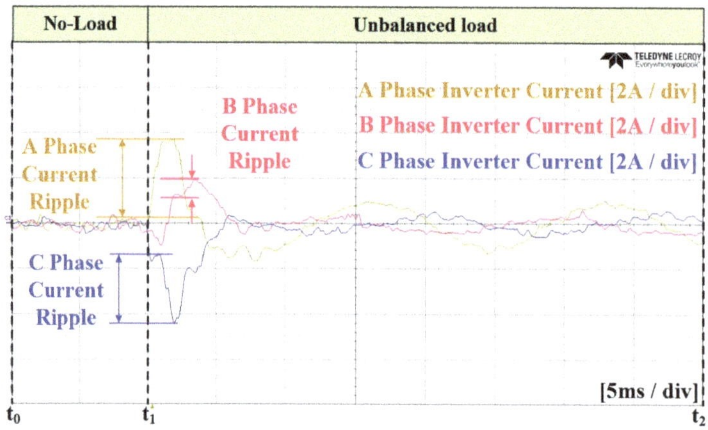

Figure 43. Enlarged waveform of the three-phase inverter current in Case 3.

6. Conclusions

This study aimed to alleviate the current imbalance in power distribution systems with significant load fluctuations in order to obtain balanced power from the power source. The conventional method, which employs an LPF-based dq0 reference generation and a PI current controller, can determine the degree of imbalance, but it requires a low-frequency LPF to extract the magnitude of the AC component. This drawback leads to an excessive time delay. In addition, the PI current controller introduces time delays through the action of the integrator. This poses challenges when applying the method to distribution systems in which loads vary in real-time. To address the aforementioned issues, this study applied an ANN-based reference generator and a current controller to an unbalanced compensation system. The ANN was trained using MATLAB, and the optimal values for the training algorithm and the number of neurons were selected through simulations. To validate the proposed method, simulations and experiments were conducted for the three cases. Case 1 applied the conventional method, Case 2 applied the ANN reference calculation with a PI current controller, and Case 3 applied the ANN reference calculation with an ANN current controller. Case 2 showed a faster estimation of the unbalance compared to Case 1, but it exhibited the drawback of an increased maximum current ripple and a degraded THD.

This problem stems from the periodic incorrect output of the ANN reference calculation. In Case 3, by training the ANN current controller with periodic incorrect output and the steady and transient states of the PI controller, the maximum current ripple and THD were improved compared to those noted in Case 1. The improvement in the current THD was confirmed, through simulations and experiments, to be comparable to that of Case 1. Therefore, the proposed ANN-based current compensation method was validated as suitable for power distribution systems with rapidly changing load conditions. In this paper, it was observed that a complete balance of the grid currents could not be achieved due to voltage distortion in the experiments. However, in the future, it will be necessary to research compensation, not only for current, but also for voltage, especially under constrained grid conditions, in order to maintain a higher level of balance.

Author Contributions: T.-G.K. and C.-G.A. conceived and designed the theory and experiment; T.-G.K. and C.-G.A. performed the experiment; T.-G.K. and C.-G.A. analyzed the theory; T.-G.K. wrote the manuscript; C.-G.A. reviewed the manuscript and searched conventional studies; C.-Y.W. and J.Y. participated in research plan development and revised the manuscript. All authors have read and agreed to the published version of the manuscript.

Funding: This work was supported by the Korea Institute of Energy Technology Evaluation and Planning (KETEP) and the Ministry of Trade, Industry, and Energy (MOTIE) of the Republic of Korea. (No. 2019381010001B), This work was supported by the BK21 FOUR Project.

Institutional Review Board Statement: Not applicable.

Informed Consent Statement: Not applicable.

Data Availability Statement: Data are contained within the article.

Conflicts of Interest: The authors declare no conflicts of interest.

References

1. Singh, B.; Al-Haddad, K.; Chandra, A. A review of active filters for power quality improvement. *IEEE Trans. Ind. Electron.* **1999**, *46*, 960–971. [CrossRef]
2. Rafi, F.H.M.; Hossain, M.J.; Rahman, M.S.; Taghizadeh, S. An overview of unbalance compensation techniques using power electronic converters for active distribution systems with renewable generation. *Renew. Sustain. Energy Rev.* **2020**, *125*, 109812. [CrossRef]
3. Inzunza, R.; Akagi, H. A 6.6-kV transformerless shunt hybrid active filter for installation on a power distribution system. *IEEE Trans. Power Electron.* **2005**, *20*, 893–900. [CrossRef]
4. Torabi Jafrodi, S.; Ghanbari, M.; Mahmoudian, M.; Najafi, A.; Rodrigues, E.M.G.; Pouresmaeil, E. A Novel Control Strategy to Active Power Filter with Load Voltage Support Considering Current Harmonic Compensation. *Appl. Sci.* **2020**, *10*, 1664. [CrossRef]
5. Herrera, R.S.; Salmeron, P.; Vazquez, J.R.; Litran, S.P.; Perez, A. Generalized instantaneous reactive power theory in poly-phase power systems. In Proceedings of the 2009 13th European Conference on Power Electronics and Applications, Barcelona, Spain, 8–10 September 2009; pp. 1–10.
6. Chen, M.; Jie, Y.; Wang, C.; Li, G.; Qiu, L.; Zhong, W. Optimized Reactive Power Control of Module Power Imbalance of Cascaded Converter. *IEEE Open J. Power Electron.* **2022**, *3*, 2–12. [CrossRef]
7. Jones, V.; Balda, J.C.; Adapa, R. Current Compensators for Unbalanced Electric Distribution Systems. In Proceedings of the 2018 IEEE Electronic Power Grid (eGrid), Charleston, SC, USA, 12–14 November 2018; pp. 1–6. [CrossRef]
8. Vijay, A.S.; Doolla, S.; Chandorkar, M.C. Unbalance mitigation strategies in microgrids. *IET Power Electron.* **2020**, *13*, 1687–1710. [CrossRef]
9. Hintz, A.; Prasanna, U.R.; Rajashekara, K. Comparative Study of the Three-Phase Grid-Connected Inverter Sharing Unbalanced Three-Phase and/or Single-Phase systems. *IEEE Trans. Ind. Appl.* **2016**, *52*, 5156–5164. [CrossRef]
10. Siddartha, A.N.R.S.; Basu, T.S.; Chakraborty, C. A simple time domain approach for harmonic, load unbalance and reactive power compensation. In Proceedings of the IECON 2013—39th Annual Conference of the IEEE Industrial Electronics Society, Vienna, Austria, 10–13 November 2013; pp. 8504–8509. [CrossRef]
11. Lyu, W.; Wang, R.; Xu, W.; Mu, S.; Guo, R.; Liang, J.; Liu, Z.A. Novel Three Phase Unbalance Detection and Compensation method of Active Power Filter. In Proceedings of the 2018 Chinese Automation Congress (CAC), Xi'an, China, 30 November–2 December 2018; pp. 418–422. [CrossRef]
12. Verdelho, P.; Marques, G.D. An active power filter and unbalanced current compensator. *IEEE Trans. Ind. Electron.* **1997**, *44*, 321–328. [CrossRef]

13. Afonso, J.; Couto, C.; Martins, J. Active Filters with Control based on the p-q Theory. *IEEE Ind. Electron. Soc. Newsl.* **2000**, *47*, 5–10.
14. Baeckeland, N.; Venkatramanan, D.; Kleemann, M.; Dhople, S. Stationary-Frame Grid-Forming Inverter Control Architectures for Unbalanced Fault-Current Limiting. *IEEE Trans. Energy Convers.* **2022**, *37*, 2813–2825. [CrossRef]
15. Roiu, D.; Bojoi, R.I.; Limongi, L.R.; Tenconi, A. New Stationary Frame Control Scheme for Three-Phase PWM Rectifiers Under Unbalanced Voltage Dips Conditions. *IEEE Trans. Ind. Appl.* **2010**, *46*, 268–277. [CrossRef]
16. Somkun, S.; Chunkag, V. Unified Unbalanced Synchronous Reference Frame Current Control for Single-Phase Grid-Connected Voltage-Source Converters. *IEEE Trans. Ind. Electron.* **2016**, *63*, 5425–5436. [CrossRef]
17. Kesler, M.; Ozdemir, E. Synchronous-Reference-Frame-Based Control Method for UPQC Under Unbalanced and Distorted Load Conditions. *IEEE Trans. Ind. Electron.* **2011**, *58*, 3967–3975. [CrossRef]
18. Wang, L.; Lam, C.-S.; Wong, M.-C. Unbalanced Control Strategy for A Thyristor-Controlled LC-Coupling Hybrid Active Power Filter in Three-Phase Three-Wire Systems. *IEEE Trans. Power Electron.* **2017**, *32*, 1056–1069. [CrossRef]
19. Wang, Z.; Wu, B.; Xu, D.; Cheng, M.; Xu, L. DC-Link Current Ripple Mitigation for Current-Source Grid-Connected Converters Under Unbalanced Grid Conditions. *IEEE Trans. Ind. Electron.* **2016**, *63*, 4967–4977. [CrossRef]
20. Herrera, R.S.; Salmerón, P.; Kim, H. Instantaneous Reactive Power Theory Applied to Active Power Filter Compensation: Different Approaches, Assessment, and Experimental Results. *IEEE Trans. Ind. Electron.* **2008**, *55*, 184–196. [CrossRef]
21. Bhattacharya, A.; Chakraborty, C.; Bhattacharya, S. Shunt compensation. *IEEE Ind. Electron. Mag.* **2009**, *3*, 38–49. [CrossRef]
22. Sun, Y.; Li, S.; Fu, X.; Dong, W.; Ramezani, M.; Balasubramanian, B. Approximate dynamic programming vector controllers for operation of IPM motors in linear and overmodulation regions. *IEEE Trans. Transp. Electrif.* **2020**, *7*, 659–670. [CrossRef]
23. Dong, W.; Li, S.; Fu, X.; Li, Z.; Fairbank, M.; Gao, Y. Control of a buck DC/DC converter using approximate dynamic programming and artificial neural networks. *IEEE Trans. Circuits Syst. I Regul. Pap.* **2021**, *68*, 1760–1768. [CrossRef]
24. Gupta, N.; Singh, S.P.; Dubey, S.P. Neural network based shunt active filter for harmonic and reactive power compensation under non-ideal mains voltage. In Proceedings of the 5th IEEE Conference on Industrial Electronics and Applications, Taichung, Taiwan, 15–17 June 2010.
25. Fang, Z.P.; Lai, J.-S. Generalized instantaneous reactive power theory for three-phase power systems. *IEEE Trans. Instrum. Meas.* **1996**, *45*, 293–297. [CrossRef]
26. George, J.; Jose, T.L.; Jacob, J. A decoupled reference generation algorithm for harmonic, reactive power and current unbalance compensation in three-phase systems. In Proceedings of the 2011 Annual IEEE India Conference, Hyderabad, India, 16–18 December 2011; pp. 1–6. [CrossRef]
27. Kim, G.-H.; Hwang, C.; Jeon, J.-H.; Ahn, J.-B.; Kim, E.S. A novel three-phase four-leg inverter based load unbalance compensator for stand-alone microgrid. *Int. J. Electr. Power Energy Syst.* **2015**, *65*, 70–75. [CrossRef]
28. Kim, S.-H. An Energy Management Method Using ANN and Hierarchical Control for Bipolar DC Microgrid Based on Droop Control. Ph.D. Thesis, Sungkyunkwan University Graduate School, Seoul, Republic of Korea, 2023.

energies

MDPI

Article

Multivariable Algorithm Using Signal-Processing Techniques to Identify Islanding Events in Utility Grid with Renewable Energy Penetration

Ming Li [1,†], Anqing Chen [1,*,†], Peixiong Liu [1,2,†], Wenbo Ren [1,†] and Chenghao Zheng [1,†]

[1] Institute of Automotive Engineers, Hubei University of Auomotive Technology, Shiyan 442002, China
[2] Hanjiang Water Resources and Hydropower (Group) Co., Ltd. Hydropower Company, Shiyan 442002, China
* Correspondence: 20180030@huat.edu.cn
† These authors contributed equally to this work.

Abstract: This paper designs a multi-variable hybrid islanding-detection method (HIDM) using signal-processing techniques. The signals of current captured on a test system where the renewable energy (RE) penetration level is between 50% and 100% are processed by the application of the Stockwell transform (ST) to compute the Stockwell islanding-detection factor (SIDF) and the co-variance islanding-detection factor (CIDF). The signals of current are processed by the application of the Hilbert transform (HT), and the Hilbert islanding-detection factor (HIDF) is computed. The signals of current are also processed by the application of the Alienation Coefficient (ALC), and the Alienation Islanding Detection Factor (AIDF) is computed. A hybrid islanding-detection indicator (HIDI) is derived by multiplying the SIDF, CIDF, AIDF, and an islanding weight factor (IWF) element by element. Two thresholds, designated as the hybrid islanding-detection indicator threshold (HIDIT) and the hybrid islanding-detection indicator fault threshold (HIDIFT), are selected to detect events of islanding and also to discriminate such events from fault events and operational events. The HIDM is effectively tested using an IEEE-13 bus power network, where solar generation plants (SGPs) and wind generation plants (WGPs) are integrated. The HIDM effectively identified and discriminated against events such as islanding, faults, and operational. The HIDM is also effective at identifying islanding events on a real-time distribution feeder. The HIDM is also effective at detecting islanding events in the scenario of a 20 dB signal-to-noise ratio (SNR). It is established that the HIDM has a small non-detection zone (NDZ). The effectiveness of the HIDM is better relative to the islanding-detection method (IDM) supported by the discrete wavelet transform (DWT), an IDM using a hybridization of the slantlet transform, and the Ridgelet probabilistic neural network (RPNN). An IDM using wavelet transform multi-resolution (WT-MRA)-based image data and an IDM based on the use of a deep neural network (DNN) were used. The study was performed using the MATLAB software (2017a) and validated in real-time using the data collected from a practical distribution power system network.

Keywords: Alienation Coefficient; fault event; Hilbert transform; islanding event; operational event; renewable energy; signal-processing technique; Stockwell transform; utility grid

1. Introduction

Globally, renewable energy (RE) is integrated into utility grids to meet future energy demand due to the depletion of fossil fuels, the price fluctuations of oil and natural gas, and the adverse impact of these energy sources on the environment. RE sources include mini or micro hydroelectric plants, wind generation plants (WGPs), solar photovoltaic (PV) plants, fuel cells, geothermal energy, and biomass energy [1]. These RE sources are integrated into the utility grid in the form of large plants and distributed generators (DGs). DGs are connected to the grid network near the load center. A condition or operating scenario in which a part of the utility network remains connected to one or more DGs where the load is still energized by the DGs and, at the same time, is isolated from the main grid is called islanding [2]. Islanding may be intentional (pre-planned) for the condition of load shedding and during the maintenance period to the ensure security and protection of the power system from cascade failure. Another form of islanding is unintentional, which is initiated because of a fault or the failure of equipment. This may create problems regarding power quality (PQ), grid safety, and stability [3]. Unintentional islanding events are undesirable and need to be identified as soon as they occur in the network. Different islanding-detection techniques have been developed, which are mainly categorized into two categories: remote (communication-based) islanding and local islanding-detection methods [4].

1.1. Review of Literature

A comprehensive review of the methods used for islanding events and a comparative study of passive, active, and hybrid islanding-detection methods (IDMs) are included in [5]. This study detailed the types of islanding methods. Local IDM methods such as passive, active, and hybrid were described. Remote IDMs, including the power line signaling scheme, transfer trip scheme, and impedance insertion method, were elaborated in detail. Different international standards for islanding detection were also discussed. A comparative study of local and remote IDMs considering the working principle, response time, non-detection zone (NDZ), maloperation possibility of IDMs, cost, impact on power quality (PQ), and multiple RE generators was presented, which is effective at selecting an IDM suitable for a particular grid topology and quantum of generation. In [6], the authors designed a predictive Goertzel model algorithm for active islanding detection with solar PV plants. This algorithm used the discrete Fourier transform (DFT) to resolve the issue of the non-detection zone (NDZ). This method obtains results at a minimum computation time considering the N+2 number of operations. However, to ensure high efficiency, additional 2N+1 operations are also required. It performed this by calculating the magnitude and phase of a particular frequency in discrete form. Hence, the performance of the IDM is dependent on the method of approximation. In [7], the authors designed an IDM that leveraged the micro phasor measurement unit ($\mu - PMU$) to detect islanding events with the minimum chance of a cyber attack. However, the working of the $\mu - PMU$ is mainly dependent on the communication network used to transmit the data. This communication network is prone to cyber attacks, which is the drawback of this method. This requires additional steps in the treatment of the data, which increase the computational time and burden. This IDM effectively used the difference in the angle between the positive and negative sequence components to detect islanding events. In [8], the authors presented a secured-communication-based passive anti-islanding technique. This IDM uses the phase angle difference of the information related to the superimposed impedance at the point of the DG connection and the point of common coupling (PCC). This method is effective at detecting islanding conditions even with low power mismatches and distinguishing islanding events from fault and switching events. This IDM is fast, free from fixed threshold settings, easy to implement, and securely discriminates the island from the fault, as well as the switching events. However, due to the use of communication channels, the cost of the IDM's implementation is high. In [9], the authors proposed a hybrid passive–active systematic methodology for the detection of islanding events. This IDM is effective at overcoming the drawbacks of the passive IDM. A fuzzy classifier for different circumstances, including the NDZ, is effectively used in this method. This IDM has high accuracy, is robust and fast, and can be easily implemented for inverter-based RE generators. This IDM is effective at maintaining the quality of power. However, in instances where the fuzzy rule fails to detect the islanding, the active mode of the inverter starts and reactive power injection is initiated. In [10], the authors presented a detailed study on the analysis and verification of IDMs for grid-integrated PV plants. This method is effective against false-positive detection due to the transients injected into the grid from outside sources.

1.2. Research Gaps

After a detailed analysis of the existing literature, it is seen that the performance of IDMs based on the use of one type of mathematical formulation suffers from any one of the demerits such as low efficiency, a high NDZ, false tripping, failure when external transients are available, etc. The performance of these IDMs might be improved by a hybrid combination of the signal analysis methods, which is the research objective for this paper.

1.3. Research Contributions

The following are the research contributions:

- An HIDM based on the hybrid combination of the ST, HT, and ALC is designed. This considers the processing of the current signals using the ST to compute the SIDF and CIDF. Currents are also processed using the HT and ALC to compute the HIDF and AIDF, respectively. The SIDF, CIDF, HIDF, AIDF, and IWF are multiplied to compute the HIDI. This HIDM combines the merits of the ST, HT, and ALC, which resulted in the following merits of the proposed HIDM:
    - The HIDM is effective at detecting islanding conditions and discriminating these events, fault events, and the operational conditions using the threshold HIDIT and HIDIFT
    - The HIDM is effective at detecting the islanding condition in the availability of the noise of 20 dB SNR.
    - The HIDM has a small non-detection zone.
    - The performance of the HIDM is not affected even with a 100% RE penetration level. However, the performance of the HIDM considering individual applications of either of the ST, HT, or ALC deteriorate when the RE penetration level increases above 50%.
- The HIDM was effectively tested on an IEEE-13 node network with a 50% and 100% contribution in the generation mix from RE plants (both solar and wind).
- The HIDM can effectively be used to identify the islanding conditions in a real-time distribution feeder of a practical utility grid.
- The performance of the HIDM is better relative to an IDM considering the DWT, an IDM using a combination of the slantlet transform and RPNN, an IDM using the wavelet transform multi-resolution (WT-MRA)-based image data, and an IDM based on the use of a Deep Neural Network (DNN).

1.4. Structure of the Paper

The contents of the paper are organized into nine sections. The Introduction, literature review, background of the research, research gaps, research contribution, and organization of the paper are discussed in Section 1. The test power grid and technical parameters are detailed in Section 2. The HIDM and related mathematical formulations are elaborated in Section 3. Section 4 elaborates on the results and their discussion for the detection of islanding events. The results and discussion for the detection of fault events are elaborated in Section 5. The results and their discussion to detect operational events are elaborated in Section 6. The real-time validation of the HIDM is included in Section 7. The relative performance comparison of the HIDM is elaborated in Section 8. The research work summary and conclusions are included in Section 9.

2. Test Utility Grid

The IEEE-13 node test system, rated at 5 MVA and 4.16 kV, was used to perform the study of islanding event detection in utility networks where a high RE generation mix is available. This was an electric power distribution system that consists of thirteen buses, twelve branches, six loads, and one generator. A generator was implemented to represent a large-area network of power. A solar generation plant (SGP) of a capacity of 1 MW and a wind generation plant (WGP) of a capacity of 1.5 MW were connected to 680 nodes of the test network using the transformers SGP-GT and WGP-GT, respectively. This ensured a 50% and 100% contribution of RE in the generation mix to meet the load demand of the test system. The test system configuration and network topology are described in Figure 1. This test system was integrated into a large-area utility network using a utility grid transformer (UG-T). A distribution feeder transformer (DF-T) was equipped between nodes 633 and 634, which helps to maintain node 634 at 0.48 kV. An island formation switch (IFS) was equipped near node 650, which was used for island formation in the test system. The details of the loads and transformers in the test system are included in Table 1 and Table 2, respectively [11,12]. The SGP was formed by a solar photovoltaic (PV) array, which was formed by the integration of 66 parallel strings. Each string consisted of five modules integrated in series. The SGP plant was integrated into node 680 of the test system using a DC/DC converter, a DC/AC converter, and the SGP-GT. The details of the various components of the SGP and associated equipment are available in [13,14]. The WGP was formed by a doubly fed induction generator (DFIG) and wind mills. The WGP plant was connected to node 680 of the test system with the help of the WGP-GT. The details of the various components of the WGP and associated equipment are available in [15,16]. This test system was modeled using the MATLAB software (2017a).

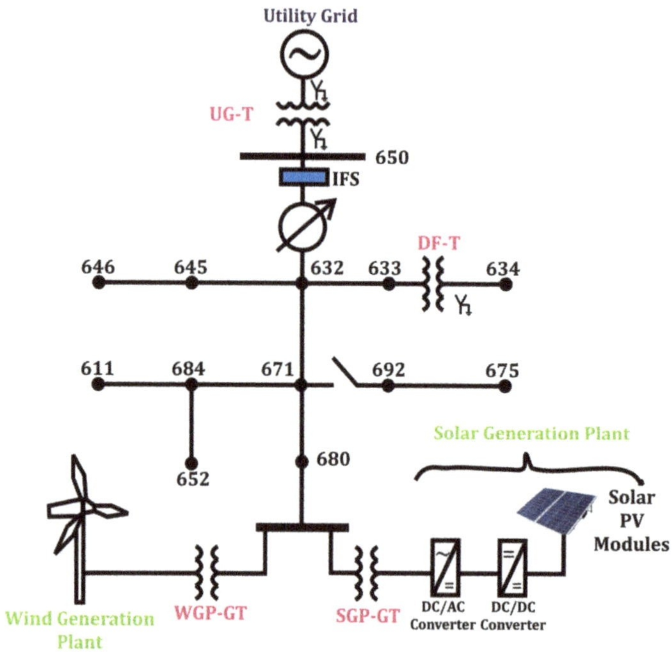

Figure 1. Test utility network with renewable energy generators.

Table 1. Load details of test system.

S. No.	Node Number	Load Detail
1	634	400 kW and 290 kVAr
2	645	170 kW and 125 kVAr
3	646	230 kW and 132 kVAr
4	652	128 kW and 86 kVAr
5	671	1255 kW and 718 kVAr
6	675	843 kW and 462 kVAr
7	692	170 kW and 151 kVAr
8	611	170 kW and 80 kVAr
9	632–671	200 kW and 116 kVAr distributed load

Table 2. Technical data of transformers equipped in test system.

S. No.	Transformer Symbol	MAV Rating	Voltage Ratio
1	UG-T	10 MVA	115 kV/4.16 kV
2	DF-T	5 MVA	4.16 kV/0.48 kV
3	WGP-GT	5 MVA	4.16 kV/0.575 kV
4	SGP-GT	1 MVA	4.16 kV/0.270 kV

3. Multi-Variable Hybrid Islanding Detection Method

This section demonstrates the various steps used for the design of the hybrid islanding-detection method (HIDM). A hybrid islanding-detection indicator (HIDI) was designed to recognize the islanding conditions in the utility grid with a high RE mix in the generation. The HIDI also discriminated between islanding events and faulty and operational events. The HIDI was computed using the Hilbert islanding-detection factor (HIDF), the Stockwell islanding-detection factor (SIDF), the covariance islanding-detection factor (CIDF), and the Alienation-Coefficient-based islanding-detection factor (AIDF). The HIDF, SIDF, CIDF, AIDF, and HIDI are elaborated in this section in detail. All the steps of the HIDM to detect and discriminate the islanding events from operational and faulty events are elaborated in Figure 2. The mathematical formulation of all the steps of the HIDM was designed using the MATLAB software.

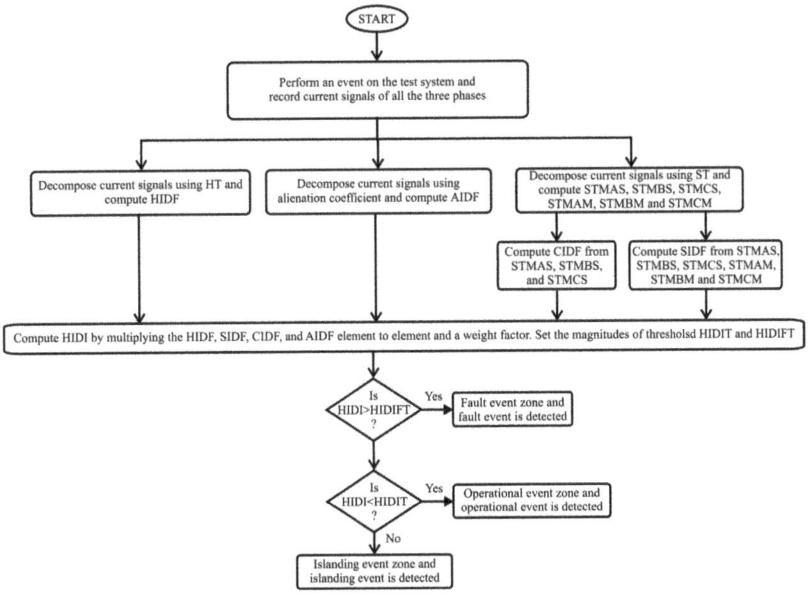

Figure 2. Hybrid islanding-event-detection algorithm.

3.1. Hilbert Islanding-Detection Factor

The HIDF was computed by processing the currents of all the phases using the Hilbert transform (HT). The current of phase A ($i_1(\tau)$) was processed using the HT to compute the HIDF for phase A (HIDFA) as detailed below [17]:

$$HIDFA = \left| \frac{1}{\pi} \int_{-\infty}^{+\infty} \frac{i_1(\tau)}{(t-\tau)} d\tau \right| \tag{1}$$

where t is the time and τ is the sampling time interval. Equation (1) is solved using the convolution of the current signal considering periodic summation [18]. Similarly, the HIDF for phase B (HIDFB) and the HIDF for phase C (HIDFC) were computed by processing the current of phase B ($i_2(\tau)$) and the current of phase C ($i_3(\tau)$), respectively. The HIDF was computed by taking the arithmetic mean of the HIDFA, HIDFB, and HIDFC as detailed below:

$$HIDF = \frac{HIDFA + HIDFB + HIDFC}{3} \tag{2}$$

3.2. Stockwell Islanding-Detection Factor

The SIDF was obtained by processing the currents of all the phases using the Stockwell transform (ST). The current of phase A ($i_1(t)$) was processed using the ST to compute the output matrix with absolute values for phase A (STMA) as detailed below [19]:

$$STMA = \left| \int_{-\infty}^{+\infty} i_1(t) \frac{|f|}{\sqrt{2\pi}} e^{-\frac{f^2(\tau-t)^2}{2}} e^{-j2\pi ft} dt \right| \tag{3}$$

where f is the frequency and t is the time. Equation (3) was solved using the discrete version, and the approximation was performed by considering $f \rightarrow n/NT$ and $\tau \rightarrow jT$. Here, n: the instantaneous sample number, N: the total number of samples, and T: the time period of the current signal. The STMA is a complex matrix, which has signal information. The rows and columns of this matrix represent the time and frequency spectrum, respectively. The matrix elements compute the time–frequency contours and can be used to compute the energy vector of the instantaneous frequency using Parseval's theorem [20,21]. The currents of phase B ($i_2(t)$) and phase C ($i_3(t)$) are also processed using Equation (3) for computing the ST output matrix with absolute values for phase B (STMB) and phase C (STMC), respectively. We compute the summation of every element of each column of matrix STMA (STMAS) as described below:

$$STMAS = \Sigma_{j=1}^{j=384} \left(\Sigma_{i=1}^{i=768} (STMA) \right) \tag{4}$$

where i and j indicate the row numbers and column numbers of the STMA matrix, respectively. Similarly, the summation of every element of each column of the matrix STMB (STMBS) and the summation of every element of each column of the matrix STMC (STMCS) are computed. The median of the STMA (STMAM) is computed by the relation below:

$$STMAM = median(STMA) \tag{5}$$

Each element of the STMAM matrix indicates the median of the respective column of the STMA matrix, which is computed as described below:

$$median = \frac{(\frac{n}{2})^{th} term + (\frac{n}{2} + 1)^{th} term}{2} \tag{6}$$

Here, n is the total row numbers in the STMA matrix, which equal 384. Similarly, the median of the STMB and STMC matrices are computed to obtain the STMBM and STMCM, respectively. The Stockwell islanding-detection factor for phase A (SIDFA)is evaluated by element by element multiplication of the STMAS and STMAM as described below:

$$SIDFA = STMAS \times STMAM \tag{7}$$

The Stockwell islanding-detection factor for phase B (SIDFB) is evaluated considering the element by element multiplication of the STMBS and STMBM. Similarly, the Stockwell islanding-detection factor for phase C (SIDFC) is evaluated considering the element by element multiplication of the STMCS and STMCM. The Stockwell islanding-detection factor (SIDF) is evaluated by taking the arithmetic mean of the SIDFA, SIDFB, and SIDFC

$$SIDF = \frac{SIDFA + SIDFB + SIDFC}{3} \tag{8}$$

3.3. Co-Variance Islanding-Detection Factor

The co-variance islanding-detection factor for phase A (CIDFA) is computed by considering samples of the current cycle (X) and samples of the previous cycle (healthy nature cycle) (Y) of the STMAS as described below:

$$CIDFA = \Sigma\left(\frac{X_i Y_i}{N}\right) \tag{9}$$

Here, N indicates the number of scores in each set of data, which is 64 samples per cycle. X represents the mean of the N scores in the current cycle data set, and Y represents mean of the N scores in the previous cycle data set. This operation is performed using the inbuilt function of MATLAB as detailed below:

$$CIDFA = cov(STMAS) \tag{10}$$

The co-variance islanding-detection factor for phase B (CIDFB) and co-variance islanding-detection factor for phase C (CIDFC) are computed from the STMBS and STMCS, respectively. The co-variance islanding-detection factor is computed by taking the arithmetic mean of the CIDFA, CIDFB, and CIDFC as detailed below:

$$CIDF = \frac{CIDFA + CIDFB + CIDFC}{3} \tag{11}$$

3.4. Alienation Islanding-Detection Factor

The Alienation islanding-detection factor (AIDF) is computed by processing the currents of all phases using the Alienation Coefficient (ALC). The Alienation islanding-detection factor for phase A (AIDFA) is computed by the sample-based Alienation Coefficients of the current of phase A taking a sampling frequency of 3.84 kHz. The AIDF for phase A (AIDFA) is evaluated considering the relation below [22]:

$$AIDFA = 1 - RA^2 \tag{12}$$

Here, RA indicates the correlation coefficient of the variables x and y for phase A. Here, the variable x represents the mean of the N scores in the current cycle data set, and the variable y represents the mean of the N scores in the previous cycle data set. The correlation coefficient RA is computed using the formulation below [22].

$$RA = \frac{N_s \sum xy - (\sum x)(\sum y)}{\sqrt{[N_s \sum x^2 - (\sum x)^2][N_s \sum y^2 - (\sum y)^2]}} \tag{13}$$

where N_s indicates the sample numbers, which equal 64. Similarly, the AIDF for phase B (AIDFB) and AIDF for phase C (AIDFC) are computed by processing the currents of phase B and phase C, respectively. The AIDF is computed by taking the arithmetic mean of the AIDFA, AIDFB, and AIDFC as detailed below:

$$AIDF = \frac{AIDFA + AIDFB + AIDFC}{3} \tag{14}$$

3.5. Hybrid Islanding-Detection Indicator

The hybrid islanding-detection indicator (HIDI) is computed by multiplying the HIDF, SIDF, CIDF, and AIDF element by element and an islanding weight factor (IWF) as detailed below:

$$HIDI = HIDF \times SIDF \times CIDF \times AIDF \times IWF \tag{15}$$

Here, the IWF is considered equal to 8×10^6. Two thresholds were selected for detecting the islanding conditions and also to discriminate such events from fault events and operational events. The first threshold is known as the hybrid islanding-detection indicator threshold (HIDIT), which is considered equal to 200. The second threshold is known as the hybrid islanding-detection indicator fault threshold (HIDIFT), which is considered equal to 10^{10}. The magnitudes of the HIDIT and HIDIFT were considered after testing the algorithm on 25 data sets of all events investigated in this research. This data set was evaluated by varying the event incidence angle, fault impedance, different levels of RE generation, events at different nodes of the test system, etc. Different zones for the detection of islanding, faults, and operational events are shown in Figure 3. For the event for which the peak magnitude of the HIDI is higher than the HIDIFT, then the event will be identified as a fault event. For the event for which the peak value of the HIDI is lower than the HIDIT, then the event will be identified as an operational event. If the peak value of the HIDI is higher than the HIDIT and less than the HIDIFT, then the event will be identified as an islanding event.

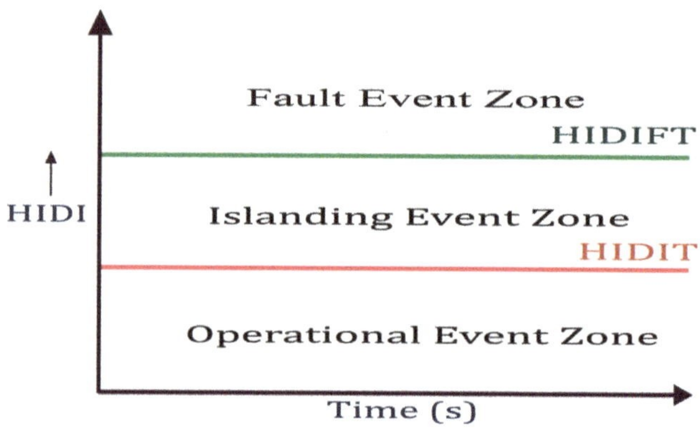

Figure 3. Different zones of events.

4. Detection of Islanding Events: Simulation Results and Discussion

The results of the simulation and their discussion for the identification of islanding events with generation from both the WGP and SGP, generation from the SGP, and generation from the WGP are discussed in this section. The results for the analysis of the impact of the islanding incidence angle on the efficacy of the HIDM are also discussed in this section. The results for islanding detection in noisy scenarios are also discussed. The results to determine the non-detection zone (NDZ) are also discussed in this section. The results are plotted for a 12-cycle time period. The current signals used to obtain the results of islanding detection demonstrated in this section were synthetically generated by simulation studies performed on the test system of a utility network with the RE generators described in Figure 1 using the MATLAB software. The results of this section are related to the event of sudden islanding, where the complete network with all generators, loads, capacitors, and reactors is in a circuit and also in an energized state.

4.1. Detection of Islanding Event with Generation from WGP and SGP

The test system was simulated with generation from both the WGP and SGP. The IFS switch was opened at 0.1 s to form the island of the test system. The currents of all phases were captured on node 632 of the test system. The current signals were processed using the HIDM to evaluate the HIDF, SIDF, AIDF, and HIDI. The current waveform of all phases, HIDF, SIDF, AIDF, and HIDI, is depicted in Figure 4a–e in the respective sequence. The maximum magnitude of the HIDI and the computational time of the HIDM are tabulated in Table 3 and Table 4, respectively.

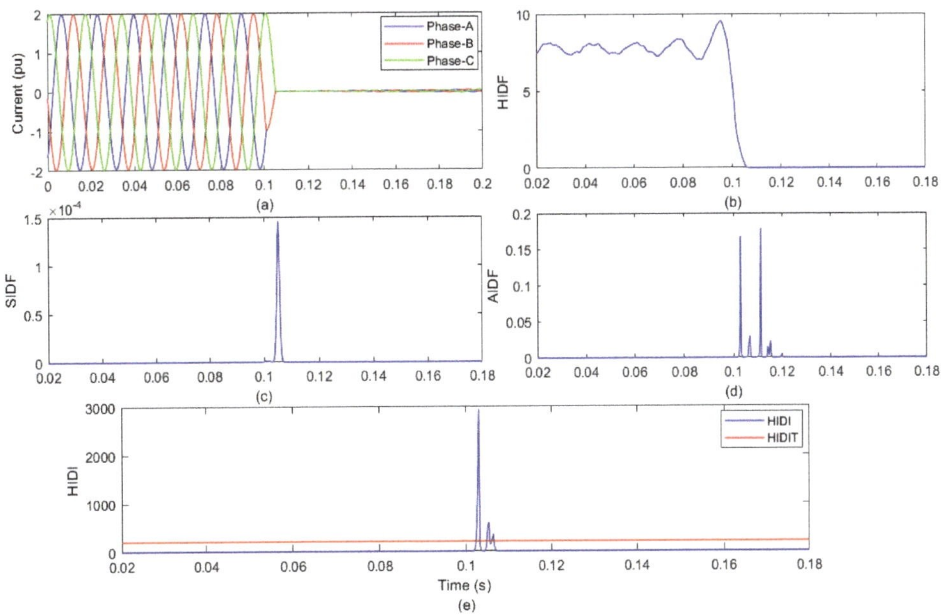

Figure 4. Detection of islanding with WGP and SGP generation: (**a**) current waveform; (**b**) HIDF; (**c**) SIDF; (**d**) AIDF; (**e**) HIDI.

Figure 4a indicates the decrease in current due to the islanding event at 0.1 s. The magnitude of the HIDF is also reduced after an islanding event, as illustrated in Figure 4b. Figure 4c elaborates that the SIDF has a zero value before and after islanding incidence and becomes high at the instant of islanding incidence. Figure 4d represents that the value of the AIDF is zero before and after islanding incidence and becomes high at the moment of islanding event occurrence. However, spikes are observed for a window of one cycle. Figure 4e elaborates that the value of the HIDI is zero prior to island formation and becomes higher than the HIDIT. However, the peak magnitude is lower relative to the HIDIFT. Hence, the peak of the HIDI falls in the islanding event zone. Hence, islanding with WGP and SGP generation is effectively detected, as well as discriminated from fault and operational events. The computational time of the HIDM for islanding detection with both the WGP and SGP is 0.278454 ms.

Table 3. Maximum magnitude of HIDI.

S. No.	Islanding/Fault/Operational Event	Maximum HIDI
1	Islanding event with WGP and SGP generation	2928.9
2	Islanding event with generation from SGP	5.9226×10^{8}
3	Islanding event with generation from WGP	5.2911×10^{5}
4	Phase to ground fault	1.9427×10^{11}
5	Phase to phase fault	2.441×10^{11}
6	Two phases to ground fault	2.3601×10^{13}
7	Three-phase fault	1.6289×10^{13}
8	Three phases to ground fault	2.6822×10^{12}
9	Feeder operation	9.028
10	Capacitor operation	49.782
11	Load operation	5.699×10^{-6}

Table 4. Islanding detection time of the HIDM.

S. No.	Islanding/Fault/Operational Event	Computational Time (ms)
1	Islanding event with WGP and SGP generation	0.278454
2	Islanding event with generation from SGP	1.650991
3	Islanding event with generation from WGP	0.443420
4	Phase to ground fault	1.320318
5	Phase to phase fault	1.631831
6	Two phases to ground fault	0.418473
7	Three-phase fault	1.397293
8	Three phases to ground fault	0.629446
9	Feeder operation	0.247305
10	Capacitor operation	0.282738
11	Load operation	0.344931

4.1.1. Testing of HIDM for Different Islanding Incidence Angles

The HIDM was tested to identify islanding conditions with generation from the WGP and SGP for occurrence angles of 0°, 30°, 60°, 90°, 120°, 150°, and 180°. The peak magnitude of the HIDI for the different occurrence angles is included in Table 5. This shows that the peak value of the HIDI is less than the HIDIFT and higher than the HIDIT. Hence, it falls in the islanding event zone, which indicates that islanding events are effectively recognized for different incidence angles, and this event is also differentiated from fault conditions and operational events.

Table 5. Peak magnitude of HIDI for different incidence angles.

S. No.	Islanding Incidence Angle	Peak Magnitude of HIDI
1	0°	2928.9
2	30°	1.5447×10^5
3	60°	1.5447×10^5
4	90°	64,472
5	120°	64,472
6	150°	4.1167×10^5
7	180°	4.1167×10^5

4.1.2. Impact of Noise on HIDM Performance

The HIDM was tested for detecting the islanding event with generation from the WGP and SGP in the scenario of 20 dB SNR noise. The islanding event was simulated, and the currents were captured on test bus 632. The noise of level 20 dB SNR is superimposed on the current waveform. The current with noise was decomposed by the application of the HIDM, and the HIDI was evaluated, which is illustrated in Figure 5. The peak magnitude is equal to 4.28326×10^6, which falls in the islanding event zone. Hence, the islanding condition with generation from the WGP and SGP in the scenario of a 20 dB SNR noise level is effectively detected using the proposed HIDM.

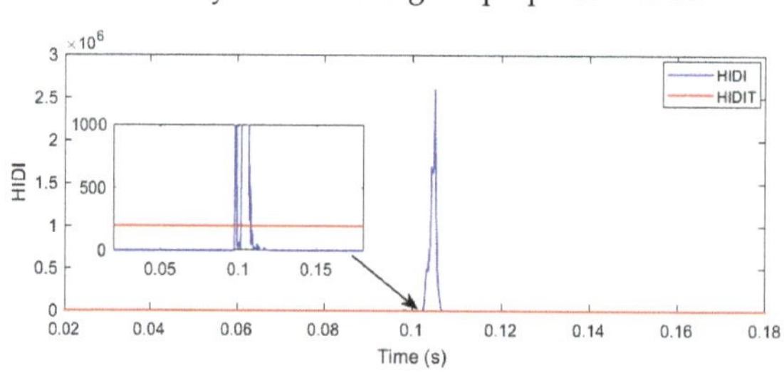

Figure 5. HIDI for islanding event with generation from both WGP and SGP in the scenario of 20 dB SNR noise.

4.1.3. Determination of NDZ of Islanding Event

A non-detection zone (NDZ) is an operational zone in which an islanding event is not detected by the HIDM. This indicates a mismatch between the power taken by the load and the generated power. The following mathematical formulation is used to detect the NDZ [23]:

$$\Delta P = 3V \times I - 3(V + \Delta V) \times I = -3\Delta V \times I \quad (16)$$

where V = 4.16 kV is the network rated voltage and I represents the current supplied by the utility grid to the test network. The NDZ range was considered to be 0.92 pu to 1.1 pu. A mismatch in real power was obtained by changing the generated power from the WGP and SGP. Various operating scenarios of power mismatch and voltage are included in Table 6. The NDZ limits where the islanding condition is not detected are marked as IND (islanding not detected), and limits where the islanding event is identified are marked as ID (islanding detected). It is observed that the NDZ limits for the HIDM in terms of power mismatch and voltage are indicated as (a) $\Delta P = -4$ kW to $\Delta P = 5$ kW at $V = 1.0$ pu and (b) $V = 0.96$ pu to $V = 1.1$ pu at $\Delta P = 0$ kW. ΔP is taken as positive for the scenario when the power generated on the island is high relative to the load on the island and vice versa. The NDZ limits of the HIDM are normal and permissible in general conditions. For these limits of the NDZ, the proposed HIDM will work well.

Table 6. Islanding NDZ of HIDM.

ΔP	Voltage (pu)								
(kW)	0.92	0.94	0.96	0.98	0.99	1.0	1.02	1.06	1.1
7	ID	ID	ID	ID	ID	ID	ID	ID	ID
6	ID	ID	ID	ID	ID	ID	ID	ID	ID
5	ID	ID	ID	ID	ID	IND	ID	ID	ID
4	ID	ID	ID	ID	IND	IND	IND	ID	ID
3	ID	ID	ID	ID	IND	IND	IND	IND	ID
2	ID	ID	ID	IND	IND	IND	IND	IND	IND
1	ID	ID	ID	IND	IND	IND	IND	IND	IND
0	ID	ID	IND	IND	IND	IND	IND	IND	IND
−1	ID	ID	ID	IND	IND	IND	IND	IND	IND
−2	ID	ID	ID	ID	IND	IND	IND	IND	IND
−3	ID	ID	ID	ID	IND	IND	IND	IND	ID
−4	ID	ID	ID	ID	ID	IND	IND	ID	ID
−5	ID	ID	ID	ID	ID	ID	ID	ID	ID
−6	ID	ID	ID	ID	ID	ID	ID	ID	ID
−7	ID	ID	ID	ID	ID	ID	ID	ID	ID

4.2. Detection of Islanding with SGP

The test system was simulated with data from the SGP. The IFS switch was opened at 0.1 s to form the island of the test network with solar generation. The currents of all phases were captured on node 632 of the test system. The current signals were analyzed using the proposed method to compute the HIDF, SIDF, AIDF, and HIDI. The current waveforms of all phases, HIDF, SIDF, AIDF, and HIDI, are depicted in Figure 6a–e in their respective sequences. The maximum magnitude of the HIDI and the computational time of the HIDM are included in Table 3 and Table 4, respectively.

Figure 6a illustrates that the current decreases and becomes zero at the moment of the islanding condition at 0.1 s. The magnitude of the HIDF is also reduced after an islanding event, as elaborated in Figure 6b. Figure 6c elaborates that the value of the SIDF becomes high at the moment of islanding condition incidence, which is zero before and after islanding condition incidence. Figure 6d elaborates that the value of the AIDF is zero before and after the islanding condition and becomes high at the instant of the islanding event occurrence. However, spikes are observed for a window of one cycle. Figure 6e elaborates that the magnitude of the HIDI is zero before island formation and becomes higher than the HIDIT. However, the peak magnitude is low relative to the HIDIFT. Hence, the peak value of the HIDI falls in the islanding event zone. Therefore, islanding conditions with the SGP are effectively detected, as well as differentiated from both faulty and operational conditions. The computational time of the HIDM for islanding detection with both the WGP and SGP is 1.650991 ms.

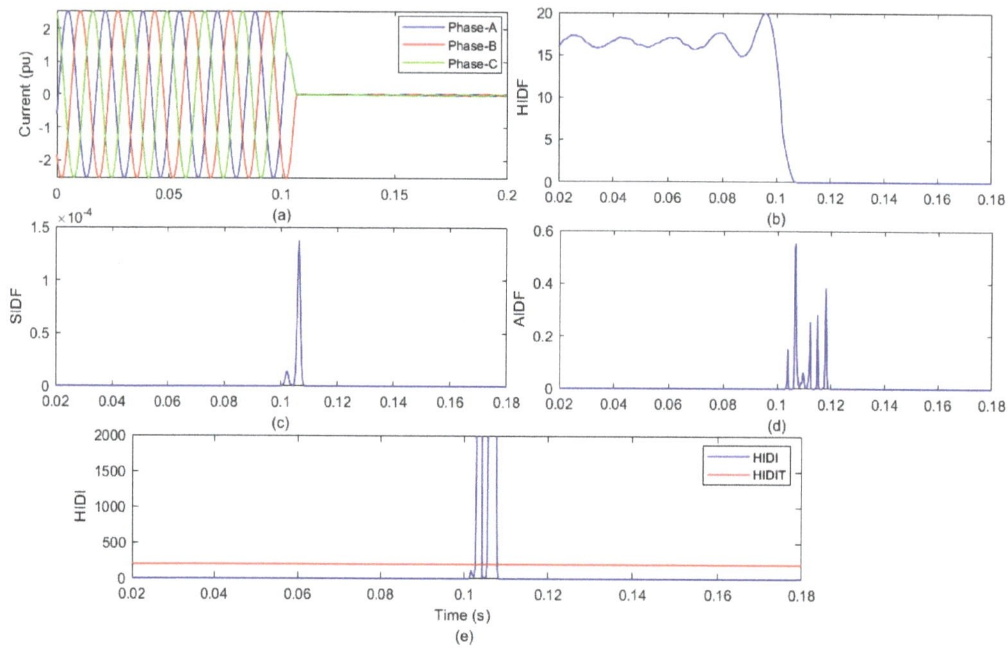

Figure 6. Detection of islanding event with generation from SGP: (**a**) current waveform; (**b**) HIDF; (**c**) SIDF; (**d**) AIDF; (**e**) HIDI.

4.3. Detection of Islanding with WGP

The test system was simulated with data from the WGP. The IFS switch was opened at 0.1 s to form the island of the test system with wind generation. The currents of all phases were captured on node 632 of the test system. The current signals were analyzed using the HIDM to compute the HIDF, SIDF, AIDF, and HIDI. The current waveforms of all phases, HIDF, SIDF, AIDF, and HIDI, are depicted in Figure 7a–e in their respective orders. The maximum magnitude of the HIDI and the computational time of the HIDM are tabulated in Table 3 and Table 4, respectively.

Figure 7a illustrates that the current decreases and becomes zero due to island formation at 0.1*s*. The magnitude of the HIDF is also reduced after an islanding event, as depicted in Figure 7b. Figure 7c elaborates that the value of the SIDF becomes high at the moment of islanding event incidence, which is zero before and after the islanding condition. Figure 7d indicates that the magnitude of the AIDF is zero before and after the islanding condition and becomes high at the moment of the islanding condition. However, spikes are observed for a window of one cycle. Figure 7e elaborates that the magnitude of the HIDI is zero before island formation and becomes higher than the HIDIT due to island formation. However, the peak magnitude is low relative to the HIDIFT. Hence, the peak value of the HIDI falls in the islanding event zone. Therefore, islanding conditions with the WGP are effectively detected, and this is differentiated from both faulty and operational events. The computational time of the HIDM for islanding detection with the WGP is 0.443420 ms.

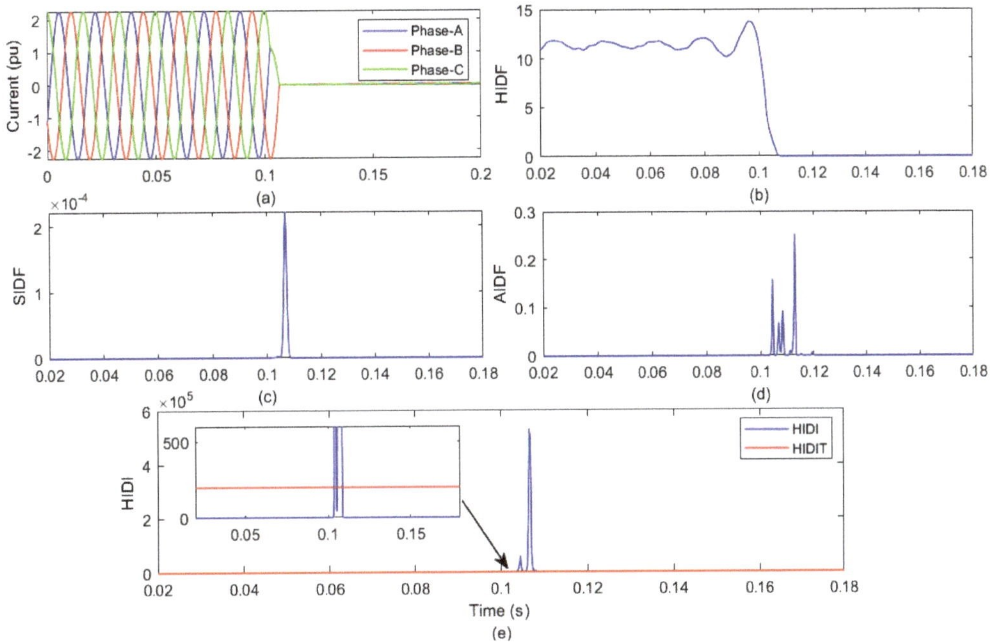

Figure 7. Detection of islanding event with generation from WGP: (**a**) current waveform; (**b**) HIDF; (**c**) SIDF; (**d**) AIDF; (**e**) HIDI.

4.4. High Penetration Level of RE (100%)

The test system was simulated considering a generation mix of 2 MW solar power and 3 MW wind power to meet the load demand of the test system of 5 MVA. Hence, an RE penetration level of 100% was achieved. The IFS switch was opened at 0.1 s to form the island of the test system. The currents of all phases were captured on node 632 of the test system. The current signals were processed using the HIDM to evaluate the HIDI, as demonstrated in Figure 8. It was inferred that the peak magnitude of the HIDI is zero before island formation and becomes higher than the HIDIT. However, the peak magnitude is low relative to the HIDIFT. Hence, the peak value of the HIDI falls in the islanding event zone. Therefore, an islanding condition with a 100% RE penetration level is effectively detected, and this is differentiated from both faulty and operational events. Further, the peak magnitude of the HIDI for 100% RE penetration is higher compared to the scenario of 50% RE penetration.

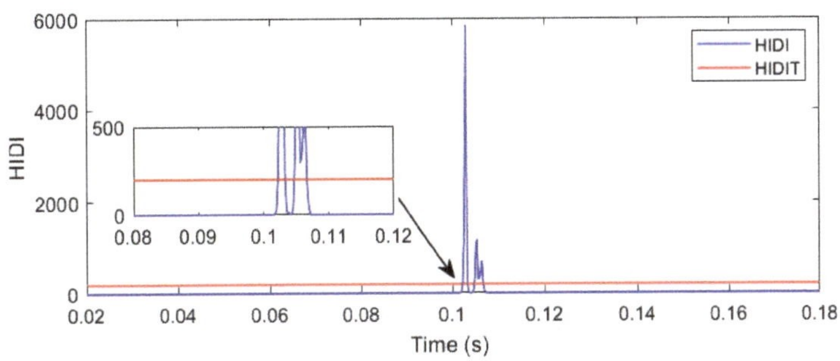

Figure 8. HIDI for islanding with 100% RE penetration.

5. Testing of HIDM to Identify and Discriminate the Fault Events from Islanding Events: Simulation Results and Discussion

In this section, the discussion of the results for the identification of fault conditions and discriminating the same from islanding events is included. The investigated fault conditions include phase to ground (LG), phase to phase (LL), two phases to ground (LLG), three phases (LLL), and three phases to ground (LLL). The current signals used to obtain the results of fault detection demonstrated in this section were synthetically generated by simulation studies performed on the test system of a utility network with the RE generators described in Figuree 1 using the MATLAB software.

5.1. Phase to Ground Fault

The test system was simulated with generation from both the WGP and SGP. An LG fault was realized at 0.1 s on junction 671 of the test system. The currents of each phase were captured at junction 632 of the test system. The current signals were processed using the proposed method to compute the HIDF, SIDF, AIDF, and HIDI. The current waveform of all phases with the LG fault, HIDF, SIDF, AIDF, and HIDI, are depicted in Figure 9a–e in their respective order. The maximum magnitude of the HIDI and the computational time of the HIDM for the condition of the LG fault are included in Table 3 and Table 4, respectively.

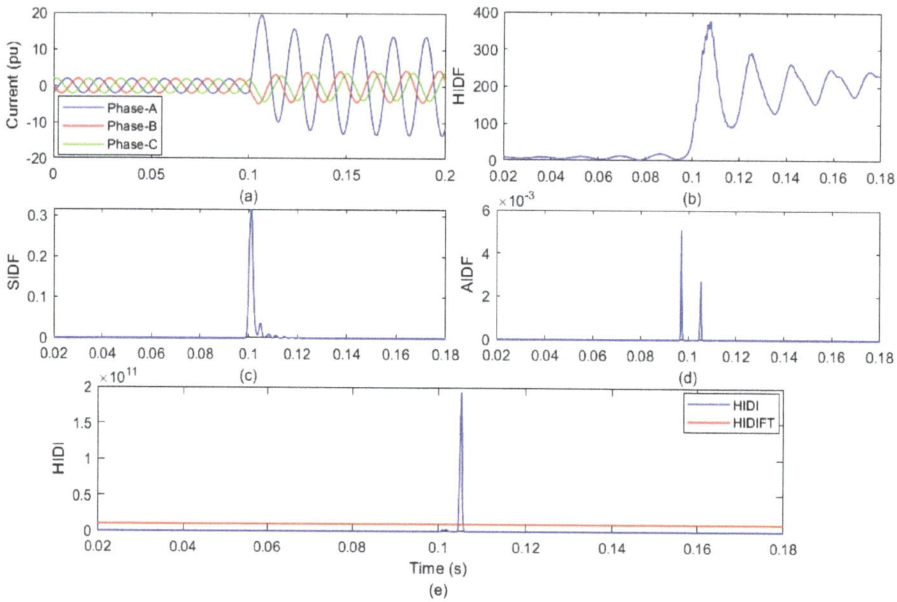

Figure 9. Testing of HIDM for LG fault: (**a**) current waveform; (**b**) HIDF; (**c**) SIDF; (**d**) AIDF; (**e**) HIDI.

Figure 9a depicts that the current of phase A increases after the LG fault at 0.1 s, while the currents of phases B and C remain at the same value. The peak of the HIDF increased after the LG fault, as elaborated in Figure 9b. Figure 9c elaborates that the value of the SIDF becomes high at the moment of LG fault occurrence, which is zero before and after the LG fault. Figure 9d illustrates that the value of the AIDF is zero prior to and after the LG fault and becomes high at the moment of the LG fault. However, spikes are observed for a window of one cycle. Figure 9e elaborates that the value of the HIDI is zero prior to the LG fault event and becomes higher than the HIDIFT at the moment of fault. Hence, the peak value of the HIDI falls in the fault event zone. Therefore, LG fault events in the presence of both the WGP and SGP are effectively detected and also differentiated from both islanding and operational conditions. The computational time of the HIDM for LG fault event detection with both the WGP and SGP is 1.320318 ms.

5.2. Phase to Phase Fault

The test system was simulated with generation from both the WGP and SGP. An LL fault was realized at 0.1 s on junction 671 of the test system. The currents of every phase were captured at junction 632 of the test system. The current signals were decomposed using the HIDM for computing the HIDF, SIDF, AIDF, and HIDI. The current waveforms of all phases with the LL fault and HIDI are elaborated in Figure 10a,b, respectively. The maximum value of the HIDI and the computational time of the HIDM for the LL fault are included in Table 3 and Table 4, respectively.

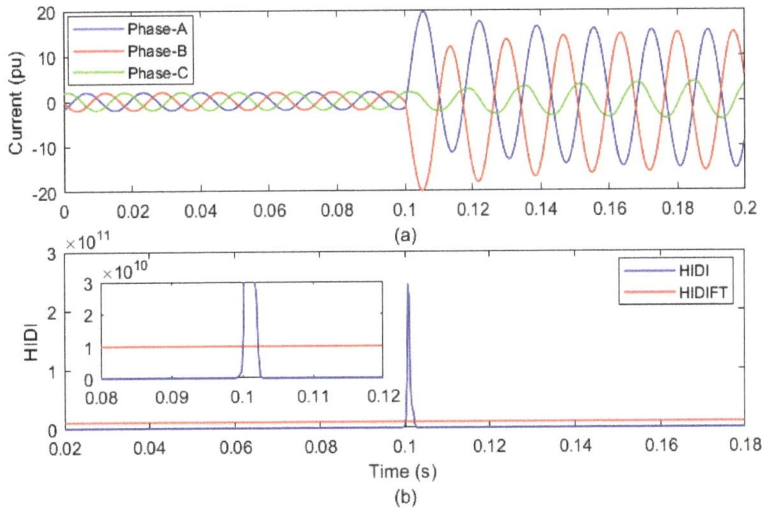

Figure 10. Testing of HIDM for LL fault: (**a**) current waveform; (**b**) HIDI.

Figure 10a depicts that the current of phases A and B becomes high due to the LL fault event at 0.1 s, while the current of phase C remains unaltered. Figure 10b elaborates that the magnitude of the HIDI is zero prior to the LL fault and becomes higher than the HIDIFT due to the LL fault. Hence, the peak value of the HIDI falls in the fault event zone. Therefore, LL fault events with both the WGP and SGP are efficiently detected and also differentiated from both the islanding and operational conditions. The computational time of the HIDM for LL fault event detection with both the WGP and SGP is 1.631831 ms.

5.3. Two Phases to Ground Fault

The test system was simulated with generation from both the WGP and SGP. An LLG fault was realized at 0.1s on junction 671 of the test system. The currents of every phase were captured at junction 632 of the test system. The current signals were decomposed using the HIDM for computing the HIDF, SIDF, AIDF, and HIDI. The HIDI for the LLG fault event is depicted in Figure 11. The maximum magnitude of the HIDI and the computational time of the HIDM for the LLG fault are included in Table 3 and Table 4, respectively.

Figure 11 illustrates that the magnitude of the HIDI is zero prior to the LLG fault event and attains values higher than the HIDIFT due to the LLG fault. Hence, the peak magnitude of the HIDI falls in the fault event zone. Therefore, LLG fault events with the WGP and SGP are efficiently detected and also differentiated from both islanding and operational conditions. The computational time of the HIDM for LLG fault event detection with both the WGP and SGP is 0.418473 ms.

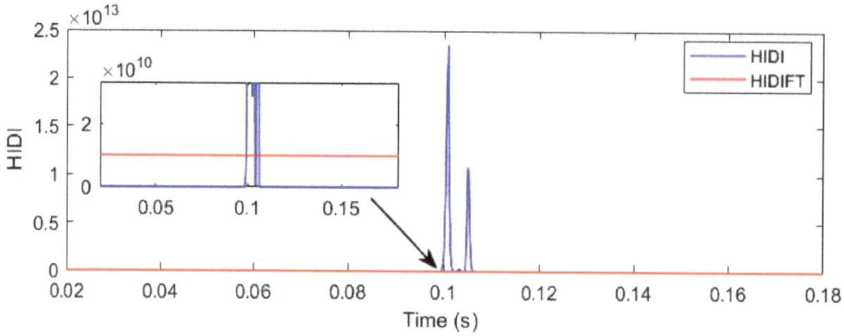

Figure 11. HIDI for testing the HIDM to identify LLG fault.

5.4. Three-Phase Fault

The test system was simulated with generation from both the WGP and SGP. An LLL fault was realized at 0.1s on node 671 of the test system. The currents of every phase were captured at junction 632 of the test system. The current signals were processed by applying the HIDM to compute the HIDF, SIDF, AIDF, and HIDI. The current waveform of all phases with the LLL fault and HIDI is depicted in Figure 12. The maximum magnitude of the HIDI and the computational time of the HIDM in the event of an LLL fault are included in Table 3 and Table 4, respectively.

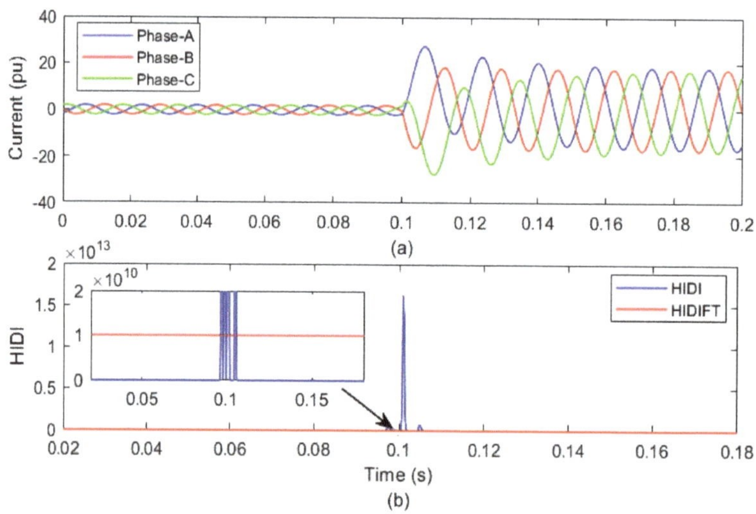

Figure 12. Testing of HIDM for LLL fault event: (**a**) current waveform; (**b**) HIDI plot.

Figure 12a depicts that the current of every phase increases due to the LLL fault at 0.1 s. Figure 12b elaborates that the magnitude of the HIDI is zero before the LLL fault event and becomes higher than the HIDIFT due to the LLL fault. Hence, the peak magnitude of the HIDI falls in the fault event zone. Therefore, LLL faults with both the WGP and SGP are efficiently detected and also differentiated from both the islanding and operational conditions. The computational time of the HIDM for LLL fault event detection with both the WGP and SGP is 1.397293 ms.

5.5. Three Phases to Ground Fault

The test system was simulated with generation from both the WGP and SGP. An LLLG fault was realized at 0.1s on junction 671 of the test system. The currents of every phase were recorded at junction 632 of the test system. The current signals were processed using the HIDM for computing the HIDF, SIDF, AIDF, and HIDI. The HIDI for the LLLG fault is depicted in Figure 13. The maximum magnitude of the HIDI and the computational time of the HIDM for the LLL fault are included in Table 3 and Table 4, respectively.

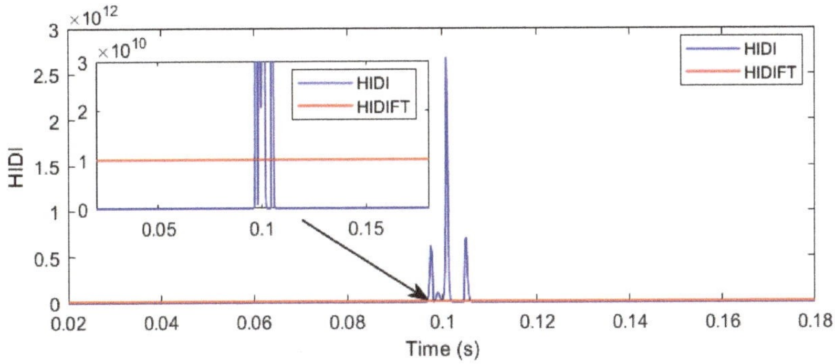

Figure 13. HIDI for the event of LLLG fault.

Figure 13 elaborates that the magnitude of the HIDI is zero prior to the LLLG fault and becomes higher than the HIDIFT due to the LLLG fault. Hence, the peak magnitude of the HIDI falls in the fault event zone. Therefore, LLLG faults with the WGP and SGP are efficiently detected and also differentiated from both islanding and operational conditions. The computational time of the HIDM for LLLG fault event detection with both the WGP and SGP is 0.629446 ms.

6. Testing of HIDM to Identify and Discriminate the Operational Events from Islanding Events: Simulation Results and Discussion

This section discusses the results to detect operational conditions and differentiate them from islanding conditions. The investigated operational events include the feeder operation, capacitor operation, and load operation. The current signals used to obtain the results of operational event detection demonstrated in this section were synthetically generated by simulation studies performed on the test system of a utility network with the RE generators described in Figure 1 using the MATLAB software. These switching events may be for maintenance purposes or for load curtailment purposes.

6.1. Feeder Operation

The test system was simulated with generation from both the WGP and SGP. A feeder opening event was simulated by opening the switch between junctions 671 and 692 at 0.0667 s and re-closing at 0.133 s. The currents of every phase were captured at junction 632 of the test system. The current signals were processed by applying the HIDM for computing the HIDF, SIDF, AIDF, and HIDI. The current waveform of all phases with feeder operation events (HIDF, SIDF, AIDF, and HIDI) is depicted in Figure 14a–e, respectively. The maximum magnitude of the HIDI and the computational time of the HIDM for the event of feeder operation are tabulated in Table 3 and Table 4, respectively.

Figure 14a illustrates that small-magnitude variations of the current in all phases are observed at the moments of feeder opening and closing. The value of the HIDF decreases at the moment of feeder opening and, again, increases at the moment of feeder closing with small-magnitude oscillations, as detailed in Figure 14b. Figure 14c elaborates that the value of the SIDF becomes high at the moment of feeder opening and closing. At all other times, the SIDF is zero. Figure 14d elaborates that the value of the AIDF becomes high at the moment of feeder opening and closing. Figure 14e elaborates that the value of the HIDI becomes high at the moment of feeder operation and closing, but the peak value is always less than the HIDIT, which falls in the operational event zone. Therefore, feeder operation events in the presence of both the WGP and SGP are effectively detected and also differentiated from both islanding and fault conditions. The computational time of the HIDM for operational event detection with both the WGP and SGP is 0.247305 ms.

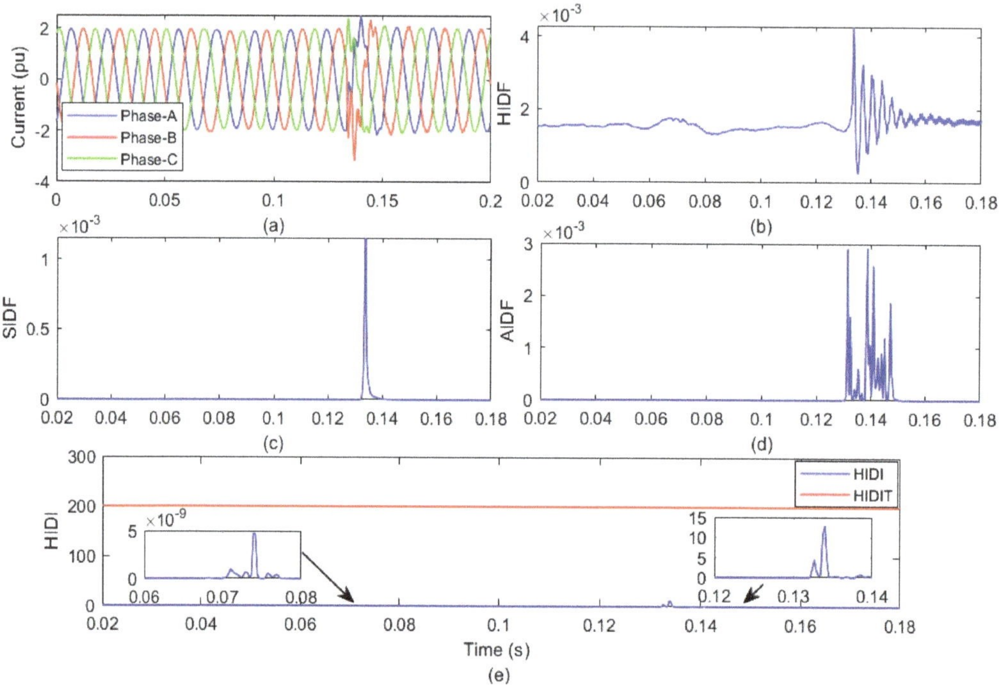

Figure 14. Testing of HIDM for feeder operation: (**a**) current waveform; (**b**) HIDF; (**c**) SIDF; (**d**) AIDF; (**e**) HIDI.

6.2. Capacitor Operation

The test system was simulated with generation from both the WGP and SGP. A capacitor operation condition was simulated by disconnecting the capacitor of rating 600 kVAR connected on junction 675 of the test system at 0.0667 s and re-connecting at 0.133 s. The currents of all phases were captured at junction 632 of the test system. The current signals were processed by the application of the HIDM for computing the HIDF, SIDF, AIDF, and HIDI. The HIDI for the capacitor operation event is depicted in Figure 15. The maximum magnitude of the HIDI and the computational time of the HIDM in the event of capacitor operation are tabulated in Table 3 and Table 4, respectively.

Figure 15 elaborates that the magnitude of the HIDI becomes high at the time of capacitor disconnecting and re-connecting, but the peak magnitude is always less than the HIDIT, which falls in the operational event zone. Therefore, capacitor operation events in the presence of both the WGP and SGP are efficiently detected and also differentiated from both islanding and fault conditions. The computational time of the HIDM for the capacitor operation event detection with both the WGP and SGP is 0.282738 ms.

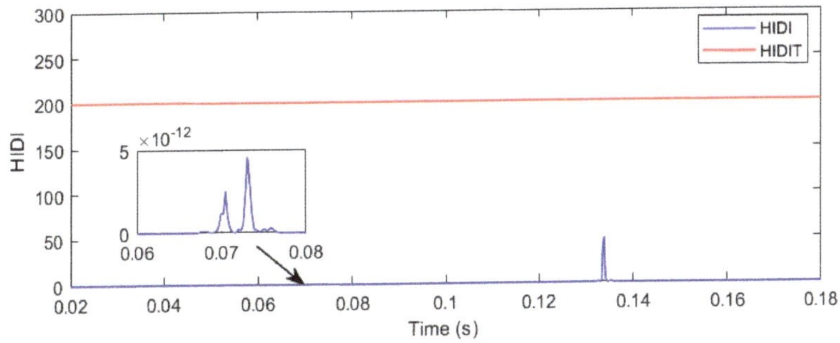

Figure 15. HIDI for the event of capacitor operation.

6.3. Load Operation

The test system was simulated with generation from both the WGP and SGP. The load operation condition was simulated by disconnecting the load of 843 kW and 462 kVAR connected on junction 675 of the test system at 0.0667 s and re-connecting at 0.133 s. The current signals were processed using the HIDM for computing the HIDF, SIDF, AIDF, and HIDI. The HIDI for the load operation event is depicted in Figure 16. The maximum magnitude of the HIDI and computational time of the HIDM for the event of load operation are tabulated in Table 3 and Table 4, respectively.

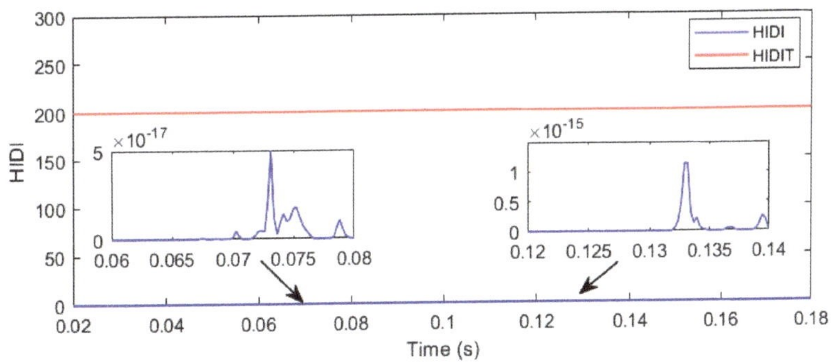

Figure 16. HIDI for the event of load operation.

Figure 16 elaborates that the magnitude of the HIDI becomes high at the time of load disconnecting and re-connecting, but the peak magnitude is always less than the HIDIT, which falls in the operational event zone. Therefore, load operation events in the presence of both the WGP and SGP are effectively detected and also differentiated from both islanding and fault conditions. The computational time of the HIDM for load operation event detection with both the WGP and SGP is 0.344931 ms.

7. Real-Time Validation of HIDM

This section validates the HIDM by extracting the current signals from a practical distribution network facility. The HIDM was tested for detecting the islanding event that occurred on a practical distribution power system network in Rajasthan State, India. The investigated distribution feeder is rated at 33 kV and emanates from a 33/11 kV grid sub-station (GSS). The details of the load and RE plants connected to the distribution feeder are included in Table 7. This feeder was manually tripped from the 33/11 kV GSS, and the current was captured for a period of 0.2 s. The current samples were recorded using a disturbance recorder (DR) installed on the feeder. The sampling rate used by the disturbance recorder is 2.5×10^4 samples per second. The main components of the DR include the central processing unit (CPU) module, measurement input module, binary input module with hysteresis of input current levels, and power supply module. The modules are integrated into internal communication buses. Communication is performed using an Ethernet network channel and various protocols. This current is processed using the HIDM, and the HIDI was evaluated, which is illustrated in Figure 17. It was observed that the peak value of the HIDI was equal to 1.5881×10^6, which falls in the islanding event zone of the proposed HIDM. Hence, it has been established that the HIDM effectively detected the islanding condition incident on the real-time distribution feeder.

Table 7. Load and RE generation details of real-time distribution feeder.

S. No.	Name of Parameter	Quantity of Parameter
1	Load	9.28 MW
2	Rooftop solar plant	0.9 MW
3	Solar PV plant	3.40 MW
4	Wind power plant	1.0 MW

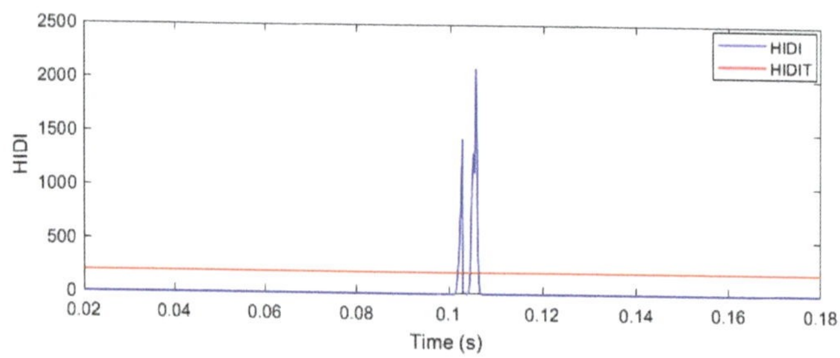

Figure 17. HIDI for islanding event on a practical distribution feeder.

8. Relative Performance of HIDM

The performance of the multi-variable HIDM was compared with the islanding-detection method (IDM) using the DWT [24], the IDM using a combination of the slantlet transform and RPNN [25], the IDM using the wavelet transform multi-resolution (WT-MRA)-based image data [26], and the IDM based on the Deep Neural Network (DNN) [27]. The performance was compared in terms of the RE penetration level for which the IDM detects events effectively, the sampling frequency, the computational time, the noise level for which the performance of the IDM is not affected, and the NDZ. A performance comparison study of the considered IDMs is provided in Table 8. This shows that, in terms of the parameters considered for the comparison study, the performance of the HIDM is better compared to that of the IDMs using DWT, a hybrid combination of the slantlet transform and RPNN.

Table 8. Islanding non-detection zone of HIDM.

S. No.	Reference	Technique	RE Penetration Level	Sampling Frequency	Computational Time	Noise Level for Performance of Algorithm Is Not Deteriorated	NDZ
1	[24]	DWT	20%	6.4 kHz	2 s	40 dB SNR	High
2	[25]	Slantlet transform+RPNN	10%	19.8 kHz	0.17 s	25 dB SNR	High
3	[26]	WT-MRA-based image data	40%	6.4 kHz	0.18 s	20 dB SNR	Moderate
4	[27]	DNN	46.67%	3.84 kHz	2 s	30 dB SNR	Not investigated
5	Proposed HIDM	ST + HT + ALC	50% and 100%	3.84 kHz	$t < 2$ ms	20 dB SNR	Moderate

9. Conclusions

A multi-variable hybrid islanding-detection method supported by a hybrid combination of the ST, HT, and ALC to process the current signals to design the HIDI has been presented. The proposed HIDM is efficient in detecting islanding conditions in the presence of generation from the WGP, SGP, and both from the WGP and SGP with a minimal NDZ. This was achieved for RE penetration levels of 50% and 100% in the test network. The HIDM is efficient for detecting islanding conditions in the scenario of a high noise level of $20dB$ SNR. The HIDM is also effective at detecting fault conditions like LG, LL, LLG, LLL, and LLLG. The HIDM effectively discriminates between islanding conditions and fault ones. The HIDM effectively discriminates between islanding conditions and operational ones. The HIDM is also effective at identifying the islanding condition on a real-time distribution feeder of a practical utility grid. The performance of the HIDM is better compared to islanding-detection methods using DWT, an IDM using WT-MRA-based image data, an IDM using DNN, and an IDM using a combination of the slantlet transform and RPNN in terms of the RE penetration level for which the IDM detects events effectively, the sampling frequency, the computational time, the noise level for which the performance of the IDM is not affected, and also the NDZ. The study was completed on an IEEE-13 bus test feeder using the MATLAB software.

Author Contributions: All authors contributed equally. All authors have read and agreed to the published version of the manuscript.

Funding: This research received no external funding.

Data Availability Statement: Data are contained within the article.

Conflicts of Interest: Author Peixiong Liu was employed by the company Hanjiang Water Resources and Hydropower (Group) Co., Ltd. Hydropower Company. The remaining authors declare that the research was conducted in the absence of any commercial or financial relationships that could be construed as a potential conflict of interest.

Abbreviations

The following abbreviations are used in this manuscript:

AC	Alternating current
AIDF	Alienation islanding-detection factor
ALC	Alienation Coefficient
CIDF	Co-variance islanding-detection factor
CPU	Central processing unit
DC	Direct current
DF-T	Distribution feeder transformer
DFIG	Doubly fed induction generator
DG	Distributed generator
DNN	Deep Neural Network
DR	Disturbance recorder
DWT	Discrete wavelet transform
GSS	Grid sub-station
HIDF	Hilbert islanding-detection factor
HIDI	Hybrid islanding-detection indicator
HIDIFT	Hybrid islanding-detection indicator fault threshold
HIDIT	Hybrid islanding-detection indicator threshold
HIDM	Hybrid islanding-detection method
HT	Hilbert transform
IDM	Islanding-detection method
IEEE	Institute of Electrical and Electronics Engineering
IFS	Island formation switch
IWF	Islanding weight factor

LG	Phase to ground fault
LL	Phase to phase fault
LLG	Two phases to ground fault
LLL	Three-phase fault
LLLG	Three phases to ground fault
MATLAB	Matrix laboratory
NDZ	Non-detection zone
PMU	Phasor measurement unit
PCC	Point of common coupling
PQ	Power quality
PV	Photovoltaic
RE	Renewable energy
RPNN	Ridgelet probabilistic neural network
SGP	Solar generation plant
SGP-GT	Solar generation plant generator transformer
SIDF	Stockwell islanding-detection factor
SNR	Signal-to-noise ratio
ST	Stockwell transform
UG-T	Utility grid transformer
WGP	Wind generation plant
WGP-GT	Wind generation plant generator transformer
WT-MRA	Wavelet transform multi-resolution

References

1. Wei, W.; Shao, C.; Huang, H.; Zhuang, W. A novel formula for plastic hinge length of reinforced concrete columns in seismic regions. *Struct. Infrastruct. Eng.* **2022**, 1–16. [CrossRef]
2. Alhasnawi, B.N.; Jasim, B.H. A new energy management system of on-grid / off-grid using adaptive neuro-fuzzy inference system. *J. Eng. Sci. Technol.* **2020**, *15*, 3903–3919.
3. Alhasnawi, B.N.; Jasim, B.H.; Rahman, Z.A.S.A.; Guerrero, J.M.; Esteban, M.D. A Novel Internet of Energy Based Optimal Multi-Agent Control Scheme for Microgrid including Renewable Energy Resources. *Int. J. Environ. Res. Public Health* **2021**, *18*, 8146. [CrossRef]
4. Timbus, A.; Oudalov, A.; Ho, C.N. Islanding detection in smart grids. In Proceedings of the 2010 IEEE Energy Conversion Congress and Exposition, Atlanta, GA, USA, 12–16 September 2010; pp. 3631–3637. [CrossRef]
5. Larik, N.A.; Tahir, M.F.; Elbarbary, Z.S.; Yousaf, M.Z.; Khan, M.A. A comprehensive literature review of conventional and modern islanding-detection methods. *Energy Strategy Rev.* **2022**, *44*, 101007. [CrossRef]
6. Kumari, J.S.; Lenine, D.; Satish, A.; Kumar, T.S.; Kalaivani, C.; Kumar, M.D.; Swamy, G.K.; Suresh, Y.V.; Reddy, J.N.; Kumar, J.K.; et al. A model predictive Goertzel algorithm based active islanding detection for grid integrated photovoltaic systems. *Microprocess. Microsystems* **2022**, *95*, 104706. [CrossRef]
7. Shukla, A.; Dutta, S.; Sadhu, P.K. An island detection approach by μ-PMU with reduced chances of cyber attack. *Int. J. Electr. Power Energy Syst.* **2021**, *126*, 106599. [CrossRef]
8. Nale, R.; Biswal, M.; Kishor, N. A passive communication based islanding-detection technique for AC microgrid. *Int. J. Electr. Power Energy Syst.* **2022**, *137*, 107657. [CrossRef]
9. Elshrief, Y.A.; Asham, A.D.; Bouallegue, B.; Ahmed, A.; Helmi, D.H.; Abozalam, B.A.; Abd-Elhaleem, S. An innovative hybrid method for islanding detection using fuzzy classifier for different circumstances including NDZ. *J. Radiat. Res. Appl. Sci.* **2022**, *15*, 129–142. [CrossRef]
10. Alsabban, M.; Bertozzi, O.; Ahmed, S. Analysis and Verification of Islanding Detection Techniques for Grid-integrated PV Systems. In Proceedings of the 2023 IEEE PES Conference on Innovative Smart Grid Technologies—Middle East (ISGT Middle East), Abu Dhabi, United Arab Emirates, 12–15 March 2023; pp. 1–5. [CrossRef]
11. Shaik, A.G.; Mahela, O.P. Power quality assessment and event detection in hybrid power system. *Electr. Power Syst. Res.* **2018**, *161*, 26–44. [CrossRef]
12. Kersting, W. Radial distribution test feeders. *IEEE Trans. Power Syst.* **1991**, *6*, 975–985. [CrossRef]
13. Mahela, O.P.; Shaik, A.G. Detection of power quality events associated with grid integration of 100kW solar PV plant. In Proceedings of the 2015 International Conference on Energy Economics and Environment (ICEEE), Greater Noida, India, 27–28 March 2015; pp. 1–6. [CrossRef]
14. Mahela, O.P.; Shaik, A.G. Comprehensive overview of grid interfaced solar photovoltaic systems. *Renew. Sustain. Energy Rev.* **2017**, *68*, 316–332. [CrossRef]
15. Mahela, O.P.; Shaik, A.G. Comprehensive overview of grid interfaced wind energy generation systems. *Renew. Sustain. Energy Rev.* **2016**, *57*, 260–281. [CrossRef]

16. Mahela, O.P.; Shaik, A.G. Power Quality Detection in Distribution System with Wind Energy Penetration Using Discrete Wavelet Transform. In Proceedings of the 2015 Second International Conference on Advances in Computing and Communication Engineering, Dehradun, India, 1–2 May 2015; pp. 328–333. [CrossRef]
17. Derviškadić, A.; Frigo, G.; Paolone, M. Beyond Phasors: Modeling of Power System Signals Using the Hilbert Transform. *IEEE Trans. Power Syst.* **2020**, *35*, 2971–2980. [CrossRef]
18. Feldman, M. Hilbert transform in vibration analysis. *Mech. Syst. Signal Process.* **2011**, *25*, 735–802. [CrossRef]
19. Arranz, R.; Ángel Paredes.; Rodríguez, A.; Muñoz, F. Fault location in Transmission System based on Transient Recovery Voltage using Stockwell transform and Artificial Neural Networks. *Electr. Power Syst. Res.* **2021**, *201*, 107569. [CrossRef]
20. Rodríguez, A.; Aguado, J.; Martín, F.; López, J.; Muñoz, F.; Ruiz, J. Rule-based classification of power quality disturbances using S-transform. *Electr. Power Syst. Res.* **2012**, *86*, 113–121. [CrossRef]
21. Ray, P.K.; Kishor, N.; Mohanty, S.R. Islanding and Power Quality Disturbance Detection in Grid-Connected Hybrid Power System Using Wavelet and *S*-Transform. *IEEE Trans. Smart Grid* **2012**, *3*, 1082–1094. [CrossRef]
22. Rathore, B.; Shaik, A.G. Wavelet-Alienation based protection scheme for multi-terminal transmission line. *Electr. Power Syst. Res.* **2018**, *161*, 8–16. [CrossRef]
23. Samet, H.; Hashemi, F.; Ghanbari, T. Minimum non detection zone for islanding detection using an optimal Artificial Neural Network algorithm based on PSO. *Renew. Sustain. Energy Rev.* **2015**, *52*, 1–18. [CrossRef]
24. Sharma, R.; Singh, P. Islanding detection and control in grid based system using wavelet transform. In Proceedings of the 2012 IEEE Fifth Power India Conference, Murthal, India, 19–22 December 2012; pp. 1–4. [CrossRef]
25. Ahmadipour, M.; Hizam, H.; Othman, M.L.; Radzi, M.A.M.; Murthy, A.S. Islanding detection technique using Slantlet Transform and Ridgelet Probabilistic Neural Network in grid-connected photovoltaic system. *Appl. Energy* **2018**, *231*, 645–659. [CrossRef]
26. Kong, X.; Xu, X.; Yan, Z.; Chen, S.; Yang, H.; Han, D. Deep learning hybrid method for islanding detection in distributed generation. *Appl. Energy* **2018**, *210*, 776–785. [CrossRef]
27. Hussain, A.; Mirza, S.; Kim, C.H. Islanding detection and classification of non-islanding disturbance in multi-distributed generation power system using deep neural networks. *Electr. Power Syst. Res.* **2023**, *224*, 109807. [CrossRef]

energies

Article

A Novel Refined Regulation Method with Modified Genetic Commutation Algorithm to Reduce Three-Phase Imbalanced Ratio in Low-Voltage Distribution Networks

Dazhao Liu [1], Zhe Liu [1], Ti Wang [1], Zhiguang Xie [1], Tingting He [2,*], Aixin Dai [2] and Zhiqiang Chen [2]

1. State Grid Beijing Electric Power Company, No. 41 Qianmen West Street, Beijing 100031, China; liudazhao@bj.sgcc.com.cn (D.L.); liuzhea@bj.sgcc.com.cn (Z.L.); wangti@bj.sgcc.com.cn (T.W.); xiezhiguang@bj.sgcc.com.cn (Z.X.)
2. School of Electrical Engineering, Beijing Jiaotong University, No. 3 Shangyuan Cun, Haidian District, Beijing 100044, China; 22110393@bjtu.edu.cn (A.D.); 22125997@bjtu.edu.cn (Z.C.)

* Correspondence: hetingting@bjtu.edu.cn

Abstract: The three-phase imbalance in low-voltage distribution networks (LVDNs) seriously threatens the security and stability of the power system. At present, a standard solution is automatic phase commutation, but this method has limitations because it does not address the branch imbalance and premature convergence or instability of the commutation algorithm. Therefore, this paper proposes a novel refined regulation commutation system, combined with a modified optimized commutation algorithm, and designs a model and simulation for feasibility verification. The refined regulatory model incorporates branch control units into the traditional commutation system. This effectively disperses the main controller's functions to each branch and collaborates with intelligent fusion terminals for precise adjustment. The commutation algorithm designed in this paper, combined with the above model, adopts strategies such as symbol encoding, cubic chaotic mapping, and adaptive adjustment based on traditional genetic algorithms. In addition, in order to verify the effectiveness of the proposed method, this paper establishes a mathematical model with the minimum three-phase imbalance and commutation frequency as objectives and establishes a simulation model. The results of the simulation demonstrate that this method can successfully lower the three-phase imbalance of the low-voltage distribution network. It leads to a decrease of the main circuit's three-phase load imbalance rate from 27% to 6% and reduces each branch line's three-phase imbalance ratio to below 10%. After applying the method proposed in this paper, the main and branches circuit three-phase imbalance are both lower than the limit ratio of the LVDNs, which can improve the quality and safety of electricity consumption. Additionally, the results also prove that the commutation algorithm under this method has faster convergence speed, better application effect, and better stability, which has promotion and application value.

Keywords: low-voltage distribution networks; three-phase imbalance; refined regulation commutation; cubic adaptive genetic algorithm

1. Introduction

The low-voltage distribution network is an important component of the power system and plays a crucial role in supplying power to end users. However, with the increase in power demand and variations in power load, the issue of three-phase imbalance in LVDNs is becoming increasingly prominent [1]. Uneven current distribution in the power grid, due to a three-phase imbalance, can cause equipment overload, damage, and low energy efficiency, all of which have a major negative impact on the dependability and quality of the power supply [2–4]. Therefore, it is of great theoretical and practical significance to solve the three-phase load imbalance problem in low-voltage distribution networks.

In China, the low-voltage feeders of LVDNs are usually three-phase, four-wire systems with two power supply modes: single-phase and three-phase [5]. At present, the main methods to solve the three-phase imbalance in LVDNs can be divided into manual phase adjustment, inter-phase reactive power compensation, and automatic phase commutation [6,7]. The method of manual phase sequence adjustment is to adjust the line load phase sequence manually regularly according to the monitoring data of the low-voltage side of the distribution transformer. This method cannot guarantee real-time and safety, and the workload is large [8]. Reactive power compensation refers to the parallel connection of capacitors between three phases to ensure a three-phase balance. This method can solve the three-phase imbalance to a certain extent, but it cannot fundamentally reduce line losses in the distribution area and has high maintenance costs [9]. The method of automatic phase commutation is based on an automatic commutation device (ACD). This device adjusts the phase sequence of some electrical loads through data collection and intelligent judgment. As a result, it fundamentally solves the problem of three-phase imbalance. This method can switch user load phases online, solve three-phase imbalance from the source, and improve power supply reliability [10,11].

At present, the method of automatic phase commutation often uses a one-multi-type commutation regulation system. In this system, a main controller monitors all intelligent commutation switches [12,13]. This method can indeed effectively balance the main circuit three-phase load of the low-voltage distribution network. However, since it is unable to monitor and adjust the current and voltage on each branch line, a serious three-phase imbalance still exists on the branch lines [6,14]. Moreover, with the continuous expansion of the power grid-scale and the continuous increase of load, the problem of three-phase load imbalance on branch lines also becomes more serious. Therefore, a more refined commutation regulation system should be designed, along with corresponding commutation algorithms with good convergence and stability.

The choice of optimization algorithm is key to reducing the degree of three-phase imbalance using ACDs [15]. Among them, artificial intelligence algorithms have significant advantages in data processing. These algorithms also have strong computational capabilities, enabling them to achieve ideal imbalance regulation through commutation. Artificial intelligence algorithms play an important role in various fields of application [16]. In terms of data processing, artificial intelligence algorithms can quickly and accurately analyze, extract, and understand a large amount of data. In terms of computing power, intelligence algorithms can utilize powerful computing resources for efficient calculations and improve the accuracy and efficiency of algorithms. In addition, artificial intelligence algorithms also have self-learning and adaptive capabilities. By continuously extracting patterns and patterns from data, artificial intelligence algorithms can self-learn and enhance their own abilities. These advantages enable intelligence algorithms to play an important role in various fields of application.

ACDs based on artificial intelligence algorithms are gradually receiving attention from researchers. For example, in [17], genetic algorithms were proposed to solve phase load balancing problems, optimize three-phase imbalance, and reduce terminal switch losses. In [18], the particle swarm optimization algorithm was employed to address the commutation scheme with optimization objectives of commutation frequency and three-phase imbalance. In [19], the optimal commutation scheme was solved using the modified Pigeon-Inspired Optimization. This approach demonstrated the advantages of introducing adaptive parameters and Cauchy perturbations in Pigeon-Inspired Optimization. In [20], the modified sparrow search algorithm was utilized to solve the multi-objective optimal model. The model aimed to minimize the commutation switch quantity and line loss cost, resulting in the derivation of an optimal commutation scheme. These methods have, to some extent, solved the three-phase imbalance on the main circuit, but the three-phase imbalance problem on the branch line still exists, and the optimization of commutation stability has not been taken into account.

Therefore, based on existing problems and references, this paper proposes a refined adjustment method with a modified genetic commutation algorithm to solve the problem of three-phase imbalance in LVDNs. This method aims to reduce the three-phase imbalance ratio on the main circuit and branch in LVDNs. It also focuses on optimizing the use of as few switch actions as possible during commutation. In addition, the commutation algorithm has good convergence and stability. This paper makes the following original contributions:

(1) A new commutation regulation mode. This paper proposes a refined regulation model to solve the issue of three-phase imbalance on branches in LVDNs. Based on traditional commutation systems, this model adds branch control units to finely adjust the three-phase load.
(2) A new commutation algorithm. To improve the commutation effect and stability, and minimize the number of commutation actions, this paper proposes a new commutation algorithm. This algorithm adopts strategies such as symbol encoding, cubic chaotic mapping, and adaptive adjustments based on a genetic algorithm to optimize algorithm performance.
(3) Simulation examples in combination with (1) and (2). To verify the effectiveness of the proposed method, this paper establishes a mathematical model with the objectives of minimizing three-phase imbalance and commutation frequency. Additionally, a simulation model is created using the data of LVDNs as an example.

2. Analysis and Modeling of Three-Phase Imbalanced System Commutation Considering Refined Regulation

2.1. Refined Regulation Commutation System

The automatic phase commutation method is one of the standard methods to solve the three-phase imbalance problem in LVDNs, with advantages such as low cost, simple, and safe use. This method does not alter the existing structure of the power grid's distribution network. It achieves an even distribution of unbalanced loads to each phase through automatic phase-shifting, resulting in a reduction of the three-phase imbalance ratio. With the continuous rise of branch loads in distribution networks and the increasing demand for electricity, researching more scientific and efficient intelligent commutation systems is the key content of automatic phase commutation methods [21–23]. Traditional commutation systems often adopt a one-multi-type regulation mode, where one master controller monitors all commutation switches, as shown in Figure 1. A △/Y-connected transformer is used as an example.

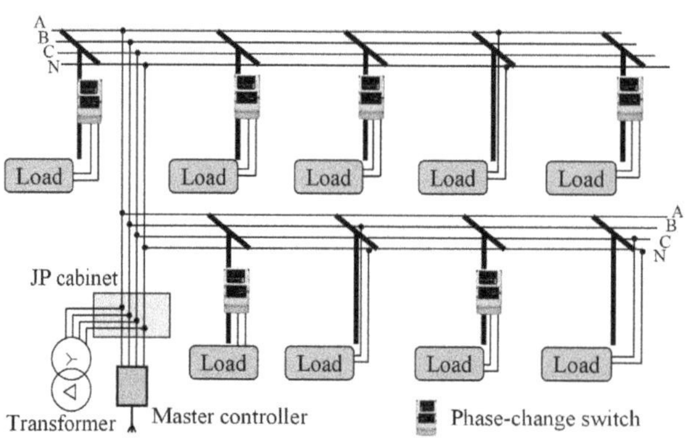

Figure 1. Traditional commutation mode diagram of LVDNs.

In traditional commutation systems, the master controller is usually installed at the low-voltage outlet line on the secondary side of the distribution transformer to collect real-time three-phase data and calculate three-phase imbalance. When the three-phase imbalance is too large, the commutation scheme is calculated according to the commutation algorithm and sent to the switches for commutation operation through wireless. Assuming there are three branches in a traditional system, where all loads on the first branch are in phase A, all loads on the second branch are in phase B, and all loads on the third branch are in phase C. Obviously, in this situation, the main circuit is in a completely three-phase balance state, but each branch is in a serious three-phase imbalance state. Therefore, the traditional commutation mode cannot solve the three-phase imbalance problem on each branch. In order to solve the three-phase imbalance problem of the main circuit and each branch in the LVDNs, this paper proposes a refined commutation mode, as shown in Figure 2.

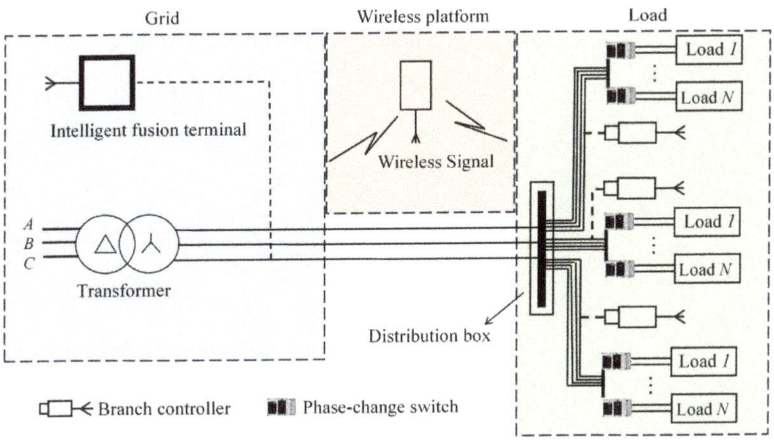

Figure 2. Three-phase balance commutation system under refined regulation mode.

The refined regulation of the commutation system involves overlaying the host functions in a traditional mode onto the unit monitoring points of each branch. This enables real-time monitoring and regulation of the load on each branch by the respective units. Then, using communication technology, the current data is uploaded to the intelligent terminal equipment on the distribution transformer side. The terminal calculates the imbalance degree based on the current situation of each branch line. Then, an improved genetic algorithm is used to optimize the commutation strategy and determine the optimal commutation switch action plan. It is then issued to the intelligent commutation switch to complete live commutation.

This method can effectively regulate the three-phase load imbalance in each branch line, which avoids the shortcomings of the traditional centralized adjustment mode. The key to the commutation system in the refined commutation mode is the commutation control terminal of the branch unit detection node. The intelligent commutation terminal is installed on the bus of each branch, and it needs to detect the total current and voltage of this branch in real-time to determine whether the branch has three-phase imbalance problems. When a three-phase imbalance problem is detected, the terminal generates a control signal based on the load current of the commutation switch collected by the terminal. The commutation switch is then driven to complete the commutation operation, enabling the intelligent management of the three-phase imbalance. Among them, the commutation switching device in the new mode still adopts the seamless commutation technology with an equal voltage of 0 milliseconds, which can switch the single-phase load's uninterrupted power supply between the three phases. Finally, the collected load current and fault signals need to be uploaded to the cloud server to achieve remote detection and control of the commutation system, ensuring the safe and stable operation of the system.

Therefore, optimizing the quality of the commutation algorithm is one of the decisive factors for intelligent commutation, and it is also the key to refining the commutation effect of the commutation system. This paper establishes a mathematical model for branch line commutation, which is used for subsequent commutation algorithm improvement.

2.2. Mathematical Model of Branch Commutation

Whether the three-phase imbalance in the distribution area can be reduced and whether the number of commutation switch actions can be minimized are two important criteria for judging the effectiveness of the phase sequence adjustment methods. In addition, considering the voltage imbalance and various indicators at the point of common coupling as a comprehensive indicator will be more comprehensive. However, considering the efficiency and stability of the commutation algorithm, this section takes the minimum three-phase imbalance and the minimum number of switch actions as the objective functions. On this basis, constraints such as commutation switch phase sequence were established, and a branch commutation mathematical model was established for the commutation algorithm model.

Three-phase imbalance generally refers to the phenomenon of three-phase load imbalance in the power system, and the degree of imbalance exceeds a certain range. The measurement of three-phase imbalance serves as an indicator for this phenomenon, yet, a universally standardized calculation method for the three-phase imbalance is currently lacking across different fields worldwide. In practical engineering, the calculation method for current unbalance is widely used and provides more accurate results [24]. Therefore, this paper calculates the three-phase current unbalance based on the distribution network operation and maintenance regulations:

$$\varepsilon = max\left\{\frac{|i_A - i_{avg}|}{i_{avg}}, \frac{|i_B - i_{avg}|}{i_{avg}}, \frac{|i_C - i_{avg}|}{i_{avg}}\right\} \times 100\% \tag{1}$$

where $i_{avg} = (i_A + i_B + i_C)/3$ represents the average current of the three phases, and i_A, i_B, and i_C represent the currents of phases A, B, and C, respectively.

Assuming there are N ACDs in the substation area, the objective function for minimizing the three-phase unbalance is shown as, as derived from Equation (1):

$$\delta_1 = min\{\varepsilon_1, \varepsilon_2, \ldots, \varepsilon_N\} \tag{2}$$

Each time a commutation command is issued, a commutation switch operation is performed, and excessive switch operations can shorten the lifespan of the switch. Therefore, it is critical to reduce how often the ACD switch actions occur. This solves practical issues like cost reductions in addition to guaranteeing the long-term use of commutation switches. The current phase sequence of a specific ACD in the substation area can be represented as:

$$M = \begin{cases} [1,0,0]^T, \ Phase\ A \\ [0,1,0]^T, \ Phase\ B \\ [0,0,1]^T, \ Phase\ C \end{cases} \tag{3}$$

During the optimization process, the switch state matrix has only three possible states, representing the connection to phases A, B, and C, respectively. The phase sequence states of each ACD in the substation area can be represented as M_x:

$$M_x = [M_1, M_2, \ldots, M_N] \tag{4}$$

By multiplying the phase sequence states of each ACD with the user-side currents, the sum of currents for each phase can be obtained:

$$[I_A, I_B, I_C]^T = M \times I \tag{5}$$

After commutation, the phase sequence states of each ACD will change, denoted as M'. A change in phase sequence for an ACD is represented as 1, while no change is represented as 0:

$$K = \begin{cases} 0, M = M' \\ 1, M \neq M' \end{cases} \tag{6}$$

The total number of commutation actions for all phasing devices can be expressed as:

$$\sigma = \sum_{i=0}^{N} K \tag{7}$$

The function for minimizing the number of commutation operations for all phasing devices can be expressed as:

$$\delta_2 = min\left\{\sum_{i=0}^{N} K\right\} \tag{8}$$

To simplify the use of the optimization commutation algorithm, weights are assigned to δ_1 and δ_2, resulting in the final objective function expression for the three-phase load imbalance commutation model established in this paper:

$$\begin{aligned} \delta &= \omega_1 \times \delta_1 + \omega_2 \times \delta_2 \\ &= \omega_1 \times min\{\varepsilon_1, \varepsilon_2, \ldots, \varepsilon_N\} + \omega_2 \times min\left\{\sum_{i=0}^{N} K\right\} \end{aligned} \tag{9}$$

3. Optimization Commutation Algorithm Based on Cubic Adaptive Modified Genetic

In order to achieve comprehensive optimization of three-phase imbalance in LVDNs, it is essential to develop an intelligent optimization algorithm. This algorithm will provide effective strategies for the commutation system to ensure optimal results. Genetic algorithm has relatively loose requirements for functions during the operation process, and there are no certain rules to constrain the algorithm. When encountering nonlinear problems, compared to other algorithms, this algorithm has obvious advantages in dealing with non-linear problems. With the highlighting of this advantage, this algorithm becomes effortless in dealing with practical problems and often appears in production optimization problems [25–27]. Therefore, this paper focuses on solving the function optimization problem with the three-phase imbalance and commutation frequency as the objectives. It utilizes the modified genetic algorithm (MGA) as the main commutation strategy. The proposed algorithm modifies the TGA, mainly using symbol encoding to represent commutation. The algorithm incorporates cubic chaotic mapping and stochastic universal selection to mitigate the risk of local optima. It also utilizes adaptive adjustment and species invasion interference to expedite the optimization process.

3.1. Encoding Method

In genetic algorithms, the encoding method directly affects the operation of genetic operators such as crossover and mutation operators, which largely determines the efficiency of genetic evolution [28]. Therefore, the encoding method should not only meet the needs of practical problems but also consider algorithm performance. In the context of this paper, the executing devices for automatic phase commutation are commutation switches, which can switch between phases A, B, and C, defined as:

$$A = \begin{bmatrix} 1 \\ 0 \\ 0 \end{bmatrix}, B = \begin{bmatrix} 0 \\ 1 \\ 0 \end{bmatrix}, C = \begin{bmatrix} 0 \\ 0 \\ 1 \end{bmatrix} \tag{10}$$

Matrix encoding is commonly employed in classical genetic algorithms. However, the high dimensionality of the matrix leads to long gene lengths, potentially hampering convergence speed and decreasing algorithm efficiency [29]. Therefore, this paper chooses symbol encoding in terms of the encoding method. Symbol encoding refers to the collection of gene values in a chromosome encoding string taken from a symbol set with no numerical meaning but only code meaning. This method offers the advantages of simple mutation and the freedom from accuracy limitations. It facilitates the integration of specialized problem-solving knowledge and the incorporation of relevant approximation algorithms. In the automatic optimization commutation problem of this paper, the symbol codes corresponding to the three phases are shown in Table 1.

Table 1. The phase sequence encoding of the optimized commutation process.

Phase Sequence	Matrix Form	Symbolic Coding
A	$[1,0,0]^T$	0
B	$[0,1,0]^T$	1
C	$[0,0,1]^T$	2

3.2. Modified Algorithm Process

The traditional genetic algorithm (TGA), also known as a simple genetic algorithm, is a heuristic optimization algorithm that simulates biological evolution. The basic process of this algorithm involves determining the encoding method and fitness calculation method. An initial population is generated and subsequent iterations involve steps such as selection, crossover, and mutation until a convergence condition is reached. The traditional algorithm flowchart is shown in Figure 3.

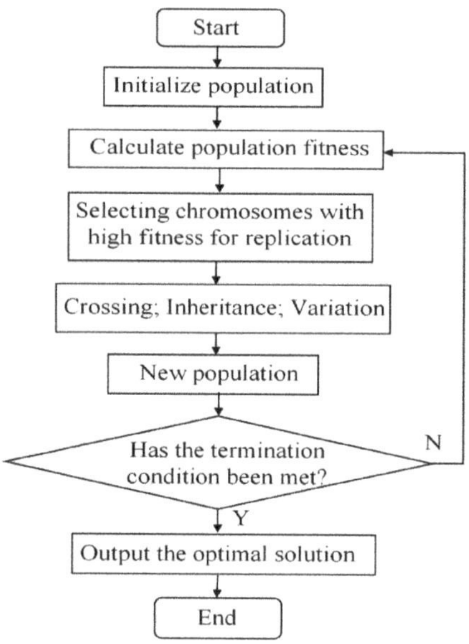

Figure 3. Flow chart of the traditional genetic algorithm.

The TGA possesses a straightforward genetic evolution process, straightforward operation, and significant application value. Nonetheless, in the context of practical problem solving, the TGA exhibits sluggish convergence speed and susceptibility to being trapped in local optima. Additionally, mutation may also lead to individuals who do not satisfy the constraint conditions [30]. In order to solve the above problems, this paper has made improvements and proposed the MGA: cubic adaptive genetic algorithm (CAGA), as shown in Figure 4.

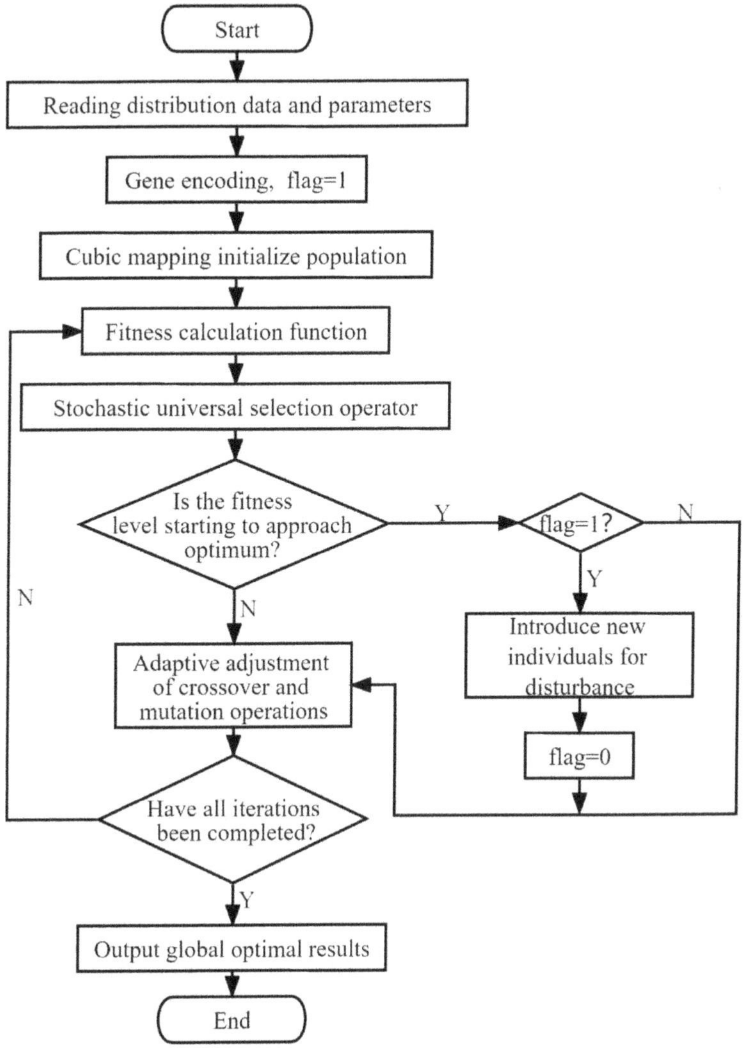

Figure 4. Flow chart of the modified genetic algorithm.

The individual genes in the population of TGA are generated randomly. This leads to uncertainty in population initialization and increases the risk of entering a local optimal state. To form an initial population with uniformity and diversity, MGA uses a cubic chaotic mapping function to initialize the population. The mathematical expression of the cubic chaotic map used in this paper is shown in Equation (11):

$$z_{k+1} = \rho z_k(1 - z_k^2) \tag{11}$$

where Z_k represents the current input value, Z_{k+1} represents the next value, and ρ is a constant.

After initializing the population through the cubic chaotic mapping function, the population is encoded according to the encoding method in Table 1 to form the initial population. Then, use the objective function established in Section 2.2 as the fitness function of the genetic algorithm for the next step of operation.

During the iteration process, due to the calculation of fitness for each generation of population individuals, the fitness matrix is constantly updated [31,32]. To obtain the optimal commutation sequence, MGA adopts a stochastic universal selection method to select excellent individuals as parents. Then, a dynamic probability strategy is adopted to meet the iterative requirements of crossover and mutation. In genetic algorithms, the crossover and mutation operations synergistically compete and cooperate, allowing the algorithm to achieve a balance between global and local search capabilities. However, in the TGA, the probabilities of crossover and mutation are constant and cannot meet the changing needs of the population during the genetic process. To solve this problem, this paper proposes modified formulas for crossover and mutation probabilities, as shown in Equations (12) and (13):

$$P_{\mathrm{c}} = \begin{cases} \frac{f - f_{\min}}{f_{\max} - f_{\min}} & f > f_{\mathrm{avg}} \\ \frac{f_{\mathrm{avg}} - f_{\min}}{f_{\max} - f_{\min}} & f \leq f_{\mathrm{avg}} \end{cases} \tag{12}$$

$$P_{\mathrm{m}} = \frac{c \times (f_{\mathrm{avg}} - f_{\min})}{f_{\max} - f_{\min}} \tag{13}$$

where P_{c} and P_{m} represent the probabilities of crossover and mutation, respectively, $f_{\max}$ and $f_{\min}$ represent the maximum and minimum fitness values of the current generation, f_{avg} represents the average fitness value of the current generation, f represents the fitness value of an individual in the current generation, and C represents a constant.

Moreover, to ensure that the algorithm does not fall into local optima, a disturbance called species invasion is introduced. When the fitness stabilizes, a fixed individual is randomly generated and added to the current population to break out of the current optimal solution and continue searching for the global optimum. In addition, MGA adopts an adaptive adjustment strategy to further adjust the proportion of outstanding individuals in the parents. In the early stages of algorithm operation, the population needs to constantly approach excellent individuals. MGA retains a small number of outstanding individuals from the current generation. The excellent offspring obtained through selection, crossover, and mutation are then used as the initial population for the next generation. As the algorithm iterates, the number of outstanding individuals increases, resulting in an increased proportion of retained parents. This effectively accelerates the optimization process of the algorithm. The Equation is as follows:

$$\eta = \begin{cases} c_1 & g < 0.1\, g_{\max} \\ c_2 \left[\frac{c_3 (g_{\max} - g)}{g_{\max}} \right] & 0.1\, g_{\max} \leq g \leq 0.9\, g_{\max} \\ c_4 & g > 0.9\, g_{\max} \end{cases} \tag{14}$$

where c_1, c_2, c_3 and c_4 are constants; g represents the current population genetic algebra; and $g_{\max}$ represents the total algebra of population genetics.

Through continuous iterative calculation, the fitness value and switch state are continuously updated until the iteration process is completed. The output of the optimal commutation switch state, commutation frequency, and current imbalance value are sent to the optimal commutation command based on the switch state under the global extreme value. Finally, combined with the optimized commutation algorithm proposed in this paper, the refined commutation mode is shown in Figure 5.

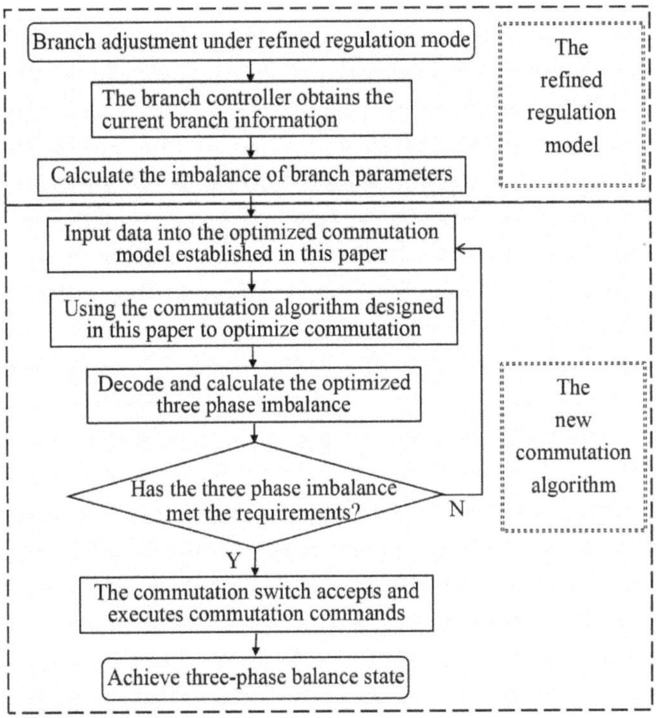

Figure 5. Proposed optimized commutation process.

4. Refined Commutation Case of LVDNs Based on Modified Genetic Algorithm

4.1. Case Data and Analysis

The distribution network structure of the LVDNs is generally a three-phase, four-wire system, and the commutation execution household wiring is generally a single-phase, two-wire system. Taking a certain distribution station area as an example, there are four branches connected to a transformer. Each branch has a varying number of commutation switches. Assuming that each branch has 30 users, and each branch is fixed with 10 commutation switches. The topology structure of this distribution substation is shown in Figure 6.

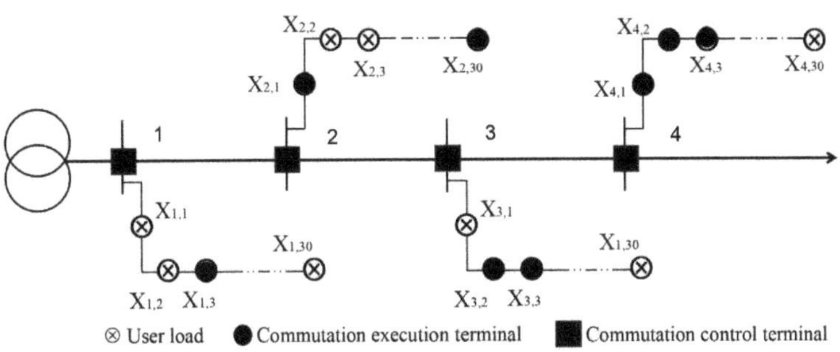

Figure 6. Topology of the distribution substation.

The execution end of the commutation device works in conjunction with the monitoring and control end of each branch. Each branch adjusts the current on the branch line by manipulating the commutation switches to achieve three-phase load balancing. The installation positions of commutation switches, the current value, and the imbalance of each branch in the distribution station area are shown in Table 2.

Table 2. Installation position of commutation switches, current value, and imbalance value for each branch line.

Branch Number	Installation Position of Commutation Switch and Current Value (A)											Imbalance Value (%)
Branch 1	Position	3	4	5	9	14	17	19	23	28	29	35%
	Current (A)	27	14	24	13	16	28	16	17	19	26	
Branch 2	Position	1	5	9	13	15	24	25	28	29	30	12%
	Current (A)	26	18	21	24	26	17	13	26	24	21	
Branch 3	Position	2	3	6	12	16	20	21	23	26	29	23%
	Current (A)	18	21	24	17	13	27	15	25	17	27	
Branch 4	Position	1	2	3	9	10	14	21	22	25	29	38%
	Current (A)	30	23	22	25	16	32	33	19	21	20	

According to Equation (1), the three-phase load imbalance of each branch in the substation is calculated as follows: 35%, 12%, 23%, and 38%. The main circuit three-phase load imbalance of the substation is as high as 27%. At this time, both the branch lines and the main circuit of the case substation have high three-phase load imbalances, which will affect the safety and economic benefits of the substation operation. The following section will use optimization algorithms to solve the three-phase load imbalance problem in the case substation.

4.2. Comparison of Optimization Commutation Algorithms

The use of a TGA to solve the three-phase load imbalance problem in the substation has drawbacks such as poor convergence, high instability of calculation results, etc. Therefore, this section conducted comparative experiments on the traditional commutation algorithm and the proposed commutation algorithm. These algorithms were used to individually solve the three-phase load imbalance problem in the branch-independent commutation station area.

The algorithm experiment is based on MATLAB R2021a, with a population size of 100, and the number of iterations set to 50. The comprehensive objective function is shown in Equation (9), where the value ω_1 is 0.95 and the value ω_2 is 0.05, and the reciprocal is used to represent the minimum value of the comprehensive objective function. We conducted algorithm comparison experiments using the classic genetic algorithm before the improvement and the CAGA modified in this paper. The comparison results of the comprehensive objective function evolution curves for each branch are shown in Figure 7.

The figure shows the comparison of algorithm improvement effects under different branches. In this embodiment, the genetic algorithm demonstrates a good optimization effect for the four branches in the station area. Although the final convergence values are consistent, the number of iterations when the improved algorithm reaches the optimal value is significantly smaller than the traditional algorithm. For example, the proposed method converges in the eighth generation in the optimization commutation of branch 2. In comparison, the traditional method converges in the 22nd generation. The efficiency of the modified iteration curve is significantly improved. Therefore, compared with traditional methods, the method proposed in this paper has a better imbalance resolution effect and the advantages of faster convergence speed and higher iteration efficiency.

To further demonstrate the stability of this algorithm in optimizing commutation problems, this paper further conducted algorithm stability verification experiments. Taking 20 populations in branch 1 as an example, compare the optimal imbalance and imbalance under the optimal fitness of each population, as well as the minimum number of commutations for each population and the number of commutations under the optimal fitness, to achieve a stability evaluation. The comparison results of the imbalance before and after the application of the modified algorithm for optimizing commutation are shown in Figure 8.

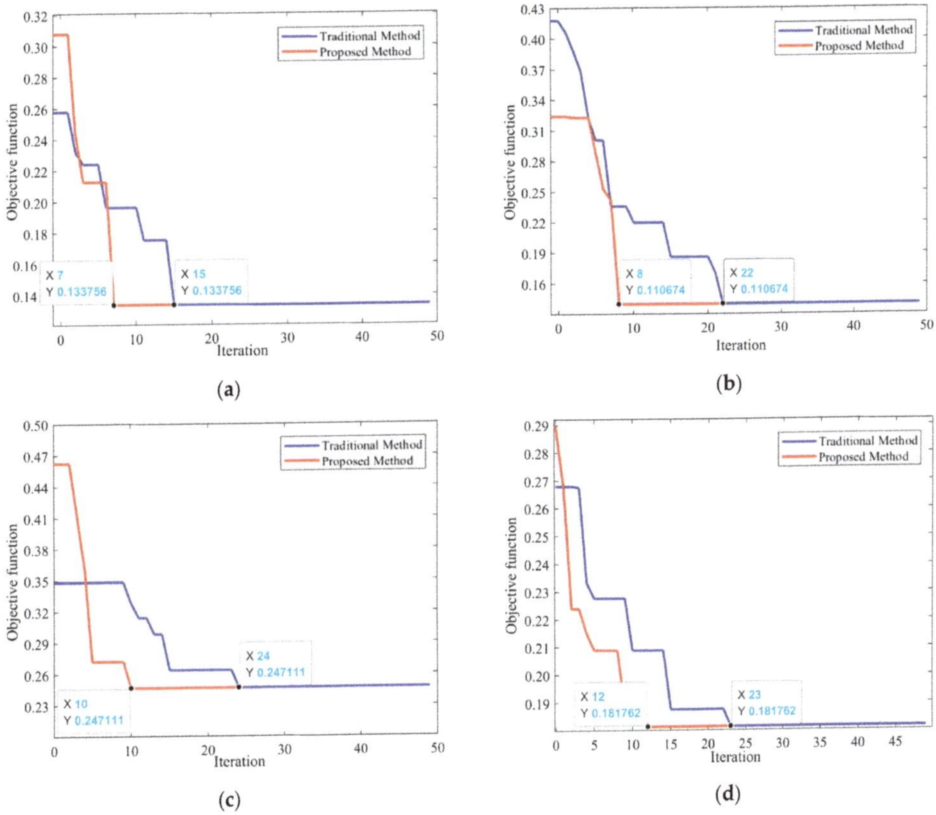

Figure 7. Comparison of optimized commutation effects for each branch using traditional and proposed methods. (**a**) Branch 1; (**b**) branch 2; (**c**) branch 3; and (**d**) branch 4.

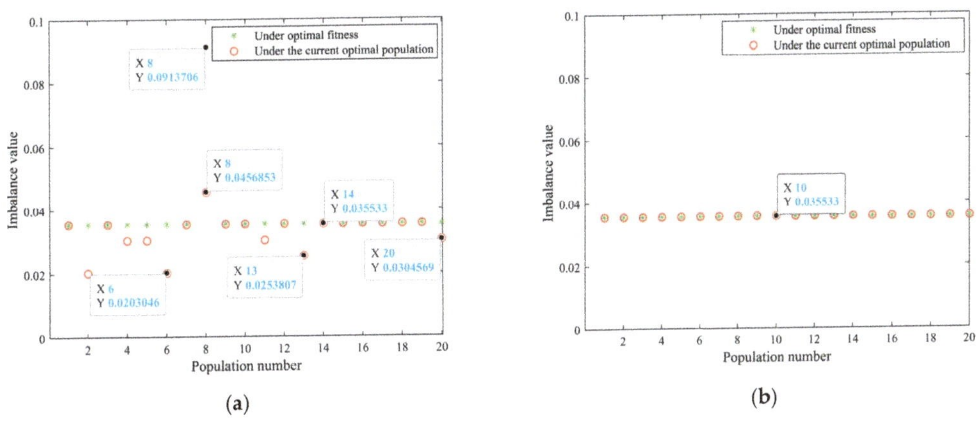

Figure 8. Stability comparison of commutation algorithms based on imbalance before and after improvement. (**a**) Traditional method; (**b**) proposed method.

From the figure, the distribution of imbalance results under traditional methods is chaotic and there are many deviations, resulting in poor stability of algorithm operations. Among them, the marked black dots represent partial deviation points. The proposed method demonstrates a high coincidence rate of two imbalances and a concentrated distribution of results. This indicates its good robustness and effectiveness in reducing three-phase imbalances. As shown in Figure 9, the comparison results of the number of commutations before and after the application of the modified algorithm for optimizing commutation are presented.

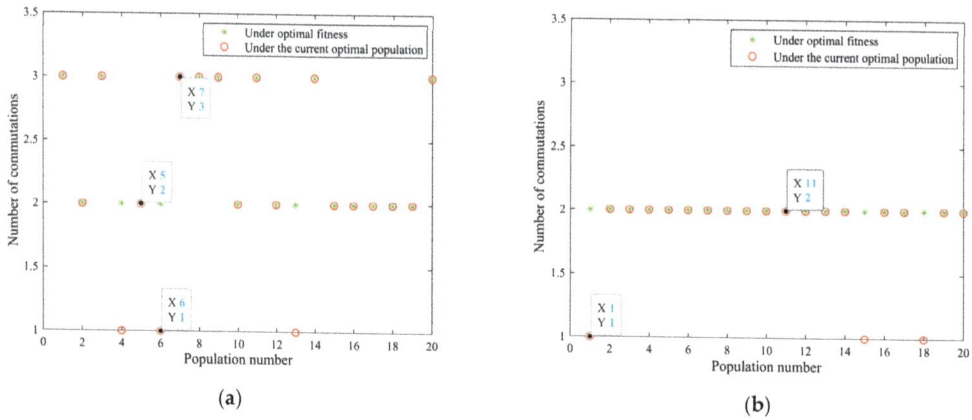

Figure 9. Comparison of the stability of the number of commutation actions between traditional and proposed methods. (**a**) Traditional method; (**b**) proposed method.

As shown in the figure, the distribution of results under traditional methods is too discrete, resulting in poor commutation effect and poor stability of the results. The two results obtained by applying the proposed method have a relatively concentrated distribution, high coincidence rate, and good robustness. Among them, the marked black dots represent partial deviation points. In summary, the CAGA designed in this paper is significantly superior to the TGA in solving the problem of optimal commutation. This algorithm can effectively reduce the imbalance of three-phase load and has both stability and convergence.

4.3. Main and Branch Circuit Commutation Results of LVDNs

The method designed in this paper is applied to optimize the commutation of each branch in the case substation. The comparative results of the commutation effects for each branch and the main circuit are shown in Table 3.

Table 3. Comparative table of optimization commutation effects for different algorithms.

Branch Number	Imbalance before Optimization (%)	Imbalance after Optimization (%)	Optimization Effect (%)	Commutation Actions (Times)
Branch 1	35%	4%	89%	2
Branch 2	12%	1%	92%	2
Branch 3	23%	10%	57%	3
Branch 4	38%	9%	76%	2
Main	27%	6%	21%	9

From Table 3, the main circuit three-phase imbalance of the substation area has decreased from 27% before governance to 6%. The imbalance of branch 1 has decreased from 35% before governance to 4%. The imbalance of branch 2 has decreased from 12% to 1%. The imbalance of branch 3 has decreased from 23% to 10%. The imbalance of branch 4 has decreased from 38% to 9%. This not only solves the three-phase imbalance of each branch but also reduces the number of commutation switches. MATLAB R2021a/Simulink is employed to facilitate the simulation and comparative analysis of the three-phase currents in the substation, both before and after the optimization commutation process. The parameters are set as follows: power supply voltage of 5 kV, transformer of 5 kV/220 V, total length of 10 km (total resistance of 2 Ω, reactance of 2 Ω), three-phase loads of 100 kW each, and neutral line resistance of 2 Ω. The comparative results of the three-phase currents before and after optimization commutation in the substation are shown in Figure 10.

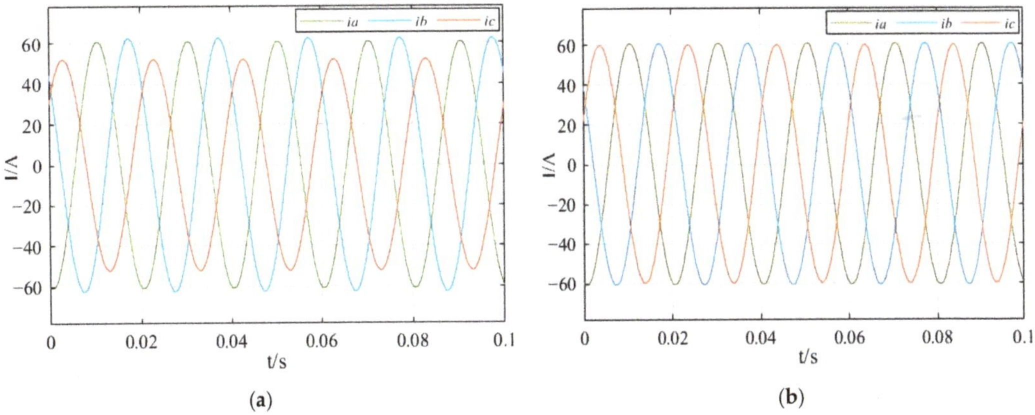

Figure 10. Comparative waveform of three-phase currents before and after optimization commutation. (**a**) Without the method proposed in this paper; (**b**) with the method proposed in this paper.

From the figure, it can be intuitively observed that the amplitude difference of the three-phase current before optimizing commutation is significant. At this time, the phase difference is not 120 degrees, and the amplitude of the current on the neutral line is also large. This will lead to significant line power consumption on the neutral line, seriously affecting the operational efficiency of the substation area. In contrast, after optimizing the commutation, the amplitude of the three-phase current is basically the same, with a phase difference of about 120 degrees. The current on the neutral line is close to zero, indicating that the three-phase current in the power system is balanced during normal operation. Currently, the distribution station area forms a relatively balanced power system. Therefore, the method proposed in this paper can effectively solve the three-phase imbalance of the main circuit, and the three-phase imbalance problem on each branch line has also been effectively solved.

5. Conclusions

This paper proposes a novel refined regulation method with faster convergence speed and better stability to solve the problem of branch three-phase load imbalance in low-voltage distribution networks. In the proposed refined regulation model, the three-phase imbalance on the branches is fully considered, and the branch control unit is added, which performs tuning in conjunction with intelligent fusion terminals. In the proposed algorithm, symbol encoding is used to represent the phase of commutation, and cubic chaotic mapping and stochastic universal selection are used to avoid local optima, and adaptive adjustment and species invasion disturbance are used to accelerate the optimization process. To test the practical application effect of the proposed algorithm, an example model was established based on the branch topology structure and branch commutation data of a certain distribution station area. In simulation, the comprehensive objective functions were set as the minimum three-phase imbalance and the minimum number of commutation actions. This method was then compared with TGA-based methods. The simulation results show that the proposed method has faster convergence speed and better stability. After independent branch commutation based on the proposed method, the three-phase load imbalance of each branch is reduced to below 10%, and the three-phase load imbalance of the main circuit is reduced from 27% to 6%. Therefore, after applying the proposed method, the main circuit three-phase imbalance in the distribution area and each branch line has been effectively solved, proving that this method has high practical value and development prospects. In future work, consideration can be given to line loss, commutation time, transient state of the circuit, etc., in order to further study the three-phase imbalance problem in LVDNs.

Author Contributions: Project administration, methodology, software, D.L.; conceptualization, investigation, data curation, Z.L. and T.W.; data curation, Z.X.; writing—original draft preparation, project supervision, T.H. and A.D.; writing—review and editing, Z.C. All authors have read and agreed to the published version of the manuscript.

Funding: This work was supported in part by the State Grid Beijing Electric Power Company under Grant 520204230002.

Data Availability Statement: Data available in a publicly accessible repository.

Conflicts of Interest: Dazhao Liu, Zhe Liu, Ti Wang, Zhiguang Xie were employed by the State Grid Beijing Electric Power Company. The remaining authors declare that the research was conducted in the absence of any commercial or financial relationships that could be construed as a potential conflict of interest.

Abbreviations

The following abbreviations are used in this manuscript:

LVDNs	Low-voltage distribution network
ACDs	Automatic commutation devices
MGA	Modified genetic algorithm
TGA	Traditional genetic algorithm
CAGA	Cubic adaptive genetic algorithm

References

1. Xiao, M.; Wang, S.; Ullah, Z. D-pmu and 5g-network-based coordination control method for three-phase imbalance mitigation units in the lvdn. *Energies* **2021**, *14*, 2754. [CrossRef]
2. Ma, K.; Fang, L.; Kong, W. Review of distribution network phase unbalance: Scale, causes, consequences, solutions, and future research directions. *CSEE J. Power Energy Syst.* **2020**, *6*, 479–488.
3. Yan, S.; Tan, S.-C.; Lee, C.-K.; Chaudhuri, B.; Hui, S.Y.R. Electric springs for reducing power imbalance in three-phase power systems. *IEEE Trans. Power Electron.* **2015**, *30*, 3601–3609. [CrossRef]
4. Heidari-Akhijahani, A.; Safdarian, A.; Lehtonen, M. Unbalance mitigation by optimal placement of static transfer switches in low voltage distribution feeders. *IET Gener. Transm. Distrib.* **2020**, *14*, 4612–4621. [CrossRef]
5. Bao, G.; Ke, S. Load transfer device for solving a three-phase unbalance problem under a low-voltage distribution network. *Energies* **2019**, *12*, 2842. [CrossRef]
6. Zeraati, M.; Golshan, M.E.H.; Guerrero, J.M. Voltage quality improvement in low voltage distribution networks using reactive power capability of single-phase PV inverters. *IEEE Trans. Smart Grid* **2018**, *10*, 5057–5065. [CrossRef]
7. Pati, U.; Mistry, K.D. Management of three-phase electrical loads using phase balancing optimization technique to reduce voltage imbalance in distribution networks. *Electr. Eng.* **2023**, *14*, 3997–4010. [CrossRef]
8. Wang, Y.; Chu, Z.; Chen, G.; Zhang, T.; Ma, Y.; Chen, C.; Lin, Z. A robust control strategy for the automatic load commutation device considering uncertainties of source and load. *Appl. Sci.* **2023**, *13*, 7390. [CrossRef]
9. Wang, X.; Dai, K.; Chen, X.; Zhang, X.; Wu, Q.; Dai, Z. Reactive power compensation and imbalance suppression by star-connected buck-type D-CAP. *Energies* **2019**, *12*, 1914. [CrossRef]
10. Fang, L.; Ma, K.; Zhang, X. A statistical approach to guide phase swapping for data-scarce low voltage networks. *IEEE Trans. Power Syst.* **2020**, *35*, 751–765. [CrossRef]
11. Ferreira, P.D.F.; Carvalho, P.M.S.; Ferreira, L.A.F.M.; Ilic, M.D. Distributed energy resources integration challenges in low-voltage networks: Voltage control limitations and risk of cascading. *IEEE Trans. Sustain. Energy* **2013**, *4*, 82–88. [CrossRef]
12. Liu, S.; Jin, R.; Qiu, H.; Cui, X.; Lin, Z.; Lian, Z.; Lin, Z.; Wen, F.; Ding, Y.; Wang, Q.; et al. Practical method for mitigating three-phase unbalance based on data-driven user phase identification. *IEEE Trans. Power Syst.* **2020**, *35*, 1653–1656. [CrossRef]

13. Wang, X.; Ji, Y.; Wang, J.; Zhao, Y.; Ye, P.; Qi, L.; Yang, S.; Liu, S. Research on reconfiguration of distribution network considering three-phase unbalance. *Wirel. Commun. Mob. Comput.* **2022**, *2022*, 9906100. [CrossRef]
14. Li, W.; Li, M.; Zhang, N.; Zhou, X.; Zhou, J.; Song, G. Optimizing allocation of distributed electric heating for large-scale access distribution considering the influence of power quality. *Energies* **2022**, *15*, 3587. [CrossRef]
15. Chen, J.; Zhou, F.; Gu, L.; Yin, H.; Zhang, L.; Gao, C. Evaluation index and evaluation method of three-phase imbalance treatment effect based on commutation. *IEEE Access* **2022**, *10*, 101913–101921. [CrossRef]
16. Ahmed, I.; Jeon, G.; Piccialli, F. From artificial intelligence to explainable artificial intelligence in industry 4.0: A survey on what, how, and where. *IEEE Trans. Ind. Inform.* **2022**, *18*, 5031–5042. [CrossRef]
17. Han, P.; Pan, W.; Zhang, N.; Wu, H.; Qiu, R.; Zhang, Z. Optimization method for artificial phase sequence based on load forecasting and non-dominated sorting genetic algorithm. *Dianll Xltong Zidonghua/Autom. Electr. Power Syst.* **2020**, *44*, 71–78.
18. Zhang, J.; Hao, S. Treatment of three-phase imbalance based on commutation switch. *Electr. Autom.* **2021**, *43*, 98–100+104.
19. Hu, M.-Y.; Zhang, H.-F.; Miao, S.-P.; Yang, F.; Qu, Z.M. Phase sequence adjustment method for three-phase load imbalance in station area based on improved pigeon-inspired optimization. *China Electr. Power* **2022**, *55*, 121–128.
20. Wang, J.-Y.; Wu, Z. Research on three-phase unbalanced load management based on improved sparrow search algorithm. *Jilin Electr. Power* **2022**, *50*, 19–24.
21. Rocha, N.; de Oliveira, Í.A.; de Menezes, C.E.C.; Jacobina, C.B.; Dias, J.A.A. Single-phase to three-phase converters with two parallel single-phase rectifiers and reduced switch count. *IEEE Trans. Power Electron.* **2016**, *31*, 3704–3716. [CrossRef]
22. Coppo, M.; Bignucolo, F.; Turri, R. Generalized transformer modelling for power flow calculation in multi-phase unbalanced networks. *IET Gener. Transm. Distrib.* **2017**, *11*, 3843–3852. [CrossRef]
23. Borges, P.R.O.; Ramos, J.E.; Carvalho, C.A.T.; Pires, V.A.; Carsoso, G.; Ramos, D.B.; de Morais, A.P. Repowering rural single-phase distribution network: A non-conventional proposal using two overhead wires and the ground as the third phase. *Electr. Power Syst. Res.* **2017**, *150*, 105–117. [CrossRef]
24. Zhang, Y.; Sun, X.; Zhu, L.; Yang, S.; Sun, Y. Research on three-phase unbalanced commutation strategy based on the spotted hyena optimizer algorithm. *Complexity* **2022**, *2022*, 2092421. [CrossRef]
25. Vasconcelos, J.A.; Ramirez, J.A.; Takahashi, R.H.C.; Saldanha, R.R. Improvements in genetic algorithms. *IEEE Trans. Magn.* **2001**, *37*, 3414–3417. [CrossRef]
26. Wen, S.; Rios, A.; Ge, Y.; Itti, L. Beneficial perturbation network for designing general adaptive artificial intelligence systems. *IEEE Trans. Neural Netw. Learn. Syst.* **2022**, *33*, 3778–3791. [CrossRef]
27. Deb, K.; Pratap, A.; Agarwal, S.; Meyarivan, T. A fast and elitist multi objective genetic algorithm: NSGA-II. *IEEE Trans. Evol. Comput.* **2002**, *6*, 182–197. [CrossRef]
28. Vairam, S.; Selladurai, V. Permutation genetic algorithm based encoding method for parallel machine scheduling and balancing. *Appl. Mech. Mater.* **2014**, *573*, 368–373.
29. Zhang, Y.; Liu, M. An improved genetic algorithm encoded by adaptive degressive ary number. *Soft Comput.* **2018**, *22*, 6861–6875. [CrossRef]
30. Jiacheng, L.; Lei, L. A hybrid genetic algorithm based on information entropy and game theory. *IEEE Access* **2020**, *8*, 36602–36611. [CrossRef]
31. Wang, C.; Liu, X.; Hu, H.; Han, Y.; Yao, M. Energy-efficient and load-balanced clustering routing protocol for wireless sensor networks using a chaotic genetic algorithm. *IEEE Access* **2020**, *8*, 158082–158096. [CrossRef]
32. Shrestha, A.; Mahmood, A. Improving genetic algorithm with fine-tuned crossover and scaled architecture. *J. Math.* **2016**, *2016*, 4015845. [CrossRef]

MDPI AG
Grosspeteranlage 5
4052 Basel
Switzerland
Tel.: +41 61 683 77 34

Energies Editorial Office
E-mail: energies@mdpi.com
www.mdpi.com/journal/energies

www.ingramcontent.com/pod-product-compliance
Lightning Source LLC
Chambersburg PA
CBHW061548220926
51850CB00005B/286
* 9 7 8 3 7 2 5 8 7 1 8 6 5 *